Procedures and Methodologies for the Control and Improvement of Energy-Environmental Quality in Construction

Procedures and Methodologies for the Control and Improvement of Energy-Environmental Quality in Construction

Editors

Benedetto Nastasi
Francesco Mancini

MDPI • Basel • Beijing • Wuhan • Barcelona • Belgrade • Manchester • Tokyo • Cluj • Tianjin

Editors
Benedetto Nastasi
Department of Planning, Design
Technology of Architecture
Sapienza University of Rome
Rome
Italy

Francesco Mancini
Department of Planning, Design
Technology of Architecture
Sapienza University of Rome
Rome
Italy

Editorial Office
MDPI
St. Alban-Anlage 66
4052 Basel, Switzerland

This is a reprint of articles from the Special Issue published online in the open access journal *Energies* (ISSN 1996-1073) (available at: www.mdpi.com/journal/energies/special_issues/procedures_methodologies_control).

For citation purposes, cite each article independently as indicated on the article page online and as indicated below:

LastName, A.A.; LastName, B.B.; LastName, C.C. Article Title. *Journal Name* **Year**, *Volume Number*, Page Range.

ISBN 978-3-0365-1780-3 (Hbk)
ISBN 978-3-0365-1779-7 (PDF)

Contents

About the Editors

Benedetto Nastasi

Benedetto Nastasi is an Assistant Professor in Building Physics and Building Energy Systems. He carries out research and teaches courses in the field of Hydrogen and Smart Energy Systems in the Built Environment. He is currently participating in the H2020 project GIFT—Geographical Islands FlexibiliTy. He is an alumnus of Sapienza University of Rome, from which he graduated Summa Cum Laude in Architectural and Building Engineering in 2011 and obtained his PhD with Honors in Energy Saving and Distributed Microgeneration in 2015. From 2015 to 2019 he worked as a Researcher at TU/e Eindhoven University of Technology, Guglielmo Marconi University and TU Delft University of Technology. He has also worked as an Energy and Sustainability Consultant at ISES International Solar Energy Society—Italy and ANEV, the Italian Wind Energy Association. He has won several awards for his research, including the Best Poster Award at SEE SDEWES, 2016, Best Invited Session Chairman at SEB, 17 and Best Senior Researcher at the Smart Energy Systems Conference, 2018.

Francesco Mancini

Francesco Mancini holds a Mechanical Engineering degree from the University of Rome La Sapienza (1999) and aPhD in Energetics (2003). He has authored 60 scientific papers in the fields of fluidodynamics, heat transfer, building services and energetics. From 2002 to 2006, he worked as a Contract Professor of "Building Services" at the University of Rome "La Sapienza"—Faculty of Architecture "Valle Giulia"; and as an Assistant Professor in Physiscs and Building Services from 2006 at the University of Rome "La Sapienza"—Faculty of Architecture "Valle Giulia". He is a a member of AiCARR (Associazione italiana Condizionamento Riscaldamento e Refrigerazione).

Editorial

Procedures and Methodologies for the Control and Improvement of Energy-Environmental Quality in Construction

Benedetto Nastasi * and **Francesco Mancini**

Department of Planning, Design and Technology of Architecture, Sapienza University of Rome, Via Flaminia 72, 00196 Rome, Italy; francesco.mancini@uniroma1.it
* Correspondence: benedetto.nastasi@outlook.com

Citation: Nastasi, B.; Mancini, F. Procedures and Methodologies for the Control and Improvement of Energy-Environmental Quality in Construction. *Energies* **2021**, *14*, 2353. https://doi.org/10.3390/en14092353

Received: 25 January 2021
Accepted: 7 April 2021
Published: 21 April 2021

Publisher's Note: MDPI stays neutral with regard to jurisdictional claims in published maps and institutional affiliations.

Overview of the Articles in This Special Issue

Building performance from an energy and an environmental point of view is fundamental due to the large amount of GHG emissions related to the building sector.

Building retrofitting is aimed at improving such performance to reduce the impact of the building. For this purpose, the Special Issue "Procedures and Methodologies for the Control and Improvement of Energy-Environmental Quality in Construction" has been launched, intended for building technology researchers, building physics experts and urban environment scholars. Among a huge number of submissions, 13 articles were accepted and published.

The first paper published within this Special Issue, authored by Ying et al. [1], deals with the natural ventilation performance in different configurations of yards in office buildings. It found that a higher comfort level corresponds to the multi-yard building type compared to the overall courtyard type. The second paper, authored by Piasecki et al. [2], is about the development of an experimental relation for predicting building users' satisfaction based on the Weber–Fechner law to provide an easy-to-use Indoor Environmental Quality (IEQ) index. The next paper, authored by Piasecki and Kostyrko [3], developed a weighting scheme for the IEQ index accounting for entropy-based and statistic-based approaches. The fourth paper authored by Mancini et al. [4] explored the potential contribution as load flexibility of dwellings in Italy for Demand Response activities, finding out that the most flexible interval is in winter season weekends, accounting for thermal power of Heat Pumps and possible heat storage.

Rosa in [5] investigated the solar energy technologies integrable in historical buildings to increase renewable energy integration by complying with architectural constraints. In [6], Battisti investigates the thermal comfort in open spaces around existing buildings in Rome and the possible improvement thanks to cool materials, greenery and permeable green surfaces. Computational Fluid Dynamics techniques are used in [7] by Gomez et al. to study the fire smoke behavior in an enclosed space, and they present an easy tool to support the design of smoke control systems.

Cieślikiewicz et al. in their study [8] monitored in situ the drying process of masonry walls and recorded the changes in the temperature and moisture as part of the renovation the historical building's basement. Vaisi et al. [9] provided a new thermal energy benchmark for university buildings focusing on monthly resolution in order to improve the accuracy of national action plans and their realization for energy-efficient built environment. In the tenth paper, Grassi et al. [10] investigate how the reduced temperature of a second-generation district heating supply can be handled despite the possible occurrence of discomfort caused by the lower output of radiators when working at reduced temperatures. In [11], Cumo et al. present a decision support tool for selecting the best energy retrofitting strategies integrated with a GIS tool for helping planners and Public Administrators.

Persiani et al. [12] authored a review article accounting for the balance between human and built environment resilience by highlighting the role of biomedical signals in indoor comfort including the use of stress research. Finally, Bourikas et al. [13] investigated through surveys and experimental measurements the impact of thermal, acoustic and air quality perception in office buildings, finding out that air quality and noise perception affects the thermal sensation.

Author Contributions: Conceptualization, B.N.; writing—original draft preparation, B.N. and F.M.; writing—review and editing, B.N. All authors have read and agreed to the published version of the manuscript.

Funding: This research received no external funding.

Institutional Review Board Statement: Not applicable.

Informed Consent Statement: Not applicable.

Conflicts of Interest: The authors declare no conflict of interest.

References

1. Ying, X.; Wang, Y.; Li, W.; Liu, Z.; Ding, G. Group Layout Pattern and Outdoor Wind Environment of Enclosed Office Buildings in Hangzhou. *Energies* **2020**, *13*, 406. [CrossRef]
2. Piasecki, M.; Kostyrko, K.; Fedorczak-Cisak, M.; Nowak, K. Air Enthalpy as an IAQ Indicator in Hot and Humid Environment—Experimental Evaluation. *Energies* **2020**, *13*, 1481. [CrossRef]
3. Piasecki, M.; Kostyrko, K. Development of Weighting Scheme for Indoor Air Quality Model Using a Multi-Attribute Decision Making Method. *Energies* **2020**, *13*, 3120. [CrossRef]
4. Mancini, F.; Romano, S.; Lo Basso, G.; Cimaglia, J.; de Santoli, L. How the Italian Residential Sector Could Contribute to Load Flexibility in Demand Response Activities: A Methodology for Residential Clustering and Developing a Flexibility Strategy. *Energies* **2020**, *13*, 3359. [CrossRef]
5. Rosa, F. Building-Integrated Photovoltaics (BIPV) in Historical Buildings: Opportunities and Constraints. *Energies* **2020**, *13*, 3628. [CrossRef]
6. Battisti, A. Bioclimatic Architecture and Urban Morphology. Studies on Intermediate Urban Open Spaces. *Energies* **2020**, *13*, 5819. [CrossRef]
7. Gomez, R.S.; Porto, T.R.N.; Magalhães, H.L.F.; Santos, A.C.Q.; Viana, V.H.V.; Gomes, K.C.; Lima, A.G.B. Thermo-Fluid Dynamics Analysis of Fire Smoke Dispersion and Control Strategy in Buildings. *Energies* **2020**, *13*, 6000. [CrossRef]
8. Cieślikiewicz, Ł.; Łapka, P.; Mirowski, R. In Situ Monitoring of Drying Process of Masonry Walls. *Energies* **2020**, *13*, 6190. [CrossRef]
9. Vaisi, S.; Mohammadi, S.; Nastasi, B.; Javanroodi, K. A New Generation of Thermal Energy Benchmarks for University Buildings. *Energies* **2020**, *13*, 6606. [CrossRef]
10. Grassi, B.; Piana, E.A.; Beretta, G.P.; Pilotelli, M. Dynamic Approach to Evaluate the Effect of Reducing District Heating Temperature on Indoor Thermal Comfort. *Energies* **2021**, *14*, 25. [CrossRef]
11. Cumo, F.; Giustini, F.; Pennacchia, E.; Romeo, C. Support Decision Tool for Sustainable Energy Requalification the Existing Residential Building Stock. The Case Study of Trevignano Romano. *Energies* **2021**, *14*, 74. [CrossRef]
12. Persiani, S.G.L.; Kobas, B.; Koth, S.C.; Auer, T. Biometric Data as Real-Time Measure of Physiological Reactions to Environmental Stimuli in the Built Environment. *Energies* **2021**, *14*, 232. [CrossRef]
13. Bourikas, L.; Gauthier, S.; Khor Song En, N.; Xiong, P. Effect of Thermal, Acoustic and Air Quality Perception Interactions on the Comfort and Satisfaction of People in Office Buildings. *Energies* **2021**, *14*, 333. [CrossRef]

Article

Group Layout Pattern and Outdoor Wind Environment of Enclosed Office Buildings in Hangzhou

Xiaoyu Ying [1], Yanling Wang [2,*] , Wenzhe Li [2], Ziqiao Liu [2] and Grace Ding [3]

1   Department of Architecture, Zhejiang University City College, No. 51, Huzhou Street, Gongshu District, Hangzhou 310015, China; yingxiaoyu@zucc.edu.cn
2   Department of Architecture, Zhejiang University, No.866, Yuhangtang Road, Sandun Town, Xihu District, Hangzhou 310027, China; 21712123@zju.edu.cn (W.L.); 21812107@zju.edu.cn (Z.L.)
3   Department of Built Environment, University of Technology Sydney, Broadway, Ultimo NSW 2007, Australia; Grace.ding@uts.edu.au
*   Correspondence: 21712105@zju.edu.cn; Tel.: +86-1576-394-4820

Received: 2 December 2019; Accepted: 9 January 2020; Published: 14 January 2020

Abstract: This paper presents a study of the effects of wind-induced airflow through the urban built layout pattern using statistical analysis. This study investigates the association between typically enclosed office building layout patterns and the wind environment. First of all, this study establishes an ideal site model of 200 m × 200 m and obtains four typical multi-story enclosed office building group layouts, namely the multi-yard parallel opening, the multi-yard returning shape opening, the overall courtyard parallel opening, and the overall courtyard returning shape opening. Then, the natural ventilation performance of different building morphologies is further evaluated via the computational fluid dynamics (CFD) simulation software Phoenics. This study compares wind speed distribution at an outdoor pedestrian height (1.5 m). Finally, the natural ventilation performance corresponding to the four layout forms is obtained, which showed that the outdoor wind environment of the multi-yard type is more comfortable than the overall courtyard type, and the degree of enclosure of the building group is related to the advantages and disadvantages of the outdoor wind environment. The quantitative relevance between building layout and wind environment is examined, according to which the results of an ameliorated layout proposal are presented and assessed by Phoenics. This research could provide a method to create a livable urban wind environment.

Keywords: CFD; enclosed building; wind environment; group layout; Hangzhou; China

---

1. Introduction

According to the statistics of the National Bureau of Statistics of China, as seen in Figure 1, the total population of the mainland increased from 1367.82 million to 1395.38 million during the five years from 2014 to 2018. From the perspective of urban and rural structures, the resident population of urban areas has increased from 749.16 million to 831.37 million. The proportion of the urban population to the total population (urbanization rate) increased from 54.77% to 59.58% [1], indicating that it is in the middle and late stages of urbanization. The urbanization process has led to a sharp increase in urban density and the scale of cities in China. Many environmental problems have become increasingly prominent. The increasingly rough urban underlying surface and highly concentrated anthropogenic heat emissions together form a special urban climate environment, mainly manifested as weak urban winds, the heat island effect, and impeded urban air circulation and pollutant diffusion [2]. Unreasonable architectural layouts or architectural forms create an outdoor static wind zone, which is not conducive to the spread of pollutants and exhaust gases during the spring and autumn, and which is also not conducive to heat dissipation in the summer. These problems have promoted the public's vision of the wind environment in urban space [3].

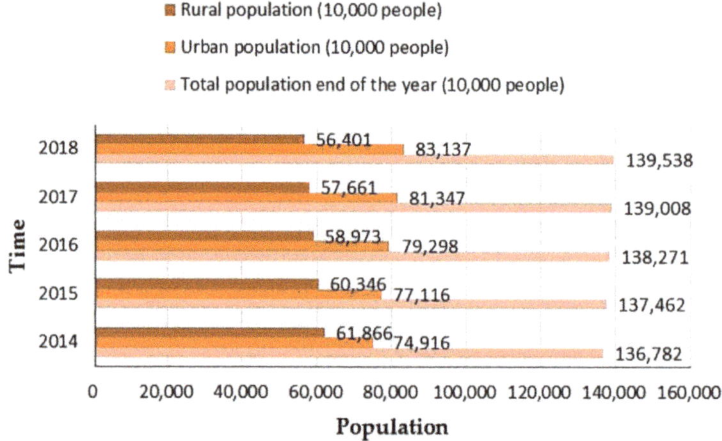

Figure 1. Mainland population statistics from China's National Bureau [1].

The office building is one of the most important types of architecture in the modern era. It accommodates a large number of working people in the city and has a profound impact on the urban spatial pattern. In addition to the place of residence, people spend the longest time in office spaces. With the development of society, people's demand for office buildings is also increasing. Office buildings are becoming more diverse, humane, low-carbon, and intelligent [4]. At present, research on office buildings focuses on form and function and lacks attention on factors affecting the outdoor environment.

The enclosed multi-story building layout can provide more natural lighting, create a better sense of territory and belonging, and reflect a traditional spatial mood. At the same time, because the enclosed space has good natural ventilation potential and is easy to create a relatively independent outdoor microclimate, it is increasingly used in urban modern architectural design [5]. The enclosed courtyard is often regarded as a microclimate modifier that improves the comfort conditions of the surrounding environment [6]. However, at present, the academic circles have rarely considered the natural ventilation performance of enclosed multi-story office buildings, and the association between typically enclosed office building layout patterns and the outdoor wind environment is still in the vague cognitive stage. In actual use, space utilization declines due to poor natural ventilation performance.

There are plenty of studies on wind motion through urban buildings [7]. Guo et al. [8] investigated the urban ventilation path, scattered morphology, and green space system that have a remarkable effect on promoting ventilation and alleviating the urban heat island effect. Enclosed city blocks are extremely unfavorable to ventilation. Kuo et al. explored the pedestrian-level wind flow characteristics inside the street canyon. The variables included the street canyon width, approaching wind direction, and podium height [9]. Guo et al. [10] used the art gallery as a case, and suggested that with comprehensive consideration given to elevation aesthetics and plane functions, one could create certain forms at suitable positions of a building to deflect wind and direct it through wind tunnels by distributing building volumes and creating open-up spaces and openings, so as to facilitate natural ventilation. Sharples et al. [11] carried out a wind tunnel study investigating the airflow through courtyard and atrium building models. The results from the study suggested that the small scale open courtyard in an urban environment had poor ventilation performance. Abdulbasit et al. [6] combined experimental and simulation methods in their research. The result verified that the manipulation of the courtyard configuration and its orientation impacted its microclimate modifying ability. Liu and Huang found that based on the computational fluid dynamics (CFD) calculation and simulation results, the growth of the height of the patio has exerted the prime influence on the natural ventilation performance of the "Yinzi" dwellings, which were enclosed buildings [12]. Jin et al., focusing on severe cold regions, studied the relationships between the mean wind velocity ratio at the pedestrian level and the residential areas' building densities and high-rise buildings' layouts [13]. Xu et al. studied the traditional courtyards in southern Jiangsu Province. They proposed three strategies, including adjusting the courtyard layout, modifying the aspect ratio, and building an ecological buffer space to maximize its benefits of regulating the microclimate in winter and summer [14]. Ying et al. [15] analyzed the wind environment around a group of six square high-rise buildings. The research models in this study were six homogeneous cube models. Their study in 2019 [16] simulated the wind environment of 12 typical single-enclosed building opening schemes, as is shown in Figure 2 [16]. This research provided new ideas for the study of the outdoor wind environment of a single enclosed multi-story building. Overall, several investigations carried out simulations of the wind in enclosed buildings such as large-scale urban blocks, actual architecture cases, and homogeneous or single-volume models. There is no quantitative research on the outdoor wind environment of enclosed multi-story building groups in complex urban environments, which should be of great concern.

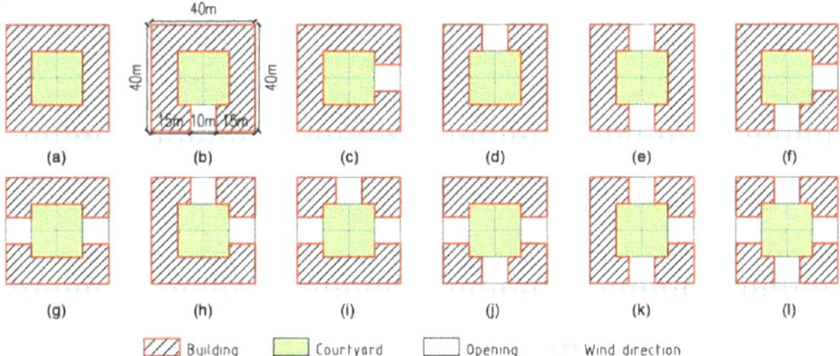

Figure 2. Twelve typical single-enclosed building opening schemes: (**a–l**), Opening scheme 1–12.

Therefore, it is necessary to research the effects of wind-induced airflows through the urban built layout pattern using statistical analysis and to investigate the association between typically enclosed office building layout patterns and the wind environment. This paper also proposes feasible improvement schemes for the unsatisfactory wind environment layout, providing a reference and design basis for architects. According to the possibility of architect design and the requirements of building fire protection [17], this article summarizes four typical enclosed office building group layouts. At the same time, for the convenience of comparative research, the model is simplified into four typical layouts: the multi-yard parallel opening, the multi-yard returning shape opening, the overall courtyard parallel opening, and the overall courtyard returning shape opening, which are explained in detail below.

At present, the research methods of the urban wind environment mainly include three methods: the base practical measurement method, wind tunnel experiments, and computer numerical simulation [18]. Computational numerical simulation is the main research method in current research. Due to the data analysis capabilities and experimental cost constraints, the base practical measurement method and wind tunnel experiments are relatively limited.

2. Materials and Methods

2.1. Methods

This paper obtained four typical multi-story enclosed office building group layouts according to Hangzhou City Planning Management Technical Regulations and the design specifications [17,19].

Then it researched the climate of Hangzhou and determined the comfortable range of the wind speed ratio, according to temperature and humidity, as the evaluation standard. Phoenics software was used to set the boundary conditions, simulate the natural ventilation performance of the four layout forms, and calculate the wind speed ratios of each measuring point. Finally, this paper used Excel software for data processing and comparative analysis. Moreover, the article also makes further research on layout forms with poor natural ventilation performance.

2.2. Boundary Conditions

The numerical simulation software Phoenics used in this paper based on Reynold's time-averaged equations can automatically select the required conditions for calculation. Boundary conditions were chosen accordingly.

2.2.1. Inlet Wind Speed Distribution

The air movement is typically horizontal, showing less vertical behavior. However, in an urban environment, "topography" affects the wind motions. When the airflow passes through different areas and topographic zone (oceans, mountains, forests, cities, etc.), friction reduces the energy of the wind; when the wind speed is reduced, the wind structure (such as the turbulence degree, the swirl scale, etc.) can also change. The degree of change decreases with the increase in height, until at a certain height, the influence of ground roughness can be ignored. This layer of the atmosphere, which is affected by the friction of the earth's surface, is called the atmospheric boundary layer. The height of the atmospheric boundary layer varies with the meteorological conditions and the topographic and surface roughness. In general, the range of 300 m above the ground (not exceeding 1000 m) is within the scope of the atmospheric boundary layer, so that the wind speed above this range is not affected by the surface but can flow freely under the action of the atmosphere gradient.

2.2.2. Mean Wind Speed Index Rate Distribution

In the atmospheric boundary layer, the average wind speed changes with altitude, and the variation is called wind shear or wind profile. At present, most countries use empirical exponential distribution to describe the change of average wind speed with altitude in near-surface layers. The velocity of approaching wind can be approximately described by a power-law profile [20] as follows:

$$U(z) = U_G \times (z/z_G)^{\alpha}. \tag{1}$$

In this formula $U(z)$ represents the average wind speed at any height z, U_G is the average wind speed at the standard height z_G, and the index α is a parameter describing the ground roughness. The building in this article complies with category C in Table 1; thus α is 0.22. The gradient height z_G was assumed to be 400 m.

Table 1. Standard four types of landforms in China [21].

Category	Underlying Surface Properties	α	z_G/m
A	Offshore, island, sea, and desert areas	0.12	300
B	Fields, villages, jungles, hills, and towns and suburbs with sparse houses	0.16	350
C	Urban districts with dense buildings	0.22	400
D	Urban areas with dense buildings and high housing	0.30	450

α—Ground roughness coefficient, z_G—Gradient wind height in meters.

2.2.3. Boundary Conditions of All the Wall Surfaces

The research assumes that the airflow on the domain outlet has fully developed, and the airflow has returned to normal flow without building obstruction. Therefore, the outlet boundary is partially unidirectional. This paper used a no-slip wall boundary condition at all the wall surfaces and a normal zero gradient boundary condition at the domain outlet and the domain top as well as symmetrical boundary conditions at the two lateral boundaries of the domain. The boundary conditions for lateral and upper surfaces do not have significant influences on the calculated results around the target building because the computational domain is large enough [22–24].

2.3. Grid Size and Independence

Franke et al. [25] suggested using at least ten cells on each side of the building and at least three cells with pedestrian wind speeds at 1.5–2 m height above the ground. This paper used three kinds of grids (coarse, medium, and fine) at the central area including the coarse grid 15 m, the medium grid 7 m, and the fine grid 3 m in the X and Y axis. The grid independence study confirmed that numerical results with medium grids change little when the grids become finer. Thus, all other models used medium grids.

2.4. Domain Size

The setting of the calculation domain is related to the credibility of wind field simulation results. For the size of the computational domain, the blockage ratio should be below 3% based on knowledge of wind tunnel experiments [26]. Baetke et al. [27] defined the blockage ratio as the ratio of the frontal area of the cube to the vertical cross-sectional area of the computational domain. On the advice of Mochida et al. [22] and Shirasawa et al. [23], the lateral and the top boundary should be set 5 H or more away from the building, where H is the height of the target building. The outflow boundary should be set at least 10 H behind the building. Where the building surroundings are considered, the height of the computational domain should be set to correspond to the boundary layer height determined by the terrain category of the surroundings [20]. Following this suggestion, for this study, the domain size was 1200 m, 1200 m, and 400 m in the longitudinal (x), lateral (y), and vertical (z) directions, respectively.

2.5. Solution Methods and Convergence Condition

This paper chose the Realizable k-ε model to solve the problem. The governing equations used were those suggested by Cheng et al. [28]. Franke et al. [29] presented best practices guidelines for the CFD simulation of flows in the urban environment, developed within the COST Action 732 framework. Almost all the research articles related to this topic considered these guidelines as the best practice reference for urban wind CFD simulations. Initializing Reynolds-averaged Navier Stokes (RANS) simulations with uniform velocity, turbulent flow energy, and energy dissipation rate fields typically requires 10^3 iterations to reach convergence. Calculation are conducted until the desired level of convergence is reached, i.e., the constant residuals of all equations are 10^{-4} or less [29,30]. The calculation conditions of the Phoenics are shown in Table 2.

Table 2. Computational condition of Phoenics.

Computational Condition	Setting
Computational domain	1200 × 1200 × 400 m
Central meshing	Grid interval of 7 m in X and Y axis, 1 m in Z axis
Turbulence model	Standard k-ε turbulence model
Incoming flow speed	2.7 m/s (summer), 3.8 m/s (winter), at the height of 10 m
Incoming flow direction	South (summer), north (winter)
Calculation rule	SIMPLEC model
Convergence condition	The maximum permissible residual error is 10^{-4}
Total number of iterations	1000

2.6. Validation of the CFD Simulation

This article is a further study based on the wind environment of the enclosed multi-story building and the courtyard space. The authors validated the implemented CFD simulation method in previous research [15,16]. By using the same boundary conditions and solutions as in Zhang's study [31], the CFD simulation results were compared with wind tunnel experiments of similar buildings for experimental verification. It also showed that the CFD simulation results had good agreement with experimental results.

2.7. Layout Model Setting

According to the possibility of architect design and the requirements of building fire protection [17], this paper divided the layout of the enclosed office building group into the following two types, as seen in Figure 3: multi-yard type and overall courtyard type, including the multi-yard parallel opening (M-p), the multi-yard returning shape opening (M-r), the overall courtyard parallel opening (O-p), and the overall courtyard returning shape opening (O-r), as is shown in Figure 4. The ideal site model was set to 200 m × 200 m (length, width). The south side was the main road with a width of 28 m where the main entrance to the site was set up. The roads on the east, west, and north sides were secondary roads with widths of 21 m, and which had secondary entrances. The building group size was 79.5 m × 79.5 m × 24 m. The building was 8 m away from the road red line, the distance between the groups was 25 m, the width of the enclosed office building group was 15 m, the green space rate was not less than 30%, the building density was not more than 40%, the plot ratio was not more than 2.2, and the room depth was 18.4 m. These indicator parameters were selected according to Hangzhou City Planning Management Technical Regulations by the Hangzhou Planning Bureau [19]. The technical specifications of the four layouts shown in Table 3 met the above requirements.

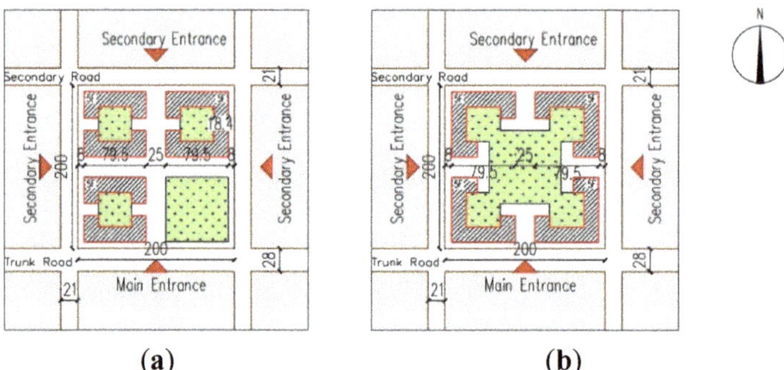

(a)   (b)

Figure 3. Two courtyard layout patterns. (**a**) Multi-yard type; (**b**) Overall courtyard type.

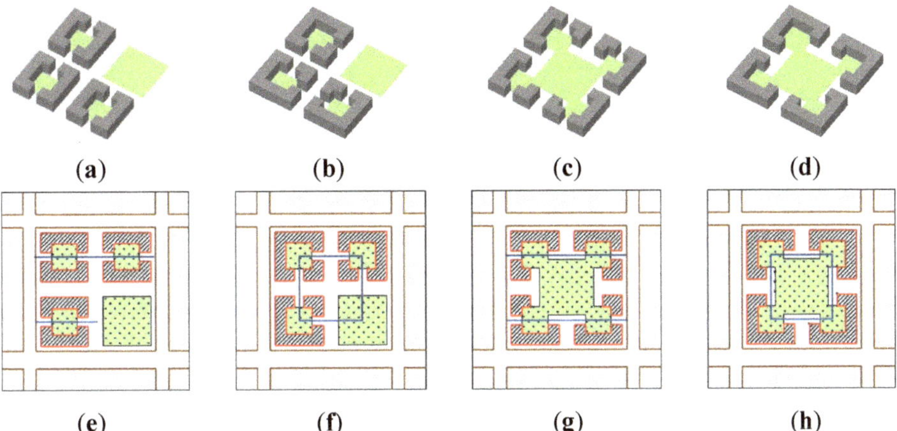

Figure 4. Four typical multi-enclosed office building group layouts. (**a**) M-p model; (**b**) M-r model; (**c**) O-p model; (**d**) O-r model; (**e**) M-p floor plan; (**f**) M-r floor plan; (**g**) O-p floor plan; (**h**) O-r floor plan.

Table 3. Technical specifications of four layouts.

Index	M-p	M-r	O-p	O-r
Building density	29.6%	29.6%	28.2%	31.0%
Floor area ratio	1.48	1.48	1.41	1.55
Green space ratio	29.5%	29.5%	36.3%	36.3%
Building story number	5	5	5	5
Building height (m)	24	24	24	24

2.8. Evaluation Criterion

This paper used Hangzhou as an example to have more practical significance. The research of the historical wind environment in Hangzhou indicated that the dominant wind direction in winter and summer in this region was obvious, and the wind frequency of the dominant wind direction was significantly higher than other wind directions. To simplify the calculation model, when simulating the wind environment around the building, the summer monsoon was set as the south wind (2.7 m/s) and the winter monsoon was set as the north wind (3.8 m/s) [32].

Many cities in the world have required evaluation of the wind environment before the construction of a building. The wind speed of the surrounding environment of the building was limited and required. Generally, the wind speed at a height of 1.5 m from the ground in the pedestrian zone is less than 5 m/s to meet the basic requirements that do not affect people's normal outdoor activities [33]. Stathopoulos et al. [34,35] suggested more data about a wider range of weather conditions and from different climates are needed to promote the new outdoor human comfort standards and described an approach towards the establishment of an overall comfort index taking into account, in addition to wind speed, the temperature and relative humidity in the urban area under consideration. The current evaluation methods mainly include relative trip comfort, wind speed probability statistics, and wind speed ratio evaluation methods. The relative trip comfort assessment method and wind speed probability statistical assessment method are both related to human subjective evaluation. The wind speed ratio is the ratio of wind velocity at each point (height = 1.5 m) to the wind velocity at the identical height at the inflow boundary, which reflects the degree of change in wind speed due to the presence of the building. The wind speed ratio equation is

$$R = V_0/V \tag{2}$$

where R is the wind speed ratio, V_0 is the velocity of a point, and V is the inflow velocity.

Kubota et al. [33] suggested that when the wind speed ratio is greater than 2.0, pedestrians will feel that the wind is too strong. On the other hand, people will not feel the presence of wind when the wind speed ratio is less than 0.5. Hyungkeun et al. found that pedestrians feel discomfort even at low wind speeds in winter. There are limits in assessing the comfort of pedestrians in winter since the existing criteria only consider the mechanical effects of the wind [36]. It is not reasonable to define pedestrian comfort value without considering other weather conditions, such as temperature and humidity. The average winter temperature of Hangzhou is −2.2 °C, the average relative humidity is 82% [32], and the climate comfort index should be greater than 25 according to the calculation method and grading principle of the climatic comfort index of the China State Meteorological Bureau [37]. Therefore, the winter wind speed should be less than 3.53 m/s, and the wind speed ratio less than 0.93 and more than 0.5. At the same time, the average summer temperature of Hangzhou is 32.4 °C, the average relative humidity is 62%, and the climate comfort index should be less than 80. Therefore, the summer wind speed should be greater than 1.25 m/s, and the wind speed ratio should be greater than 0.5 and less than 2.0.

3. Results and Discussion

Since wind is faster at the corner of buildings due to the less amount of topography, existing flow increases in speed at the corners in order to connect to the streamlines [38]. Therefore, the measurement points selected in the simulation were all locations with a large pedestrian flow and unfavorable wind speed, as is shown in Figure 5. Point D_1 was the measuring point in the courtyard where the summer monsoon may have adverse effects. Point D_5 was the measuring point in the courtyard where the winter monsoon may have adverse effects. D_2 and D_6 were measuring points with the concentrated pedestrian flow on the axis parallel to the wind direction and located at the corner of the building. D_3 and D_4 were the measuring points of concentrated human flow on the axis perpendicular to the wind direction.

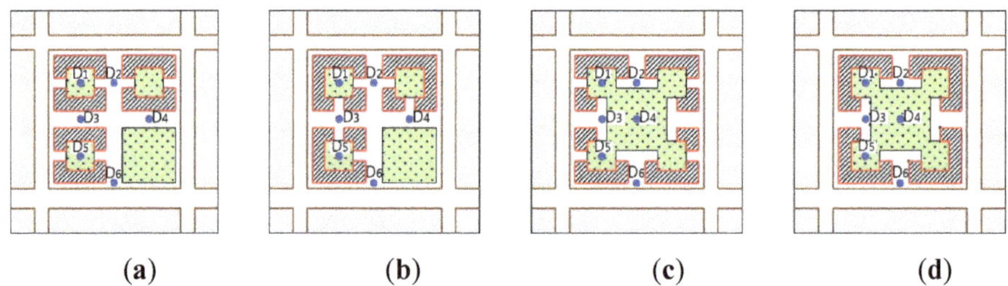

(a)   (b)   (c)   (d)

Figure 5. Distribution of the measuring points of four layouts. (**a**) Distribution of the measuring points of M-p; (**b**) Distribution of the measuring points of M-r; (**c**) Distribution of the measuring points of O-p; (**d**) Distribution of the measuring points of O-r.

3.1. Analysis of Wind Simulation Results in Summer

Figure 6 shows the simulation results of the wind environment at an outdoor pedestrian height (1.5 m) under the influence of the south wind in summer. The direction of the wind was south, and the wind speed was 2.7 m/s. The data of 6 measuring points of each scheme were statistically analyzed.

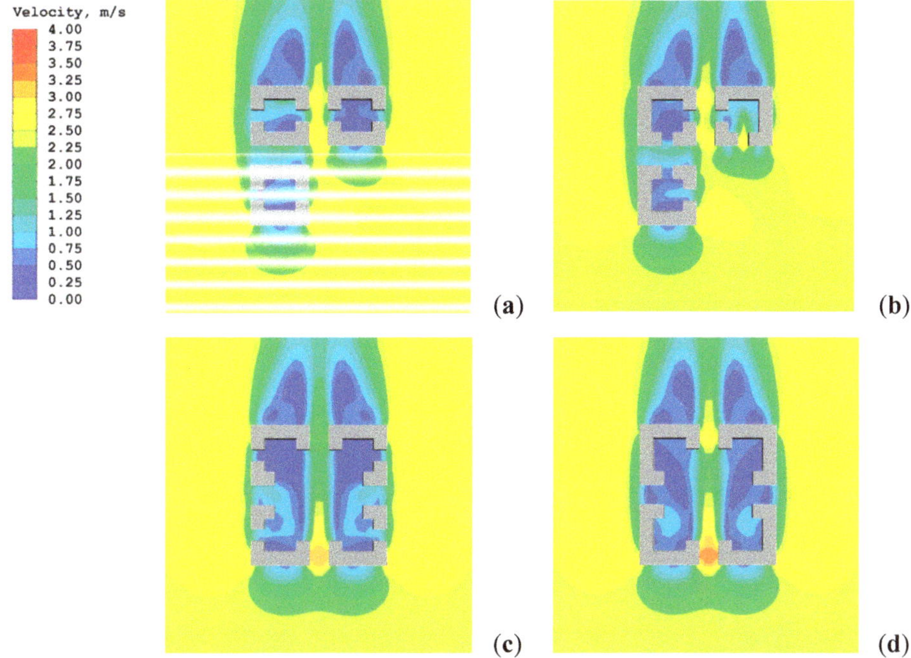

Figure 6. Summer simulation results of four layouts. (**a**) Summer wind simulation results of the M-p scheme; (**b**) Summer wind simulation results of the M-r scheme; (**c**) Summer wind simulation results of the O-p scheme; (**d**) Summer wind simulation results of the O-p scheme.

Figure 7 shows the contour map of the wind speed ratio at the outdoor pedestrian height (1.5 m) of each layout scheme.

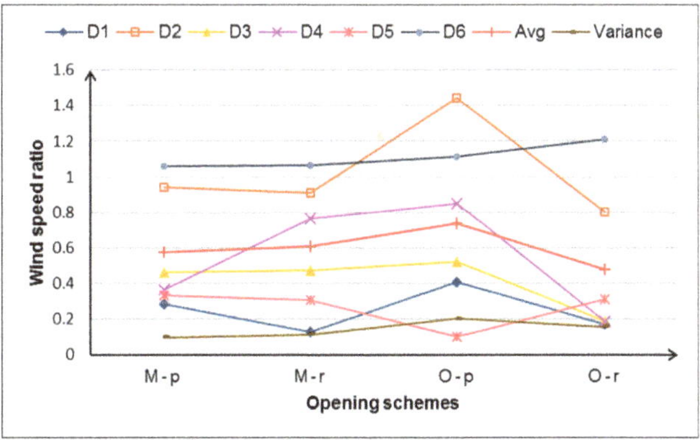

Figure 7. Measuring points' wind velocity ratio of summer wind in four schemes.

It can be seen intuitively from the figure that the wind speed of the measuring points on the axis parallel to the inflow wind direction was relatively large in the four layout schemes. They were the D_2 and D_6 measuring points in the multi-yard type and the D_2, D_4, and D_6 measuring points in the overall courtyard type. The second was the windward measurement point of the building, which was the multi-yard type measurement point D_4. The measured wind speeds in the courtyard and the measured wind speeds between the two buildings (north and south) were relatively small, namely the measuring points D_1, D_3, and D_5. This also verified that it was easy to form tunnel wind on an axis parallel to the wind direction, so the wind speed at the measuring point on the axis was relatively large.

Besides, the wind speed ratio on the central axis consistent with the incident wind direction tended to decrease along the wind direction in general consideration. However, simulation results and data showed that only the O-p layout scheme met this situation, and other schemes were irregularly distributed, which indicates that the layout of the enclosed building group had a greater impact on the wind environment.

Among the four layout forms, the variance of measuring points of the O-p type was the largest, which was 0.20, and the wind speed ratio varied from 0.10 to 1.44. The variance of measuring points of the M-p type was the smallest, which was 0.10, and the wind speed ratio varied from 0.28 to 1.06. This indicated that the natural ventilation performance of each area of the O-p layout was very different, and pedestrians may feel uncomfortable. In Figure 8, the natural ventilation performance of the M-p layout was relatively uniform, and the wind speed ratio was closest to the comfort range (0.5–2.0), so the ventilation performance was excellent.

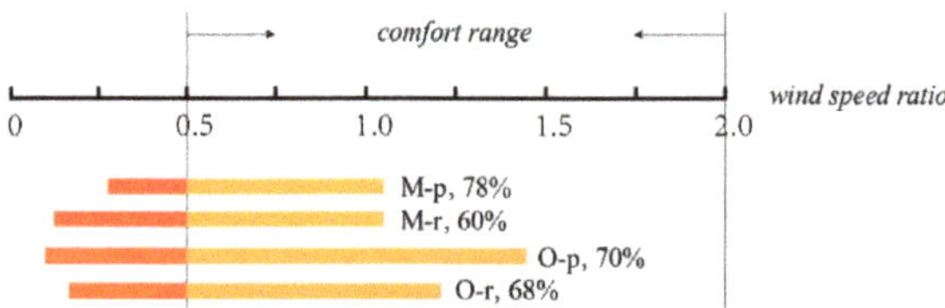

Figure 8. Comfort ratio of the four layouts (summer wind).

As can be seen from the Avg curve in Figure 7, the average wind speed ratio of the O-p type was the largest at 0.74, and the average wind speed ratio of the O-r type was the smallest at 0.48. The M-p and M-r types were 0.57 and 0.60, respectively. Therefore, the O-r layout was not conducive to outdoor ventilation because of the weak outdoor airflow. The evaluation of summer ventilation performance was as follows: M-p > M-r > O-p > O-r.

3.2. Analysis of Wind Simulation Results in Winter

Figure 9 shows the simulation results of the wind environment at an outdoor pedestrian height (1.5 m) under the influence of the north wind in winter. The direction of the wind was north, and the wind speed was 3.8 m/s. The data of 6 measuring points of each scheme were statistically analyzed.

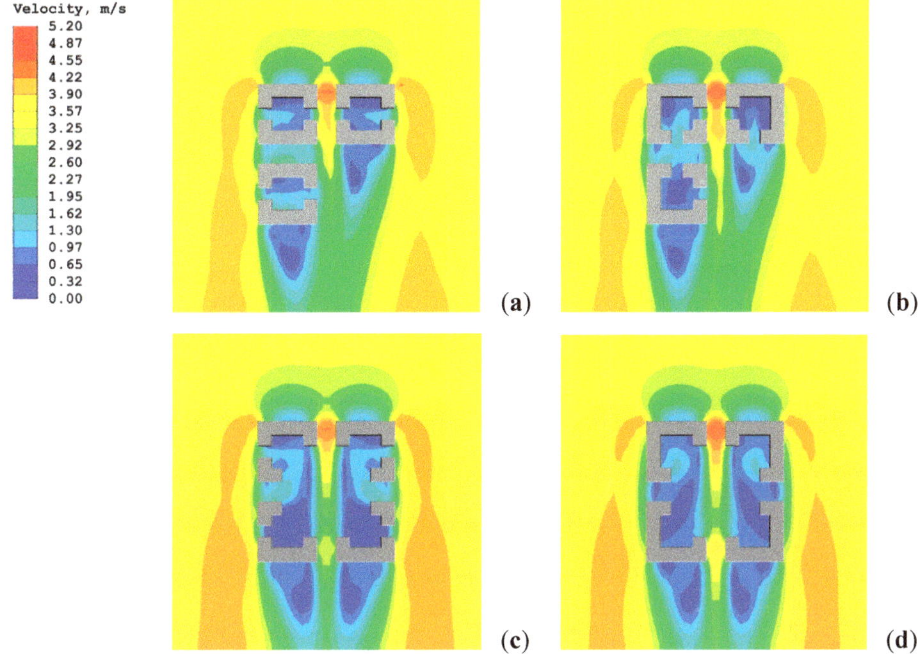

Figure 9. Winter simulation results of four layouts. (**a**) Winter wind simulation results of M-p scheme; (**b**) Winter wind simulation results of M-r scheme; (**c**) Winter wind simulation results of O-p scheme; (**d**) Winter wind simulation results of O-p scheme.

Figure 10 shows the contour map of the wind speed ratio at the outdoor pedestrian height (1.5 m) of each layout scheme.

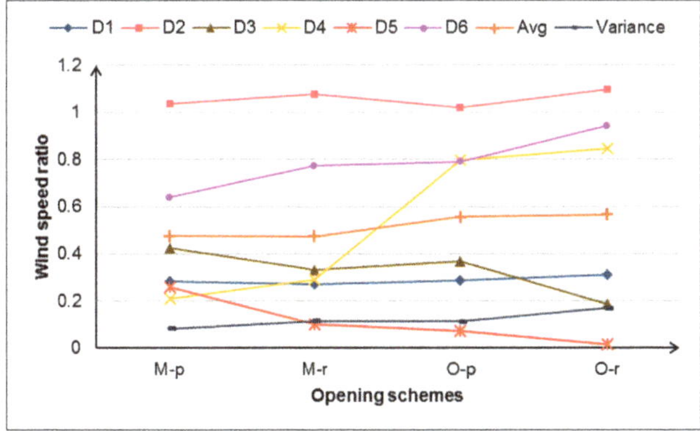

Figure 10. Measuring points' wind velocity ratio of winter wind in four schemes.

Similar to summer monsoon simulation results, the wind speed of the measuring points on the axis parallel to the inflow wind direction was relatively large in the four layout schemes. The actual wind speed exceeded 3 m/s in many places, so winter shelter measures should be considered. Among the four layout forms, the variance of measuring points of the O-r type was the largest, which was 0.17, and the wind speed ratio varied from 0.02 to 1.10. The variance of measuring points of the M-p type was the smallest, which was 0.08, and the wind speed ratio varied from 0.21 to 1.04. This indicates that the natural ventilation performance of each area of the O-r layout was very different, and the natural ventilation performance of the M-p layout was relatively uniform, and the wind speed ratio was closest to the comfort range (0.5–0.93), as is shown in Figure 11.

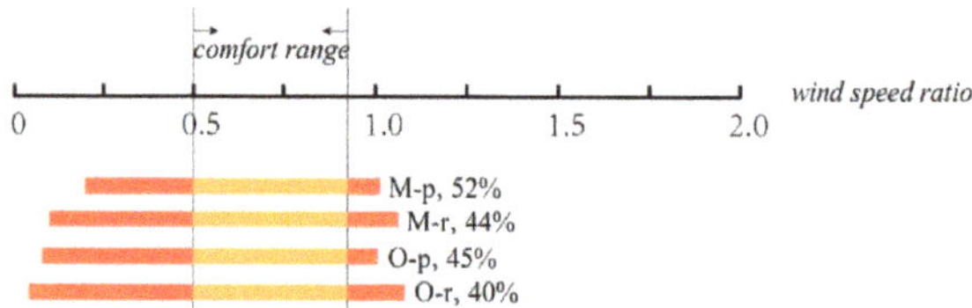

Figure 11. Comfort ratio of the four layouts (winter wind).

As can be seen from the Avg curve in Figure 10, the average wind speed ratio of the ratios of the four layout forms was between 0.47 and 0.57, which meets the comfort requirements of the winter monsoon. The evaluation of winter ventilation performance was as follows: M-p > M-r > O-p > O-r.

3.3. Influence of Overhead Ratio on Outdoor Wind Environment of Enclosed Building Group

From the analysis of the simulation results, results indicated that the outdoor wind environment of the O-r type was the most unfavorable. Therefore, taking the O-r type as an example, the influence of the overhead ratio on the outdoor wind environment of an enclosed building group was discussed. The model and measuring points were set, as shown in Figure 12 (the overhead ratio in this paper refers to the ratio of the ground floor area to the floor space of the building).

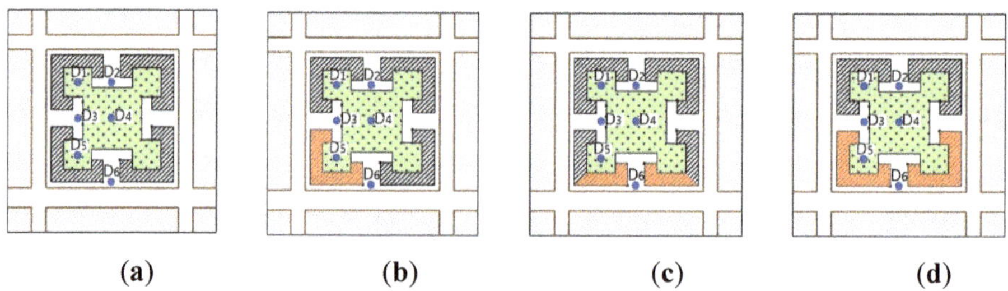

(a)　　　　**(b)**　　　　**(c)**　　　　**(d)**

Figure 12. Distribution of the overhead measuring points. (**a**) Overhead ratio = 0; (**b**) Overhead ratio = 25% (one side); (**c**) Overhead ratio = 25% (middle); (**d**) Overhead ratio = 50%, the orange part represents the overhead position.

Figure 13 shows the outdoor wind environment under the influence of summer and winter winds when the ground floor of the southwest side was overhead. This indicates that when the ground floor was overhead, it had a great influence on the summer wind airflow and little influence on the winter wind airflow. It could meet the needs of summer ventilation and winter shelter that could create a comfortable outdoor wind environment.

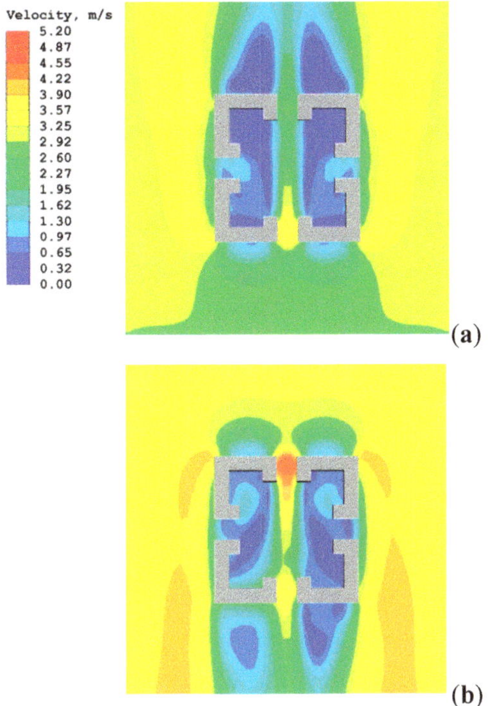

Figure 13. Simulation results with overhead layers using the overhead ratio of 25% on one side as an example. (**a**) Summer wind simulation result with overhead layers; (**b**) Winter wind simulation result with overhead layers.

Figure 14a is a contour map of wind speed ratios of four layout schemes under the influence of the summer monsoon. The overhead layer was on the side of the inflow direction, and the overhead ratio was 0, 25% (one side), 25% (middle), and 50%. The figure indicates that when the ground floor was overhead, it was more comfortable because of more airflow in summer. However, with the same overhead ratio, the outdoor wind environment was more comfortable when the overhead layer was in the middle (Figure 12c) than on one side (Figure 12b). As for the O-r layout, the simulation results of the overhead layers in the middle (Figure 12c) and the wind direction (Figure 12d) were very different, indicating that the wind conditions in the two cases were similar.

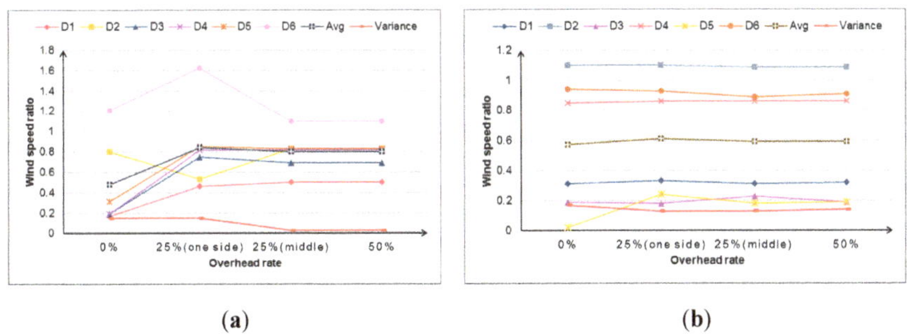

Figure 14. Wind velocity ratios when the overhead layers are in the inflow wind direction. (**a**) Wind velocity ratio of summer wind in O-r type; (**b**) Wind velocity ratio of winter wind in O-r type.

Figure 14b is a contour map of wind speed ratios of four layout schemes under the influence of winter monsoon. The overhead layer was on the side of the outlet direction, and the overhead ratio was 0, 25% (one side), 25% (middle), and 50%. This indicates that when the ground floor was overhead, the natural ventilation performance of the winter monsoon did not change much, and the effect of the overhead ratio on the outdoor wind environment was also small. This means that the impact of the overhead rate on the outdoor wind environment in winter can be ignored at this time.

When the overhead layer was located in the summer inflow direction, it could improve the problem of the unfavorable outdoor wind environment in the enclosed building group in the summer. However, it had little effect on the winter monsoon airflow. Therefore, setting up an overhead layer in the summer inflow direction is an effective measure to improve the outdoor wind environment of the O-r layout.

4. Conclusions

The enclosed space has good natural ventilation potential and application prospects, but the current actual situation is not very optimistic. Existing research lacks consideration of the natural ventilation performance of enclosed spaces and fails to solve the problem of low space utilization. Furthermore, the existing wind environment evaluation standards rarely consider the impact of temperature and humidity. The innovation of this article is to clarify the relationship between the layout of four typical enclosed office building group layouts (the M-p, M-r, O-p, O-r types) in Hangzhou and the comfort of the outdoor wind environment. At the same time, this article proposes improvement measures for the layout of poor natural ventilation performance. It studies the relationship between the overhead rate and the outdoor wind environment taking the O-r layout type as an example. Further research directions should clarify the parameter relationship between the two. Furthermore, the research method in this paper takes into account regional climate characteristics in more depth based on the best practice guidelines for CFD simulations. The conclusions of this paper can be summarized as follows:

The layout of the enclosed building group has a great impact on the outdoor wind environment, and the natural ventilation performance of the multi-yard type is better than the overall courtyard type. Among the four (M-p, M-r, O-p, O-r) layouts, the natural ventilation performance of the M-r layout type is the best, and the natural ventilation performance of the O-r layout type is the worst. The evaluation of natural ventilation performance is M-p > M-r > O-p > O-r.

In the enclosed building group it is easy to create a more comfortable outdoor wind environment in the courtyard space under the influence of the winter monsoon. However, appropriate measures should be taken where tunnel wind may be generated.

However, the courtyard space needs to consider measures to promote airflow in the courtyard space during the summer monsoon. It is an effective measure to use overhead layers locally. When the overhead ratio is 25% (middle), it not only has a high plot ratio but also can create a comfortable outdoor wind environment. The overhead ratio may have a certain functional relationship with the natural ventilation performance of the enclosed office building group, but more data is needed to verify this.

Author Contributions: X.Y. and Y.W. conceived the study, performed simulations and data analysis; X.Y. and Y.W. wrote the paper; W.L., Z.L. and G.D. provided constructive comments. All authors have read and agreed to the published version of the manuscript.

Funding: This research was funded by National Natural Science Foundation of China grant number 51878608; Natural Science Foundation of Zhejiang Province grant number LY18E080025; and Hangzhou Social Development self-declaration project grant number 20180533B08.

Conflicts of Interest: The authors declare no conflicts of interest.

References

1. National Bureau of Statistics of China. Available online: http://www.stats.gov.cn/tjsj/ndsj/ (accessed on 5 November 2019).
2. Wang, J.G. Research on large-scale urban spatial form based on urban design. *Chin. Sci. (Ser. E Tech. Sci.)* **2008**, *39*, 830–839.
3. Ye, Z.N. Review and prospect of research status of urban wind environment in China. *Planner* **2015**, *31*, 236–241.
4. Yuan, M. Preliminary analysis on design principles and development trends of modern office buildings. *Anhui Archit.* **2015**, *202*, 19–20.
5. Liu, X.X.; Liu, T. Optimized comparison of enclosing arrangement on sunshine condition. *J. Qingdao Technol. Univ.* **2014**, *35*, 61–70.
6. Abdulbasit, A.; Ibrahim, N.; Ahmad, S.S.; Yahya, J. Courtyard design variants and microclimate performance. *Procedia-Soc. Behav. Sci.* **2013**, *101*, 170–180.
7. Hang, J.; Li, Y.; Sandberg, M. Experimental and numerical studies of flows through and within high-rise building arrays and their link to ventilation strategy. *J. Wind Eng. Ind. Aerodyn.* **2011**, *99*, 1036–1055. [CrossRef]
8. Guo, F.; Zhu, P.S.; Wand, S.; Duan, D.; Jin, Y. Improving natural ventilation performance in a high-density urban district: A building morphology method. *Procedia Eng.* **2017**, *205*, 952–958. [CrossRef]
9. Kuo, C.-Y.; Tzeng, C.-T.; Ho, M.-C.; Lai, C.M. Wind Tunnel Studies of a Pedestrian-Level Wind Environment in a Street Canyon between a High-Rise Building with a Podium and Low-Level Attached Houses. *Energies* **2015**, *8*, 10942–10957. [CrossRef]
10. Guo, W.H.; Liu, X.; Yuan, X. A case study on optimization of building design based on CFD simulation technology of wind environment. *Procedia Eng.* **2015**, *121*, 225–231. [CrossRef]
11. Sharples, S.; Bensalem, R. Airflow in courtyard and atrium buildings in the urban environment: A wind tunnel study. *Sol. Energy* **2001**, *70*, 237–244. [CrossRef]
12. Liu, S.; Huang, C. Optimization of Natural Ventilation of "Yinzi" Dwellings in Western Hunan Based on Orthogonal Experiment and CFD. *Buildings* **2016**, *6*, 25. [CrossRef]
13. Jin, H.; Liu, Z.; Jin, Y.; Kang, J.; Liu, J. The Effects of Residential Area Building Layout on Outdoor Wind Environment at the Pedestrian Level in Severe Cold Regions of China. *Sustainability* **2017**, *9*, 2310. [CrossRef]
14. Xu, X.; Luo, F.; Wang, W.; Hong, T.; Fu, X. Performance-Based Evaluation of Courtyard Design in China's Cold-Winter Hot-Summer Climate Regions. *Sustainability* **2018**, *10*, 3950. [CrossRef]
15. Ying, X.Y.; Zhu, W.; Hokao, K.; Ge, J. Numerical research of layout effect on wind environment around high-rise buildings. *Archit. Sci. Rev.* **2013**, *33*, 1097–1103. [CrossRef]
16. Ying, X.Y.; Wang, Y.L.; Li, W.Z.; Liu, Z.Q.; Ding, G. Enclosed multi-story buildings' opening and wind environment of courtyard space. *Huazhong Archit.* **2019**, *37*, 49–52.
17. China Institute of Building Standard Design & Research. *Architectural Design Code for Fire Protection*; China Planning Press: Beijing, China, 2015; pp. 1–74.
18. Yang, J.Y.; Zhang, T.; Tan, Y. Research on urban wind environment: Technology evolution and integration of evaluation systems. *South Archit.* **2014**, *3*, 31–38.
19. *Hangzhou City Planning Management Technical Regulations*; Hangzhou Planning Bureau: Hangzhou, China, 2008.
20. *Recommendations for Loads on Buildings*; Architectural Institute of Japan: Tokyo, Japan, 2004.
21. Chinese Academy of Architecture. *The National Standard of the People's Republic of China*; Lode for Load of Building Structures; China Building Industry Press: Beijing, China, 2006.
22. Mochida, A.; Tominaga, Y.; Murakami, S.; Yoshie, R.; Ishihara, T.; Ooka, R. Comparison of various k-ε models and DSM applied to flow around a high-rise building—Report on AIJ cooperative project for CFD prediction of wind environment. *Wind Struct.* **2002**, *5*, 227–244. [CrossRef]
23. Shirasawa, T.; Tominaga, T.; Yoshie, R.; Mochida, A.; Yoshino, H.; Kataoka, H.; Nozu, T. Development of CFD method for predicting wind environment around a high-rise building part 2: The cross comparison of CFD results using various k-ε models for the flow field around a building model with 4:4:1 shape. *AIJ J. Technol. Des.* **2003**, *18*, 169–174. [CrossRef]

24. Yoshie, R.; Mochida, A.; Tominaga, Y.; Kataoka, H.; Harimoto, K.; Nozu, T.; Shirasawa, T. Cooperative project for CFD prediction of pedestrian wind environment in Architectural Institute of Japan. *J. Wind Eng. Ind. Aerodyn.* **2005**, *95*, 1551–1578. [CrossRef]

25. Franke, J.; Hirsch, C.; Jensen, A.G.; Krüs, H.W.; Schatzmann, M.; Westbury, P.S.; Miles, S.D.; Wisse, J.A.; Wright, N.G. Recommendations on the use of CFD in wind engineering. In *COST Action C14, Impact of Wind and Storm on City Life Built Environment, Proceedings of the International Conference on Urban Wind Engineering and Building Aerodynamics, Rhode-St-Genèse, Belgium, 5–7 May 2004*; Van Beeck, J.P.A.J., Ed.; vonKarman Institute, Sint-Genesius-Rode: Sint-Genesius-Rode, Belgium, 2004.

26. Tominaga, Y.; Mochida, A.; Yoshie, R.; Kataoka, H.; Nozu, T.; Yoshikawa, M.; Shirasawa, T. AIJ guidelines for practical applications of CFD to pedestrian wind environment around buildings. *J. Wind Eng. Ind. Aerodyn.* **2008**, *96*, 1749–1761. [CrossRef]

27. Baetke, F.; Werner, H.; Wengle, H. Numerical simulation of turbulent flow over surface-mounted obstacles with sharp edges and corners. *J. Wind Eng. Ind. Aerodyn.* **1990**, *35*, 129–147. [CrossRef]

28. Cheng, Y.; Lien, F.S.; Yee, E.; Sinclair, R. A comparison of large Eddy simulations with a standard k–ε Reynolds-averaged Navier–Stokes model for the prediction of a fully developed turbulent flow over a matrix of cubes. *J. Wind Eng. Ind. Aerodyn.* **2003**, *91*, 1301–1328. [CrossRef]

29. Franke, J.; Hellsten, A.; Schlünzen, H.; Carissimo, B. Best practice guideline for the CFD simulation of flows in the urban environment. In *COST Action 732, Quality Assurance and Improvement of Microscale Meteorological Models*; COST Office: Brussels, Belgium, 2007.

30. Moonen, P.; Dorer, V.; Carmeliet, J. Effect of flow unsteadiness on the mean wind flow pattern in an idealized urban environment. *J. Wind Eng. Ind. Aerodyn.* **2012**, *104–106*, 389–396. [CrossRef]

31. Zhang, A.; Gao, C.; Zhang, L. Numerical simulation of the wind field around different building arrangements. *J. Wind Eng. Ind. Aerodyn.* **2005**, *93*, 891–904. [CrossRef]

32. Meteorological Data Department of Meteorological Information Center; Department of Building Science of Tsinghua University. *Special Meteorological Data Set for Building Thermal Environment Analysis in China*; China Building Industry Press: Beijing, China, 2005; p. 126.

33. Kubota, T.; Miura, M.; Tominaga, Y.; Mochida, A. Wind tunnel tests on the relationship between building density and pedestrian-level wind velocity: Development of guidelines for realizing acceptable wind environment in residential neighborhoods. *Build. Environ.* **2008**, *43*, 1699–1708. [CrossRef]

34. Stathopoulos, T. Pedestrian level winds and outdoor human comfort. *J. Wind Eng. Ind. Aerodyn.* **2006**, *94*, 769–780. [CrossRef]

35. Stathopoulos, T.; Wu, H.; Zacharias, J. Outdoor human comfort in an urban climate. *Build. Environ.* **2004**, *39*, 297–305. [CrossRef]

36. Hyungkeun, K.; Kyungsoo, L.; Taeyeon, K. Investigation of Pedestrian Comfort with Wind Chill during Winter. *J. Sustain.* **2018**, *10*, 274. [CrossRef]

37. Wang, X.L.; Wei, X.D.; Wu, Y.Q.; Wang, X.F. The change of climate comfortable index in Hangzhou city during the past 50 years. *J. Baoji Univ. Arts Sci. (Nat. Sci.)* **2014**, *34*, 31–35.

38. Baghaei Daemei, A.; Khotbehsara, E.M.; Nobarani, E.M.; Bahrami, P. Study on wind aerodynamic and flow characteristics of triangular-shaped tall buildings and CFD simulation in order to assess drag coefficient. *Ain Shams Eng. J.* **2019**, *10*, 541–548. [CrossRef]

Article

Air Enthalpy as an IAQ Indicator in Hot and Humid Environment—Experimental Evaluation

Michał Piasecki [1],* , **Krystyna Kostyrko [1]**, **Małgorzata Fedorczak-Cisak [2]** and **Katarzyna Nowak [2]**

[1]   Building Research Institute, Filtrowa 1, 00-611 Warszawa, Poland; k.kostyrko@itb.pl
[2]   Faculty of Civil Engineering, Cracow University of Technology, 24 Warszawska Street, 31-150 Krakow, Poland; mfedorczak-cisak@pk.edu.pl (M.F.-C.); knowak@pk.edu.pl (K.N.)
*   Correspondence: m.piasecki@itb.pl; Tel.: +48-22-5664-352

Received: 17 February 2020; Accepted: 16 March 2020; Published: 20 March 2020

Abstract: The authors studied the impact of indoor air humidity in the range of 60% to 90% on building user perception in the temperature range of 26 to 28 °C. The research thesis was put forward that the impact of humidity on indoor air quality dissatisfaction of building users in a warm and humid indoor environment is greater than that indicated in thermal comfort models. The presented experiment examined the indoor air quality perception of $n = 28$ subjects in the test chamber of a nearly zero energy building under ten environmental conditions, together with a thermal comfort assessment. The authors developed an experimental relation for predicting building users' satisfaction based on the Weber–Fechner law, where the predicted percentage of dissatisfied users (PD) is determined by means of air enthalpy (h), $PD = f(h)$. The obtained results confirmed the sated thesis. Additionally, the intersection points of the experimental function and isotherms resulting from the Fanger model are presented, where the thermal comfort assessment starts to indicate lower user dissatisfaction results than experimental values. The authors recommend the experimental equation for humid air enthalpies in the range of 50 to 90 kJ/kg. The indoor air quality assessment based on the enthalpy value is simple and can be used to determine the overall Indoor Environmental Quality index of a building (IEQ_{index}).

Keywords: indoor air quality; IAQ; enthalpy; humidity; thermal comfort; TC; dissatisfaction; panel tests; nearly zero energy building; NZEB; indoor environmental quality; IEQ

---

1. Introduction

1.1. Literature Review

People are constantly exposed to the indoor environment of buildings, which is crucial for human thermoregulation and respiratory process; consequently, people's reactions reflect the level of indoor air parameters. The impact of the indoor environment is responsible for people's health, psychophysical state and influences behavioural change, concentration and work efficiency. As early as 1936, Yaglou [1] considered the effect of temperature and humidity on people in a study for American Society of Heating, Refrigerating and Air Conditioning Engineers (ASHRAE) on ventilation requirements. To date, both parameters are considered to be the most important elements impacting the satisfaction of building users. Strategies based on the physical measurements of the indoor environment allow to take the necessary steps to ensure adequate indoor human comfort [2]. Since the beginning of the twentieth century, many environmental variables such as temperature, black ball temperature, relative humidity, air velocity, radiation and others have helped determine various indoor thermal comfort indicators [3]. Each variable, however, can show a dominant effect in certain situations, not necessarily additive or linear. For example, humidity is indicated as a determinant of user satisfaction in hot and humid environments [4]. There are differences between the values selected by various authors as comfort conditions and the indicators used, e.g., operative temperature and humidity, effective temperature (ET*), Standard Effective Temperature (SET*), as well as black globe probe and humidity, black globe humidity index [5] and others. Also, the final parameters expressing the comfort of users due to the temperature and humidity in rooms in various publications are different. These include: thermal sensation vote (TSV), predicted mean vote (PMV), humidity sensation vote (HSV), thermal comfort index (TC), thermal comfort vote (TCV), thermal/humidity acceptance (THAV), percentage dissatisfied (*PD*), perceived indoor quality (PAQ), indoor air acceptability (ACC) and indoor air quality index (*IAQ*). The authors, like most researchers, tend to express building comfort parameters in % of satisfaction. The difficulty in transferring theoretical models to real conditions in nearly zero energy buildings (NZEB) creates a constant need to validate various comfort models taking into account very specific parameters of building, indoor condition specifications and scenarios of use [6].

The impact of humidity on human thermal and indoor air quality perception in hot and humid environments has been studied in several works in the previous years with various parameters and indicators. The use of enthalpy in studies of indoor perceptions has been carried out so far by using two main separate approaches. Most scientists have focused on the effects of temperature and humidity on people's thermal perception, where enthalpy was only a side indicator. Other researchers have analysed the direct impact of humidity and temperature on air quality perception. In fact, any research known to us did not study these two perception effects simultaneously in NZEBs with determination of the applicability ranges of these two approaches. Fang [7] argued that in higher temperatures and humidity, the respiratory cooling system is insufficient, so the air can be perceived as stuffy and uncomfortable. The linear correlation between air acceptability and enthalpy observed by Fang indicates that respiratory cooling is necessary for an acceptable perception of air quality. Also, Toftum [8] studied the thermal perception of comfort felt due to the human respiratory cooling system on a group of panellists in order to test air quality. His experiment led to almost the same conclusions. The correlation between air freshness and humidity was also found by Berglund and Cain [9]. When the respiratory cooling effect drops to a certain level (humidity and temperature, in practice enthalpy), the air will be perceived as bad, regardless of whether it is clean or contaminated. What is interesting is that Berglund even considered the possibility of supplying cool and dry air to alleviate the perception of poor air quality, without removing contaminants. Fang's team [7] found a correlation between the acceptance of air pollution by panel members and enthalpy, which represents the energy content of air humidity. In the five tested levels of selected pollutants in the air (temperature ranges: 18, 23 and 28 °C, and relative humidity (RH) ranges: 30%, 50% and 70%), there were highly significant linear regressions of acceptability against the enthalpy of the evaluated air at five levels of pollution. At low enthalpy (low temperature and humidity), the level of contamination was a key factor in the perception of air quality, and air pollution was less important for the perceived air quality as air enthalpy increased. In addition to some enthalpy level, for example at 28 °C and relative humidity 70%, temperature and humidity synergy were found to be the key determinants of perceived air quality. In this case, air was perceived as unacceptable, regardless of whether it was clean or not. In our article, we are interested in humid air without pollutants. A simple linear model was presented based on the results of Fang. The linear model of indoor air acceptability as a function of enthalpy using the following Equation (1) is:

$$ACC = a{\cdot}h + b \tag{1}$$

where, ACC is indoor air acceptability (takes values from sensory test −1 to 1, where −1 is completely unacceptable, 1 is fully acceptable), h is air enthalpy (kJ/kg), and a and b are linear regression coefficients different for specific air pollution levels. For clean air, the relationship was found to be as follows (2):

$$ACC = -0.033{\cdot}h + 1662 \tag{2}$$

Air acceptability (2) can be transformed using the Wargocki transformation into the percentage dissatisfied $PD_{\_Fang}$ in % using the formula provided in Reference [10] (3):

$$PD_{\_Fang} = \frac{100}{1 + \frac{1}{\exp(-4.28 \cdot ACC + 0.42)}} \tag{3}$$

According to authors, analysing a building's comfort using a percentage of dissatisfied *PD* offers many benefits. Various elements of the assessed indoor comfort can be integrated and the indicator is easily understood with the model IEQ developed by our team in Building Research Institute (ITB) [11], with the logistic regression IEQ model verified in Hong Kong [12] and with a literature review of the IEQ models creation [13]. The *PD* indicator can be used to designate indoor environmental quality index (IEQ) and for a building's certification [14].

Toftum's team [15,16] investigated the effect of humidity and temperature of inhaled air on perceived air acceptability. The air inhaled by subjects was rated as warmer and less acceptable with increasing air humidity and temperature. They developed a model that predicts the percentage of people dissatisfied with insufficient respiratory cooling depending on the actual evaporative and convective cooling of the airways (see Equation (4)). Both the temperature and humidity of the inhaled air had an impact on human perception of thermal breath sensitivity, freshness and acceptability. Respondents perceived inhaled air as cooler and more acceptable at lower temperatures and humidity. The results of this study confirmed the hypothesis that the perception of excess humidity is associated with the respiratory cooling system. Toftum's work resulted in a model for predicting the percentage of dissatisfaction with reduced respiratory cooling. The model is based on assessments of air at temperatures (t_a) in the range of 20 to 29 °C and vapour pressure (p_a) in the range of 1000 Pa to 3000 Pa. Enthalpy was in the range of 50 to 80 kJ/kg. The results, in accordance to Equation (4), indicate a rapid increase in dissatisfaction level with an increase in enthalpy above the 55 kJ/kg value:

$$PD_{\_Toftum} = \frac{100}{1 + \exp[-3.58 + 0.18 \cdot (30 - t_a) + 0.14 \cdot (42.5 - 0.01 \cdot p_a)]} \tag{4}$$

Another direction of using enthalpy for indoor comfort tests has been determined partly by the researchers dealing with the classic perception of thermal comfort. Thermal comfort is a condition in which a person feels that their body is in a state of natural heat balance, i.e., it feels neither warmth nor cold. The main research on thermal comfort was conducted by Fanger [17,18]. The results of his research became the basis for the development of the international standard (ISO) 7730 standard [19] on the analytical determination and interpretation of thermal comfort using the calculation of PMV and *PD* indicators and criteria of local thermal comfort and any other interesting articles validating the model, e.g., Reference [6]. *PD* is an indicator related to thermal comfort in indoor environments [20], finding use in the engineering assessment of the thermal comfort of rooms, this is the expected percentage of dissatisfaction with the thermal conditions in the room. People who choose −3, −2, +2, +3 on the predicted mean vote scale (PMV) during the experiment are considered dissatisfied with the thermal comfort in the room. PMV for nearly zero energy buildings is converted into *PD* in % according to the author's empirical formula [6] from the experiment validating the Fanger model:

$$PD = 100 - 99.9 \cdot \exp(-0.0355 \cdot PMV^4 - 0.242 \cdot PMV^2) \tag{5}$$

The PMV model is based on the identification of skin temperature and sweating rate required for optimal comfort conditions, based on experimental data and literature e.g., from Rohles and Nevins [21]. Thermal comfort has been characterised by taking into account the parameters of the environment and the human body, using models of extended heat transfer. Increased dissatisfaction according to model (4) is significantly higher than in classic models of thermal comfort based on Fanger's model (5). These indicate the considerable importance and potential of including this approach for planning nearly zero energy buildings assessments.

Further research on the Fanger model included the process of adapting people to thermal conditions. Van Hoof et al. [22] have reviewed thermal comfort models for indoor applications from the second half of the 1990s to 2010. Djongyang [23] reviewed the contribution of the adaptive model approach, addressing the behavioural and psychological adaptation of people in an indoor environment. Halawa and van Hoof [24] have reviewed the adaptive model approach. Croitoru [25] refers in more detail to human thermophysiological models and adaptive psychological models, again promoting the combination of the human body thermoregulation model with the numerical approach as the most effective tool for assessing thermal comfort in an indoor environment. Air humidity was addressed by ASHRAE, which developed a standard for building comfort requirements. The standard is known as ASHRAE Standard 55: 2017 [26]. The purpose of this standard is to define a combination of indoor thermal environmental and personal factors that create thermal environmental conditions, which are acceptable to most residents in the building space. One of the most recognisable features of the Standard 55 is the "ASHRAE Comfort Zone" presented on a modified psychrometric chart. The standard allows the use of thermal comfort charts in places where people have certain levels of activity that cause a metabolic rate in the range of 1.0 to 1.3 met (kcal/kg/hour), and where clothes provide thermal insulation from 0.5 clo to 1.0 clo (1 clo = 0.155 m^2 K/W). The comfort zone is based on PMV values from −0.5 to +0.5. Enthalpy recognised by ASHRAE as comfortable is in the range of 35 to 55 kJ/kg in winter and 40 to 60 kJ/kg in summer.

The results on human perception to high humidity in higher temperatures, as provided in Reference [27], indicated that the impact of humidity on human responses was significant and increased with an increase in air temperature when the relative humidity was above 70%. The indoor comfort in hot–humid conditions was also studied by Kleber and Wagner [28]. A total of 136 subjects were tested in a climate chamber with specific hot–humid conditions in a test facility at the Karlsruhe Institute of Technology. Nine experimental conditions with high operative temperature and different relative humidity levels (26, 28 and 30 °C combined with 50%, 65% and 80% humidity) were studied. The significant influence of air humidity on indoor air quality and thermal perception was found. The acceptability of indoor air quality under temperatures of 26 to 30 °C and RH of 60% to 80% was also studied by He et al. from Hunan University [29]. The authors confirmed a significant increase in dissatisfaction level for higher humidity. In Reference [30], Jing et al. studied the influence of relative humidity on thermal comfort in an environmental chamber. Twenty subjects were exposed to nine combinations of humidity and temperatures. Once again, higher humidity had a negative effect on the subjects' thermal comfort. Zhai [31] examined the effects of air movement from ceiling fans on subjective thermal comfort and perceived air quality for hot–humid environments. In a climate chamber controlled at three temperatures (26, 28 and 30 °C) and two relative humidity levels (RH 60% and 80%), sixteen subjects dressed in summer clothing (0.5 clo) were exposed to seven levels of air speed ranging from 0.05 m/s to 1.8m/s. The subjects were asked to evaluate thermal sensation, comfort and perception of indoor air quality. Without air movement, the unacceptable limit established by the ASHRAE standard 55 was reached very quickly even with moderate humidity. In Reference [32], Buonocore studied naturally ventilated building environments to evaluate the influence of relative humidity and air speed on the occupants' thermal perception. Indoor environmental variables were measured alongside questionnaires, focusing on thermal environment and air movement evaluation. The results indicated that relative humidity had a significant negative impact on thermal perception. Rana [33] used subjective responses in surveys as grounds to validate the performance of the thermal comfort prediction. The results confirm that humidity index may be an important predictor of indoor comfort at high humidity. The impact of humidity on the comfort of building users is also analysed in the literature on human comfort [34].

The relationship between indoor air humidity and the humidity of the partitions and walls that are not discussed in this publication have been presented and discussed by Kaczorek [35] and

Krause [36,37]. Air humidity also has a significant impact on the building's energy consumption, which was, for example, discussed by Gawin [38].

In the Discussion Section, the authors refer to studies focused on sensory comfort evaluation tests, and the results are compared.

1.2. Research Hypothesis

In Reference [39], research focused on tropical climates found that the International standard for indoor climate, ISO 7730 based on Fanger's predicted mean vote (PMV/*PD*) equations, does not adequately describe comfortable conditions in a wide spectrum of temperatures and humidity. This paper presents some of the evidence and suggests ways in which ISOs are failing in determining the implications of air humidity. The direct impact of air humidity on air quality has been studied in a relatively small number of papers. The perception of IEQ in hot humid conditions was studied in References [7,15] in ventilated ecological houses [40] and in a climate test facility [28], where for higher temperatures and humidity, the dissatisfaction of users with indoor air quality (respiratory cooling) is higher than the discomfort associated with the thermal sensation. Taking this into consideration, the authors' intention was to analyse the actual humidity impact on the perception of building occupants in an experimental study, taking place in a test chamber located in a low-energy building. The authors believe that air enthalpy is the most suitable indicator for determining the effects of humidity on user comfort (as a percentage of dissatisfaction with IAQ) at selected conditions. Related studies on the impact of humidity on air quality have usually been carried out using a method in which users inhaled air with specified pollutants, vapour gradients and varying temperatures (also without indoor pollutants) [7,15]. The authors of this article put forward the thesis that one should not conduct experimental tests of actual thermal comfort and air quality of the indoor environment separately because indoor air discomfort in hot and humid environments may be higher than thermal discomfort. Example results of the percentage of dissatisfied *PD* estimated in the enthalpy function resulting from the Fanger model and the ISO 7730 standard based on general human thermal balance ($PD\_{ISO7730}$, Table 1) and results obtained in models based on thermal sensations resulting from respiratory cooling comfort models (PD_{Fang}, PD_{Toftum}, Table 1) [7,15] differ significantly (especially for higher enthalpies), as shown in Table 1 for selected temperatures, humidity and calculated enthalpy values (clothing level = 0.6 clo, metabolic rate = 1.1 met). Enthalpy (h) and *PD* for sample temperatures and humidity are obtained by the Monte Carlo simulation and related model implementation. Based on Table 1, it can be concluded that for rooms with increased air enthalpy, the Fanger thermal comfort model may not be suitable for the overall indoor human comfort assessment. Significant discomfort occurs due to the comfort associated with air quality. This research subject was analysed in this paper by the experimental approach.

Taking into consideration the differences of up to several dozen percent in the results of the expected percentage of indoor environment dissatisfaction (Table 1, air quality versus thermal comfort), the authors decided to conduct an experiment in order to empirically analyse the humidity impact on human comfort under comparable boundary conditions. In the indoor environment of the building, almost zero energy-specific conditions with increased humidity and temperature were identified, enabling the use of surveys to determine the percentage dissatisfied.

The authors hypothesise that it is possible to express the percentage of dissatisfaction in conditions of increased humidity and temperature as the function, probably logarithmic as the Weber–Fechner law, of the enthalpy stimulus. The hypothesis originates in the science of the physiology of the human body, for which the universal Weber–Fechner law should apply [41,42]:

$$R = k \cdot \ln(S) \tag{6}$$

where R is the human perceptual variable related to stimulus, S is the stimulus of the environment causing the response and k is the ratio of proportionality. The authors' intention is to check experimentally whether in an indoor hot and humid environment Equation (6) can be used where enthalpy is an important stimulus (Equation (7)):

$$PD_{\_exp} = k \cdot \ln \left(\frac{h}{h_{th}} \right)$$ (7)

where $PD_{\_exp}$ is the percentage of dissatisfaction with the enthalpy (%), h is the actual enthalpy (kJ/kg) and h_{th} is the border neutral enthalpy for user perception (kJ/kg).

Table 1. Percent dissatisfied (*PD*) estimated for three human comfort models for selected indoor environment data (Monte Carlo method), where $PD_{\_ISO7730}$ values are estimated using the Fanger thermal comfort model, *PD*_Toftum is estimated using the indoor air acceptance model (4) and *PD*_Fang is estimated using the indoor air acceptance model—Equations (2) and (3) (under the assumptions: air speed < 0.1 m/s, 0.6 clo, 1.1 met, $t_a = t_{mr}$).

t_a °C	RH %	h [1] kJ/kg	$PD_{\_ISO7730}$ %	$PD_{\_Toftum}$ %	$PD_{\_Fang}$ %
26	50	53	8	32	39
27	50	56	15	40	50
28	50	58	26	48	58
26	60	58	9	43	58
27	60	61	18	52	65
28	60	65	30	61	69
26	70	64	11	55	68
27	70	67	21	64	73
28	70	71	34	73	76
26	80	70	13	66	75
27	80	73	24	75	78
28	80	77	39	82	80
26	90	75	15	76	79
27	90	79	27	83	81
28	90	83	44	88	83
29	90	88	62	92	85

[1] For enthalpy calculation, the Magnus–Tetens and Clausius–Clapeyron approximation was used as the accepted method in climatology.

Hypothetically, it is assumed that the enthalpy neutral for users (h_{th}) is in the range of 50 to 55 kJ/kg, which has to be confirmed by tests. The authors hypothesise that the Fanger model applies only to the threshold/neutral enthalpy value, and that the above models similar to Fang and Toftum are valid for a general comfort assessment. Also, Jokl [43] conducted research on the introduction of the Weber–Fechner law to assess thermal comfort but he did not confirm the relationship experimentally.

2. Materials and Methods

The main research objective is to express the percentage of user dissatisfaction in conditions of increased humidity and temperature in indoor building environment as the function of air enthalpy, $PD = f(h)$, by experimental evaluation. Based on the hypothesis, the authors set out to empirically determine the role of enthalpy in indoor comfort models by sensory evaluation tests undertaken in the NZEB building.

2.1. Research Scheme and Approach

For illustrative purposes, the authors provide a research scheme in Figure 1.

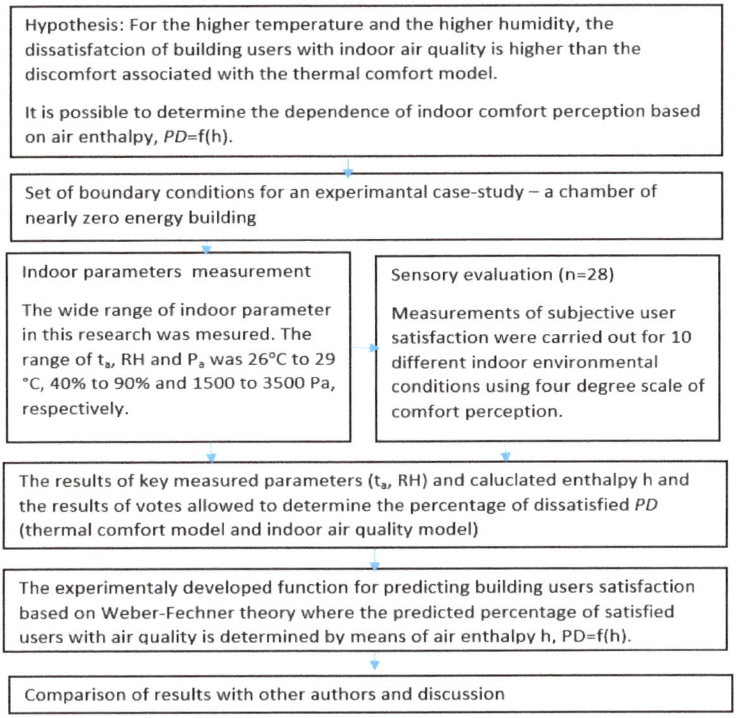

Hypothesis: For the higher temperature and the higher humidity, the dissatisfatcion of building users with indoor air quality is higher than the discomfort associated with the thermal comfort model.

It is possible to determine the dependence of indoor comfort perception based on air enthalpy, *PD*=f(h).

Set of boundary conditions for an experimantal case-study – a chamber of nearly zero energy building

Indoor parameters measurement

The wide range of indoor parameter in this research was mesured. The range of t_a, RH and P_a was 26°C to 29 °C, 40% to 90% and 1500 to 3500 Pa, respectively.

Sensory evaluation (n=28)

Measurements of subjective user satisfaction were carried out for 10 different indoor environmental conditions using four degree scale of comfort perception.

The results of key measured parameters (t_a, RH) and caluclated enthalpy h and the results of votes allowed to determine the percentage of dissatisfied *PD* (thermal comfort model and indoor air quality model)

The experimentaly developed function for predicting building users satisfaction based on Weber-Fechner theory where the predicted percentage of satisfied users with air quality is determined by means of air enthalpy h, PD=f(h).

Comparison of results with other authors and discussion

Figure 1. The assumptions and steps of the experiment.

2.2. Experimental Facilities

The test on users' perception of indoor air quality in hot and humid environments at NZEB office buildings was carried out in the experimental chamber of a nearly zero energy building (Central Europe), designed for physics tests in "in situ" conditions. Figure 2 shows a test functionality of the test chamber. The test room is thermally and acoustically insulated from the rest of the building. It is mainly warmed by heated walls and floors, and by a fan coil unit, which may additionally heat air if necessary. Building Management System (BMS) allows to control the room and wall temperatures, humidity, the number of exchanges and the concentration of CO_2 in the air. Additional humidity generating devices (M1–M3) were located in the room to help maintain a high level of humidity (RH > 70%). A total of 14 subjects could be in the room at each test.

The main measurement system (MC point at Figure 2) of the indoor environment was set up in front of the panel group. In addition, several testing sets were used by the authors to determine the homogeneity of thermal conditions in the room. The ventilation rate was about one air exchange per hour. The respondents answered questions (votes) in artificial light: 450 ± 50 lux. The measured CO_2 concentration level (for tests) did not exceed 1000 ± 30 ppm. The measured Total Volatile Organic Compounds (TVOC) air concentration levels did not exceed 150 ± 36 μg/m^3. Throughout the test, the authors took the necessary measures and actions to remove the potential asymmetry of the temperature field, air flow and humidity in the chamber, "quasi-stabilising" the indoor conditions before each test and maintaining the continuity of conditions during the test. The air speed was set at 0.1 ± 0.05 m/s. The difference between radiant temperature and air temperature was controlled "online" by a building management system (BMS) as less than ±1 °C. The measured vertical air temperature gradient was less than ±1 °C between the floor and the head of a seated person. The temperatures in the laboratory room changed by no more than ±1 °C at each test point. This was ensured by the heating of the walls and floor in the test room.

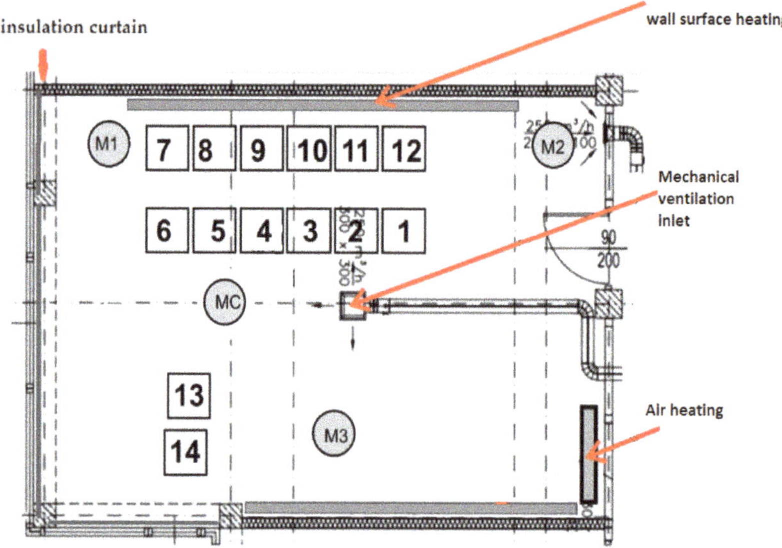

Figure 2. Chamber for measuring indoor comfort (where 1–14 are seats for participants in the sensory tests, M1–M3 are devices for generating high humidity, and MC is the location of microclimate sensors).

Outdoor conditions were monitored throughout the experiment, but the authors made an assumption that the variability of atmospheric conditions does not influence the specific results of the experiment in a statistically significant way and a test chamber of NZEB is sufficiently well insulated from the outdoor environment.

2.3. Enthalpy Determination Method

The enthalpy value, h, in kJ/kg was determined for all temperature and humidity conditions (ten tests). Enthalpy, h, was calculated on the basis of measured air temperature (t_a) and humidity (RH) using the ideal gas law, as follows:

$$h = h_a + w \cdot h_g \tag{8}$$

where:

$h_a = 1.006 \cdot t_a$, dry air enthalpy,

h_g = water vapour enthalpy,

t_a = measured air temperature,

$w = 0.622 \cdot P_a/(P_0 - P_a)$, humidity factor,

$w \cdot h_g = w \cdot (2501 + 1.805 \cdot t_a)$, water vapour enthalpy multiplied by the humidity factor,

$P_a = RH/100 \cdot P_S$, partial pressure of water vapour,

$P_s = 610.94 \cdot \exp(17.625 \cdot t_a/(t_a + 243.04))$, saturation vapour pressure (Pa), and

P_0 = atmospheric pressure.

In the designed test condition, the calculated enthalpy values were in the range of 50 to 90 kJ/kg.

2.4. Thermal Comfort Model-Measurements

The level of thermal discomfort was also determined in the experiment using a measuring device. PMV and *PD* were calculated in accordance to ISO 7730 for each sensory test. PMV/PD are reference parameters for thermal environmental assessment, as provided in standard EN 16798-1 [44]. PMV is a function of measured physical parameters as presented:

$$PMV = f(t_a, t_{mr}, v_a, p_v M, I_{cl})$$ (9)

where t_a is the measured indoor air temperature (°C), t_{mr} is the radiant temperature (°C), v_a is the measured air velocity (m/s), p_v is the water vapour partial pressure (Pa), M is the human metabolic rate (W/m^2) and I_{cl} is the human clothing insulation (m^2 K/W). All these parameters were measured and determined. The clothing level of was calculated at 0.6 clo, and metabolic rate was set at 1.1 met. The measurement methodology was based on ISO 7726. In practice, the values of PMV and $PD_{\_ISO7730}$ were obtained by measurement equipment but for validation purposes, they were re-calculated using the web tool found at http://comfort.cbe.berkeley.edu/EN of the Centre for the Built Environment, University of California, Berkeley. The obtained $PD_{\_ISO7730}$ results using the normative ISO method formed the basis for comparing the thermal comfort model with experimental results obtained from the assessment of indoor air quality perception.

2.5. Air Perception Sensory Evaluation Tests—Vote

The survey evaluation involved 28 subjects, students of the University of Technology, in two sessions (day 1 and day 2) on 7 November and 9 December 2019. The panellists group was ethnically homogeneous, with variation 100% white Caucasian. 20% were men and 80% women; however, gender differences were not included in the results discussion. The authors assume that the perception of comfort depends on body parameters and not on gender directly. The participants signed their consent to participate in the tests and declared their health state and anthropometric parameters prior to the tests. The anthropometric data characterising the panel group is provided in Table 2.

Table 2. Anthropometric data of tested panel groups with expanded uncertainty at the confidence level of $1 - \alpha = 0.95$.

Group	Gender	Group Size	Age (years)	Height (cm)	Body Weight (kg)	Skin Surface "DuBois" (m^2)	Body Mass Index	Clothing Insulation (clo)
Academic youth	Mean	28	23 ± 1	167 ± 8	62 ± 10	1.8 ± 0.3	22 ± 4	0.6 ± 0.1

The panellists group had a BMI of 22 ± 4. The neutral limit of body weight is in the range of 18.5 < BMI < 24.9. The value of clothing's thermal resistance (clo) between women and men was averaged and calculated. The fact that some women have long hair and wear extra underwear (like bras) was included in the uncertainty estimation. Subjects were wearing long trousers, short-sleeved shirts and shoes, which provides an insulation of clothes (I_{clo}) at 0.6 ± 0.1 clo. The performed physical activity was set as 1.1 ± 0.15 met (semi-active sitting/working in a seated position). The group remained air-conditioned in neutral conditions (N, Figure 3) before each test at PMV = −0.1; t_a = 23 °C, RH = 40%. The respondents evaluated their comfort perception in writing using the air acceptance vote considering temperature and humidity. The provided question was: determine using a 4-degree scale whether prevailing air conditions including actual humidity and temperature are comfortable for work, where 0 = neutral (comfortable), 1 = just comfortable, 2 = just not comfortable, 3 = not comfortable. Participants knew that air quality would be tested for different temperatures and humidity, but no values were given for actual parameters. Voting took place after about 15 minutes of being in the tested conditions in accordance to the experimental timetables (Figures 3 and 4). After each vote, the subjects returned to thermally neutral conditions (second room). Due to the number of voting places in the test room (maximum of 14), the panel group was divided into two smaller A and B groups voting at 15-minute intervals under the same temperature conditions. On day 1, in group A, there were 12 panellists, while in group B there were 11. On day 2, there were 14 panellists in groups A and B (n =

28). Figures 3 and 4 present the timetable of experiments. At day 1, six panellists were excluded from the experiment for being too late, having cold symptoms or wearing unsuitable clothing.

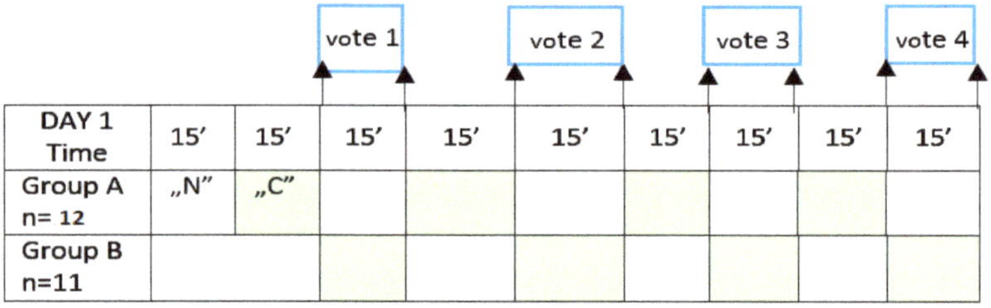

			vote 1		vote 2		vote 3		vote 4
DAY 1 Time	15′	15′	15′	15′	15′	15′	15′	15′	15′
Group A n= 12	„N"	„C"							
Group B n=11									

Figure 3. Experimental timetable (day 1, n = 23, "N"—neutral conditions, "C"—test chamber).

			vote 5		vote 6		vote 7		vote 8		vote 9		vote 10
DAY 2 Time	15′	15′	15′	15′	15′	15′	15′	15′	15′	15′	15′	15′	15′
Group A n=14	„N"	„C"											
Group B n=14													

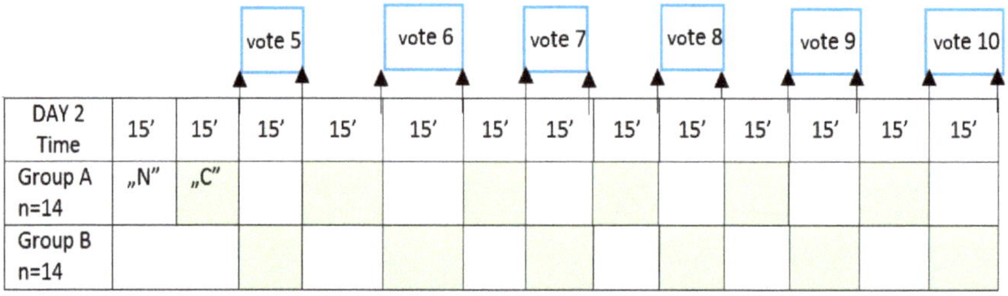

Figure 4. Experimental timetable (day 2, n = 28, "N"—neutral conditions, "C"—test chamber).

The experiment was conducted for ten specific conditions with different values of air enthalpy. While conducting the experiment, the authors took into account the level of energy required for metabolism, i.e., the demand for food (decided on a 3-hour maximum). The group did not consume meals for up to two hours before the study and during the tests. During the test, students were allowed to drink water (0.5 l maximum) to supplement their needs related to the secretion of sweat. A longer experiment could cause disturbances in the concentration of young people, and the aim was to maintain activity on the same level for 2 to 3 hours.

The authors did not take into consideration other human factors that may affect the results of comfort tests including: psycho-conditions, physiological circadian (day) rhythm and the level of nutrition before tests. Aware of the limitations, the authors argued that these conditions had a statistically minor significant impact on the results obtained, well within the calculated expanded uncertainty of 26%.

2.6. The Measuring Equipment

The range of indoor parameters was 26 to 29 °C, humidity was 40% to 90% and vapour pressure was 1500 to 3500 Pa. In such conditions, the enthalpy values were in the range of 45 to 90 kJ/kg. The indoor air parameters measurement device (MC) was located in the middle of the test area. Figure 5 shows the device used for measuring thermal indoor parameters: t_a = actual air temperature, t_g = temperature of black globe probe (heat radiation meter), t_{nw} = wet-bulb temperature, RH = relative air humidity and v_a = air flow speed. The measurements of physical indoor parameters were provided at three height levels: 0.05, 1 and 1.6 m above floor level in parallel. Only the chest level (1 m) of seated participants was considered for further calculations.

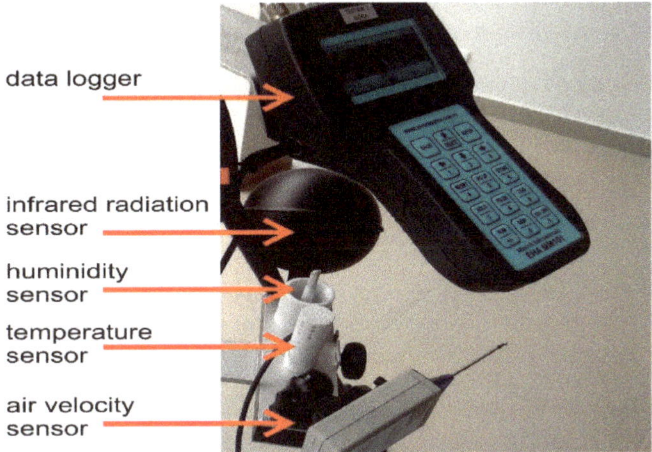

Figure 5. Sensors of the EHA-MM101 device for indoor environment tests (MC).

Measurements at three heights allowed for the reduction of any possible negative gradient of vertical temperature. The technical data and sensor resolution are presented in Table 3. All measuring sensors were calibrated by an accredited certifying laboratory.

Table 3. Sensors information.

Sensor	Measurement Range	Resolution	Accuracy
Temperature	−20 to 50 °C	0.01 °C	0.5 °C
Humidity	0% to 100%	0.1% RH	5%
Air speed	0.01 to 10 m/s	0.01 m/s	2%
Radiant temperature	0 to 50 °C	0.01 °C	2%
Carbon dioxide	0 to 5000 ppm	0.1 ppm	1 ppm

Other assumptions of the assessment methodology for determining thermal comfort were based on EN ISO 7730 [19]. The authors' intention was to maintain indoor air conditions in which the main pollutants are on a neutral level for the perception of users, as enthalpy is the only variable studied. Continuous CO_2 measurement was carried out during the experiment by a FYAD00 sensor and other side devices integrated with the building management system (BMS). The mean carbon dioxide concentration during tests was 650 ± 15 ppm, which corresponds to a neutral percentage of dissatisfaction ($PD_{CO2} < 10\%$) in accordance to Reference [45] and [46]. During day 1 and day 2, a measurement of volatile organic compounds (VOCs) air pollution was carried out. The air was collected on Tenax adsorbent samples and transported to the laboratory, then tested using the ISO 16000-6 and ISO 16000-3 provisions. Air samples were thermally desorbed and analysed in a Shimadzu QP2010 chromatograph. The VOCs were identified by the mass spectral database. The mean of total VOCs concentration (TVOC) was 120 μg/m^3 ± 18%, which corresponds to a neutral percentage of dissatisfaction ($PD_{TVOC} < 5\%$) in accordance to Reference [47].

As part of the calculations, the realistic uncertainty of measurement for all measuring devices was determined Table 4. The standard deviation of panel 'votes' was 12.9%. Uncertainties have been determined using the recommendations: for a model IEQ reliability analysis provided in Reference [48], for thermal comfort subjective test vote uncertainty analysis [49,50]. The specified expanded uncertainty of *PD_exp* assessment for the provided experiment was 26%. The uncertainty for enthalpy calculation considers the provisions in Reference [51]. The calculated effect of humidity on the enthalpy value in our research range had a standard deviation of 2.38%, and the effect of temperature on the enthalpy value determination had a standard deviation of 0.96%. The actual standard deviation of indoor air enthalpy determination in our case was 2.70%.

Table 4. Overall uncertainty (U) of experimental indoor perception evaluation (PD_exp = f(h)) based on enthalpy determination uncertainties (SD_h) and vote results (SD_{vote}).

Parameter	SD_h %	SD_{vote} %	U %
PD_exp	2.7	12.9	$2 \cdot (7.29 + 166.4)^{-2} = 26.4$

The standard deviations of the estimated thermal insulation of clothes were 0.6 ± 0.1 clo, and the metabolic rate of workers was 1.1 ± 0.15 met. The calculated uncertainty for the $PD_{ISO7730}$ determination for Fanger thermal comfort was 3.22%, considering References [48,52].

2.7. The Boarder Assumptions

The results refer to the buildings with a mechanical ventilation. The experiment results refer to the range of indoor parameters: temperature 26 to 29 °C, relative humidity range 40% to 90% and the enthalpy range of 45 to 90 kJ /kg. The following assumptions were used for all tests: air speed < 0.1 m/s, 0.6 clo, 1.1 met, $t_a = t_{mr}$. These assumptions are valid for the experiment as well as all thermal comfort calculations.

The panellists group was ethnically homogeneous, Caucasian. Results may not be representative for other ethnic groups. 20% were men and 80% women; however, gender differences were not included in the results discussion. Due to the fact that the research was conducted on students, the results may not be representative for older people.

The authors assume that other potential air pollution factors than CO_2 and TVOC (determined during the experiment) did not affect the subjects' satisfaction results obtained.

The subject group size (n = 28) affects the significance of the data analysis. In practice, the authors considered the sample size issue with Raosoft calculator (http://www.raosoft.com/samplesize.html) to set a minimum number of subjects. The authors adopted the following assumptions: the expected margin of error, the amount of error that we were able to tolerate, in our experience in sensory evaluations with panellists (students) is 20%. The confidence level was 95%. The global population size of university students was 14,000. The minimum recommended size of our survey calculated was 24. The authors set *n* = 28 test subjects for practical and technical reasons. With the expanded uncertainty of 26%, a sample of 28 (k = 95%) ensures reproducibility and representativeness of the study.

3. Results

3.1. Experimental Relation of Dissatisfaction with Perception of Indoor Air Condition

The respondents evaluated their comfort perception in writing using the air acceptance vote (considering temperature and humidity). As part of the experiment, measurements of user satisfaction were carried out for ten different indoor environment conditions using a 4-degree scale (where "0" is neutral air conditions (comfortable), "1" is just comfortable, "2" is just not comfortable, and "3" is not comfortable). The results of key measured parameters (t_a, RH) and calculated enthalpy (h), and the results of votes are presented in Table 5. Similar to other studies, like Fanger's, the number of dissatisfied participants was counted, including those who answered "2" or "3" in the survey.

Table 5. Vote results—number of panellist votes using a 4-degree scale (whether prevailing air conditions are comfortable for work, where 0 is neutral (comfortable), 1 is just comfortable, 2 is just not comfortable, 3 is not comfortable) for 10 selected indoor conditions and the calculated percentage of dissatisfied (PD_exp).

Test Number	Number of Votes	t_a	RH	h	Number of Panellists Voting for Degree in Comfort Scale "0"–"3"				PD_exp
–	–	°C	%	kJ/kg	"0"	"1"	"2"	"3"	%
1	23	27.2	44.9	53	11	8	3	1	17
2	23	27.0	45.8	53	12	7	3	1	17
3	23	26.8	44.8	52	13	8	2	0	9
4	23	26.6	43.4	52	17	3	1	0	5
5	28	28.3	56.8	63	0	16	8	4	43
6	28	28.6	59.7	64	2	14	8	4	43
7	28	28.5	62.3	68	2	13	10	3	46
8	28	28.3	71.8	73	2	10	8	8	57
9	28	27.6	80.5	77	1	3	6	16	79
10	28	27.4	86.0	79	1	2	10	15	89

The authors present the obtained results in the form of a relation of dissatisfaction with the perception of indoor conditions, *PD_exp* = f (*h*). In order to evaluate the shape of the curve, logarithmic regression was used to determine the experimental equation, consistent with the hypothesis that the *PD* results should correspond to Weber–Fechner's law, where indoor air enthalpy (h) is a stimulus (Figure 6).

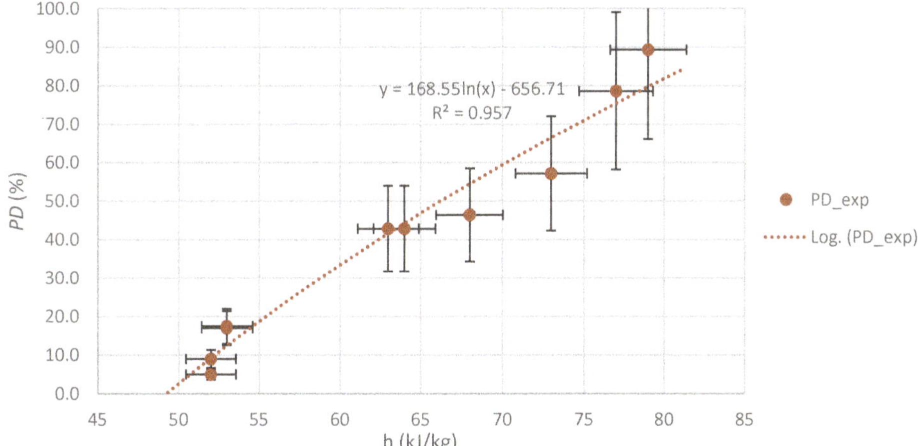

Figure 6. Percent of dissatisfied (subjects) in enthalpy function (*PD_exp*), results of experimental research and regression line (ln.(*PD_exp*)) under the following assumptions: air speed < 0.1 m/s, 0.6 clo, 1.1 met, $t_a = t_{mr}$, $c_{CO2} < 600$ ppm, $c_{TVOC} < 150$ μg/m³.

There is an assumption that the human perception neutral enthalpy is a hypothetical point at percentage dissatisfied (*PD*) = 0%. This enthalpy, h_{th}, indicates a border neutral perception for an unpolluted hot and humid air quality, the border point of the TC comfort zone defined by the Fanger equation. From this point, the impact of higher enthalpy on users can be calculated from the converted regression equation. The converted experimental dependence of users' dissatisfaction *PD* in % in the enthalpy function with $R^2 = 0.957$ takes the following form Equation (10):

$$PD\_exp = 168.55 \cdot ln\left(\frac{h}{49.22}\right) \tag{10}$$

According to the results obtained, neutral enthalpy, h_{th}, is 49.2 kJ/kg. Fang obtains a corresponding value of neutral acceptance for h = 45 kJ/kg, while Toftum obtains a value of 55 kJ/kg [15]. In a thermal comfort assessment, a neutral value of enthalpy cannot be determined.

3.2. Enthalpy Prediction for which the Thermal Comfort Model Gives Understated Results

The method of enthalpy prediction for a given temperature, at which the results from the thermal comfort model are starting to be lower than the actual dissatisfaction (as shown by the experimental model), is provided. Figure 7 shows a comparison of the authors' experimental indoor air perception function (*PD_exp*) and the estimated results of the predicted percentage of satisfaction with the thermal comfort model for the same enthalpy range for the indoor conditions under the following assumptions: air speed < 0.1 m/s, 0.6 clo, 1.1 met, $t_a = t_{mr}$. PD_ISO values for thermal comfort were calculated using the Fanger model from ISO 7730. The thermal dissatisfaction results were estimated for three constant temperatures 26, 27 and 28 °C. For a constant temperature value, 26 °C, and maximum relative humidity, RH = 100%, the maximum value of the dissatisfaction percentage is 17% (Figure 7). For a constant temperature value, 27 °C, and a maximum relative humidity, RH = 100%, the maximum value of the dissatisfaction percentage is 30% (Figure 7). For a constant temperature value, 28 °C, and a maximum relative humidity, RH = 100%, the maximum value of the dissatisfaction percentage is 47%. The comparison of the thermal comfort and indoor air quality models shows that *PD* experimental values start to be higher from the intersection points A to C in Figure 7.

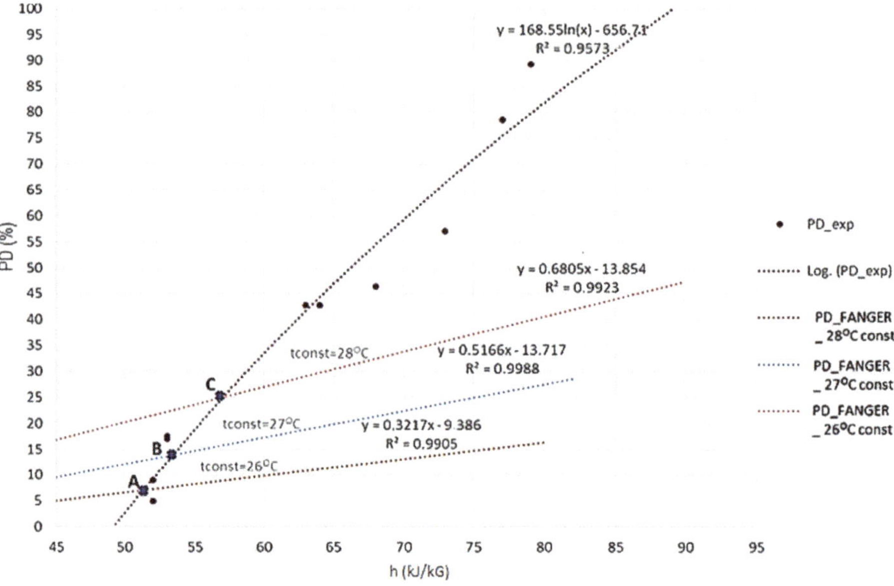

Figure 7. Comparison of the panel test results (black dots) of indoor air perception (*PD_exp*) with the results estimated using the thermal comfort model (*PD_ISO7730*) for three constant temperatures: 26, 27 and 28 °C (under the assumptions: air speed < 0.1 m/s, 0.6 clo, 1.1 Met, $t_a = t_{mr}$). Intersection points are presented in the graph with dots marked A, B and C. The *PD_Fanger* lines are interrupted in points where RH reaches 100%.

The real (experimental) occupants' dissatisfaction level is higher than the *PD* level predicted with the help of the Fanger equation curves.

Enthalpy values (*h*) for intersection points A, B and C are determined for three temperatures. The estimated humid air enthalpy points (at the three fixed temperatures) set the boundaries of the non-binding Fanger equation zone for human thermal comfort assessment. In practice, the intersection of logarithmic and linear functions can be solved by the numerical method. The mathematic solution involves the W Lambert Function [53]. Equation (11) takes the form:

$$m_1 \cdot h + c_1 = m_2 \cdot \ln(h) + c_2 \tag{11}$$

where coefficients m_1 and c_1 describe the linear function of dissatisfied occupants based on Fanger's model, and m_2 and c_2 describe the experimental logarithmic relationship $PD = f(h)$. A function (Equation (11)) can be rewritten as Lambert Function $x = W(x) \cdot e^{W(x)}$ in the form of Equation (12):

$$-\frac{m_1}{m_2} e^{\frac{c_1 - c_2}{m_2}} = -\frac{c_1 \cdot h}{m_2} e^{\frac{-h \cdot c_1}{m_2}} \tag{12}$$

which has the solution of Equation (13):

$$h = -\frac{m_2}{m_1} \cdot W\left(-\frac{m_1}{m_2} \cdot e^{\frac{c_1 - c_2}{m_2}}\right) \tag{13}$$

For value x calculated from the left side of Equation (12), value W(x) is read from Lambert's Function and used for h in Equation (13). For instance (point B, Figure 7), if $m_1 = 0.5166$, $c_1 = -9.386$ and $m_2 = 168.55$, $c_2 = 656.71$, $x = -0.139$ gives $W(x) = -0.164$, so (12) indicates h = 53.4 kJ/kg. The enthalpy values obtained in accordance to Equations (11)–(13) are presented in Table 6. Equation (11) can be used to determine the enthalpy for a specific temperature at which the Fanger equation begins to indicate understated results to the actual dissatisfaction of users.

Table 6. The estimated humid air enthalpy points A, B and C calculated with Equation (13) (at three fixed temperatures, see Figure 7) set the boundaries of the non-binding Fanger equation zone for human thermal comfort assessment under the following assumptions: air speed < 0.1 m/s, 0.6 clo, 1.1 met and $t_a = t_{mr}$.

Point	t_a °C	Parameter m_1	Parameter c_1	x g_w/kg_{dryair}	h kJ/kg	PD_exp %
A	26	0.3217	−9.386	9.9	51.3	7
B	27	0.5166	−13.717	10.2	53.4	14
C	28	0.6805	−13.854	11.3	57.1	25

4. Discussion

In the presented experiment, the authors confirmed that the impact of air humidity on user dissatisfaction related to indoor air quality has a greater impact on perception than a thermal sensation estimated based on the ISO 7730 model [19] and the ASHRAE Standard 55-2017 [26] for a hot and humid environment. Considering the significant increase in user dissatisfaction with indoor air quality for temperatures of 26 to 28 °C with enthalpy higher than 51.3, 53.4 and 57.1 kJ/kg, in relation to the *PD* value resulting from the thermal comfort model, the authors state that due to the global user satisfaction and indoor environmental quality index, the thermal assessment model based on the ISO7730 standard should not be used, as it gives underestimated results. The authors state that the perception of thermal- and air-related comfort dominated on comfort thermal perception and cannot be separately perceived "in situ" by users. The authors recommend using the indoor air quality model instead of the thermal model for high enthalpies. The expected total percentage of dissatisfied users (*PD* = 100%) by experimental function is h = 88 kJ/kg. Above this, there is a 95% probability that all users are dissatisfied. The presented isotherm based on the Fanger model indicates that the dissatisfaction percentage is two times lower than experimental *PD* for t = 28 °C and three times lower for 27 °C. The experimental results find confirmation in some papers. In Reference [40], Simonson stated that humidity is exactly twice as important for IAQ than for thermal comfort. Investigation of indoor thermal comfort in hot and humid conditions in a German climate test facility was analysed by Kleber et al. [28]. The tests were conducted in a similar temperature and humidity range and the impact of humidity and temperature on the air quality perceived by subjects was taken into account. The results of Kleber, $PD_{\_Kleber}$ (for t_a 26–28 °C, RH 60% to 80%, subjects n = 136) and those published by Simonson, $PD_{\_Simonson}$ (for 28 °C, RH 60%–80%) are compared with our experimental results (*PD_exp*) in Figure 8.

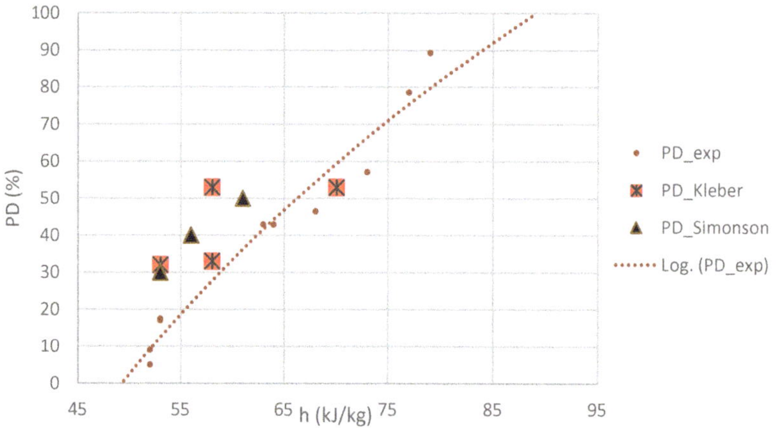

Figure 8. Comparison of the experimental results of indoor discomfort perception (log.PD_exp) with the former research presented by Kleber (*PD_Kleber*) [28] and Simonson (*PD_Simonson*) [40].

The results of Kleber show even higher dissatisfaction of air quality perception values for the same enthalpies. The results are correlated and follow the same trend. They indicate a higher level of dissatisfaction than the thermal comfort model in similar conditions. The key point of discussion is to compare the experimental results with the results obtained by Fang [7] and Toftum [15], who conducted studies on the impact of temperature and humidity on the IAQ perception of subjects (n = 38 and 40). The results obtained by Toftum, presented in Figure 9 (*PD_Toftum*), were calculated based on Formula (4). Fang used linear regression to describe acceptability of air enthalpy at the selected pollution levels (including clean air). Acceptability function (2) was transformed by the authors using the Wargocki Equation (3) into the percentage of satisfied users (*PD_Fang*). It is presented in Figure 9 as *PD_Fang*.

The results presented in Figure 9 are also correlated and have the same logarithmic dissatisfaction trend. The differences may result from a slightly different panel test method, but they also show that the dissatisfaction of users from indoor air quality for enthalpy above about 55 kJ/kg is higher than the one derived from the thermal comfort model. The acceptability of indoor air quality under temperatures of 26 to 30 °C and RH = 60% to 80% was studied by researchers from Hunan University [29]. The authors also confirmed a significant increase in dissatisfaction for high humidity. For example, for a temperature of 28 °C and RH humidity 80% (i.e., enthalpy of about 80 kJ/kg), they obtained a dissatisfaction value of 95%, which is even greater than the value that would result from our experimental function, by up to 10%.

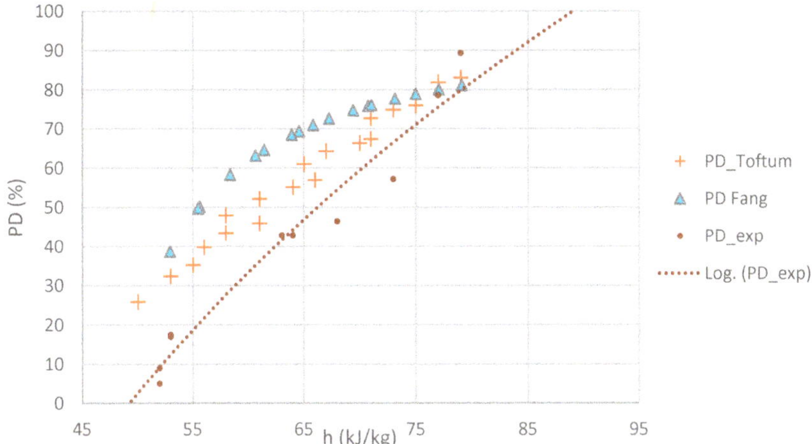

Figure 9. Comparison of experimental results of percentage dissatisfied in function of the air enthalpy (log.*PD-exp*) to the research results obtained by Fang (*PD_Fang*) [7] and Toftum (*PD_Toftum*) [15].

The authors believe that the proposed function, *PD* = f(h), is representative for the assessment of the indoor comfort of rooms equipped with a mechanical ventilation, such as that of nearly zero energy buildings, and is innovative in addressing the actual indoor environmental comfort for hot and humid conditions. As the calculation of enthalpy and the percentage of dissatisfaction on the basis of Relationship (9) is easier for non-experts than the calculation of PMV and *PD* from any thermal comfort model, the results obtained may find a wider practical application in the design of HVAC control BMS systems or the planning of heat and ventilation levels in existing buildings. There are examples in the literature of using enthalpy to control HVAC systems, especially when building users need to dry or humidify the indoor air [54]. For the estimation of indoor human comfort for mechanical ventilation eligibility, it can hardly be evaluated by indoor-outdoor temperature difference, only as used in conventional methods. The indoor temperature alone is obviously not sufficient to evaluate the indoor air enthalpy. A possible approach to address this problem is to use the spectrum of factors affecting the indoor condition, which was an example presented in reference [55]. The dual enthalpy control adds another enthalpy characteristic parameter sensor in the return air. The air with the lower enthalpy is brought into the conditioning section of the air handler. This is an efficient method of control that can be used with the Earth Air Heat Exchanger, as presented in Reference [56].

5. Conclusions

Numerous publications in which indoor environmental quality of buildings is assessed most often use the Fanger thermal model [57]. As shown in the article, under specific conditions such as increased humidity and temperature, this model will not give the correct results. It is proposed to evaluate the user satisfaction based on the air enthalpy that can be easily determined as a basic thermodynamic parameter. The authors presented the experimental curve of physical dependence (model) for predicting building occupants' dissatisfaction in hot and humid environments, *PD* = f(*h*). This relationship is primarily based on the Weber–Fechner law and the predicted percentage of dissatisfied users by air quality can be determined by means of air enthalpy (*h*). The presented experiment has examined the indoor air quality (IAQ) perception of a panel group (*n* = 28) in the experimental NZEB building under ten environmental conditions (humid air but unpolluted). The obtained results indicate a much higher level of dissatisfaction of subjects' perception with indoor air quality in a warm and humid environment than that resulting from the Fanger thermal comfort model (TC). The authors suggest using the proposed model instead of the thermal one for the range of enthalpy between 50 and 90 kJ/kg to assess the overall indoor environmental quality level of a building. Providing assessment with this method is simple and practical because enthalpy depends mainly on two parameters: temperature and humidity. Authors believe that the presented conclusions are important for building comfort prediction and modelling and prove the general thesis that in a hot indoor environment, air humidity (in practice, air enthalpy) is more important for an IAQ model than for the TC model.

Author Contributions: Conceptualisation, M.P. and K.K.; methodology, M.P.; software, M.P.; validation, M.P., K.K., M.F.-C. and K.N.; formal analysis, M.P.; investigation—test, M.F.-C., K.N.; resources, M.P. and M.F.-C.; data curation, M.P. and M.F.-C.; writing—original draft preparation, M.P.; writing—review and editing, M.P., K.K., M.F.-C., K.N.; visualisation, M.P.; supervision, M.P.; project administration, M.P.; funding acquisition, M.P. and M.F.-C. All authors have read and agreed to the published version of the manuscript.

Funding: This research received no external funding.

Conflicts of Interest: The authors declare no conflict of interest.

Abbreviations

The following abbreviations are used in this manuscript:

ACC	indoor air acceptability index (-)
c_{CO2}	concentration of carbon dioxide (ppm)
h	air enthalpy (kg/kJ)
h_{th}	the border neutral enthalpy for user perception (kg/kJ)
IAQ_{index}	indoor air quality index (percentage of persons satisfied with indoor air quality) (%)
I_{cl}	clothing thermal insulation (m^2 K/W or clo, where 1 clo = 0.155 m^2 K/W)
IEQ_{index}	Indoor Environmental Quality index (combined value of percentage of persons satisfied) (%)
LCI	lowest concentration interest
m_1 and c_1	equation coefficients of linear function of persons dissatisfied with defined temperature based on Fanger (-)
m_2 and c_2	equation coefficients of function of percentage dissatisfied with enthalpy (%)
M	metabolic rate (met)
p_a	vapour pressure (Pa)
PD	percentage dissatisfied (%)
$PD_{(IEQ)}$	percentage dissatisfied with IEQ (%)
$PD_{\_exp}$	percentage dissatisfied with indoor perception by experimental evaluation (%)
$PD_{\_ISO7730}$	estimated percentage dissatisfied with thermal comfort by ISO 7730 (%)
PD_{Fanger}	estimated percentage dissatisfied with thermal comfort by Fanger model by ISO 7730 (%)
$PD_{\_Fang}$	estimated percentage dissatisfied based on Fang model (%)
$PD_{\_Toftum}$	estimated percentage dissatisfied based on Totftum model (%)
PMV	predicted mean vote—Thermal Sensation Scale (ISO 7730)
PPD	predicted percentage dissatisfied (ISO 7730)
RH	relative humidity of air (%)
SD_h	experimental standard deviation of enthalpy determination (%)
SD_{vote}	experimental standard deviation of panel votes (%)
t_a	indoor air temperature (°C)
TC	thermal comfort index (%)
TVOC	total volatile organic compounds
t_g	black globe temperature (°C)
t_{mr}	mean radiant temperature (°C)
U	overall uncertainty (%)
v_a	air velocity (m/s)
VOC	volatile organic compounds

References

1. Yaglou, C.P. Sanitary Aspects of Air Conditioning. *Am. J. Public Heal. Nations Heal.* **1938**, *28*, 143–147. [CrossRef]
2. Chu, C.M.; Jong, T.L. Enthalpy estimation for thermal comfort and energy saving in air conditioning system. *Energy Convers. Manag.* **2008**, *49*, 1620–1628. [CrossRef]

3. Medeiros, C.M.; Baêta, F.D.C.; de Oliveira, R.F.; Tinôco, I.D.F.; Albino, L.F.; Cecon, P.R. Índice térmico ambiental de produtividade para frangos de corte. *Rev. Bras. Eng. Agrícola e Ambient.* **2005**, *33*, 19–27. [CrossRef]

4. Chandra, S.; Fairey, P.W.; Bowen, A. *Passive Cooling by Natural Ventilation: A Literature Review*; Final Report, Task 1; Solar Energy Center: Cocoa, FL, USA, 1982; Volumes 1 and 2.

5. Buffington, D.E.; Collazo-Arocho, A.; Canton, G.H.; Pitt, D.; Thatcher, W.W.; Collier, R.J. Black Globe-Humidity Index (BGHI) as Comfort Equation for Dairy Cows. *Trans. ASAE* **1981**, *24*, 711–714. [CrossRef]

6. Piasecki, M.; Fedorczak-Cisak, M.; Furtak, M.; Biskupski, J. Experimental confirmation of the reliability of fanger's thermal comfort model-Case study of a near-zero energy building (NZEB) office building. *Sustainbility* **2019**, *11*, 2461. [CrossRef]

7. Fang, L.; Clausen, G.; Fanger, P.O. Impact of temperature and humidity on the perception of indoor air quality. *Indoor Air* **1998**, *9*, 193–201. [CrossRef]

8. Toftum, J. Human response to combined indoor environment exposures. *Energy Build.* **2002**, *34*, 601–606. [CrossRef]

9. Berglund, L. Thermal and non-Thermal Effects of Humidity on Comfort. *J. Hum. -Environ. Syst.* **1997**, *97*, 239–246. [CrossRef]

10. Lan, L.; Wargocki, P.; Wyon, D.P.; Lian, Z. Effects of thermal discomfort in an office on perceived air quality, SBS symptoms, physiological responses, and human performance. *Indoor Air* **2011**, *21*, 376–390. [CrossRef] [PubMed]

11. Piasecki, M.; Kostyrko, K. Indoor Environmental Quality Assessment Model IEQ Developed in ITB. Part 1. Choice of the Indoor Environmental Quality Sub-Component Models. *Ciepłownictwo Ogrzew. Went.* **2018**, *49/6*, 223–232. [CrossRef]

12. Lai, A.C.K.; Mui, K.W.; Wong, L.T.; Law, L.Y. An evaluation model for indoor environmental quality (IEQ) acceptance in residential buildings. *Energy Build.* **2009**, *41*, 930–936. [CrossRef]

13. Heinzerling, D.; Schiavon, S.; Webster, T.; Arens, E. Indoor environmental quality assessment models: A literature review and a proposed weighting and classification scheme. *Build. Environ.* **2013**, *70*, 210–222. [CrossRef]

14. Piasecki, M. Practical implementation of the indoor environmental quality model for the assessment of nearly zero energy single-family building. *Buildings* **2019**, *9*, 214. [CrossRef]

15. Toftum, J.; Jørgensen, A.S.; Fanger, P.O. Upper limits of air humidity for preventing warm respiratory discomfort. *Energy Build.* **1998**, *28*, 15–23. [CrossRef]

16. Toftum, J.; Fanger, P.O. Air humidity requirements for human comfort. *ASHRAE Trans.* **1999**, *105*, 81–86.

17. Fanger, P. Calculation of Thermal Comfort, Introduction of a Basic Comfort Equation. *ASHRAE Trans.* **1967**, *73*, 1–20.

18. Fanger, P.O. Assessment of man's thermal comfort in practice. *Occup. Environ. Med.* **1973**, *30*, 313–324. [CrossRef] [PubMed]

19. ISO. *ISO 7730: Ergonomics of the Thermal Environment Analytical Determination and Interpretation of Thermal Comfort Using Calculation of the PMV and PPD Indices and Local Thermal Comfort Criteria*; International Organization for Standardization: Geneva, Switzerland, 2005.

20. Olesen, B.W.; Parsons, K.C. Introduction to thermal comfort standards and to the proposed new version of EN ISO 7730. *Energy Build.* **2002**, *34*, 537–548. [CrossRef]

21. Rohles, F.H., Jr.; Nevins, R.G. Thermal comfort: New directions and standards. *Aerosp. Med.* **1973**, *44*, 730–738.

22. Van Hoof, J. Forty years of Fanger's model of thermal comfort: Comfort for all? *Indoor Air* **2008**, *18*, 182–201. [CrossRef]

23. Djongyang, N.; Tchinda, R.; Njomo, D. Thermal comfort: A review paper. *Renew. Sustain. Energy Rev.* **2010**, *4*, 2626–2640. [CrossRef]

24. Halawa, E.; Van Hoof, J. The adaptive approach to thermal comfort: A critical overview. *Energy Build.* **2012**, *51*, 101–110. [CrossRef]

25. Croitoru, C.; Nastase, I.; Bode, F.; Meslem, A.; Dogeanu, A. Thermal comfort models for indoor spaces and vehicles—Current capabilities and future perspectives. *Renew. Sustain. Energy Rev.* **2015**, *44*, 304–318. [CrossRef]

26. ANSI/ASHRAE. *Standard 55-2017: Thermal Environmental Conditions for Human Occupancy*; ASHRAE Inc.: Atlanta, GA, USA, 2017.

27. Jin, L.; Zhang, Y.; Zhang, Z. Human responses to high humidity in elevated temperatures for people in hot-humid climates. *Build. Environ.* **2017**, *114*, 257–266. [CrossRef]

28. Kleber, M.; Wagner, A. Investigation of indoor thermal comfort in warm-humid conditions at a German climate test facility. *Build. Environ.* **2018**, *128*, 216–224. [CrossRef]

29. He, M.; Li, N.; He, Y.; He, D.; Wang, K. Influences of Temperature and Humidity on Perceived Air Quality with Radiant Panel Workstation. *Procedia Eng.* **2017**, *205*, 765–772. [CrossRef]

30. Jing, S.; Li, B.; Tan, M.; Liu, H. Impact of relative humidity on thermal comfort in a warm environment. *Indoor Built Environ.* **2013**, *22*, 4. [CrossRef]

31. Zhai, Y.; Zhang, Y.; Zhang, H.; Pasut, W.; Arens, E.; Meng, Q. Human comfort and perceived air quality in warm and humid environments with ceiling fans. *Build. Environ.* **2015**, *90*, 178–185. [CrossRef]

32. Buonocore, C.; De Vecchi, R.; Scalco, V.; Lamberts, R. Influence of relative air humidity and movement on human thermal perception in classrooms in a hot and humid climate. *Build. Environ.* **2018**, *156*, 233–242. [CrossRef]

33. Rana, R.; Kusy, B.; Jurdak, R.; Wall, J.; Hu, W. Feasibility analysis of using humidex as an indoor thermal comfort predictor. *Energy Build.* **2013**, *64*, 17–25. [CrossRef]

34. Frontczak, M.; Wargocki, P. Literature survey on how different factors influence human comfort in indoor environments. *Build. Environ.* **2011**, *46*, 922–937. [CrossRef]

35. Kaczorek, D. Moisture buffering of multilayer internal wall assemblies at the micro scale: Experimental study and numerical modelling. *Appl. Sci.* **2019**, *9*, 3438. [CrossRef]

36. Nowoświat, A.; Skrzypczyk, J.; Krause, P.; Steidl, T.; Winkler-Skalna, A. Estimation of thermal transmittance based on temperature measurements with the application of perturbation numbers. *Heat Mass Transf.* **2018**, *54*, 1477–1489. [CrossRef]

37. Orlik-Kożdoń, B.; Steidl, T. Experimental and analytical determination of water vapour transmission properties of recyclable insulation material. *Constr. Build. Mater.* **2018**, *192*, 798–807. [CrossRef]

38. Gawin, D.J.; Koniorczyk, M.; Wieckowska, A.; Kossecka, E. Effect of moisture on hygrothermal and energy performance of a building with cellular concrete walls in climatic conditions of Poland. *ASHRAE Trans.* **2004**, *110*, 795–803.

39. Nicol, F. Adaptive thermal comfort standards in the hot-humid tropics. *Energy Build.* **2004**, *36*, 628–637. [CrossRef]

40. Simonson, C.J. Moisture, thermal and ventilation performance of Tapanila ecological house. In *VTT Tiedotteita—Valtion Teknillinen Tutkimuskeskus*; VTT TIEDOTTEITA; Technical Research Centre of Finland: Espoo, Finland, 2000.

41. Dehaene, S. The neural basis of the Weber-Fechner law: A logarithmic mental number line. *Trends Cogn. Sci.* **2003**, *12*, 244–246. [CrossRef]

42. Goldstein, E.B.; Humphreys, G.W.; Shiffrar, M.; Yost, W.A. *Blackwell Handbook of Sensation and Perception*; Blackwell Publishing: Oxford, UK, 2008; ISBN 0631206841.

43. Jokl, M.V. A methodology for the comprehensive evaluation of the indoor climate based on human body response: Evaluation of the hygrothermal microclimate based on human psychology. *Energy Build.* **2014**, *85*, 458–463. [CrossRef]

44. CEN. *EN 16798 Energy Performance of Buildings—Ventilation of Buildings—Part 1: Indoor Environmental input Parameters for Design and Assessment of Energy Performance of Buildings Addressing indoor Air Quality, Thermal Environment, Lighting and Acoustics*; CEN: Brussels, Belgium, 2019.

45. Piasecki, M.; Kostyrko, K.B. Combined Model for IAQ Assessment: Part 1—Morphology of the Model and Selection of Substantial Air Quality Impact Sub-Models. *Appl. Sci.* **2019**, *9*, 3918. [CrossRef]

46. Piasecki, M.; Kostyrko, K.; Pykacz, S. Indoor environmental quality assessment: Part 1: Choice of the indoor environmental quality sub-component models. *J. Build. Phys.* **2017**, *41*, 264–289. [CrossRef]

47. Piasecki, M.; Kozicki, M.; Firlag, S.; Goljan, A.; Kostyrko, K. The approach of including TVOCs concentration in the indoor environmental quality model (IEQ)- case studies of BREEAM certified office buildings. *Sustainbility* **2018**, *10*, 3902. [CrossRef]

48. Piasecki, M.; Kostyrko, K.B. Indoor environmental quality assessment, part 2: Model reliability analysis. *J. Build. Phys.* **2018**, *5*, 1–28. [CrossRef]

49. Wang, J.; Wang, Z.; de Dear, R.; Luo, M.; Ghahramani, A.; Lin, B. The uncertainty of subjective thermal comfort measurement. *Energy Build.* **2018**, *181*, 38–49. [CrossRef]

50. Alfano, A.; Palella, B.I.; Riccio, G. The role of measurement accuracy on the thermal environment assessment by means of PMV index. *Build. Environ.* **2011**, *46*, 1361–1369. [CrossRef]
51. Lira, I.; Taylor, J.R. Evaluating the Measurement Uncertainty: Fundamentals and Practical Guidance. *Am. J. Phys.* **2003**, *71*, 409–410. [CrossRef]
52. Ribeiro, A.S.; Alves e Sousa, J.; Cox, M.G.; Forbes, A.B.; Matias, L.C.; Martins, L.L. Uncertainty Analysis of Thermal Comfort Parameters. *Int. J. Thermophys.* **2015**, *36*, 2124–2149. [CrossRef]
53. Corless, R.M.; Gonnet, G.H.; Hare, D.E.G.; Jeffrey, D.J.; Knuth, D.E. On the Lambert W function. *Adv. Comput. Math.* **1996**, *5*, 329–359. [CrossRef]
54. Mazzei, P.; Minichiello, F.; Palma, D. HVAC dehumidification systems for thermal comfort: A critical review. *Appl. Therm. Eng.* **2005**, *25*, 677–707. [CrossRef]
55. Ali, M. Efficient indoor thermal comfort control via TSK adaptive control of HVAC systems. In Proceedings of the Conference Paper on "Computer, Control and Communication Engineering" (KIC4E-19), Kuala Lumpur, Malaysia, 23–25 September 2019.
56. Romanska-Zapala, A.; Bomberg, M.; Dechnik, M.; Fedorczak-Cisak, M.; Furtak, M. On Preheating of the Outdoor Ventilation Air. *Energies* **2020**, *13*, 15. [CrossRef]
57. Manfren, M.; Nastasi, B.; Piana, E.; Tronchin, L. On the link between energy performance of building and thermal comfort: An example. *AIP Conf. Proc.* **2019**, *2123*, 020066.

 *energies*

Article

Development of Weighting Scheme for Indoor Air Quality Model Using a Multi-Attribute Decision Making Method

Michał Piasecki * and **Krystyna Kostyrko**

Thermal Physics, Acoustic and Environment Department, Building Research Institute, Filtrowa 1, 00-611 Warsaw, Poland; k.kostyrko@itb.pl
* Correspondence: m.piasecki@itb.pl; Tel.: +48-22-5664-352

Received: 21 April 2020; Accepted: 10 June 2020; Published: 16 June 2020

Abstract: When planning the energy demand of ventilation, proper consideration should be given to the possible scenarios of indoor air quality and pollutant concentrations. The purpose of the present research is to create a practical method of prioritising indoor air pollutants, considering technical, economical and health aspects, in the Indoor Air Quality model (IAQ). In order to find the global weights for the combined IAQ_{index} model sub-elements (in practice, air pollutant concentrations), the Multi-Criteria Decision Making (MCDM) approach is used. The authors have approached the problem of a weighting scheme in a model such as the complex model of the IAQ related to making decisions with many criteria and with the Multi-Attribute Decision Making MADM approach (specifically MCDM). The basis of the MADM method is a decision matrix constructed rationally by the authors, which includes six attributes: actual indoor air carbon dioxide concentration, total volatile organic compounds (TVOCs) and formaldehyde HCHO concentration, and their anthropogenic and construction product emissions to the indoor environment. The decision model of IAQ_{index} includes five alternatives (possible situations), and the combination of pollutant concentration attributes with additional emission attributes is related to the indoor environment under specific situation. For defining the weights of criteria, the authors provide objective approaches: (i) entropy-based approach considering measuring the amount of information, and (ii) CRITIC, a statistic-based approach. The value of the presented method, i.e., the determination of global weights for IAQ components, is shown as a practical application to determine IAQ and the Indoor Environmental Quality (IEQ) index for an office building used as a case study.

Keywords: indoor air quality models; indoor air quality; indoor environmental quality; sustainable building; multi-criteria decision analysis; MCDM; MADM; user dissatisfaction; weights system; building comfort; PD; IAQ; IEQ

---

1. Introduction and Literature Review

1.1. Research Problem

In literature on the Multi-Attribute Decision Making (MADM)method, the combined Indoor Air Quality ($\sum$IAQ) model optimisation problem has not been addressed thus far to set up the weights of model sub-components (in practice due to excess indoor pollutant concentrations and risk to occupants). One of the main criteria for optimising such a decision problem should be, directly or indirectly, the cost of ventilation for the elimination of excess pollution concentrations. For the purposes of the decision problem, it is assumed that the general criterion for optimising such a weighing system is to include a range of IAQ indicators and components, i.e., types of contaminants that are most important due to health and toxicity or the amount of air to be removed by the ventilation system.

The MADM method, as a specific kind of Multi-Criteria Decision Making (MCDM) approach, has emerged as a formal methodology for assessing available information and providing values supporting decision-makers in many areas, including environmental engineering. The published results suggest that, over the past decade, there has been a significant increase in the use of MCDM in technical areas for civil engineering applications, the design of indoor environments and the improvement of outdoor environments [1]. MCDM makes it possible to assess complicated decision problems limited by numerous and divergent criteria based on expert subjective assessments (subjective weights) and decision models, the purposes of which are to calculate objective weights according to the rules of decision theory and order multiple parameters of the analysed problem. Recently, the construction industry has become one of the main stressors to the environment and society globally [2]. In the context of the continuous decrease in energy consumption by new buildings and the fundamental need to provide comfortable and healthy conditions for the users in indoor environments, there is a growing urgency to use IAQ models that meet certain criteria. IAQ models are the main components of building Indoor Environmental Quality models [3] and increasingly used for building design software (such as Building Information Modelling, BIM) or as part of building managing systems (BMSs) controlling heat, cooling and ventilation systems (HVACs).

The basic aim of this study was to develop a new approach which combines the previously developed ΣIAQ model [4], containing loosely connected components (indoor air pollution elements and their impact on user perception), with a new weighting system based on MADM that would produce a practical tool supporting choices of the best possible indoor environment and air control strategy. As a case study, a large office building was used, for which the authors had already assessed the indoor environmental parameters [4,5], and the authors recalculated the ΣIAQ and Indoor Environmental Quality (IEQ) indices using the new weighting scheme. A similar issue for the external environment was presented in [6]. This team built the MCDM decision model to study the impacts of outdoor air pollution on the economic development of selected zones in China.

How humans perceive indoor comfort is a complex phenomenon. Authors have analysed the problem from a holistic point of view following current developments described widely [3–5]. Researchers assessing indoor comfort in buildings have focused primarily on optimizing individual components of the indoor environment (thermal, light and acoustic comfort, and air quality), as these four parameters connected to the human senses are considered by occupants to be the most important in determining their comfort [3]. Our research work is aimed at identifying the relationship between the indoor parameters and the resulting reactions of building occupants (in particular, in the area of IAQ). This multi-component approach to indoor comfort focuses on the 'human stimuli response' and has led to well-known indoor comfort models such as the thermal comfort model developed by Fanger. In recent years, this holistic approach to the indoor environment has been presented in the European standard EN 15251 and later in EN 16798-1: 2018 [7] where, for the first time, a wider number of indoor comfort criteria were considered. These standards suggest an approach to the classification of IAQ and IEQ models and support certification of buildings taking into account specific components of indoor environment, however, they do not provide a practical guidance or an approach on how to combine them into one indicator that could be useful to classify indoor environmental conditions. The authors solved this problem in a previous publication [4]. When the predominant indoor pollutants have different characteristic values, there is a large problem regarding the choice and weighting method of IAQ sub-components. The problem is considered in this article with use of the MADM approach. Taking into consideration the characteristics and correlations of the selected pollutants, IAQ may be characterized by representative indicators. Our studies on Building Research Establishment Environmental Assessment Method (BREEAM)certified buildings [5] and Leadership in Energy and Environmental Design (LEED) system shown out that carbon dioxide, total volatile organic compounds (TVOCs) and HCHO are the main worldwide, independent and representative environmental indicators. They are being used worldwide as an evaluation index of IAQ in buildings (mainly offices with mechanical ventilation). Since each of these indicators represents a pollutant class with comparable indoor sources and characteristic dissemination and the indexed group avoids unreliable measurements. This is based on the fact that indicators are "too small" because of decreased critical concentrations. A method of data pre-treatment is proposed by us in the procedures, which reflects concentration differences between pollutants and considerstheir influence on the satisfaction of the building users. Taking into consideration the existing knowledge gap, the authors proposed an objective method to determine weights in theIAQ model in this article. The article provides the procedure to calculate global weights for the IAQ model sub-components. The ΣIAQ model considers the impact of air concentrations of selected air pollutants (CO_2, TVOCs, HCHO, etc.) on the building users' perceptions. The weighted IAQ model was developed by the authors considering the standard EN 16798-1:2019 [7] provisions with the intention to create it as a main sub-component of the overall Indoor Environmental Quality (IEQ) model [8,9] with assumptions described in references [3,8]. The IAQ model and weight system applies to office buildings with mechanical ventilation. Quasi-static indoor environment conditions and full air mixing in the building are assumed.

1.2. Basic Information on the MADM Approach

Some analogies to the presented IAQ weighting problem can be found in the literature, and the studies described so far have led to the establishment of rankings of alternative solutions to the decision problem. The MADM method has beenused as an effective tool for choosing the optimal composition of construction materials [10,11], estimating their weights [12] for the selection of the optimal air filtration method to remove pollutants [13], and choosing the optimal set of indoor environment parameters in an office building [14] when the alternatives are differently dated combinations of these parameters (attributes). Additionally, the MADM technique solved the problem of the impact of air pollution on the level of economic development of Chinese cities (for example, the city of Wuhan) [6]. This project is interesting to the authors because it concerns the impact of air pollution. The research involved determining the weights of air pollutants of various types of VOCs, and of the physical parameters affecting these pollutants (attributes of the decision problem), and establishing a ranking of the specific systems of these environmental parameters occurring in different years (alternatives to the decision problem). In these simulations, a modified Technique for Order of Preference by Similarity to Ideal Solution (TOPSIS) was used. This approach uses neural networks to determine weights.

MCDM was also proposed inchoosing a strategy to design sustainable buildings [15]. The purpose of this study was to provide a method combiningclimate changeeffects, adaptive thermal satisfaction, life cycle assessment (LCA), life cycle cost (LCC) analysis and multi-criteria decision making in order to set the most valuable design strategy that would improve building object sustainability. For the assessment and selection of the most appropriate design strategies for buildings, attributes were analysed using a multi-disciplinary approach based on sustainability criteria. In order to find the best building design strategy and make decisions based on several criteria, the Complex Proportional Assessment (COPRAS) method was used [16,17]. This method [15] assumes proportional and direct dependence of the significance of the usability degree of the tested versions of buildings (alternatives) in the system of criteria (attributes) described by the parameters: (1) hours of comfort in the room, (2) primary energy demand, (3) CO_2 emissions, and (4) costs). This method was successfully used to deal with the complex selection of problems. Based on this study, by using the COPRAS method, the authors provided a ranking of the most valuable structural solutions for buildings that meet four characteristic parameters/criteria and their weights built within the decision model.

Most often, MCDM decision models have been used by researchers in the process of selecting components of construction materials by methods such as the "preference technique in classification according to the similarity to the ideal solution" (TOPSIS) [18,19], which uses the principles of analytical hierarchy process (AHP) [17] and various versions of Elimination and Choice Expressing the Reality (ELECTRE) [1]. The method using entropic weights, created on the basis of Shannon's information theory [20] for the selection of materials [10,11], was also used. Additionally, the method of material selection from many alternatives based on Jahan's research [12] and the analysis of correlation between Criteria Importance Through Intercriteria Correlation (CRITIC) criteria [12] was used. Alemi-Ardakani [10] analysed the possibilities of optimising the composition of composite materials using the Adjustable Mean Bars (AMB), Modified Digital Logic (MDL), Numeric Logic (NL) and CRITIC [10] methods, as well as the Peng method [21] of creating rankings of preference functions using composition enrichment technology (PROMETHEE—Preference Ranking Organization Method for Enrichment Evaluations) or a preference selection indicator [1].

MCDM methods have now become so popular for extending knowledge of environmental impacts and sustainable construction that several groups of scientists reviewed the applications and development trends of decision models in "environmental sciences" (Cegan, Linkov et al., 2011 [22] and 2017 [23]) and in sustainable construction (Navarro et al. 2019 [1]).MADM methods are a special case of MCDM methods in which decision-making criteria have been replaced by attributes, i.e., features of problems or processes that are evaluated by analysts. Each MADM technique has its own specifics, assumptions and principles. Almost all of the above-mentioned MCDM methods are characterised by the complexity of the mathematics of their moderate to extreme models. Application of these techniques is difficult, requiring advanced mathematical skills and knowledge. Therefore, the creation of undemanding MADM methods is highly desirable. Multi-purpose optimisation based on proportion analysis, named Multi-Objective Optimization by Ratio Analysis(MOORA)was proposed by Brauers and Zavadskas (2006) [24] and is used for the selection of construction materials [25], as it involves uncomplicated mathematics. Therefore, it can be used effortlessly and effectively to choose the best materials or solutions.

To solve the weight problem in the IAQ model, the authors used two methods: entropy and CRITIC. The objective weighting measure was proposed with the Shannon entropy concept provided in [20]. In MCDM, the entropy relates to the diversity degree within the attribute data set. The higher this diversity degree, the higher the attribute weight. The smaller the entropy within the associated data to the attribute, the higher the power of discrimination of the attribute in changing ranks of alternatives. The entropy weights are calculated with only one set of values dedicated to the same attribute j in all alternative levels of the decision matrix. Entropy weight is a typical attribute weight, but it is always introduced to the global weight as a factor in all decision matrix indices. In the entropy method, an attribute with a performance rating that is very different from the others has more importance in the problem because it has a greater impact on ranking outcomes. The attribute has lesser importance if all attributes have comparable performance ratings for that attribute. An objective method to weigh criteria importance through inter-criteria correlation (CRITIC) based on the SD approach was proposed by Diakoulaki and later explored by Jahan [12]. The higher the level of interdependency between attributes, the larger the ranking outcome error. The CRITIC approach is a popular objective method to calculate weights, which solves the problem of interdependencies between the attributes in all combinations by considering the correlation between the sets of variables from various alternative levels, while calculating the weights. It is one of the objective methods of weight determination that belongs to the class of correlation methods. It uses the information contained in the data matrix in the form of the degree of deviation of a variant value from a given mean value of the criteria. The CRITIC method is particularly useful in the IAQ model, as we show later, because the parameters adopted as attributes in the decision model of this problem are the concentrations of selected pollutants and the emissions of the same pollutants; therefore, in the IAQ decision matrix there are correlating attributes.

1.3. Basic Information on the ΣIAQ Model

The ΣIAQ model, in which the authors are looking for objective weight sub-elements, uses a selected number of indoor air pollutants ($P_1, \ldots$ j) and their impact on user dissatisfaction (Percentage

Dissatisfied as PD = f(c_j) in %). The combined $\sum\text{IAQ}_\text{index}$ (cumulative percentage of satisfied users with indoor air quality, including selected pollutant impacts of user perception) Equation is

$$\sum\text{IAQ}_\text{index} = \text{WP}_1\cdot\text{IAQ}(\text{P}_1)_\text{index} + \text{WP}_2\cdot\text{IAQ}(\text{P}_2)_\text{index}, \dots, \text{WP}_j\cdot\text{IAQ}(\text{P}_j)_\text{index} \tag{1}$$

where sub-indices $\sum\text{IAQ}(\text{P}_j)$ are the percentage of users satisfied with the pollutant concentration; $WP_1, \dots P_j$ are weights for each IAQ sub-component for groups of air pollutants with comparable concentrations.

There is a difference in Δc_j concentration between the measured air concentration of pollutant c_j and the recommended "reference" concentration c_{ref} (reference by the European Commission "safe" values such as c_{LCI} or c_{ELV}), which may be below the actual air concentration in the contaminated rooms. Thus, the concentration excess related to the ventilation rate is

$$\Delta c_j = c_j - c_{ref} \tag{2}$$

The excess concentration weights WP_j for the IAQ model until now have been calculated based on the arithmetic mean or by adjusting (normalization) all attribute Δc_j values using Equation (3):

$$W_{P,j} = \frac{\Delta c_j}{\sum_{j1\dots7}\Delta c_{j..7}} \tag{3}$$

where the sum of the adjusted weights W_{Pj} of all pollutants to be removed by ventilation should be in unity. This weighting scheme has been used by the authors in previous years, but in this article, we propose a better and more objective approach based on the MADM approach.

Three levels of complexity of the IAQ model were proposed, and in each, pollutants have been defined that were included in the IAQ model. Figure 1 shows the extended $\text{IAQ}_\text{quality}$ model with its sub-indices and also with the sub-components of the $\text{IAQ}_\text{comfort/health}$ model type, i.e., indoor air pollutants important to health with an impact on the energy balance of a building using a mechanical ventilation system.

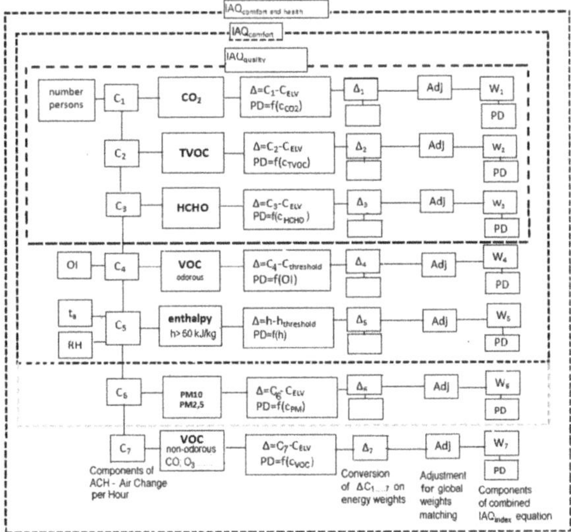

Figure 1. Combined Indoor Air Quality ($\sum$IAQ) model with weighting scheme; horizontally, the model is bound by the weighing processes of individual sub-models; vertically, the model is bound by a chain of input components (pollutant concentrations in ventilated indoor air) and a chain of output components (elements of the combined IAQ_index equation).

The experimental dependencies of the percentage of persons dissatisfied (*%PD*, where *%PD = 1-IAQ*) and the concentrations values of air pollutants, c_j, perceived in the adequate ranges are of fundamental significance to the sub-components relevant to the IAQ and IEQ models [9]. In the ΣIAQ model proposed by the authors, the accepted common approach is engaged to transform individual pollutant concentrations into sub-components before these are combined into a single index. The aggregated summary of sub-components, however, may lead to a situation in which all components are below the individual health threshold, but the final index shows that the health threshold is exceeded. Conversely, the averaging of partial sub-components may lead to an outcome whereby the overall index shows an acceptable IAQ value, but one or more partial indicators are higher than their individual health threshold. The solution proposed is to use all sub-component maximum values to provide a final form of the ΣIAQ. Considering this, we proposed the ΣIAQ model with three possible levels of model complication due to the application potential, as we show in Figure 2:

i.   building sustainable assessment schemes (like BREEAM assessment)—three sub-components, named quality, ΣIAQ_{quality};

ii.   design purpose, considering the perceptible pollutants affecting indoor satisfaction and comfort, with the intention to use this index for calculating the IEQ—five sub-components, called comfort, ΣIAQ_{comfort}; and

iii.   complex design purpose, where ΣIAQ_{index} represents health and comfort (seven sub-components, or more if necessary), called comfort and health, ΣIAQ_{comfort/health}.

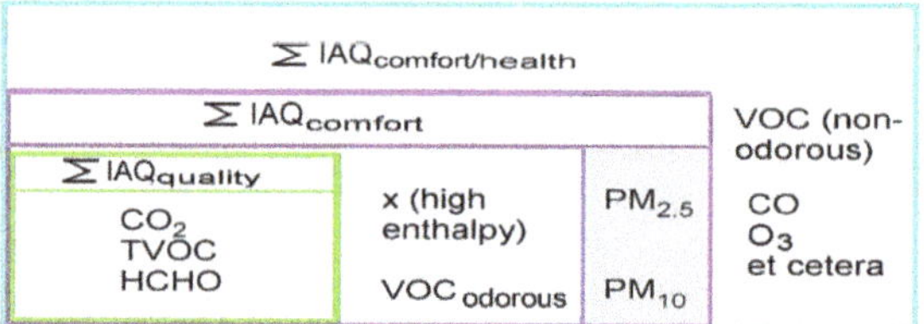

Figure 2. The IAQ model with three complexity levels reflecting the application potential.

The simplest IAQ index level can be used with the main purpose of supporting a sustainable building assessment and certification by using only three sub-components: CO_2, HCHO and TVOC. ΣIAQ_{quality} model is used in this article to assess the office building (case study).

From the dependencies, expressed as the curves for the percentage of satisfied users with pollutant IAQ(CO_2) or IAQ(VOC_{odorous}), the following Equations (4)–(6) of the models are provided:

$$\sum IAQ_{quality} = W_1 \cdot IAQ(CO_2) + W_2 \cdot IAQ(TVOC) + W_3 \cdot IAQ(HCHO) \tag{4}$$

$$\sum IAQ_{comfort} = W_1 \cdot IAQ(CO_2) + W_2 \cdot IAQ(TVOC) + W_3 \cdot IAQ(HCHO) + W_4 \cdot IAQ(VOC_{odorous}) + W_{4a} \cdot IAQ(VOC_{odorous}) + W_5 \cdot IAQ(h) \tag{5}$$

$$\sum IAQ_{comfort/health} = W_1 \cdot IAQ(CO_2) + W_2 \cdot IAQ(TVOC) + W_3 \cdot IAQ(HCHO) + W_4 \cdot IAQ(VOC_{odorous}) + W_{4a} \cdot IAQ(VOC_{odorous}) + W_5 \cdot IAQ(h) + W_6 \cdot IAQ(PM2.5, PM10) + W_7 \cdot IAQ(VOC_{non\text{-}odorous}) + \\ + W_{7a} \cdot IAQ(VOC_{non\text{-}odorous}) + W_8 \cdot IAQ(CO) + W_9 \cdot IAQ(NO_2) + W_{10} \cdot IAQ(O_3) \tag{6}$$

The structure of ΣIAQ_{comfort/health} is made of seven (or possibly more) components or IAQ sub-models, which are occupants satisfaction functions for the various types of indoor air pollutants: IAQ(TVOC), IAQ(HCHO),IAQ(CO_2), IAQ(VOC_{odorous}), IAQ(PM2.5, PM10), IAQ(enthalpy, h) and the selected IAQ(VOC_{non-odorous}). The IAQ(VOC_{non-odorous}) and IAQ(VOC_{odorous}) models should be multiplied, depending on the number of dominant VOC substances; hence, the ΣIAQ_{comfort/health} index may have more than 11 sub-components in practice.

Combined $\sum\text{IAQ}_{\text{index}}$ is an element of the IEQ index in Equation (7) [8], where the authors adopted a crude weighting system in which all elements are weighted identically (0.25 for weights W_1–W_4):

$$\text{IEQ}_{\text{index}} = 0.25 \cdot \text{TC}_{\text{index}} + 0.25 \cdot \Sigma\text{IAQ}_{\text{index}} + 0.25 \cdot \text{ACc}_{\text{index}} + 0.25 \cdot \text{L}_{\text{index}} \tag{7}$$

where TC_{index} is the thermal comfort index, expressed as the number of users satisfied with an indoor thermal comfort (in %); $\text{ACc}_{\text{index}}$ is the acoustic comfort index, i.e., the number of users satisfied with sound level (in %); and L_{index} is Daylight Quality, i.e., the number of users satisfied with daylight in the building (in %).

2. Materials and Methods

2.1. Research Procedure Diagram

Figure 3 shows the necessary research phases to calculate the global weights of the combined $\sum\text{IAQ}$ model with MADM and further usage of these weights to calculate the occupant satisfaction $\text{IAQ}_{\text{index}}$ and $\text{IEQ}_{\text{index}}$ for the case study building.

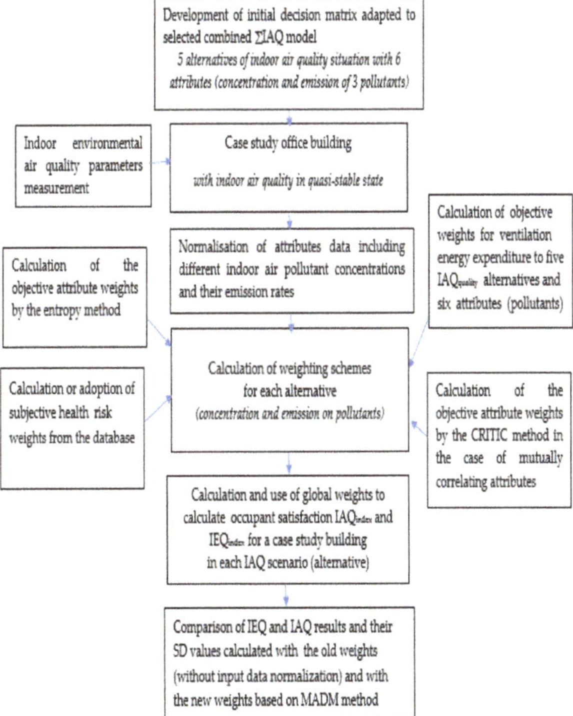

Figure 3. The research phases to calculate the global weights for the ΣIAQ model (based on the case study building).

2.2. MADM Method Applied to the ΣIAQ Model

The basis for applying the MADM method is a decision matrix, as shown in Table 1. The issue of a weighting scheme is a problem related to making decisions with several criteria, which can be analysed using the model [26]. For every type of $\sum\text{IAQ}$ model with Equations (4), (5) or (6), the decision matrix should be created separately.

Table 1. Decision matrix scheme in the MADM model for the concordance of economic and human comfort criteria for each $\sum$IAQ scheme described by Equations (4)–(6) separately with alternatives (e.g., $\sum$IAQ$_{comfort}$ scheme configurations) and with attributes (air pollution components in quasi-stable states and emissions).

Alternatives	Attributes $X_{1...m}(j = 1 .... m)$								
$a_{i....n}(i = 1 ... n)$	X_1	X_2	X_3	X_4		$X_5 (e)$	$X_{6\,(e)}$	$X_7(e)$	$X_m(e)$
a_1	x_{11}	x_{12}	x_{13}	x_{14}	......	X_{15}	X_{16}	X_{17}	..... x_{1m}
a_2	x_{21}	x_{22}	X_{23}			X_{25}	X_{26}		x_{2m}
a_3	x_{31}	X_{32}				X_{35}			x_{3m}
....	....	....	....	....	....	.....		......	......
a_n	x_{n1}	X_{n2}	X_{n3}	X_{n4}		X_{n5}	X_{n6}	X_{n7}	...... x_{nm}

In Table 1, alternatives a1, a2 ... a(n) for $\sum$IAQ$_{comfort}$ configurations are defined further (in Methods). Attributes X_1, X_2 and X_j are the indoor air pollution concentrations of CO_2, HCHO, TVOC and VOC$_{odorous}$ in a quasi-stable state of the indoor environment. $X_5(e)$, $X_6(e)$ and $X_7(e)$ are indoor emission processes and their rates per emission surface of CO_2 (e), HCH(e), introduced by an thropogenicemissions or directly emitted from construction products. In the decision matrix of the IAQ weighting scheme according to Table 1, the flow chart presented in Figure 3 provides the necessary steps and, therefore, includes MADM elements:

(1) attributes in the number j = 1 ... 3 ... , which are indoor air pollution concentrations covered by the models $\sum$IAQ$_{quality}$, $\sum$IAQ$_{comfort}$ and $\sum$IAQ$_{comfort\,/\,health}$ described by Equations (4)–(6) and Figure 2; however, the number of attributes beyond j = 7 (e.g., for $\sum$IAQ$_{comfort/health}$) is not limited because model components like VOC$_{odorous}$ or VOC$_{non-odorous}$ and other pollutants can be multiplied.

(2) alternatives i = 1 ... n, which are the $\sum$IAQ models described by Equations (4)–(6) and Figure 2 assigned to the quasi-stable state, or defined by emissions. The selected decision model alternatives correspond to three combined indoor air quality models in various attribute set configurations (i.e., changes in attribute sets, IAQ sub-models). Alternative models, i.e.,IAQ$_{comfort}$ and IAQ$_{comfort/health}$ models, have more attributes than the IAQ$_{quality}$ model; moreover, the number of these attributes is variable in both the pollutant and emission groups. In the group of alternatives, IAQ model systems not disturbed by emissions, e.g., IAQ$_{quality}$, and models distorted by emission attributes, (e) for example, were considered. The number of alternatives (Table 1) cannot be too small because the standard deviations for the set of alternatives assigned to a given attribute are calculated (i.e., for the attribute column in the decision matrix).

The assumptions in the decision model are adequate for the task of measuring the indoor air ventilation rate to obtain the assumed air quality, setting IAQ model alternatives so they meet the criteria and taking into account their importance ranking and possibility for ongoing diagnosis of alternatives to the assumed overall IEQ assessment. The model should provide a set of attributes that are important for maintaining hygienic and efficient ventilation, both of which are connected to variable air pollutants. This is important because the decisions relate to the scope of complex indoor air quality models, while planning the ventilation system depends on the amount of "excess air pollution" including VOC and SVOC compounds.

In the Materials and Methods and Results sections, an example of the implemented IAQ$_{quality}$ model is provided, as well as a shortened Table 5 decision matrix with a set of attributes of type (e) (emission), which includes additional sources of bio-emissions (CO_2 emissions), emission sources from additional equipment and finishing products (HCHO emissions), and emissions from building materials (TVOC). The choice of the attributes was facilitated by the fact that the effects of air pollution, in another open environment and system of conditions and tasks, on the economy were provided by Wang et al., 2017 [6]. This team built a Smart MCDM decision model to study the impact of outdoor air pollution on the economic development of selected zones in China. In addition, looking at the goal of the present paper to rank the importance of selection parameters for human comfort, insights could be derived from already established research procedures for the selection of construction materials [12] or waste disposal (Savic [27]). Therefore, the attributes of the decision matrix (Tables 1 and 2) include the most typical and expected air pollutants belonging to successively developed ΣIAQ systems. The selection of attributes, i.e., types of pollution, consistent with Equations (4)–(6) also corresponds to the choice of Air Quality Index (AQI) [28].

2.3. Flow Chart of Weighting Scheme Determination

In the analysed IAQ model, the authors considered excess concentrations that can be measured in a quasi-steady state, and emissions from additional and sometimes temporary sources that periodically increase in the ΣIAQ system of excess concentrations. The number of pollution type attributes is as strictly defined as the ΣIAQ specified composition. The number of emission type attributes is variable, adapted to the selected set of decision pollutants, and does not usually exceed the number of pollution type attributes.

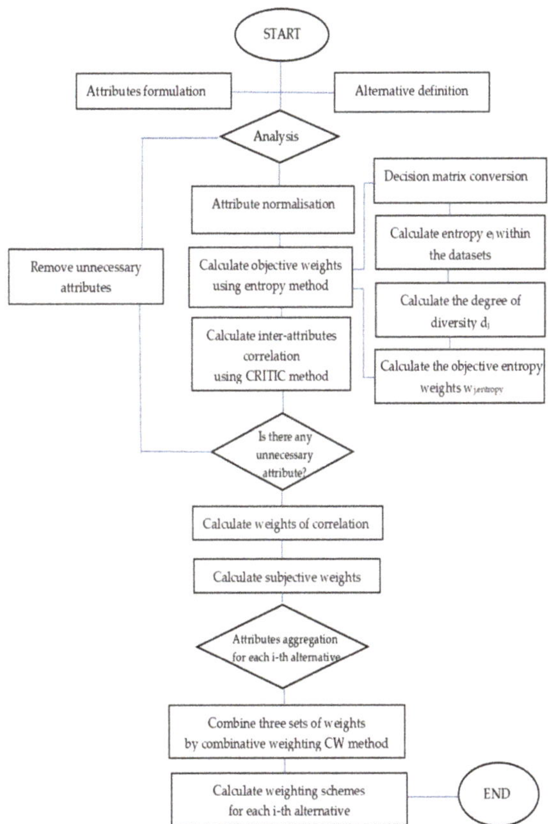

Figure 4. Chart for subjective and objective weight determinations for the combined ΣIAQ model.

The process of determining the weight for the IAQ_{index} model, seen in a number of variants of the composition of its sub-indexes, i.e., in various alternatives, involves weighting scheme calculations adequate to the level of a particular alternative. Each weighting scheme contains relative values of global weights. These weights represent the energy expenditure of the ventilation system on the elimination of excess concentration of a given pollutant (excess concentration (2) [4]) in ventilated air. The process for determining the weighting scheme for the ΣIAQ model is shown in the flow chart in Figure 4. It should be noted that the authors considered it appropriate, after specifying the technical conditions and physics of the building for which the decision model is used, to add the costs of ventilating the anticipated excess pollutant concentrations. Based on the literature of this multi-parameter problem, however, the authors found it advisable not to insert this attribute at the current level of the generalized decision model for selecting the ΣIAQ weighting scheme.

The idea of the correlation effect on weight $w_{j,CRITIC}$ stems from the fact that when the criteria correlation with other attributes is significant, it should have lower importance due to the roles of other criteria. An increase in the value of the concentration of a given indoor air pollutant, with a sudden (dramatic) increase in emissions, is accompanied by an increase in global weight because it must include an additional component, which is the correlation weight.

2.4. General Principles of Estimating IAQ Model Weighting Schemes

In this section, the authors provide a method with a step-by-step procedure for setting priorities in the group of dominant indoor air pollutants, considering technical, economic and health aspects, by global weights for each component of the IAQ model. Developed for the ΣIAQ$_{quality}$ model, the MADM decision matrix includes six attributes: carbon dioxide, total VOC (TVOC) and formaldehyde concentrations (HCHO) in the indoor environment as well as additional carbon dioxide bio-emissions and total VOCs and formaldehyde emissions. All six attributes are associated with cost criteria, and in future the decision model can be developed along this trend. The MADM decision matrix of the IAQ$_{quality}$ model includes five alternatives (combination of various IAQ$_{quality}$ sub-components) for defining indoor air quality in a building. The informationcontained in every attribute is related to the contrast of each criterion. Standard deviation (SD) and entropy are the measures of intensity and ways to present objective criteria weights. In order to calculate the weights of criteria, the authors use objective methods: entropy and CRITIC, presented in Figures 3 and 4.

Adaptation of the IAQ model with the "new" weighting procedure based on the MDMA approach to our office building case study (see Results and Discussion) was performed mainly for illustrative purposes in the context of the presented MDMA method for IAQ model development. The authors did not focus in depth on any technical issues of the building. Other Indoor Environmental Quality (IEQ) sub-components indexes, such as thermal, acoustic and visual satisfaction (in %), are presented in a paper [5] in order to provide background to the actual research problem.

The authors present the order of steps and mathematical operations leading to the determination of weighting schemes of ΣIAQ models (4), (5) and (6) separately. All model decision matrix variants are covered inan overview (Table 2), which presents decision matrices for the three types of IAQ models: ΣIAQ$_{quality}$, ΣIAQ$_{comfort}$ and ΣIAQ$_{comfort/health}$, covering various types of contaminants.

Table 2. Decision matrices for three ΣIAQ models with attribute data.

Alternatives $a = 1 \ldots n$	Attributes $X_{1 \ldots m}$ ($j = 1 \ldots m$)						Emissions		
	CO_2	TVOC	HCHO	VOC Odorous	x High Enthalpy	PM2.5	CO_2 (e)	TVOC (e)	HCHO (e)
IAQ$_{quality}$ in quasi-stable state with load by emissions									
IAQ$_{quality}$	x_{11}	x_{12}	x_{13}						
IAQ$_{quality}$(e_{min})	x_{21}	x_{22}	x_{23}				x_{27}	x_{28}	x_{29}
IAQ$_{quality}$+($e_{25\%}$)	x_{31}	x_{32}	x_{33}				x_{37}	x_{38}	x_{39}
IAQ$_{comfort}$ in quasi-stable state with load by emissions									
IAQ$_{comfort}$	x_{11}	x_{12}	x_{13}	x_{14}	x_{15}	x_{16}	x_{17}	x_{18}	x_{19}
IAQ$_{comfort}$ (e_{min})	x_{21}	x_{22}	x_{23}	x_{24}	x_{25}	x_{26}	x_{27}	x_{28}	x_{29}
IAQ$_{comfort}$+($e_{25\%}$)	x_{31}	x_{32}	x_{33}	x_{34}	x_{35}	x_{36}	x_{37}	x_{38}	x_{39}
IAQ$_{comfort/health}$ in quasi-stable state with load by emissions									
IAQ$_{comfort/health}$	x_{11}	x_{12}	x_{13}	x_{14}	x_{15}	$\ldots \ldots$	x_{17}	x_{18}	x_{19}
IAQ$_{comfort/health}$ (e_{min})	x_{21}	x_{22}	x_{23}	x_{24}	x_{25}	$\ldots \ldots$	x_{27}	x_{28}	x_{29}
IAQ$_{comfort/health}$ +($e_{25\%}$)	x_{31}	x_{32}	x_{33}	x_{34}	x_{35}	$\ldots \ldots$	x_{37}	x_{38}	x_{39}

The first step for estimating the MADM framework is normalisation of data (henceforth, the decision matrix indicators), which belong to different types and have different excess concentrations of the most important air pollutants and VOC emissions affecting IAQ indoor environments. In an indoor quasi stable state, attribute parameters include the concentration of these pollutants and additional increases in their concentration when the level is disturbed by additional sources of their emission or bio-pollutant emissions (1). Normalisation of attributes in various units (values of pollutant concentration and emission rate) in large ranges (attribute) is performed to obtain dimensionless quantities. Attribute normalisation is possible in several ways; however, for the research problem discussed here, the most common methods are as follows:

(a) Linear normalisation of attributes belonging to the decision matrix carried out according to the pattern of elements in the normalised decision matrix, belonging to the set of "cost attributes" according to Equation (8):

$$n_{ij} = \min{(x_{ij})}/x_{ij} \tag{8}$$

where $j \in N_c$ and N_c represents a set of criteria or attributes. The minimum value of the indicator x_{ij} of the decision matrix, i.e., the normalisation of attributes in the form of the levels of air pollutants or their emissions in the decision matrix of a complex IAQ model, can be more conveniently determined using Equation (9) provided by Körth [29]:

$$n_{ij} = 1 - \frac{x_{ij}}{x_j^{\max}} \tag{9}$$

if air pollution is a negative value indicator.

(b) Normalisation of attributes according to Wang's scheme [6] already applied in relation to the description of air pollution. In addition, positive and negative indicator values represent different meanings (10) and (11). Air pollution is a negative value indicator (negative values), while profit from the absence of pollutants would be a positive indicator. Similarly, the values of the emission rate of individual pollutants are also negative indicators, which can be read from the emission curves, both max (x_{ij}) and min (x_{ij}) values. Therefore, in order to normalise the attributes, the authors chose negative values according to Equation (10). Positive values (type benefit) are

$$n_{ij} = \frac{x_{ij} - \min(x_{ij})}{\max(x_{ij}) - \min(x_{ij})} \tag{10}$$

Negative values (type costs) are

$$n_{ij} = \frac{\max(x_{ij}) - x_{ij}}{\max(x_{ij}) - \min(x_{ij})} \tag{11}$$

where x_{ij} represents the original value and n_{ij} represents the value after normalisation. However, this way of normalising the IAQ decision matrix in the actual system of its application, in which some configurations (alternatives) have some attributes values equal to 0, proved to be impracticable. The same group also uses the transformer version of Equation (10) provided by Zavadskas [30] and Vujcic [31]:

$$r_{ij} = \frac{x_{ij} - \max x_{ij}}{\min x_{ij} - \max x_{ij}} \tag{12}$$

The stage of data normalisation of the decision matrix is carried out after it is built, and at the same time, the max (x_{ij}) and min (x_{ij}) values are determined in the attribute columns j. These max and min values of the x_{ij} indicators are inserted into the formula for normalisation of the input data. Therefore, prior to data normalisation, the maximum and minimum values x_{ij} adopted for the attributes in the studied conditions of the IAQ model should be compiled from the literature. These values will be used to test the normalisation of the input dataset adopted in the decision matrix, while the test normalisation will also address the question of whether a given normalisation method (formulas) normalises the input dataset correctly (the normalisation result cannot be negative) [31] and whether the set of alternatives (combinations of attributes) in the decision matrix has been determined correctly. Literature [2,5,7,32–40] on the minimum pollutant values (min x_{ij}) include the long-term concentrations [32] or c_{LCI} concentration values. The maximum values [2,7,36–38] of the max (x_{ij}) pollutant concentrations are either short-term values or allowable concentration values c_{ELV} [32] according to EN 16798-1: 2019, World Health Organization WHO guidelines [33] and EU recommendations [34]. The results of maximum and minimum concentration tests in representative groups of UK, Danish and Polish building tests published by Shrubsole [2] and Johnston [35] were used. Attributes are initially normalised assuming that, for attributes $X_1 \ldots m$, the values from Table 3 can be inserted into Equations (9) and (12) to obtain normalised attributes n_{ij}.

Table 3. Example data for the ΣIAQ model decision matrix. Attribute indices x_{ij} and normalisation results n_{ij} with two methods calculated by Equations (9) and (12), as example alternatives.

Parameters				Attributes, $X_{1 \ldots m}$					
	CO_2 ppm	TVOC µg/m^3	HCHO µg/m^3	VOC$_{odorous}$ µg/m^3 e.g., NO$_2$	x Enthalpy kJ/kg	PM2.5 µg/m^3	Emission Bio-CO_2 e mg/m^3 h.pers.	Emission TVOC e mg/m^3 h	Emission HCHO e µg/m^2 h
max(x_{ij})	1000	1000	160	660	80	100	64.84 persons	57.1	68.2
min(x_{ij})	200	154	5.8	20	34	10	16.2 one person	13.6	4.26
min x_{ij}–max x_{ij}	−800	−846	−194.2	−640	−46	−90	−48.6	−43.5	−63.94
x_{ij} lit.value	450	787	18	40	45	35	32.4 mean value	28.6 mean value	36
Normalised attribute values n_{ij} calculated according to Equation (9)									
	0.55	0.213	0.88	0.94	0.44	0.65	0.5	0.5	0.47
Normalised attribute values n_{ij} calculated according to Equation (12)									
	0.68	0.25	0.73	0.97	0.76	0.72	0.67	0.65	0.50

2.5. Calculation of the Objective Attribute Weights by the Entropy Method

Step 1. Normalisation of indicators

Since the analysed indicators of indoor air contamination may be provided in different units, authors normalised them as a first step. Normalised decisions, however, depend on the attribute's nature. In the entropy weights procedure, all attributes must be beneficial (positive values), but in the decision matrix all attributes are of a non-beneficial type. Therefore, the conversion of normalised indicators of "cost" type into indicators of "profit" requires building a matrix of "profit type". As such, the higher their indicator values are, the higher the assessment of the decision variant will be. The indicators for which the criteria are based on the assessment of current air pollutants, and the assessment of the emission rate belong to the cost type indicators, are designated by non-beneficial or cost sets. In order to determine weights using the Shannon entropy method, a specific data normalisation method is used [20]. We built a matrix $Y = (y_{ij})$, in which all criteria will be of "profit type". Then, the elements of the matrix are determined from Equation (13):

$$y_{ij} = 1/x_{ij} \tag{13}$$

The normalised decision matrix $Z = [z_{ij}]_{n \times m}$ with the elements z_{ij} for determining entropy weights is determined with Equation (14):

$$z_{ij} = \frac{y_{ij}}{\sum_{i=1}^{n} y_{ij}} = \frac{1/x_{ij}}{\sum_{i=1}^{n} 1/x_{ij}} \tag{14}$$

for various alternatives (more or less complex indoor air quality models) and their respective attribute sets (selected IAQ sub-models to assess various pollutants and emission rates).

Step 2. Calculation of the entropy measure

The entropy vector e_j for a given attribute j from the sum of components z_{ij} and $\ln(z_{ij})$ calculated for all alternatives $i = 1 \ldots n$ is determined as

$$e_j = -\frac{1}{\ln n} \sum_{i=1}^{n} z_{ij} \ln z_{ij} \tag{15}$$

where $z_{ij} \cdot \ln z_{ij}$ is equal to 0 for $z_{ij} = 0$ and the variation(divergence) level vector d_j calculated for each attribute as

$$d_j = 1 - e_j \tag{16}$$

Step 3. Defining the objective weight of each j as follows, based on the entropy concept

$$w_{jentropy} = \frac{d_j}{\sum_{j=1}^{m} d_j} \tag{17}$$

2.6. Calculation of Subjective Attribute Weights Depending on the Toxicity of Air Pollution to Humans

The weights of IAQ_{index} models must also contain, in addition to the objective entropy weights, the subjective attribute weights that describe factors related to health human toxicity from indoor air pollution.

Variant 1. S_j^{health} subjective weights from databases for criteria of "air pollution specified chemical compound" with tested chemical properties, toxicity, and "health level" and "health scores" assigned to air concentrations of this compound are treated as reference data taken directly from the literature [36] and [40–44]. For example, Yuan [40] adopted the following weights for selected pollutants as part of the research program:"French permanent survey on Indoor Air Quality" [43], US-EPA and WHO databases (Table 4).In some cases when detectinga larger number of indoor air contamination compounds with air concentration >5 µg/m^3, additive health effects are assumed, and the subjective weight S_j^{health} can be created with the EU-LCI approach.

Table 4. Subjective weights of ten pollutants (plus hot and humid air enthalpy).

Different Types of Air Pollutants and Subjective Weights									
CO	CO_2	NH_3	O_3	HCHO *	PM10	TVOC	Rn	microbes	benzene
0.21	0.01556	0.01556	0.158	0.21	0.0525	0.21	0.105	0.0233	0.19

The levels of indoor air enthalpy and related user dissatisfaction (%)					
Enthalpy kJ/kg	34.5	44.5	59.5	70.5	80.0
Dissatisf. %	10	20	40	60	81.9

* Based on the recommendation of the European Collaborative Action expert group. This group focuseson the harmonisation ofindoor emission assessments by means of developoing Lowest Concentration Interest (LCI) values (since 2011). This expert group developed a list of relevant substances and described the procedure of LCI calculation. For assessing combination effects of substances, the R-value as the health hazard index was proposed. The R_i value was established as $R_i = C_i/LCI_i$, where C_i is the concentration of compound in a test chamber i. For $R_i < 1$, it is assumed that there will be no health effects. R = sum of all R_i = sum of all ratios $(C_i/LCI_i) \leq 1$.

Variant 2. The subjective s_j^{health} weights determined by health experts depend on the toxicity of pollutants based on a number of expert opinions r greater than the number of experts k. In the case of not listing compounds in databases, the relative value (signification) of the s_j^{health} weight for the jth criterion (with a total number of m criteria) can be calculated by the SWARA method presented by Kersuliene et al. [45] and Zavadskas [46] using weight estimation procedures in several steps as follows.

Step 1. The relative importance of the criterion is determined by reference to the weight of s_j and the weight of the previous criterion s_{j-1}. Then, the weights of s_j^{health} based on the opinions of all experts are calculated. An example of a simple algorithm for determining the average subjective weight of attributes s_j is based on the average of opinions (weight proposals) of attributes collected from k experts, i.e., from the formulas given by [46]. The average value of expert attribute reviews is calculated using Equation(18):

$$\bar{t}_j = \frac{\sum_{k=1}^{r} t_{jk}}{r} \tag{18}$$

where t_{jk} is the ranking of attribute j according to respondent k, and r is the number of respondents.

Step 2. The subjective weights can be adjusted by dividing the average values of each attribute by the sum of the average values of the attributes (19):

$$s_j^{health} = \frac{\bar{t}_j}{\sum_{j=1}^{m} \bar{t}_j} \tag{19}$$

Step 3. Having determined the average values of attribute reviews from Equation (18) and knowing the established ranking of these opinions, you can calculate the standard deviation of attribute j in the ranking of opinions ("dispersion of expert ranking values") by Equation (20):

$$\sigma_j^2 = \frac{1}{r-1} \cdot \sum_{k=1}^{r} (t_{jk} - \bar{t}_j)^2 \tag{20}$$

Value σ_j^2 is useful for calculating the weights of the effects of attribute correlation $w_{j,CRITIC}$ formula.

2.7. Calculation of the Objective Weights of Correlation between Attributes by the CRITIC Method

Use of the CRITIC (Criteria Importance Through Inter-criteria Correlation) method in the case of mutually correlating attributes (e.g., attributes constituting "quasi-fixed" impurities and attributes constituting their emissions) to estimate the correlation weights between the criteria is justified. Additionaly to the intensity contrast of dataset attributes in the decision matrix, there was another concept more recently taken into consideration by researchers using MCDM. Diakoulaki's team [47] found that the greater the level of inter-dependency between attributes, the higher the error ranking outcome. CRITIC was analysed in [47] as an objective and new weighting system that took into consideration criteria correlations. The approach included contrast intensities (by means of SD criteria) and combined them with correlations weights. All material selection studies assume that the criteria are independent of each other, while they are also likely to influence each other. Some approaches proposed in recent years for weighting criteria when choosing material not only bypassed this approach, but they did not take into account the interdependence of criteria and proposed methods of objective weighting, which have relevant shortcomings. When correlation between criteria is suspected, the validity of the criteria can be adjusted by entering correlation weights. Calculating the weights of correlations between the criteria of a decision model using Pearson's correlation coefficients is recommended [9,11]. The CRITIC weights marked $w_{j,CRITIC}$ specify a correlation between the criteria, which can include all criteria (or attributes) or only criteria that are influenced. The CRITIC method by Diakoulaki [47], Jahan et al. [12] and Ardakani et al. [10] includes estimations of differences in the "intensity" of criteria values (expressed by the value of their standard deviations) and combines them with weights based on correlation coefficients (step 4). Alternatively, the modified model in Jahan [12] uses only Pearson's correlation coefficients and does not include standard deviation criteria in the equations. As justification, it was stated that the differences resulting from changes in the "intensity" of the criteria were already taken into account by the entropy method, and the correlation weights were calculated accordingly. Therefore, the values of the correlation weights between the criteria are calculated by [10,12] using the values of Pearson coefficients R_{jk}, which represent the correlation between the two selected criteria j and k (correlated criteria) and which can be calculated easily, even in MS Excel. R_{jk} values are read from the symmetrical linear criteria correlation matrix or calculated from Equation (21):

$$R_{jk} = \frac{\sum_{i=1}^{n} (x_{ij} - \bar{x}_j)(x_{ik} - \bar{x}_k)}{\sqrt{\sum_{i=1}^{n} (x_{ij} - \bar{x}_j)^2 \cdot \sum_{i=1}^{n} (x_{ik} - \bar{x}_k)^2}} \quad j \text{ and } k = 1 \ldots m \tag{21}$$

where x_{ij} and x_{ik} are n total number of alternatives (alternative IAQ models) and matrix elements (Table 1) x_{ij} (with a peak) of average values of the j criterion and x_{ij} of average values of k criteria (emission criteria $k_{1..3}$ correlating with other criteria, e.g., emission rate of pollutants, which include selected criteria pollutant concentrations). If R_{jk} is close to $+1$ or -1, this means that the criterion is correlated highly, while R_{jk} close to 0 means there is no correlation. Then, the correlation weights are calculated by the CRITIC method using R_{jk} determined by Equation (21). The effects of the correlation impact of all criteria on other criteria can be calcualted, or the impact of selected k criteria (e.g., emissions) on related criteria of the decision model j (e.g., polluting emissions), and only then are selected correlation coefficients used. A common form of the formula for correlation weights using the CRITIC method is Equation (22) [10]:

$$w_{j,CRITIC} = \frac{\sum_{k=1}^{m} (1 - \beta \cdot R_{jk})}{\sum_{j=1}^{m} (\sum_{k=1}^{m} (1 - \beta \cdot R_{jk})} \quad j \text{ and } k = 1 \ldots m \tag{22}$$

where β is equal to $+1$ if two objective criteria are of the same type (either greater is better) or $\beta = -1$ otherwise. In Equation (22), Equation (23) denotes a conflict of measure created by criterion k in relation to the decision:

$$Conflict = \sum_{k=1}^{m} \left(1 - R_{jk}\right) \tag{23}$$

According to Zawadskas [30], the sum (23) ought to be multiplied by σ_j^-, $-$ the standard deviation (SD) of the jth criterion. The inclusion of SD follows a similar approach to the entropy approach, assigning a weight to the attribute if it has a corresponding attribute value across alternatives. A "symmetrical matrix" is developed with dimensions $n \times m$ and a generic element R_{jk}. The higher the divergence in the scores of alternatives in criteria j and k, the lower the value of R_{jk}. This step determines the conflict and then the weight of each decision criterion. The sum provided below denotes the measure of conflict created by criterion j with respect to the decision.

The modified and simplified CRITIC method, which has not been used in this study, calculates the correlation weights between criteria without calculating the symmetric linear correlation matrix R_{ij}. The method determines the objective weights $w_{j,CRITIC}$ on a different principle. It determines the objective weights of the attributes by using contrast intensities for each measure, considered as SD, conflict and correlation coefficient between criteria by Zavadsas et al. [30], Vujicic et al. [31] and Adali et al. [48].

Step 1. Calculation of the normalised decision matrix for $w_{j,CRITIC}$

For each criterion x_{ij}, function r_{ij} which transforms all the values of criteria into the interval [0, 1], is provided:

$$r_{ij} = \frac{x_{ij} - \min x_{ij}}{\max x_{ij} - \min x_{ij}} \tag{24}$$

The basis of the transformation uses the concept of an ideal point. The initial matrix is converted into generic element matrix r_{ij}. Here, it should be noted that normalisation does not take account of the type of criteria.

Step 2. Calculation of standard deviations in attribute column j for all alternatives $1 \ldots n$.

Step 3. Calculation of C_j specifying the amount of information contained in the given attribute j.

In Equation (25), SD represents the deviation of various values for a given criterion of a mean value. The information of C_j contained in criterion j is calculated by Equation (25):

$$C_j = \sigma_j \sum_{j=1}^{m} (1 - R_{jk}) \tag{25}$$

where σ_j is the SD of the jth criterion or alternative, and the expression r_{jk} is considered equivalent to the correlation coefficient between the two attributes: selected attribute j and correlation attribute k.

Step 4. Calculation of the weight of the correlation of the criteria correlating k with a given criterion j is carried out using Equation(26):

$$w_{j,CRITIC} = C_j / \sum_{k=1}^{m} C_k \quad j = 1, \dots m \tag{26}$$

Calculation of correlation coefficients should not be ignored; therefore, this study adopted a modified method of data normalisation and calculation of correlation weights between attributes based on the symmetric linear correlation matrix using the full CRITIC method. Calculation of correlation weights between criteria based on the symmetric linear correlation matrix by the CRITIC method uses a modified data normalisation method. In summary, the calculations were based on the determined correlation coefficients for attribute pairs related to the same impurities $w_{j,CRITIC}$ by using the following steps.

Step 1. Normalisation of initial decision matrix (Table 1)

The recommended [6] equation for normalisation of the original decision matrix for cost criteria is Equation (27)

$$r_{ij} = \frac{\max(x_{ij}) - x_{ij}}{\max(x_{ij}) - \min(x_{ij})} \tag{27}$$

but the authors used the transformer version [32,33] in the form:

$$r_{ij} = \frac{x_{ij} - \max x_{ij}}{\min x_{ij} - \max x_{ij}} \tag{28}$$

Step 2. Calculation of standard deviations in data sets assigned to attributes j from the value r_{ij} for all alternatives $1 \dots n$.

Step 3. Calculation of symmetric linear correlation matrix R_{ij}.

The primary matrix (Table 1) is converted to the normalized decision matrix and later to symmetric linear correlation matrix of generic R_{ij} elements, which are Pearson's correlation coefficients (21) determined for all pairs of normalised sets on data r_{ij} assigned to selected attributes, at all levels of alternatives n. Then, the most likely correlations corresponding to attribute pairs (e.g., concentrations and emissions of the same compound) were selected from the matrix, and weights were calculated for them $w_{j,CRITIC}$. A coefficient of linear correlation between each pair of measures is estimated using the following formula that allows to quantify the conflict occurring among various criteria. It is obvious that more discordant the scores of the alternatives in both matrix vertical i and horizontal j sections, the lower the value R_{ij}.

Step 4. Calculation of objective weights using the CRITIC method. The weight of the jth criterion is obtained using Equation (29) presented as:

$$w_{j,CRITIC} = C_j / \sum_{k=1}^{m} C_k \tag{29}$$

where

$$C_i = \sigma_j \cdot \sum_{k=1}^{m} \left(1 - R_{jk}\right) \tag{30}$$

A higher value of C_j indicates a more considerable amount of information contained in a particular criterion. Therefore, its weight is higher. The CRITIC method assigns greater weight to criteria that have high standard deviations and low correlations with other criteria. A higher C_j value means that more information can be obtained from this criterion, which increases the relative importance of the criterion for the decision problem. The study of correlation effects by the CRITIC method is carried out by calculating the weight $w_{j,CRITIC}$ for the j attribute that requires it due to the anticipated impact of the k attribute, e.g., when it is known that the validity of the attribute "TVOC pollution" in a "quasi-stable state" will increase under the influence of the attribute correlating with"TVOC emission".

2.8. Calculation of the Objective Weights for Ventilation Energy Expenditure

The objective weights for ventilation energy refer to the combined ΣIAQ model described by polynomial (1), whose words have been assigned weights described by Equation (3), based on the concept of "excess concentration", creating the energy expenditure of ventilation. These are the relative weights calculated for each alternative of the IAQ model. The values Δc_j for pollutants adopted inthe decision matrix differ by several orders. In the IAQ index model decision matrix (Table 1), it is necessary to normalise the set of values of "excess concentrations" for the first three attributes in the $\Sigma IAQ_{quality}$ system described by Equation (4). The basic problem indata normalisation for determining weights $w_{ij,energy}$ based on the concept of "excess concentration" is the selection of formulas that can be useful for this purpose. Referring to Körth et al. [29] and Vojcica [31], it seems that the following equation can be used to normalise datasets of "type cost", which after conversion to attributes of "type profit" (or beneficial) adjust for IAQ systems corresponding to subsequent alternatives a1 to a5. Normalised values Δc_j will correspond to "energy load" weights $w_{ij,energy} = n_{ij}$:

$$n_{ij} = \left[1 - \frac{x_{ij}^{\min}}{x_{ij}}\right]^{-1} \tag{31}$$

Taking into account the input data of the decision model summarised in Tables 1 and 5, a dimensionless set of values corresponding to excess concentrations according to (2), normalised for weights, is calculated for the alternatives of this model $w_{ij,energy}$. for non-zero attributes $j = 1, 2, 3$ in the form of CO_2, TVOC and HCHO pollutant concentrations.

2.9. Combining the Weights by Calculation of Global Weights

The global weights $w_{ij,global}$ contain specified weight multiplication products of various types, adjusted to the sum of these products according to the widely used procedure for building weighting development. In the flow chart (Figure 3), the $w_{j,global}$ weights include basic entropy weights $w_{j,entropy}$, subjective weights related to health human toxicity determined using "health-related" databases s_j^{health} and correlation weights $w_{j,CRITIC}$, and ventilation energy weights $w_{ij,energy}$. At the same time, correlation weights occur only in those alternatives (ΣIAQ systems) in which decision matrices contain correlation criteria (this criterion is "pollutant emissions" in Table 1). The weights are assigned to individual attributes $j = 1 \ldots 6$ of the decision matrix, and the weighting schemes for five alternative IAQ levels $i = 1 \ldots n$ are composed of weights assigned to specific attributes at a given level of alternatives from a1 to a5, which are adjusted to the sum of weights at a given alternative level i. A modified integrated weighting method, presented by Deepa et al. [49] and [15], proposed an approach to combine all weights calculated using different weight assignment approaches into a single set of weights. The equation to calculate the weights of attributes in alternative levels is

$$w_{ij.global} = \frac{\prod_{g=1}^{n} w_g}{\sum_{g=1}^{n}\left(\prod_{g=1}^{n} w_g\right)} \tag{32}$$

where $w_{ij,global}$ represents thecomined weights of the criteria, w_g denotes the criteria weights calculated by using one of the weight calculation methods, and $g = 1, 2, \ldots, n$, where n is the number of methods used.

If a decision is made to assign a given attribute to the objective weight $w_{j,entropy}$ and the subjective weight associated with the risk to s_j^{health} health, but with adjastmentto the sum of weights at a given alternative level i, then the global weight of toxic pollution is

$$w_{ij.global} = \frac{[s_j^{health} \cdot w_{j.entropy}]}{\sum_{j=1}^{m}[s_j^{health} \cdot w_{j,entropy}]} \tag{33}$$

When the alternative and the attribute are assigned the relative weight $w_{ij,energy}$ and when the alternative indicators are compared to the attributes of type k correlating with, e.g.,"pollution emission", then the composite weight of the attribute correlating with the attribute "emission" also includes the weight of the correlation effect $w_{j,\text{CRITIC}}$ and weight $w_{j,energy}$ for the given attribute j at the level of the i-th alternative:

$$w_{ij,global} = \frac{[s_j^{health} \cdot w_{j,entropy} \cdot w_{j,CRITIC} \cdot w_{ij,energy}]}{\sum_{j=1}^{m}[s_j^{health} \cdot w_{j,entropy} \cdot w_{j,CRITIC} \cdot w_{ij,energy}]} \tag{34}$$

The denominators of the formulas for global attribute weights $j = 1 \ldots m$ at alternative levels $i = 1 \ldots n$ are, therefore, calculated as the sum of the products of entropy weights $w_{j,entropy}$ multiplied by s_j^{health} and $w_{j,\text{CRITIC}}$ and other weights, e.g.,$w_{j,energy}$ representing the importance of the attribute for the energy load.

It should be noted that the use of the $\sum$IAQ model (for calculating the total energy expenditure for ventilation and using the global weight value in the IEQ inEquation (7)) requires to determine the global weight values only for the attributes of "pollution concentrations" (in the case of the IAQ$_{quality}$ model, only for $j = 1, 2, 3$) in the form of relative global weights adjusted at all levels. The measures of the significance of the other attributes (emissions) are determined by the CRITIC method and are given by correlation weights that additionally impact the global weights of the "concentration" attributes. Therefore, although entropic weights are calculated just like CRITIC weights for all attributes, taking advantage of the fact that they are weights characteristic for given attributes (they are weights calculated in columns j, not relative weights), they can be "selected" and adjusted in the range of attributes to calculate relative global weights for all levels of alternatives.

2.10. Characteristics of the Case Study Building

In order to determine the weights of the IAQ model, the authors present a casestudy building and use the real scenarios inindoor environment quality. Thecasestudy building is a newly built high tower office with 49 floors and a net area of 59,000 m^2 made of steel and concrete materials with a heavy glass facade. The buildingis located in Warsaw. The experimental part (measurement in situ) included analysis of CO_2, HCHO and VOCs indoor concentrationsin the building interiors. Provided in [5], datasets on the pollutants supported the authors' decision matrix development including scenarios/alternatives. At the time of the test, the building was in the pre-occupancy phaseandhadempty spaces with no furniture inside. The indoor walls were plaster covered and freshly painted. Suspended ceilings were mounted, and thefloors were coveredwith a synthetic carpet. The building installationswere active during tests. The building was tested threedays after the internal work was finished. Measurements were made on the 47th floor, covering area of almost 3000 m^2. The main focus was on the IAQ_{index} for selected open spaces in which alargenumber of usershave to reside, and which represent the largest occupied floor space. ISO and CENinternational standard based analytical methods were engaged to assess"in situ"HCHO, TVOC and CO_2 concentration values. Theair samples for VOCs tests were collected viaactivesampling with anelectronic mass flow controller that controlled the air flow via test probes (10 dm^3/h for VOCs and 30 dm^3/h for HCHO). Indoor air samples were taken from selected representative office locationsand tested later in an accredited laboratory in accordance with ISO 16000-3:2011 and ISO 16000-6:2011. VOCs were sampled on the probe tubes filled with Tenax adsorbent. Later, they were thermally desorbed using a TD-20 device (Shimadzu, Kyoto, Japan). The process of separation and analysis of VOCs was achieved with aGC/MS gas chromatograph equipped with a mass spectrometer (model GCMS-QP2010, Shimadzu, Japan). VOCs were identified by comparing the retention times of chromatographic peaks with the retention times of reference compounds, by searching the NIST mass spectral database (USA). Identified VOC compounds were quantified using a relative identification factor from standard solution calibration curves. TVOC concentration was calculated by summing identified and unidentified VOCs eluting between n-hexadecane and n-hexane.

2.11. Building-Related Research Limitations

Building-related research limitations were also recognized because the ΣIAQ model and provided weighting scheme applied mainly to office buildings with mechanical ventilation. Quasi-static indoor environment conditions and full air mixing in the building wereassumed by the authors (perfect mix of indoor air). The IAQ model is based on sensory curves that were created by examining people's response to a stimulus with analmost neutral influence of other indoor environment factors. This means that the applicability of the presented IAQ model is limited to indoor thermal conditions in the range of 20–26 °C and relative humidity of 30–70%. The valid clothing level of occupants was 0.5–0.8 clo, and metabolic rate accepted was 1–1.2 m. The number of air pollutants as sub-components in the IAQ model was limited to the sub-models presented in [4]. In the example case study building, we analysed building pollution situations based on sixattributes from which we built alternative situations (a1–a5) for the building. These attributes are mostcommonly used in commercial environmental building assessment systems like BREEAM or LEED. The range of applicability of the developed weighing system wasvalid up to the maximum pollution values: 1800 mg/m^3 for ΔCO_2, 1000 μg/m^3 for TVOC and 160 μg/m^3 for HCHO.

Health risk in the context of human exposure to a given pollutant is an issue usually analysed inoccupational medicine research. In this context, we consciously limited our research scope and didnot focus directly on human health risk.

3. Development of $\Sigma IAQ_{quality}$ Model Weighting Schemes for the Case Study Building

3.1. Development of the Initial Decision Matrix Adapted to the Combined $\sum IAQ_{quality}$ Model

This section providesastep-by-step approach for a numerical calculation of combined $\sum IAQ_{quality}$ model sub-element weights presented in the example case study building.

The results of the casestudy building identified pollutants [5] and were the basis inbuilding a decision matrix. Table 5 represents the initial decision matrix (the input data table) for the defined scenario (fitting of the weighting schemes to various $IAQ_{quality}$ quasi-stable states). The matrix consisted of six attributes (concentrations and emissions of three indoor air pollutions) and five alternatives, where:

a1—alternative based on the case study specifics presented in a previous report in [4,5]. In the building under investigation, with quasi-stable air quality, for the indoor pollutant concentrations of CO_2, TVOC and HCHO, the assumption was made that the additional emissions of these three air pollutants were not measurable;

a2—alternative based on the assumption of minimum air pollution concentration levels changing very little because the emissions rate atthe minimum level does not affect the quasi-stable state of IAQ; for the CO_2 bio-pollution rate and the CO_2 concentration minimum level value, the bio-effluents rate from a single person during one hour was accepted [42];

a3—alternative predicting the measured air pollution quasi-stable concentration at aminimum level, but under the influence of CO_2 (e), TVOC (e) and HCHO (e) emission rates, it did so at a higher level (addition under minimum level waswithin the 0.25 measuring range); the TVOC or HCHO minimum levels of emission rates were calculated as 0.25 of the (max–min) range estimated by the literature on thepollution emission rate;

a4—alternative is similar to a3, but the higher influence of the emission rate of CO_2 (e) does not exist (there is no addition under the minimum level of CO_2 emission rate);

a5—alternative is similar to a3, but the higher influence of emission ratesof TVOC and HCHO does not exist (there is no addition under the minimum level of TVOC and HCHO emission rates).

Table 5. Initial decision matrix with the data assigned to various $IAQ_{quality}$ component sets and action scenariospresented by individual alternatives.

lternatives	Attributes					
	$\Delta\ CO_2$ [1],[2]	TVOC	HCHO	CO_2 e [2]	TVOC e [3]	HCHO e [4]
	ppm mg/m^3	µg/m^3	µg/m^3	mg/m^{3}h	µg/m^{3}h	µg/m^2 h
a1 test study [5] (e) emission rate = 0	450 ppm 810 mg/m^3	787	18	0	0	0
a2 minimum [3] (e) minimum TVOC and HCHO; (e) minimum CO_2 one person	360 + 16.2 = 376.2 mg/m^3	154 + 13.6 = 167.6	5.8 + 4.26 = 10.6	16.2 one person	13.6	4.26
a2-2 test study [5] (e) minimum TVOC and HCHO; (e) minimum CO_2 one person	450 ppm 810 mg/m^3	787	18	16.2 one person	13.6	4.26
a3 minimum + 1h (e) minimum + 25% concentration (e) CO_2 two persons[2]	200ppm 360 + 32.4= 392.4 mg/m^3	154 + 24.3 = 178.3	5.8 + 20.2 = 26.0	32.4 two persons	10.8 + 13.6 = 24.3	15.9 + 4.26 = 20.2
a5 minimum+ 1h(e) minimum TVOC, HCHO; (e) CO_2 minimum + 25% concentration two persons	200ppm 360 + 32.4 = 392.4 mg/m^3	167.6	10.6	32.4 two persons	13.6	4.26
Maximum x_{ij} level of ventilation load (lit.)	1000 ppm 1800 mg/m^3	1000	160	64.8 mg/m^{3}h four persons	57.1	68.2
Maximum x_{ij} value in attribute j column	810	787	26.0	32.4	24.3	20.2
Minimum x_{ij} level of ventilation load (lit.)	200 ppm 360 mg/m^3	154	5.8 10.6	16.2 mg/m^3 hone person	13.6	4.26
Minimum x_{ij} value in attribute j column	376.2	167.6	10.6	0	0	0

[1] General conversion of units, e.g., CO_2 from ppm to mg/m^3 in accordance with [50]. [2] Assumptions for the attribute of CO_2 concentration—(1) minimum 1 person max. 4 people; (2) in a room of 20 m^3 per person; (3) the CO_2 concentration caused by occupant exhalation is approximately 400–1700 ppmv,while the average volume is 20–150 m^3·person^{-1} and the air change ACH is 0.45–0.70 h^{-1} [42]. The time for indoor concentration of CO_2 to reach a quasi-steady state is between 6 and 8 happroximately; (4) the mass balance method [42] can be used to link indoor CO_2 emission rate concentration, as shown in equation m = p·c·Q, where m is the emission rate of the target compound, µgs^{-1}, p is occupant number, c is the CO_2 concentration in occupant exhaled air (400 ppmv [42]), µg m^{-3}, and Q is the respiratory rate of humans, m^3 s^{-1}, and is assumed as7.5l min^{-1} (1.25 × 10^{-4} m^3 s^{-1}). [3] TVOC—Hori [38] study ACH is ca 0.5 h^{-1}.[4] HCHO—[35] ACH is set to 0.5 h^{-1}.

The set of alternatives is used for checking how the weighting scheme changes when the emission rate of a particular air pollutant is absent or increasing.

Simultaneously, the authors calculated all weights for the additional alternative version a2-2, which differs from alternative a2. This is based on the assumption of air pollution concentration levels, taken from the case study, practically not changing the quasi-stable state because the emissions rate is at a minimum level, like in alternative a2. A minimum level for the CO_2 bio-pollution rate, and TVOC (e) and HCHO (e) minimum level emission rates were established. It can, therefore, be argued that the weighting scheme for the IEQ model with the a2-2 alternative will have a screen emission (more probable than for the a2 alternative). The decision model and its weighting schemes with alternative a2-2 and the same set of attributes havedifferent structures than alternative a2. However, as the values of indoor pollution concentrations are the same in this study (with the assumption of minimum levels of emissions alone), version a2-2 can be used for comparingthe impact of the two alternatives a2 and a2-2. Maximum x_{ij} levels of ventilation load are based on the research [2,7,32,35,38,42] and Minimum x_{ij} levels by [2,35,38,39,42].

3.2. Attribute Weights Obtained by the Entropy Method

Determination of objective attribute weights (only for attributes 1, 2 and 3, named "pollution concentration" sub-indices in $IAQ_{quality}$) from Equation (4), calculated in attribute columns by all levels of alternatives belonging to the decision matrix in accordance to the entropy approach, is based onuncertain information measurements contained in the decision matrix and directly provides a set of weights for criteriabased on mutual individual criteria values contrastedwithvariants for each criterionalone and then for all criteria at the same moment. The following steps are suggested in this direction.

Step 1. Normalisation of indicators for the entropy method—since the indicators of air pollutants are in different units, and since the attributes are of a non-beneficial type, the most recommended method to normalise them is used first. In particular, the transformation method according to Equation (34) solves this issue:

$$z_j = \frac{y_{ij}}{\sum_{i=1}^{n} y_{ij}} = \frac{1/x_{ij}}{\sum_{i=1}^{n} 1/x_{ij}} \tag{35}$$

Table A1 (in Appendix A) presents the normalised decision matrix (according to Equation (35)) with dimensionless numbers z_j representing the normalised response of alternative i on attribute j. The structure of the decision matrix contains the input parameters of atribute $j = 1, 2, 3$ determined on the assumption of an indoor quasi-stable state environment with minimum level of air pollution concentration and minimum emission rates.

Table 6a presents the shortage of normalised decision matrices (according to Equation (35)). The structure of alternative a2-2 (presented in Table 6b) contains the input parameters of atribute $j = 1, 2, 3$ based on the current experiment [5] with the assumption of an indoor quasi-stable state environment with the experimentaly determined level of air pollution concentration and with minimum emission rates.

Step 2. The entropy measure calculation. The measure of the jth attribute under indicator i (i ranges from 1 to n) is calculated using the following Equation (36):

$$e_j = -k \cdot \sum_{i=1}^{n} z_{ij} \ln z_{ij} = -\frac{1}{\ln n} \sum_{i=1}^{n} z_{ij} \ln z_{ij} \tag{36}$$

where constant $k = 1/\ln n$ guarantes that e_j ($j = 1, 2 \ldots, n$) belongs to the interval [0;1]. A constant k in the matrix with five alternatives can be calculated by Equation (37):

$$k = \frac{1}{\ln(n)} = \frac{1}{\ln(5)} = 0.621335 \tag{37}$$

Table 6. Normalised decision matrices with dimensionless numbers z_j for (**a**) alternative a2 based on the minimum level of air pollution in a quasi-stable state with minimum emission rates for (**b**) alternative a2-2 based on air pollution values from the case study in a quasi-stable state with emissions rates at a minimum.

Alternatives	Attribute X = 1 … m					
	ΔCO_2	TVOC	HCHO	CO_2 (e)	TVOC (e)	HCHO (e)
	(a)					
a1	0.1043	0.0531	0.175	0	0	0
a2	0.2304	0.2449	0.2937	0.3333	0.3201	0.4134
a3	0.2174	0.2286	0.1187	0.1666	0.1798	0.0866
a4	0.2304	0.2286	0.1187	0.3333	0.1798	0.0866
a5	0.2174	0.2449	0.2937	0.1666	0.3201	0.4134
	(b)					
a1	0.1188	0.0656	0.1986	0	0	0
a2	0.1188	0.0656	0.1986	0.3333	0.3201	0.4134
a3	0.2475	0.2828	0.1347	0.1666	0.1798	0.0866
a4	0.2624	0.2828	0.1347	0.3333	0.1798	0.0866
a5	0.2475	0.3030	0.3333	0.1666	0.3201	0.4134

Therefore, it is possible to calculate the value of k for all attributes $j = 1 \ldots 6$ and alternatives $n = 5$. With the help of Table 6a,b, it was possible to calculate e_j, d_j and entropy weights of criteria values $w_{j,entropy}$, which are presented for Variant 1 with alternative a2 and Variant 2 with a2-2 in Table 7. The degree of divergence (d_j), where d_j ($j = 1, 2 \ldots , m$), is the inherent intensity of criteria contrast C_j. The value (d_j) of the average intrinsic information contained in each criterion is calculated as Equation (38):

$$d_j = 1 - e_j \tag{38}$$

Step 3. Calulation of the entropy weights of attributes

Next, the entropy weights of each j are calculated as follows. The final criteria weight, in the third step of the approach, may be calculated by an additive normalisation (39):

$$w_{j,entropy} = (d_j)/\sum (d_j) \tag{39}$$

Table 7 presents the results of calculations using Equations (38) and (39).

Table 7. Entropy, degree of divergence and the relative weights of attribute j.

Attributes	ΔCO_2	TVOC	HCHO
Decision matrix with alternative a2			
Entropy e_j	0.9790	0.9444	0.9509
Divergence d_j	0.0210	0.0556	0.0491
Relative weight $w_{j,entropy}$	0.16706	0.44232	0.39061
Decision matrix with alternative a2-2			
Entropy e_j	0.9620	0.8907	0.9619
Divergence d_j	0.0381	0.1093	0.0380
Relative weight $w_{j,entropy}$	0.2055	0.5895	0.20496

3.3. Attribute Weights by the CRITIC Method

Step 1. Normalisation of indicators for the CRITIC method

For each criterion$j$ and indicator x_{ij}, membership function r_{ij}, which transforms the values of attributes into the interval [0,1], is provided(40):

$$r_{ij} = \frac{x_{ij} - \min\ x_{ij}}{\max\ x_{ij} - \min\ x_{ij}}$$

(40)

and the transformer version of formulas is [30,31]:

$$r_{ij} = \frac{x_{ij} - \max\ x_{ij}}{\min\ x_{ij} - \max\ x_{ij}}$$

(41)

The transformation considers the concept of an ideal point. With Equation (41), the initial matrix is converted into the normalised decision matrix with the generic elements r_{ij} (Table 8a,b).

Table 8. Normalised decision matrices. (**a**) (with alternative a2) based on minimum air pollution levels in a quasi-stable state with minimum emission rates; with zero-dimensional generic elements r_{ij}. (**b**) (with alternative a2-2) based on air pollution test study levels in a quasi-stable state with minimum emission ratesand zero-dimensional generic elements r_{ij}.

Alternatives	ΔCO_2	TVOC	HCHO	CO_2 e	TVOC e	HCHO e
			(**a**)			
σ_j	0.393	0.394	0.447	0.374	0.369	0.428
a1	0	0	0.519	1	1	1
a2	1	1	1	0.5	0.440	0.789
a3	0.963	0.967	0	0	0	0
a4	1	0.967	0	0.5	0	0
a5	0.963	1	1	0	0.440	0.789
			(**b**)			
σ_j	0.478	0.479	0.376	0.374	0.369	0.428
a1	0	0	0.519	1	1	1
a2	0	0	0.519	0.5	0.440	0.789
a3	0.963	0.967	0	0	0	0
a4	1	0.967	0	0.5	0	0
a5	0.963	1	1	0	0.440	0.789

Table A2 (located in Appendix A) presents the normalised decision matrix Equation (35) with dimensionless indicators r_{ij} representing the normalised response of alternative i on attribute j.

Step 2. Calculation of symmetric linear correlation matrix R_{jk} based on all indicators of normalised decision matrices (Table 9) and shortages of symmetric linear correlation matrices R_{jk} with alternatives a2 (Table 9a) and a2-2 (Table 9b).

Step 3. Determining criteria weights for "pollutant concentration" attributes using the CRITIC method.

The amount of information necessary to calculate C_j (42) includes measures of conflict C_j determined with Equations (30) or (42). Attribute (criteria) weights $w_{j,\text{CRITIC}}$ were obtained from Equation (43) (Table 10).

Table 9. Symmetric linear correlation matrix R_{ij} (with alternative a2) calculated for all correlation possibilities. (**a**) Shortage of symmetric linear correlation matrix R_{ij} (with alternative a2) calculated for all correlation possibilities (in this shape of the matrix, it can be assumed that only six attributes are burdened with additional correlation weights, but in the recognised attribute set only the first three will be used). (**b**) Shortage of symmetric linear correlation matrix R_{ij} (with alternative a2-2) calculated for all correlation possibilities.

Attributes	ΔCO_2	TVOC	HCHO	CO_2 e	TVOC e	HCHO e
ΔCO_2	1.0000	0.9984	−0.0170	−0.7759	−0.8451	−0.5654
TVOC	0.9984	1.0000	0.0205	−0.8012	−0.8252	−0.5346
HCHO	−0.0170	0.0205	1.0000	0.0136	0.5477	0.8340
CO_2 e	−0.7759	0.0136	0.0136	1.0000	0.6782	0.4537
TVOC e	−0.8451	−0.8252	0.5477	0.6782	1.0000	0.9185
HCHO e	−0.5654	−0.5346	0.8340	0.4537	0.9185	1.0000
(a)						
ΔCO_2	1			−0.775907		
TVOC		1			−0.8252373	
HCH0			1			0.83395897
CO_2 e	−0.775907			1		
TVOC e		−0.825237			1	
HCHO e			0.833959			1
(b)						
ΔCO_2	1			−0.7496659		
TVOC		1			−0.750258	
HCH0			1			0.8279349
CO_2 e	−0.7496659			1		
TVOC e		−0.750258			1	
HCHO e			0.8279349			1

Table 10. Results of the CRITIC method application for decision matrix with alternatives a2 and a2-2.

	$(1 - R_{jk})$		
Attributes	CO_2	TVOC	HCHO
Standard deviation	0.393	0.394	0.447
CO_2	0		
TVOC		0	
HCHO			0
CO_2 e	$1 + 0.7759 = 1.7759$		
TVOC e		$1 + 0.8252 = 1.8252$	
HCHO e			$1 - 0.83396 = 0.16604$
$SUM_{attribute}$	1.7759	1.8252	0.16604
C_j—with Equation (42)	0.6979	0.7191	0.0742
SUM C_j	1.4912 (for three attributes) (100%)		
	and with alternative a2 (Table 9a)		
$w_{j,CRITIC}$—with Equation (43)	0.46801	0.48223	0.04976
	and with alternative a2-2 (Table 9b)		
$w_{j,CRITIC}$—with Equation (43)	0.48083	0.48198	0.0372

Aconflict measure created by criterion j due to the decision defined by the rest of the criteria can be calcualted, which determinesa quantity of the information in relation to each criterion:

$$C_j = \sigma_j \sum_{j=1}^{m} (1 - R_{jk}) \tag{42}$$

Objective CRITIC criteria weights for a chosen attribute are obtained by normalising (adjusting) the values of C_j and, in particular, alternative levels (43):

$$w_{j,CRITIC} = C_j / \sum_{k=1}^{m} C_k \quad j = 1, \dots m \tag{43}$$

3.4. Relative Weights Obtained by Normalisation of the "Excess Concentration"

Taking into account the input data of the decision model compiled in the input decision matrix with alternatives (Table 5), alternatives a1 to a5 (with values of excess concentrations Δc_j according to (2) [4]) were calculated with weights $w_{ij,energy}$ for non-zero attributes $j = 1, 2, 3$. In order to determine the "energy load" weight, the following steps are recommended.

Step 1. Assuming that the minimum concentration value c_{ref} in the definition of "excess concentration" given in Equation (2) may be the minimum value min (x_{ij}), then the formula for normalisation (43) can be determined, based on the data for the alternative a1 obtained in real case studies [5] and data from the literature [2,51,52] regarding the minimum values of pollutants (TVOC and HCHO) occurring in a building. Justification for the manner of assuming the minimum and maximum values of pollution in a building with CO_2 bio-pollution is discussed in the notes of Table 5.

Step 2. Normalisation of weights (energy load) $w_{ij,energy}$ using Equation (9) transformed to (44) for attributes of type "cost". According to [29,31], it seems that the equation can be used to normalise non-beneficial datasets, which after conversion to profit (or beneficial) attributes can be adjusted for IAQ systems corresponding to subsequent alternatives from a1 to a5. Transformation of the excess concentration values, which correspond to the indicators x_{ij} at the levels of subsequent alternatives a1 to a5 taken from Table 5, gives the weight of "energy load" $w_{ij,\,energy}$:

$$w_{ij,energy} = \left[1 - \frac{x_{ij}^{min}}{x_{ij}} \right]^{-1} \tag{44}$$

Normalisation of weights based on the concept of "excess concentration"— correct applicationof weights of Equation (44) is possible on the assumption that the min (x_{ij}) value of indoor pollution determined on the basis of insitu measurements and given in the input decision matrix (Table 5) corresponds to the standard value c_{ref} from Equations (2) and (3).

Step 3. Calculation of the relative values of "energy load" weights assigned to attributes of CO_2, TVOC and HCHO pollutant indoor concentrations (Table 11).

Table 11. Normalisation of the energy load relative weights calculated with Equation (44).

Alternatives	ΔCO_2	TVOC	HCHO
	alternative a1		
x_{ij}	810 mg/m^3	787 µg/m^3	18 µg/m^3
min(x_{ij})	360 mg/m^3	154 µg/m^3	5.8 µg/m^3
$w_{ij,energy}$	0.3984	0.2744	0.3265
	alternative a2		
x_{ij}	376.2 mg/m^3	167.6 µg/m^3	10.6 µg/m^3
min(x_{ij})	360 mg/m^3	154 µg/m^3	5.8 µg/m^3
$w_{ij,energy}$	0.6148	0.3357	0.0585

Table 11. *Cont.*

Alternatives	ΔCO_2	TVOC	HCHO
	alternative a3		
x_{ij}	392.4 mg/m^3	178.3 µg/m^3	26.0 µg/m^3
$\min(x_{ij})$	360 mg/m^3	154 µg/m^3	5.8 µg/m^3
$w_{ij,energy}$	0.5841	0.3538	0.0621
	alternative a4		
x_{ij}	376.2 mg/m^3	178.3 µg/m^3	26.0 µg/m^3
$\min(x_{ij})$	360 mg/m^3	154 µg/m^3	5.8 µg/m^3
$w_{ij,energy}$	0.7290	0.2305	0.0404
	alternative a5		
x_{ij}	392.4 mg/m^3	167.6 µg/m^3	10.6 µg/m^3
$\min(x_{ij})$	360 mg/m^3	154 µg/m^3	5.8 µg/m^3
$w_{ij,energy}$	0.454567	0.462546	0.082887

4. Results

4.1. Global Weights Calculated for the IAQ$_{quality}$ Model Alternatives

The final step of the weight development methodin Section 3 is to aggregate the calculated set of "component" weights and obtain the resultant global weights. Table 12 shows the determination of global weights calculated according to Equation (33), when the attributes do not take part in the emission process (attribute type *j*),and Equation (34) when attributesdo take part in the emission process (attribute type *k*). Weight values adequate for the numerical example are taken for the set of alternatives a2and a2-2.

Table 12. Relative global weights in the weighting schemes calculated for five alternatives of the IAQ model (a1 to a5); global weights obtained with Equation (32).

	Attributes (Pollution Concentration Levels)					
	ΔCO_2	TVOC	HCHO	ΔCO_2	TVOC	HCHO
	(alternative a1) test study [5] concentrationlevel without emission			(alternative a2) minimum concentration level with (e) minimum emission rate levels		
$w_{j,entropy}$	0.16706	0.44232	0.39061	0.16706	0.44232	0.39061
s_j^{health}	0.01556	0.21	0.21	0.01556	0.21	0.21
$w_{j,CRITIC}$	–	–	–	0.46801	0.48223	0.04976
$w_{ij,energy}$	0.3984	0.2744	0.3265	0.6148	0.3357	0.0585
$w_{j,entropy} \times s_j^{health} \times w_{j,CRITIC} \times w_{ij,energy}$	0.00103562	0.025488	0.026782	0.000747946	0.0150370	0.00023878
$w_{ij,global}$	0.0194279	0.478148	0.5024236	0.0466775	0.938422	0.0149017
	(alternative a2-2) test study [5,8] concentration level with minimum emission rate levels			(alternative a3) minimum concentration level; (e) minimum +25% range of emission rate		
$w_{j,entropy}$	0.2055	0.5895	0.20496	0.16706	0.44232	0.39061
s_j^{health}	0.01556	0.21	0.21	0.01556	0.21	0.21
$w_{j,CRITIC}$	0.48083	0.48198	0.0372	0.46801	0.48223	0.04976
$w_{ij,energy}$	0.3984	0.2744	0.3265	0.5841	0.3538	0.0621
$w_{j,entropy} \times s_j^{health} \times w_{j,CRITIC} \times w_{ij,energy}$	0.000612536	0.0163725	0.0005227	0.0007106	0.0158478	0.00025347
$w_{ij,global}$	0.0349865	0.935155	0.0298593	0.0422678	0.9426554	0.0150769
	(alternative a4) minimum concentration level; (e) minimum +25% range of TVOC and HCHO emission rate			(alternative a5) minimum concentration level; (e) minimum +25% range of CO$_2$ emission rate		
$w_{j,entropy}$	0.16706	0.44232	0.39061	0.16706	0.44232	0.39061
s_j^{health}	0.01556	0.21	0.21	0.01556	0.21	0.21
$w_{j,CRITIC}$	0.46801	0.48223	0.04976	0.46801	0.48223	0.04976
$w_{ij,energy}$	0.7290	0.2305	0.0404	0.5841	0.3538	0.0621
$w_{j,entropy} \times s_j^{health} \times w_{j,CRITIC} \times w_{ij,energy}$	0.000886878	0.010324785	0.000164901	0.000710597	0.01584780	0.000251997
$w_{ij,global}$	0.07795658	0.9075486	0.014494798	0.0422712976	0.942738165	0.014990546

4.1.1. The Results of the New Weighting Method Applied on IAQ_{index} and IEQ_{index} Results for theCaseStudy Building

A numerical example of the combined $\sum IAQ_{quality}$ model with the new weighting scheme, developed according to Figure 3, is presented in Table 13 in the context of the overall Indoor Environmental Quality (IEQ) asessement of thecasestudy office building (IEQ is the predicted percentage of users satisfied with all components of indoor environmental quality). The input indoor parameters for the case study (in order to assess IEQ sub-components) are taken from previous papers [4,5], where the measured pollutant concentrations were $c_{CO2} = 450$ ppm, $c_{TVOC} = 787$ $\mu g/m^3$ and $c_{HCHO} = 18$ $\mu g/m^3$ in the case study building a week after completion of the finishing works.

Table 13. The physical parameters representing the indoor environment of the case study and the IEQ_{index} results calculated from Equation (7) with sub-component $\sum IAQ$ calculations (old approach and new); assuming realistic uncertainty of measurements for a case study (old and new weight systems applied and compared).

Sub-Index	IEQSub-Component Models	Input Values	Sub-Index (Satisfied Users) ±SD
Thermal comfort TC_{index}	PMV (Fanger and ISO 7730) $PMV = f(M, I_{cl,dyn}, t_a, t_r, v_a, p_a)$ $PD_{TC} = f(PMV)$	$I_{cl} = 0.55$ clo $t_a = 24\,°C$ $t_r = 24.5\,°C$ $v_a = 0.15$ m/s RH = 45% M = 1.1 met	90.0% ± 3.2%
$\sum IAQ_{index}$ sub-indices	$PD_{IAQ(CO2)} = 395 \cdot \exp(-15.15 \cdot C_{CO2}^{-0.25})$ $PD_{IAQ(TVOC)} = 405 \cdot \exp(-11.3 \cdot C_{TVOC}^{-0.25})$ $PMV_{HCHO} = 2 \cdot \log(C_{HCHO}/0.01)$ $PD_{HCHO} = 100 - 95 \cdot \exp(-0.03353 \cdot PMV^4 - -0.2179 \cdot PMV^2)$	$c = 450$ ppm $c = 787$ $\mu g/m^3$ $c = 18$ $\mu g/m^3$	85.2% ± 0.6% 52.0% ± 18.0% 65.8% ± 10.7%
$\sum IAQ_{index}$ (new) without emission	$\sum IAQ_{index} =$ $W_1 \cdot (CO_2) + W_2 \cdot (TVOC) + W_3 \cdot (HCHO)$ $= 0.0194279 \cdot (85.2) + 0.478148 \cdot (52.0) + 0.5024236 \cdot (65.8) =$ $= 1.65525708 + 24.863696 + 33.05947288 = 59.58$		59.58% ± 10.2%
$\sum IAQ_{index}$ (new) with emission on minimum level	$\sum IAQ_{index} = W1 \cdot (CO_2) + W2 \cdot TVOC) + W3 \cdot (HCHO) =$ $= 0.0349865 \cdot (85.2) + 0.935155 \cdot (52.0) + 0.0298593 \cdot (65.8)$ $= 2.980885 + 48.62806 + 1.96474194 = 53.57$		53.57% ± 16.8%
$\sum IAQ_{index}$ (old)	$IAQ_{VOC} = 0.96 \cdot IAQ(TVOC) + 0.04 \cdot IAQ(HCHO)$ $\sum IAQ_{index} = 0.5 \cdot IAQ(CO_2) + 0.5 \cdot IAQ(VOC)$		53.0% ± 17.3% 69.1% ± 9.0%
Acoustic comfort ACc_{index}	$PD_{ACc} = 2 \cdot (Actual_{Sound\_Pressure\_Level}(dB(A) - Design_{Sound\_Pressure\_Level}(dB(A))$ Measured actualsound level Designed sound level	55 dB(A) 45 dB(A)	80.0% ± 6.7%
Daylight L_{index}	$PD_L = -0.0175 + 1.0361/\{1 + \exp(+4.0835 \cdot (\log_{10}(E_{min}) - 1.8223))\}$	455 lux	98.4% ± 9.0%
$\sum IEQ_{index} ±$ SD (new)	$W_1 \cdot TC_{index} + W_2 \cdot \sum IAQ_{index} + W_3 \cdot ACc_{cindex} + W_4 \cdot L_{index}$		
no emission	$0.25 \cdot 90.0 + 0.25 \cdot 59.6 + 0.25 \cdot 80.0 + 0.25 \cdot 98.4 = 82.0\%$		82.0% ± 3.9%
with emission	$0.25 \cdot 90.0 + 0.25 \cdot 53.6 + 0.25 \cdot 80.0 + 0.25 \cdot 98.4 = 80.5\%$		80.0% ± 5.12%
$\sum IEQ_{index} ±$ SD (old)	$IEQ_{index} = W_1 \cdot TC_{index} + W_2 \cdot \sum IAQ_{index} + W_3 \cdot ACc_{cindex} + W_4 \cdot L_{index}$		84.4% ± 3.7%

The uncertainty of user dissatisfaction with pollutant concentration is determined by a combination of two groups of components. The first measurement uncertainty is related to the real uncertainty of parameters biases when there are measured in indoor environments. The second group concerns the uncertainty of votes in the sensory tests (reliability of responses to a given stimulus of the indoor environment are discussed in literature). In practice, a metrological analysis of the IEQ model (where $\sum IAQ$ index is the sub-model) was provided in our previous paper [8] considering provisions of the international metrological guideline: ISO BIPM JCGM 100:2009 Guide to the expression of uncertainty in measurement. For all other possible scenarios, the uncertainty for determining the $\sum$IAQ index would be up to ±10% to ±30% (depending on each pollution model uncertainty assessment and the number of IAQ model components taken into account).

5. Discussion

5.1. New Weighting Method Applied on IAQ$_{index}$ and IEQ$_{index}$ Estimation for a CaseStudy Building and Comparison of Its Results with Old Crude Weighting SystemResults

The previous approach by their weighting scheme (Table 13) was based on the treatment of all main pollutantsin ununiform way. The weights were previously set as equal (for pollutants with disproportionately different concentrations) or as dependent only on the energy expenditure on ventilation (for pollutants with similar) concentrations. The new IAQ and IEQ index results based on MDMA-developed weights werecompared to the old method. The system of objectified weights used for a numerical example of IAQ and IEQ building assessment leads to important conclusions. As expected, the use of objectified weights means that the results of the IAQ and IEQ indices weredifferent than those with subjective weights. For example, at low CO_2 indoorconcentrations, the role of this parameter in the IAQ$_{index}$ assessment was overstated. This is the case for most BMS systems controlling building ventilation, where currently the number of air exchanges (ACH) depends mainly on CO_2 concentration measurements. At relatively low CO_2 air concentrations <1500 ppm, the CO_2 priority compared to HCHO or TVOC attributes is much lower. The cumulative weight of the HCHO and TVOC arguments compared to the "old" IAQ calculation method increased by 48%. This is important from the point of view of planning and managing ventilation systems. According to calculations, the number of users satisfied with the level of CO_2 (85% satisfied) wasa good result. Nevertheless, at the same time, the theoretical number of users satisfied with TVOC concentration wasonly 52%. With novel objective weights for the TVOC argument, the number of users satisfied with IAQ will decrease. In the case of the present studyand the new weight system, the value of IAQ$_{index}$ decreased about 10% compared to previous calculations using the old method. Further, if one considers the existence of additional emissions from the building, finishing materials and furniture, the value of satisfied users will decrease by another 5.5%.

The authors strongly believe that the new approach to weights gives objective and more adequate results for the IAQ and IEQ models. Wider use of the new weighting method can be effective upon supplementing databases with the subjective weights of health risks for various volatile substances in the air.

5.2. The WeightingProblem in Indoor Air Quality Component Sets Framework

The weights in any decision-making model should not be arbitrary [53]. Some researchers still use weights in modelling without justification. A solution to the weight problem in indoor air quality modelling is proposed. Global weights are not only measures of the importance of attributes in the human–indoor building environment relationship, health risks and possible disturbances of the quasi-stable state due to correlations between attributes, but they are also measures of the potential load of ventilation impacted by emissions and excess concentrations of additional costs for energy use. In this context, the adoption of objectified weights in the IAQ model (also any other model, where the proposed method can be transferred) has significant practical contribution. It is suggested, knowing the specific air pollution values (attributes) for the analyzed building, to calculate the weights for a new alternative created by the user in accordance with the method proposed in article. For less insightful research, the authors suggest using a set of weights developed by us (Table 12) for scenarios/alternatives that aremost similar to the scenario under consideration. According to the adopted MADM method for each analyzed alternative (a1–a5 scenarios for the indoor environment described by concentration/emission levels for attributes or types of pollutants adopted in the IAQ model), global weights for each attribute (CO_2 or TVOC or HCHO concentration) are calculated separately as the numerical coefficients of Equations (4) or (6). Thisis because the component weights (e.g., CRITIC correlation weight) of global weights depend on the values of the attributes (air pollution concentrations and emissions). Table 12 shows the calculated global weights $w_{ij,global}$ for each attribute for each alternative a1–a5. For alternative a1 (measured concentrations in thecase study building with an assumption of no additional emissionsform construction products and athropogenic CO_2), global weights of the IAQ_{index} model are 0.0194279 for CO_2, 0.478148 for TVOC and 0.5024236 for HCHO. For alternative a2 (minimum concentration levels with minimum emission rates), global weights are 0.0466775 for CO_2, 0.938422 for TVOC and 0.0149017 for HCHO. For alternative a2-2 (case study result concentrations with minimum emission rates), global weights are 0.0349865 for CO_2, 0.935155 for TVOC and 0.0298593 for HCHO. For alternative a3 (minimum concentration level; (e) minimum + 25% range of emission rate), global weights are 0.0422678 for CO_2, 0.9426554 for TVOC and 0.0150769 for HCHO. For alternative a4 (minimum concentration level; (e) minimum +25% range of TVOC and HCHO emission rates), global weights are 0.07795658 for CO_2, 0.9075486 for TVOC and 0.014494798 for HCHO. For alternative a5 (minimum concentration level; emission minimum +25% range of CO_2 emission rate), global weights are 0.0422712976 for CO_2, 0.942738165 for TVOC and 0.014990546 for HCHO. The obtained global weight results clearly indicate which pollutant attributes have priority. In ranges up to about 1500 ppm CO_2 in indoor air, the weight of this attribute in the IAQ model is very small. This means that the ventilation system should take into account, in this respect, the removal of excess TVOC concentration first and, in the second place, excess HCHO. For all alteratives, the weight of the TVOC attribute is significant (for a2–a5, weight is over 90%). The weights for TVOC and HCHO are significantly higher than for CO_2 (one order of magnitude). A 25% increase in CO_2 emissions will not increase its weight in the model.

In this article, the authors have described in detail a procedure for the subjective and objective weight determination for a combined ΣIAQ model, which is composed of the following basic steps:

i.   Development of an initial decision matrix adapted to the type of the combined ΣIAQ model;

ii.   Normalisation of attribute data including different indoor air pollutant concentrations and their emission rates;

iii.   Calculation of the objective attribute weights by the entropy method;

iv.   Calculation or adoption of subjective health-connected weights from the database;

v.   Use of the CRITIC (Criteria Importance Through Inter-criteria Correlation) method in the case of mutually correlating attributes;

vi.   Calculation of objective weights for ventilation energy expenditure to various IAQ sub-components (pollutants);

vii.   Calculation of relative global weights for all significant air pollutants of the ΣIAQ model, considered in various schemes (various alternatives in the framework of decision model).

In our case study building (Table 13), the IAQ model determines global weight values for the attributes of "pollution concentrations" in the form of relative global weights adjusted at all alternative levels. The measures of the significance of the other attributes (additional emissions) are determined by the CRITIC method and are given by correlation weights that additionally impact the global weights of the "concentration" attributes. Therefore, although entropy weights are calculated just like CRITIC weights for all attributes, taking advantage of the fact that they are weights characteristic for given attributes, they can be "selected" and adjusted in the range of the first "concentration" attributes in order to calculate relative global weights for all levels of alternatives.

Thus, the determined global weights are valid within the $IAQ_{quality}$ model (4) treated as a sub-expression in the IEQ equation model described by Equation (7). It seems that the approach presented in this study may easily be replicated by other similar IAQ models.

In our future research, the authors will try to unambiguously link the proposed IAQ model and weight system with the demand for ventilation air for various alternatives and the ventilation systems used in case study buildings.

6. Conclusions

Currently, the issue of comfort assessement, based on user satisfaction models, is discussed in numerous research works, but in most of them indoor comfort is analyzed in terms of thermal comfort only [54–56].Fortunately, the recently published european standrd EN 16798-1: 2018 [7] included a wider number of indoor comfort criteria. This standard provides an approach to the classification of IAQ and IEQ models and supportscertification of buildings taking into account specific components of the environmental comfort (IEQ sub-models) including $\sum$IAQ (which is also composed of sub-components in functions of pollutant concentrations). This standard and any relatedreseach does not provide practical guidance on how to combine $\sum$IAQ model components into one indicator that could be used to classify indoor air conditions and does not provide information of how to aggregate components by an objective weighting system.With the research challenge in mind, the authors presented a relevant IAQ model in [4].In this paper, the authors provided astep-by-step procedure for determining the combined weighting scheme for the IAQ index equation, using the MADMdecision model including calculation of the objective attribute weights by the entropy method, calculation or adoption of subjective health-connected weights from the reference database, use of the CRITIC method in the case of mutually correlating attributes and calculation of objective weights for ventilation energy expenditure to various IAQ sub-components (pollutants). The assumption of the decision model is adequate for the task of rating indoor air ventilation to determine the assumed air quality (our actual work), and setting IAQ model alternatives so that they can meet the criteria taking into account their importance ranking and possibility of ongoing diagnostics of alternatives connected to the perceived actual indoor air situation. The presented weigthing scheme approach for the IAQ model may be regarded as objective, and it generates weighted criteria values (for each analysed attribute and indoor scenario) directly from the variations in criteria values. The provided method eliminates the problems of IAQ model weights with using the subjectivepoint of viewas well as the incompetence orabsence of a responsible decision-maker. The MADM approach, as shown in thiscase study, is based on simple mathematics and can, therefore, be used effectively for choosing the best IEQ and ventilation strategies.

Author Contributions: Conceptualisation, M.P. and K.K.; methodology, M.P. and K.K.; software, M.P. and K.K.; validation, M.P. and K.K.; formal analysis, M.P.; investigationtest, resources, M.P.; data curation, M.P.; writing—Original draft preparation, M.P. and K.K.; writing—Review and editing, M.P. and K.K.; visualisation, M.P.; supervision, M.P.; project administration, M.P.; funding acquisition, M.P. Both authors have read and agreed to the published version of the manuscript.

Funding: This research did not receive external funding.

Conflicts of Interest: The authors do not declare any conflicts of interest.

Appendix A

Table A1. Normalised decision matrix for the entropy method based on Equation (35).

Alternatives	Attribute $X = 1 \ldots m$					
	ΔCO_2 mg/m^3	TVOC $\mu g/m^3$	HCHO $\mu g/m^3$	CO_2 (e) $mg/m^3{\cdot}h$	TVOC (e) $\mu g/m^3 h$	HCHO (e) $\mu g/m^2 h$
a1 test study [5} +(e) emission rate = 0	$810\ mg/m^3$ $1/x_i = 0.0012$ $z_{ij} = 0.1043$	$787\ \mu g/m^3$ $1/x = 0.0013$ $z_{ij} = 0.0531$	$18\ \mu g/m^3$ $1/x_{ij} = 0.056$ $z_{ij} = 0.175$	0	0	0
a2 minimum; (e) minimum TVOC and HCHO emission +1h; (e) minimum CO_2 one person	$376.2\ mg/m^3$ $1/x_{ij} = 0.00265$ $z_{ij} = 0.2304$	$167.6\ \mu g/m^3$ $1/x = 0.0060$ $z_{ij} = 0.2449$	$10.6\ g/m^3$ $1/x_{ij} = 0.094$ $z_{ij} = 0.2937$	$16.2\ mg/m^3 h$ $1/x_{ij} = 0.062$ $z_{ij} = 0.3333$	$13.6\ \mu g/m^3 h$ $1/x_{ij} = 0.073$ $z_{ij} = 0.3201$	$4.26\ \mu g/m^2 h$ $1/x_{ij} = 0.234$ $z_{ij} = 0.4134$
a3 minimum+1h; (e) minimum TVOC HCHO +25% concentration range (e) minimum CO_2 +25% concentration two persons[2)]	$392.4\ mg/m^3$ $1/x = 0.0025$ $z_{ij} = 0.2174$	$178.3\ \mu g/m^3$ $1/x = 0.0056$ $z_{ij} = 0.2286$	$26.0\ \mu g/m^3$ $1/x = 0.038$ $z_{ij} = 0.1187$	$32.4\ mg/m^3$ $\cdot h\ 1/x_{ij} = 0.031$ $z_{ij} = 0.1666$	$\Sigma\ 24.3\ \mu g/m^3 h$ $1/x_{ij} = 0.041$ $Z_{ij} = 0.1798$	$\Sigma\ 20.2\ \mu g/m^2\ h$ $1/x_{ij} = 0.049$ $z_{ij} = 0.0866$
a4 minimum+1h; (e) minimum TVOC, HCHO +25% concentration range; (e) CO_2 minimum one person	$376.2\ mg/m^3$ $1/x = 0.00265$ $z_{ij} = 0.2304$	$178.3\ \mu g/m^3$ $1/x = 0.0056$ $z_{ij} = 0.2286$	$26.0\ \mu g/m^3$ $1/x_{ij} = 0.038$ $z_{ij} = 0.1187$	$16.2\ mg/m^3 h$ $1/x_{ij} = 0.062$ $z_{ij} = 0.3333$	$\Sigma\ 24.3\ \mu g/m^3 h$ $1/x_{ij} = 0.041$ $z_{ij} = 0.1798$	$\Sigma\ 20.2\ \mu g/m^2\ h$ $1/x_{ij} = 0.049$ $z_{ij} = 0.0866$
a5 minimum+1h; (e) minimum TVOC, HCHO; (e) CO_2 minimum + 25% concentration range; two persons	$392.4\ mg/m^3$ $1/x = 0.0025$ $z_{ij} = 0.2174$	$167.6\ \mu g/m^3$ $1/x = 0.0060$ $z_{ij} = 0.2449$	$10.6\ g/m^3$ $1/x = 0.094$ $z_{ij} = 0.2937$	$32.4\ mg/m^3 h$ $/x_{ij} = 0.031$ $z_{ij} = 0.1666$	$13.6\ \mu g/m^3 h$ $1/x_{ij} = 0.073$ $z_{ij} = 0.3201$	$4.26\ \mu g/m^2 h$ $1/x_{ij} = 0.234$ $z_{ij} = 0.4134$
$\Sigma\ 1/x_{ij}$	0.0115	0.0245	0.32	0.186	0.228	0.566

Table A2. Normalisation of decision matrix with alternative a2 (Table 5) for the CRITIC method.

Alternatives	Attribute $X = 1 \ldots m$					
	ΔCO_2	TVOC	HCHO	CO_2 (e)	TVOC (e)	HCHO (e)
	mg/m^3	$\mu g/m^3$	$\mu g/m^3$	$mg/m^3 \cdot h$	$\mu g/m^3 h$	$\mu g/m^2 h$
a1 Test report [5] ; +(e) emission rate = 0	810 mg/m^3 810–810/ −433.8 = 0	787 μg/m^3 787–787/ −619.4 = 0	18 μg/m^3 18–26/ −15.4 = 0.519	0 0–32.4/ −32.4 = 1	0 0–24.3/ −24.3 = 1	0 0–20.2/ −20.2 = 1
a2 minimum; (e) minimum TVOC, HCHO +1h; (e) minimum CO_2 one person	376.2 mg/m^3 376.2–810/ −433.8 = 1	167.6 μg/m^3 167.6–787/ −629.4 = 1	10.6 μg/m^3 10.6–26/ −15.4 = 1	16.2 mg/m^{3}h 16.2–32.4/ −32.4 = 0.5	13.6 μg/m^{3}h 13.6–24.3/ −24.3 = 0.440	4.26 μg/m^{2}h 4.26–20.2/ −20.2 = 0.789
a2-2 Test study [5] ; CO_2 TVOC, HCHO; (e) minimum emission rate; (e) CO_2 one person	810 mg/m^3 810–810/ −433.8 = 0	787 μg/m^3 787–787/ −619.4 = 0	18 μg/m^3 18–26/ −15.4 = 0.519	16.2 mg/m^{3}h 16.2–32.4/ −32.4 = 0.5	13.6 μg/m^{3}h 13.6–24.3/ −24.3 = 0.440	4.26 μg/m^{2}h 4.26–20.2/ −20.2 = 0.789
a3 minimum +1h; (e) minimum +25% TVOC, HCHO concentration range; (e) CO_2 two persons	392.4 mg/m^3 392.4–810/ −433.8 = 0.963	178.3 μg/m^3 178.3–787/ −629.4 = 0.967	26.0 μg/m^3 26.0–26.0/ −15.4 = 0	32.4 mg/m^{3}h two persons 32.4–32.4/ −32.4 = 0	24.3 μg/m^{3}h 24.3–24.3/ −24.3 = 0	20.2 μg/m^2 h 20.2–20.2/ −20.2 = 0
a4 minimum+1h; (e) minimum +25% TVOC, HCHO concentration range (e) minimum CO_2 one person	376.2 mg/m^3 376.2–810/ −433.8 = 1	178.3 μg/m^3 178.3–787/ −629.4 = 0.967	26.0 μg/m^3 26.0–26.0/ −15.4 = 0	16.2 mg/m^{3}h 16.2–32.4/ −32.4 = 0.5	24.3 μg/m^{3}h 24.3–24.3/ −24.3 = 0	20.2 μg/m^2 h 20.2–20.2/ −20.2 = 0
a5 minimum+1h; (e) minimum +25% CO_2 2 persons; (e) minimum TVOC and HCHO	392.44 mg/m^3 392.4–810/ − 433.8 = 0.963	167.6 μg/m^3 167.6–787/ −629.4 = 1	10.6 μg/m^3 10.6–26/ −15.4 = 1	32.4] mg/m^{3}h 32.4–32.4/ −32.4 = 0	13.6 μg/m^{3}h 13.6–24.3/ −24.3 = 0.440	4.26 μg/m^{2}h 4.26–20.2/ −20.2 = 0.789
Maximum x_{ij} value of concentration range lit.	1800 mg/m^3	1000 μg/m^3	160 μg/m^3	64.8 mg/m^3·h four persons	57.1 μg/m^{3}h	68.2 μg/m^{2}h
Maximum x_{ij} value in j attribute column	810 mg/m^3	787 μg/m^3	26.0 μg/m^3	32.4 μg/m^3	24.3 μg/m^{3}h	20.2 μg/m^{2}h
Minimum x_{ij} level of ventilation load (lit.)	360 mg/m^3	154 μg/m^3	5.8 μg/m^3	16.2 mg/m^{3}h	13.6 μg/m^{3}h	4.26 μg/m^{2}h
Minimum x_{ij} value in attribute j column	376.2	167.6	10.6	0	0	0
Minimum x_{ij}–maximum x_{ij}	−433.8	−619.4	−15.4	−32.4	−24.3	−20.2

References

1. Navarro, I.J.; Yepes, V.; Martí, J.V. A Review of Multicriteria Assessment Techniques Applied to Sustainable Infrastructure Design. *Adv. Civ. Eng.* **2019**, *2019*, 1–16. [CrossRef]
2. Shrubsole, C.; Dimitroulopoulou, S.; Foxall, K.; Gadeberg, B.; Doutsi, A. IAQ guidelines for selected volatile organic compounds (VOCs) in the UK. *Build. Environ.* **2019**, *165*, 106382. [CrossRef]
3. Piasecki, M.; Kostyrko, K.; Pykacz, S. Indoor environmental quality assessment: Part 1: Choice of the indoor environmental quality sub-component models. *J. Build. Phys.* **2017**, *41*, 264–289. [CrossRef]
4. Piasecki, M.; Kostyrko, K.B. Combined model for IAQ assessment: Part 1—Morphology of the model and selection of substantial air quality impact sub-models. *Appl. Sci.* **2019**, *9*, 3918. [CrossRef]
5. Piasecki, M.; Kozicki, M.; Firlag, S.; Goljan, A.; Kostyrko, K. The approach of including TVOCs concentration in the indoor environmental quality model (IEQ)—Case studies of BREEAM certified office buildings. *Sustainability* **2018**, *10*, 3902. [CrossRef]
6. Wang, Q.; Dai, H.N.; Wang, H. A smart MCDM framework to evaluate the impact of air pollution on city sustainability: A case study from China. *Sustainability* **2017**, *9*, 911. [CrossRef]
7. CEN. *Energy Performance of Buildings—Ventilation of Buildings—Part 1: Indoor Environmental Input Parameters for Design and Assessment of Energy Performance of Buildings Addressing Indoor Air Quality, Thermal Environment, Lighting and Acoustics*; EN 16798; CEN: Brussels, Belgium, 2019.
8. Piasecki, M.; Kostyrko, K.B. Indoor environmental quality assessment, part 2: Model reliability analysis. *J. Build. Phys.* **2018**, *5*, 1–28. [CrossRef]
9. Wong, L.T.; Mui, K.W.; Hui, P.S. A multivariate-logistic model for acceptance of indoor environmental quality (IEQ) in offices. *Build. Environ.* **2008**, *43*, 1–6. [CrossRef]
10. Alemi-Ardakani, M.; Milani, A.S.; Yannacopoulos, S.; Shokouhi, G. On the effect of subjective, objective and combinative weighting in multiple criteria decision making: A case study on impact optimization of composites. *Expert Syst. Appl.* **2016**, *46*, 426–438. [CrossRef]
11. Shanian, A.; Savadogo, O. A methodological concept for material selection of highly sensitive components based on multiple criteria decision analysis. *Expert Syst. Appl.* **2009**, *36*, 1362–1370. [CrossRef]
12. Jahan, A.; Mustapha, F.; Sapuan, S.M.; Ismail, M.Y.; Bahraminasab, M. A framework for weighting of criteria in ranking stage of material selection process. *Int. J. Adv. Manuf. Technol.* **2012**, *58*, 411–420. [CrossRef]
13. Moridi, P.; Atabi, F.; Nouri, J. Weighting and prioritizing of air pollutant filtration technologies for controlling NH3, PM and VOCs by fuzzy TOPSIS method. *Iran Occup. Heal.* **2015**, *12*, 1–9.
14. Keshavarz Ghorabaee, M.; Zavadskas, E.K.; Turskis, Z.; Antucheviciene, J. A new combinative distance-based assessment (CODAS) method for multi-criteria decision-making. *Econ. Comput. Econ. Cybern. Stud. Res.* **2016**, *50*, 25–44.
15. Invidiata, A.; Lavagna, M.; Ghisi, E. Selecting design strategies using multi-criteria decision making to improve the sustainability of buildings. *Build. Environ.* **2018**, *139*, 58–68. [CrossRef]
16. Zavadskas, E.K.; Kaklauskas, A.; Turskis, Z.; Kalibatas, D. An approach to multi-attribute assessment of indoor environment before and after refurbishment of dwellings. *J. Environ. Eng. Landsc. Manag.* **2009**, *17*, 5–11. [CrossRef]
17. Zavadskas, E.K.; Cavallaro, F.; Podvezko, V.; Ubarte, I.; Kaklauskas, A. MCDM assessment of a healthy and safe built environment according to sustainable development principles: A practical neighborhood approach in vilnius. *Sustainability* **2017**, *9*, 702. [CrossRef]
18. Noryani, M.; Sapuan, S.M.; Mastura, M.T. Multi-criteria decision-making tools for material selection of natural fibre composites: A review. *J. Mech. Eng. Sci.* **2018**, *12*, 3330–3353. [CrossRef]
19. Issa, U.H.; Miky, Y.H.; Abdel-Malak, F.F. A decision support model for civil engineering projects based on multi-criteria and various data. *J. Civ. Eng. Manag.* **2019**, *25*, 100–113. [CrossRef]
20. Lotfi, F.H.; Fallahnejad, R. Imprecise shannon's entropy and multi attribute decision making. *Entropy* **2010**, *12*, 53–62. [CrossRef]
21. Peng, A.H.; Xiao, X.M. Material selection using PROMETHEE combined with analytic network process under hybrid environment. *Mater. Des.* **2013**, *47*, 643–652. [CrossRef]
22. Huang, I.B.; Keisler, J.; Linkov, I. Multi-criteria decision analysis in environmental sciences: Ten years of applications and trends. *Sci. Total Environ.* **2011**, *19*, 3578–3594. [CrossRef]

23. Cegan, J.C.; Filion, A.M.; Keisler, J.M.; Linkov, I. Trends and applications of multi-criteria decision analysis in environmental sciences: Literature review. *Environ. Syst. Decis.* **2017**, *37*, 123–133. [CrossRef]

24. Brauers, W.K.M.; Zavadskas, E.K. Robustness of MULTIMOORA: A method for multi-objective optimization. *Informatica* **2012**, *23*, 1–25. [CrossRef]

25. Hafezalkotob, A.; Hafezalkotob, A. Extended MULTIMOORA method based on Shannon entropy weight for materials selection. *J. Ind. Eng. Int.* **2015**, *12*, 1–13. [CrossRef]

26. Xu, Z. *Uncertain Multi-Attribute Decision Making: Methods and Applications*; Springer: Berlin/Heidelberg, Germany, 2015. ISBN 9783662456408.

27. Jovanovic, S.; Savic, S.; Jovicic, N.; Boskovic, G.; Djordjevic, Z. Using multi-criteria decision making for selection of the optimal strategy for municipal solid waste management. *Waste Manag. Res.* **2016**, *34*, 884–895. [CrossRef] [PubMed]

28. Mintz, D. *Technical Assistance Document for the Reporting of Daily Air Quality—The Air Quality Index (AQI)*; EPA 454/B12-001; U.S. Environmental Protection Agency, Office of Air Quality Planning and Standards, Air Quality Assessment Division: Research Triangle Park, NC, USA, 2013.

29. Körth, H. Zur Berücksichtigung mehrerer Zielfunktionen bei der Optimierung von Produktionsplänen. *Math. Wirtsch.* **1969**, *6*, 184–201.

30. Zavadskas, E.K.; Stević, Ž.; Turskis, Z.; Tomašević, M. A novel extended EDAS in Minkowski space (EDAS-M) method for evaluating autonomous vehicles. *Stud. Inform. Control* **2019**, *28*, 255–264. [CrossRef]

31. Vujicic, M.; Papic, M.; Blagojevic, M. Comparative analysis of objective techniques for criteria weighing in two MCDM methods on example of an air conditioner selection. *Tehnika* **2017**, *67*, 422–428. [CrossRef]

32. Rode, C. *IEA Indoor Air Quality Design and Control in Low-Energy Residential Buildings—Annex 68—Subtask 1: Defining the Metrics. In the Search of Indices to Evaluate the Indoor Air Quality of Low-Energy Residential Buildings*; AIVC Contributed Report 17; Technical University of Denmark: Copenhagen, Denmark, 2017.

33. WHO Regional Office for Europe. *Evolution of WHO Air Quality Guidelines: Past, Present and Future*; WHO Regional Office for Europe: Copenhagen, Denmark, 2017.

34. Carrer, P.; de Oliveira Fernandes, E.; Santos, H.; Hänninen, O.; Kephalopoulos, S.; Wargocki, P. On the development of health-based ventilation guidelines: Principles and framework. *Int. J. Environ. Res. Public Health* **2018**, *15*, 1360. [CrossRef] [PubMed]

35. Johnston, C.J.; Andersen, R.K.; Toftum, J.; Nielsen, T.R. Effect of formaldehyde on ventilation rate and energy demand in Danish homes: Development of emission models and building performance simulation. *Build. Simul.* **2020**, *13*, 197–212. [CrossRef]

36. Piasecki, M.; Kostyrko, K.; Fedorczak-Cisak, M.K.N. The Air Enthalpy as IAQ Indicator in Hot and Humid Environments—Experimental Evaluation. *Energies* **2020**, *13*, 1481. [CrossRef]

37. Teleszewski, T.; Gładyszewska-Fiedoruk, K. The concentration of carbon dioxide in conference rooms: A simplified model and experimental verification. *Int. J. Environ. Sci. Technol.* **2019**, *16*, 8031–8040. [CrossRef]

38. Hori, H.; Ishimatsu, S.; Fueta, Y.; Ishidao, T. Evaluation of a real-time method for monitoring volatile organic compounds in indoor air in a Japanese university. *Environ. Health Prev. Med.* **2013**, *18*, 285–292. [CrossRef] [PubMed]

39. U.S. Green Building Council. *USGBC LEED v4 for Building Design and Construction—Current Version*; U.S. Green Building Council: New York, NY, USA, 2019.

40. Yuan, J.; Han, J.H.; Xie, X.D.; Qi, Z.Y.; Yang, X.; Zhao, B.; Zhao, R.; Ye, J.J.; Xie, X.D.; Qi, Z.Y.; et al. A new indoor air quality assessment system and its software tool. In Proceedings of the Indoor Air 2005, 10th International Conferenceon Indoor Air Quality and Climate, Beijing, China, 4–9 September 2005; Volume 1–5.

41. Tao, H.; Fan, Y.; Li, X.; Zhang, Z.; Hou, W. Investigation of formaldehyde and TVOC in underground malls in Xi'an, China: Concentrations, sources, and affecting factors. *Build. Environ.* **2015**, *85*, 85–93. [CrossRef]

42. Ye, W.; Zhang, X.; Gao, J.; Cao, G.; Zhou, X.; Su, X. Indoor air pollutants, ventilation rate determinants and potential control strategies in Chinese dwellings: A literature review. *Sci. Total Environ.* **2017**, *586*, 696–729. [CrossRef]

43. Péry, A.; Bonvallot, N.; El Yamani, M.; Boulanger, G.; Karg, F.; Mosqueron, L.; Ismert, M.; Guillossou, G.; Glorennec, P. French occupational exposure limit and toxicological reference values: Objectives and methods. *Environ.Risques Sante* **2013**, *12*, 442–449.

44. Mosqueron, L.; Kirchner, S.; Nedellec, V. Bilan des études françaises sur la mesure de la qualité de l'air à l'intérieur des bâtiments (1990–2001). *Environ. Risques Sante* **2002**, *1*, 31–41.

45. Keršulienė, V.; Zavadskas, E.K.; Turskis, Z. Selection of rational dispute resolution method by applying new step-wise weight assessment ratio analysis (swara). *J. Bus. Econ. Manag.* **2010**, *21*, 32–42. [CrossRef]

46. Zavadskas, E.K.; Turskis, Z.; Ustinovichius, L.; Shevchenko, G. Attributes weights determining peculiarities in multiple attribute decision making methods. *Eng. Econ.* **2010**, *21*, 32–43.

47. Diakoulaki, D.; Mavrotas, G.; Papayannakis, L. A multicriteria approach for evaluating the performance of industrial firms. *Omega* **1992**, *20*, 467–474. [CrossRef]

48. Adalı, E.A.; Işık, A.T. Critic and Maut Methods for the Contract Manufacturer Selection Problem. *Eur. J. Multidiscip. Stud.* **2017**, *2*, 93–101. [CrossRef]

49. Deepa, N.; Ganesan, K.; Srinivasan, K.; Chang, C.Y. Realizing sustainable development via modified integrated weighting MCDM model for Ranking Agrarian Dataset. *Sustainability* **2019**, *11*, 6060. [CrossRef]

50. American Industrial Hygiene Association AIHA. *Odor Thresholds for Chemicals with Established Health Standards*; AIHA: Fairfax, VA, USA, 2013.

51. Noguchi, M.; Mizukoshi, A.; Yanagisawa, Y.; Yamasaki, A. Measurements of volatile organic compounds in a newly built daycare center. *Int. J. Environ. Res. Public Health* **2016**, *13*, 736. [CrossRef] [PubMed]

52. Wu, Y.; Lu, Y.; Chou, D.-C. Indoor air quality investigation of a university library based on field measurement and questionnaire survey. *Int. J. Low Carbon Technol.* **2018**, *13*, 148–160. [CrossRef]

53. Fedorczak-Cisak, M.; Kotowicz, A.; Radziszewska-Zielina, E.; Sroka, B.; Tatara, T.; Barnaś, K. Multi-Criteria Optimisation of an Experimental Complex of Single-Family Nearly Zero-Energy Buildings. *Energies* **2020**, *13*, 1541. [CrossRef]

54. Manfren, M.; Nastasi, B.; Piana, E.; Tronchin, L. On the link between energy performance of building and thermal comfort: An example. *AIP Conf. Proc.* **2019**, *2123*, 020066.

55. Yun, G.Y. Influences of perceived control on thermal comfort and energy use in buildings. *Energy Build.* **2018**, *158*, 822–830. [CrossRef]

56. Bomberg, M.; Kisilewicz, T.; Mattock, C. *Methods of Building Physics*; Krakow University: Kraków, Poland, 2017; pp. 1–301.

Article

How the Italian Residential Sector Could Contribute to Load Flexibility in Demand Response Activities: A Methodology for Residential Clustering and Developing a Flexibility Strategy

Francesco Mancini [1,*], Sabrina Romano [2], Gianluigi Lo Basso [3], Jacopo Cimaglia [3] and Livio de Santoli [3]

1 Department of Planning, Design and Technology of Architecture, Sapienza University of Rome, Via Flaminia, 72-00197 Rome, Italy
2 Energy Technologies Department (DTE), Italian National Agency for Technologies, Energy and Sustainable Economic Development (ENEA), Via Anguillarese, 301-00123 Rome, Italy; sabrina.romano@enea.it
3 Department of Astronautics, Electrical Energy Engineering, Sapienza University of Rome, Via Eudossiana, 18-00184 Rome, Italy; gianluigi.lobasso@uniroma1.it (G.L.B.); jacopocimaglia.ingegneria@gmail.com (J.C.); livio.desantoli@uniroma1.it (L.d.S.)
* Correspondence: francesco.mancini@uniroma1.it; Tel.: +39-06-4991-9172

Received: 28 April 2020; Accepted: 29 June 2020; Published: 1 July 2020

Abstract: This work aims at exploring the potential contribution of the Italian residential sector in implementing load flexibility for Demand Response activities. In detail, by combining experimental and statistical approaches, a method to estimate the load profile of a dwelling cluster of 751 units has been presented. To do so, 14 dwelling archetypes have been defined and the algorithm to categorise the sample units has been built. Then, once the potential flexible loads for each archetype have been evaluated, a control strategy for applying load time shifting has been implemented. That strategy accounts for both the power demand profile and the hourly electricity price. Specifically, it has been assumed that end users access a pricing mechanism following the hourly trend of electricity economic value, which is traded day by day in the Italian spot market, instead of the current Time of Use (TOU) system. In such a way, it is possible to flatten the dwellings cluster profile, limiting undesired and unexpected results on the balancing market. In the end, monthly and yearly flexibility indexes have been defined along with the strategy effectiveness parameter. From calculations, it emerges that a dwelling cluster for the Italian residential sector is characterised by a flexibility index of 10.3% and by a strategy effectiveness equal to 34%. It is noteworthy that the highest values for flexibility purpose have been registered over the heating season (winter) for the weekends.

Keywords: Residential users; Demand Response; Flexible loads; Dwellings clustering

---

1. Introduction

The European Union established the ambitious net zero greenhouse gas emissions target by 2050. Actions heading towards the energy systems decarbonisation can be implemented by the application of energy efficiency measures and by increasing Renewable Energy Sources (RES) penetration [1]. However, large-scale RES integration within electrical systems shows important technical and safety issues, due to the RES non-programmable nature. As a consequence, figuring out, effectively, the energy supply and demand matching will play a key role in the near future [2]. In order to stabilize the grid, several technical options are currently available. Among these, a growing level of system flexibility will become necessary to reduce the purchase cost of electricity in the spot market. Adding new flexibility sources to the traditional regulation for the electricity offer will be feasible to perform that task. Basically, allowing end-users to actively participate to the electricity price formation mechanism by load time shifting leads to the implementation of so-called Demand Side Management (DSM) strategies [3,4].

DSM activities imply an energy consumption model modification and they can be mainly classified as "energy efficiency (EE)" and "demand response (DR)" [5]. The EE is for reducing the demand without dealing with the renewable energy fluctuations; conversely, DR proves to be more promising [6] since it is based on adapting the user demand profiles to the grid requirements by increasing, reducing, or moving the energy consumption.

Several research activities were focused on the evaluation of technical and safety issues associated with a growing RES share in current energy systems. In reference [7], an accurate analysis on the technical feasibility of electrical systems characterised by 100% RES was reported. The authors highlighted the importance of flexible power plants, storage technologies and Demand Response activities. Zappa et al. [8] analysed several scenarios related to whole European energy system to explore the feasibility of 100% RES electrical system. The authors performed their simulations accounting for different operating conditions, including also a significant load shifting capacity. That capacity can be provided by either industrial sector or the tertiary one, as well as by the residential sector. The interest in DR activities implementation hails from the huge value of energy consumption related to the building sector. According to data for 2018 reported in [9], this value is equal to 26.1% of total primary energy need in the EU. In Italy, the residential sector share is higher than the average value in the EU and it is equal to 28.0% (i.e., Heating 68.6%; Electrical Appliances 13.1%; Water Heating 11.5%; Cooking 6.2%, Cooling 0.6%, referred to the national primary energy consumptions) [10].

Moreover, several investigations to identify the buildings' potential of flexibility can be found in literature [11]. For instance, Rahmani-Andebili [12] built a predictive model to properly schedule the deferrable users and the energy resources. In addition, the use of buildings' thermal mass can be considered as an effective storage medium [13,14]. Indeed, that mass, different for each building, is able to store heat by anticipating or postponing the heating systems (or cooling systems) switch-on, without affecting the indoor thermal comfort conditions [15]. Other available options to manage the load flexibility consist of applying the so-called Power-to-X or Power-to-What strategies [16–18]. This is the general terminology meant for the electricity conversion into other useful energy forms. As regards the building sector, Power-to-Heat, Power-to-Power and Power-to-Gas seems to be the most promising and suitable solutions [19–22]. Moreover, specific storage devices, such as Phase change Materials (PCMs), batteries, pressurised gas vessels, and electro-fuels injection into Natural Gas (NG) network must be properly integrated into the existing energy systems [23]. Lezama et al. [24] propose an evolutionary algorithm for the management of DR activities in residential homes equipped with photovoltaics (PV) and battery systems. Having said, the recent literature on that topic has strongly focused on the electric heat pumps use which serve the end-user for space heating and Domestic Hot Water (DHW) production [25].

A holistic approach must be taken to address the building load flexibility to better understand how the application of such strategy at small-scale can actually affect the whole energy system performance.

Many studies focused on the analysis of a single dwelling instead of considering a wide group of them. Where this latter group was studied, the modelling was often oversimplified. Indeed, those models did not account for the several thermos-physical characteristics of buildings as well as the occupancy profiles [26]. Adhikari et al. [27] developed in their work an algorithm for the optimal management of an HVAC systems aggregate for residential users. Similarly, Cai et al. [28] analysed the role and implications of a commercial/residential users aggregate in the optimised management of a district heating network. Scaling down the level of analysis, D'hulst et al. [29] presented their estimation on the flexibility amount associated to a group of smart household appliances in the residential sector.

In this paper, the authors propose and describe their approach for simulating a generic dwellings aggregate based on real data extrapolation taking into account whole electricity consumptions.

To the best of the authors' knowledge, currently, there does not exist an available model characterised by high quality data for generating realistic profiles, to evaluate the potential of flexibility for the Italian residential sector. That high accuracy is required in order to properly build affordable low voltage grid models. Several studies dealing with the loads profiling activity can be found in literature for Italy and some southern Europe countries, but they are based either on benchmarks or simulations [30,31] referred to typical buildings [32]. Additionally, the most updated data, relative to the real electricity consumptions, date back to 2002 [33]. Therefore, one of the aims of this study consists of fulfilling that gap by proposing an empirical-probabilistic model of household electric loads. Such a model has been designed to generate demand side profiles based on a bottom-up approach, using also on field measurements, which have been recently registered.

Indeed, those real data has been collected by an experimental campaign over a two-year period, 2018–2019, on 14 selected dwellings [34]. That sample has been chosen as the most representative dwelling typologies of the Italian middle regions' building stock. To do so, a combined approach based on simulations and real monthly power consumption data collection of 751 dwellings, determined from utility bills has been used. Thus, by post-processing all the outcomes, it has been possible to build the electric load profile of some typical Italian dwellings.

The hourly electricity price on the Italian spot market has been recorded to plot its profile over the year. Thereafter, a decisional algorithm to shift the end-user loads has been implemented and discussed, identifying the potential opportunities the residential sector can offer to the whole Italian electric system.

In the end, from literature review, it emerged how the most investigated sectors for DR application are the industrial and tertiary ones. For that reason, the authors believe that their contribution to the knowledge in this topic consists of: (i) providing the actual electric load profile along with the amount of flexibility potential for the most common Italian dwelling typologies; (ii) elaborating a simplified methodology to create a buildings' cluster; (iii) evaluating the potential limitations related to the implementation of DR activities into a country with the lowest electrification degree of heating across the European Union; (iv) defining useful indicators in order to assess the loads shifting strategies effectiveness once they are applied. Indeed, the DR programs suitability was generally investigated paying mostly attention to the achievable results in terms of grid stability [35] and benefits for the end users [36,37]. Differently, quantifying the adopted strategies effectiveness and viability have not been widely addressed up to date.

2. Materials and Methods

This study is part of a research project aimed at characterizing residential users to assess the effectiveness of a Demand Response (DR) program application in the residential sector.

In previous works by some of the authors, a wide data collection campaign was carried out to build a database consisting of 751 typical Italian dwellings. For that purpose, an on-line tool, based on an in-house code, was provided to "interview" several end-users. In detail, this code is able to collect all the actual dwelling consumptions allowing to compare the billing data with the estimated ones hailing from a simplified dynamic simulation. To do so, a collection of the dwelling's characteristics has been carried out by means of an on-line survey dedicated to non-expert users (see Appendix A, Table A1). The typical dwelling technical parameters, such as U-value, heat capacity etc., have been assumed from the construction age, the climate zone, and the potential energy retrofitting measures. Furthermore, the solar gains have been assessed accounting for the walls and roofs colour (i.e., very light colour, light colour, medium colour, dark colour, very dark colour) as well as the shading degree in terms of time periods over the day [38].

That database was already used to perform the following tasks: (i) characterising the energy use of residential end-users and identifying the flexible loads [38]; (ii) assessing how the energy retrofitting interventions influence the energy performance [39]; (iii) estimating the potential benefits that arise from the installation of Building Automation Control systems [40]; and, simulating the climate changes implications on dwellings energy needs [41]. Thus, simulations results match positively the real collected data [38], but they refer to a standard scheduling [42]. In addition, the assumed occupants' number over the day is the declared one in the questionnaire (i.e., from 8:00 a.m. to 1:00 p.m.; from 1:00 p.m. to 7:00 p.m.; from 7:00 p.m. to 12:00 p.m.; from 12:00 p.m. to 8:00 a.m.). As demonstrated in literature, unfortunately, people have a weak perception of their own energy consumptions [43].

For those reasons, the authors believed to deepen the knowledge of residential users' behaviour, in order to acquire a more realistic schedule. In such a way, better forecast values of electricity consumptions can be computed to implement the subsequent aggregation process.

For that purpose, the 751 questionnaire respondents were asked to participate in an experimental campaign aiming at measuring the electricity consumption at their homes (Figure 1). Based on their feedback, 14 households were selected to become archetypes. In order to maximize the representativeness of the dwelling sample—the average floor surface, the number of occupants, the age, and the financial and educational conditions have been considered. In terms of geographic location, the 14 selected households are located in the Rome Municipality and in the neighbouring provinces. Notwithstanding, this is not a limit for the purpose of this study, since it refers only to the dwellings' electricity consumption. Indeed, it has to be noted that in Italy, only 0.4% of heating consumptions are due to electric-driven devices [44]. Furthermore, the electricity consumptions for cooling purpose in the Italian residential sector are very low; even though the Italian territory is characterised by different climate zones, the electricity consumptions for the inner space cooling are not strongly affected by the geographical location. As reported in [41], the normalised yearly electricity consumptions for Milan, Rome, and Naples (i.e., northern, middle, and southern regions) are very similar and they are equal to 1.3 kWh/m^2 year, 1.8 kWh/m^2 year and 1.3 kWh/m^2 year, respectively. Since the scope of the article is to propose a method to create a residential cluster, it is possible to extrapolate data, referred to the Italian middle regions, to assess, in a first approximation, the whole Italian building stock.

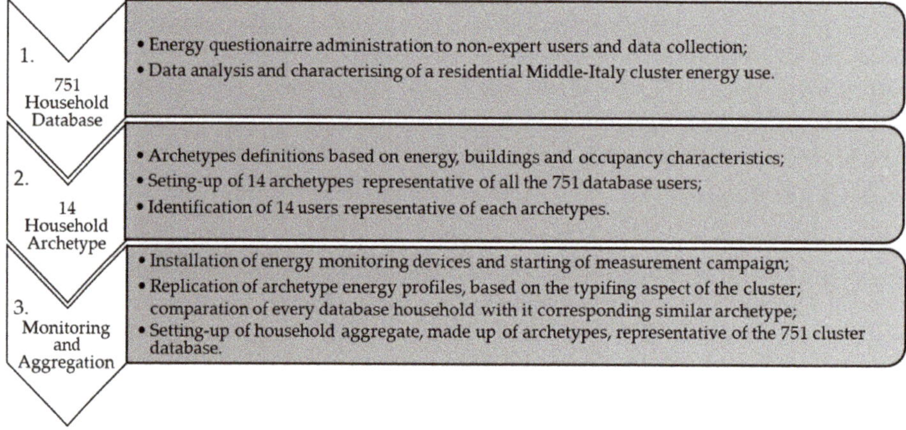

Figure 1. Flow chart related to the methodological approach.

Thus, once the occupants' consent was acquired, the selected 14 dwellings were equipped with sensors and probes for monitoring the electricity uptakes. Thereafter, data collection was carried out over two years. By processing all of those measurements, the electricity time scheduling has come out, identifying the average daily profile for each month, distinguishing also three different day types (i.e., weekdays, non-working days, and Saturdays).

First, the sample households have been considered one by one and, then, the virtual aggregate of 751 units, which has been built on the 14 archetypes. Indeed, the building aggregate modelling process combines statistical and measured data very often [45] for the estimation of hourly electricity consumption for a large group of residential buildings [46]. Moreover, the archetype-based approach is widely applied for taking into account building diversity, in terms of architectural shape, construction year, urban context, orientation and materials [47]. In so doing, the aggregate power demand as well as the energy demand (ED_{agg}) can be easily calculated by multiplying each archetype demand (ED_i) by the dwellings number (n_i) belonging to that one [48].

$$ED_{agg} = \sum_{i=1}^{N} ED_i \cdot n_i \tag{1}$$

The archetype categorization has been drawn up according to the following peculiarities [49]:

- electricity consumptions (storable loads, shiftable loads, non shiftable loads);
- electric-driven heating systems and/or Domestic Hot Water (DHW);
- PV array installation;
- dwelling size;
- the occupancy modelling (occupant number, time scheduling).

The reference archetype identification the sampled dwellings belong to has been done by means of a grade assignment according to the criteria reported in Table 1. Specifically, the grade calculation has been performed by comparing simultaneously the typifying aspects of a selected dwelling to the archetypes reference values. These latter will be presented extensively in the Section 3.2 and they have been outlined in Table 5.

Table 1. Typifying Aspects and criteria outline for the Grade calculation.

Typifying Aspects (A)	Criterium	Max Grade * (G_{max})	Grade *
Storable Loads	Relative deviation	0.15	$(A/A_{ref}) * G_{max}$ or $(A_{ref}/A) * G_{max}$
Deferrable Loads	Relative deviation	0.15	$(A/A_{ref}) * G_{max}$ or $(A_{ref}/A) * G_{max}$
Non-deferrable Loads	Relative deviation	0.15	$(A/A_{ref}) * G_{max}$ or $(A_{ref}/A) * G_{max}$
Heating or DHW **	Energy carrier	0.05	Electricity = 0.05; NG ** = 0
PV ** array	Installation/lack	0.05	Installed = 0.05; Missing = 0.00
Dwelling floor surface	Relative deviation	0.10	$(A/A_{ref}) * G_{max}$ or $(A_{ref}/A) * G_{max}$
Occupants Number	Relative deviation	0.10	$(A/A_{ref}) * G_{max}$ or $(A_{ref}/A) * G_{max}$
Occupancy in time span 8–13	presence/absence	0.10	Present = 0.10; Missing = 0.00
Occupancy in time span 13–19	presence/absence	0.10	Present = 0.10; Missing = 0.00
Occupancy in time span 19–0	presence/absence	0.025	Present = 0.025; Missing = 0.00
Occupancy in time span 0–8	presence/absence	0.025	Present = 0.025; Missing = 0.00
TOTAL		1.00	

* G_{max} is the maximum assignable grade for the selected typifying aspect; A is the actual numerical value corresponding to the selected typifying aspect; A_{ref} is the numerical value corresponding to the typifying aspect associated to the reference archetype, which is used for comparing each building; ** DHW: Domestic Hot Water; NG: Natural Gas; PV: Photovoltaic.

Subsequently, the electricity price trend has been analysed to identify the current cost criticalities of the Italian electricity system, using the open data provided by the Energy Market Manager (GME) [50]. Finally, the study has been completed superimposing the electricity price profile on the aggregate consumption curve to highlight the potential opportunities emerging from a different behaviour of residential sector. Indeed, rather than replicating the most common mechanism, where consumers optimise their purchase costs in accordance with the current tariff scheme [51,52], it has been assumed that the flexibility can be integrated in the market. In detail, by gathering the flexibility capacity of each dwelling to get the energy player scale, that energy amount can be effectively programmed for a flexible use matching the spot market profile as much as possible [53]. In this way, it is possible to create a Residential Cluster (RC), which is able to offer directly, in the day-ahead market, packs of flexible energy characterised by a specific time scheduling. In so doing, the energy traders can choose the best profiles for maximising the social wellbeing. Nonetheless, that approach presume that utilities keep completely under control the flexible loads amount, to formulate the best offer sets matching the market outcomes. They can do it directly or by means of an intermediary subject, such as an aggregator able to gather homogeneous end users.

For that purpose, a statistical approach has been applied also on the monthly National Unique Price trend (that price is commonly also known as the PUN Index by Italian Market operators). Specifically, the PUN Index is defined as the average of zonal prices in the day-ahead market, weighted by total purchases, net of purchases for pumped-storage units and of purchases by neighbouring countries' zones [54]. Thereafter, all those periods where the residential sector could either reduce or increase its electricity uptakes have been defined. Figure 2 briefly depicts the logical pathway for the decision making.

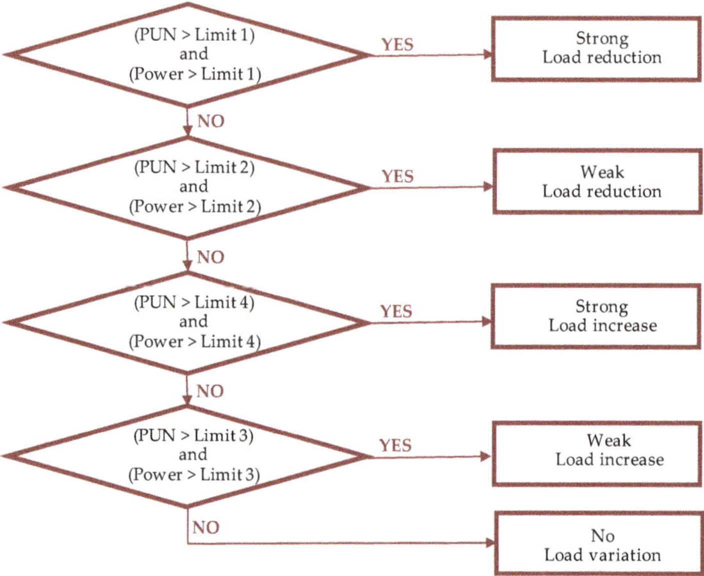

Figure 2. Flow chart related to the consumptions' optimization process (PUN: National Unique Price).

Database Description and Sample Users

As stated previously, the present study refers to a database consisting of 751 households characterised by nonhomogeneous size, technical systems, and occupancy. The average floor surface is equal to 120.4 m^2 (see Figure 3a) and it ranges between the minimum and maximum values of 22.5 m^2 and 648.0 m^2, respectively. In regards to the occupant number, the lower and upper limits are 1 and 9, respectively, while the average value is 3.4 (see Figure 3b). Thus, in terms of frequency, the 4-people dwelling is the most common (41.8%), while the 3-people and 2-people ones have a share equal to 26.1% and 17.8%, respectively. All of the samples are equipped with a heating system (see Figure 4a), showing a clear prevalence of NG-based plants (98.5%), compared to the electrically powered ones (1.5%). Moreover, a DHW device is installed in all sample dwellings (see Figure 4b), consisting mainly of instantaneous boilers (77.5%), instead of other typologies, such as condensing boiler with integrated storage (7.7%), electric water heaters (12.5%), and heat pump water heaters (2.1%).

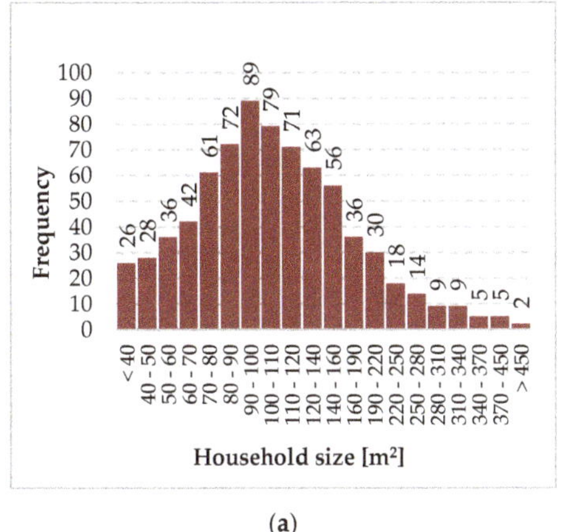

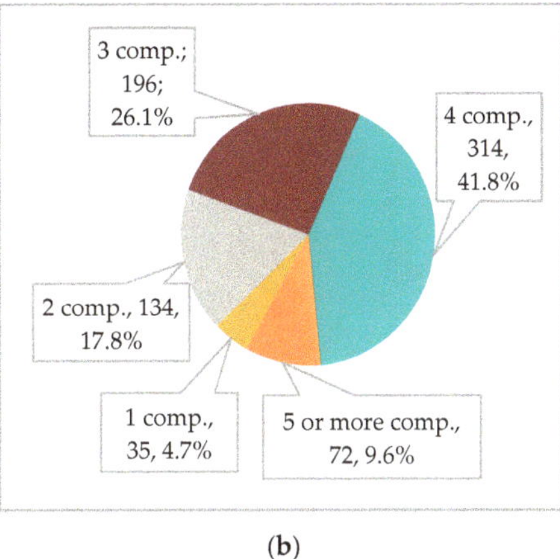

(a)   **(b)**

Figure 3. Building subdivision. (**a**) Household size. (**b**) Family occupancy.

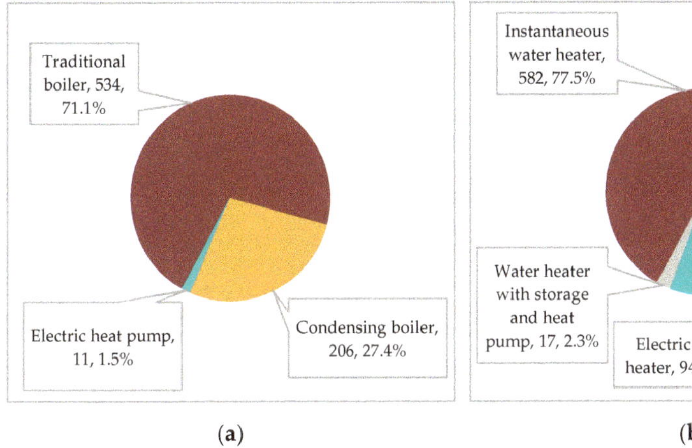

(a)   **(b)**

Figure 4. Heating system typologies: (**a**) inner Space heating; (**b**) DHW.

In 385 homes (51.3%), there is at least one fixed electric air conditioner and the average number of chilled rooms is equal to 2.9. The 751 dwellings energy characterisation has been performed by running the online simulation tool [38–41]. In that way, it has been possible to evaluate the peculiarities of each sample household in terms of management flexibility. Once the most common characteristics have been identified, 14 significative archetypes have been defined according to what is summarised in Table 2.

From the data, it emerges how all the archetypes are equipped with a NG boiler (9 non-condensing boilers and 5 condensing boilers); cooling appliances serve only a few rooms and they are installed only in 9 of the archetypes. Washing Machines (WM) are available in all archetypes with an average operating time equal to 4 cycles per week; only 11 archetypes are equipped with Dishwasher (DW), running 5 cycles per week and only 4 of them also have a Tumble Dryer (TD). Finally, the archetype lighting system is composed by 80% of fluorescent lamps or LED, and 20% by filament lamps and halogen.

Table 3 reports the Archetypes occupancy along with the family demographic composition.

Table 2. Archetypes appliances and characteristics.

Archetype	Floor Surface [m²]	Heating and DHW *	Cooling *	PV Array	WM **	DW **	TD **
#1	49	NCB	2 HP		7; 5; A+	6; 7; A	
#2	101	NCB	1 HP		10; 2.5; A		
#3	100	NCB	1 HP		7; 5; A+		
#4	50	NCB	1 HP		7; 1.5; A+		5; 0.5; A
#5	100	CB + HP	4 HP		7; 4; A++	5; 4; A	5; 4; A
#6	65	CB	3 HP		7; 6; A	12; 3.5; A	7; 0.5; B
#7	65	NCB	1 HP		7; 5; A+	6; 7; A	
#8	60	CB			7; 2; A++	12; 1.5; A+	
#9	95	NCB	2 HP		7; 5; A+++	12; 8; A+	
#10	102	NCB	1 HP		7; 3; A+	14; 5; A	
#11	67	CB			10; 5; B	6; 5; B	
#12	134	CB			7; 6; A	14; 7; A	6; 3; B
#13	124	CB			5; 4; A	12; 7; A+	
#14	123	NCB + solar collectors		3.9 kW	5; 4; A	12; 7; A+	

* NCB: Non-Condensing Boiler; CB: Condensing Boiler; HP: Heat Pump; ** WM: Washing machine; DW: Dishwasher; TD: Tumble dryer; Capacity, cycles per week, Energy Class.

Table 3. Family composition of each Archetype.

Archetype	Occupants *	Description
#1	4; (1; 3; 4; 4)	Family with two teenage children and one unemployed parent
#2	2; (0; 0; 2; 2)	Commuter Workers
#3	4; (0; 3; 4; 4)	Family with school-aged children, and one part-time working parent
#4	1; (0; 0; 1; 1)	Commuter Worker
#5	4; (1; 3; 4; 4)	Family with school-aged children, and one home parent
#6	4; (1; 3; 4; 4)	Family with school-aged children and babies, and one unemployed parent
#7	3; (0; 0; 3; 3)	Family with a baby and commuter parents
#8	2; (1; 1; 2; 2)	Commuter worker, awaiting employment
#9	3; (1; 2; 3; 3)	Family with a school-aged child, and one commuter worker
#10	2; (0; 1; 2; 2)	Family of commuter workers
#11	3; (0; 2; 3; 3)	Family with a school-aged child, and two commuter workers
#12	4; (0; 1; 4; 4)	Family with two adult children, two commuter parents
#13	2; (0; 1; 2; 2)	Family with a school-aged child, and two commuter workers
#14	2; (2; 2; 2; 2)	Two Pensioners

* Number of occupants; (8 a.m. to 1 p.m.; 1 p.m. to 7 p.m.; 7 p.m. to 12 p.m.; 12 p.m. to 8 a.m.).

Additionally, several wireless sensors and actuators have been installed in the selected households as archetypes for monitoring and keeping under control the global energy consumptions, as well as the single appliance uptakes and the indoor thermal comfort (see Figure 5).

The number of sensors to be installed has been decided according to the dwelling shape and household appliances typology, as reported in Table 4. All sensors adopt the Z-Wave communication protocol for interacting with the Energy Box (EB). This latter has the function to manage the peripheral devices and it is able to exchange data with tertiary parts, such as the utilities, by the internet connection. The Electricity Meter is responsible for monitoring the user main meter, while the Smart Plugs supervise appliances such as refrigerators, washing machines and other electrical devices. Finally, Smart Switches control DHW preparation and, where they are installed, the electric heat pumps too, for air conditioning. Thus, the Energy Box sampling time for the electrical data collection is 5 s. Post-processing was performed to calculate average values over 15 min, according to the common meters provided by Distributors. Therefore, in the present analysis, the dwellings load profiles have been built by on field measurements, on the basis of a quarterly resolution.

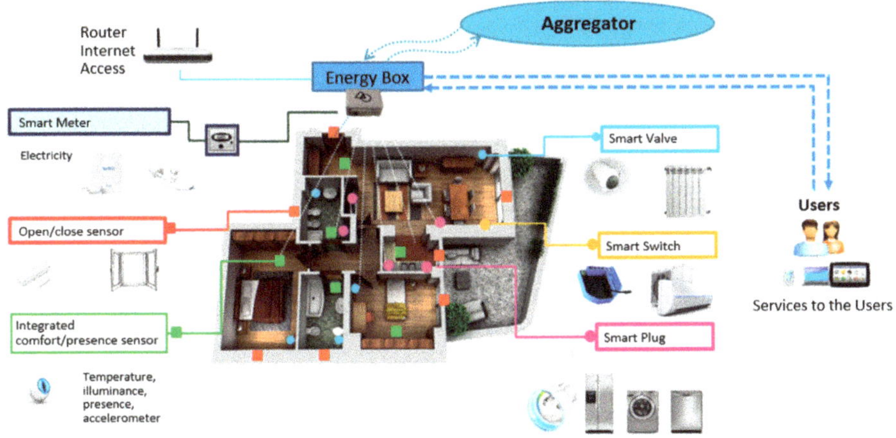

Figure 5. Control kit layout.

Table 4. Control kits configuration for the archetypes.

Function	Device	Archetype													
		#1	#2	#3	#4	#5	#6	#7	#8	#9	#10	#11	#12	#13	#14
Energy box	Gateway	1	1	1	1	1	1	1	1	1	1	1	1	1	1
	Electricity meters	1	1	1	1	2	1	1	1	1	1	1	1	1	2
Monitoring	Multi-sensors (temperature, presence, brightness)	5	6	6	4	6	6	4	4	7	6	3	9	7	7
	Windows/doors opening and closing detectors	7	8	6	5	8	8	5	5	10	10	6	9	12	9
Control	Smart Valves	6	5	0	4	3	6	5	3	8	6	0	0	7	0
	Smart Plugs	4	3	4	4	3	4	4	3	3	4	3	5	3	6
	Smart Switches	1	0	0	0	1	1	1	1	1	1	0	1	0	0

3. Results and Discussions

3.1. Electric Consumption Time Scheduling of Selected Archetypes

In this section, the results of real data processing associated to the archetype monitoring activities have been presented. By registering the actual power profile of each archetype over the years 2018 and 2019, the average daily trends have been deduced and plotted, categorising them into weekday, Saturday, and non-working day.

Figure 6 depicts clearly the average daily profile in the weekdays associated to each one of the 14 archetypes, together with their average trend, which is plotted in red line.

It can be noticed how all 14 archetypes show very similar profile trends, with a first slight peak occurrence in the early morning hours (between 6:00 a.m. and 8:00 a.m.), and a second one (much more prominent) close to the evening (between 7:00 p.m. and 10:00 p.m.). This trend can be correlated with all those periods of the day in which the occupant presence within the dwelling is the greatest. That generally occurs before and after the working time activities; for the archetypes characterized by permanent occupants inside (i.e., # 11; # 14), a third peak is observed close the central hours of the day (between 1:00 p.m. and 3:00 p.m.).

Similarly, Figure 7 shows the average daily profile in the Saturdays for each archetype. In those days, due to a higher occupancy level, slightly different trends have been registered, compared to the observed ones for weekdays: (i) the morning peak is greater and it is moved forward by two hours, approximately; (ii) the peak in the central hours of the day is higher; (iii) the average power is generally increased owing to the greater activity inside the households (i.e., cleaning, washing, etc.); (iv) there are wider differences in the shape profiles associated to each archetype.

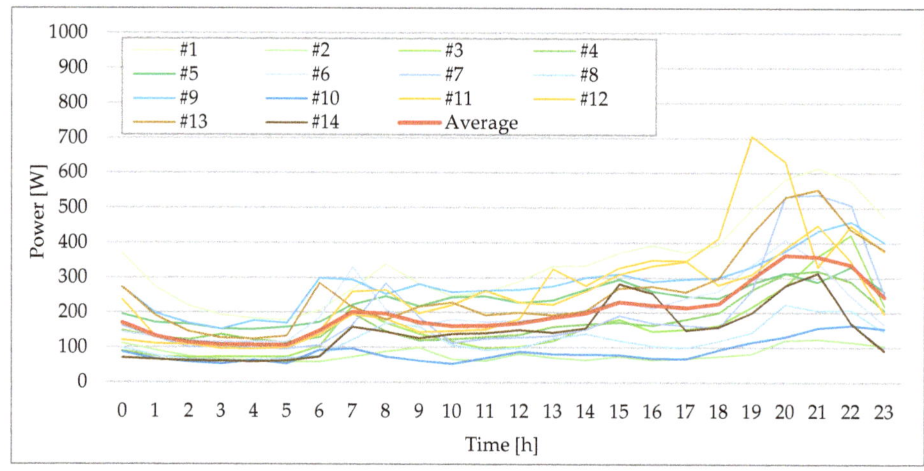

Figure 6. Archetypes average daily profiles over the weekdays.

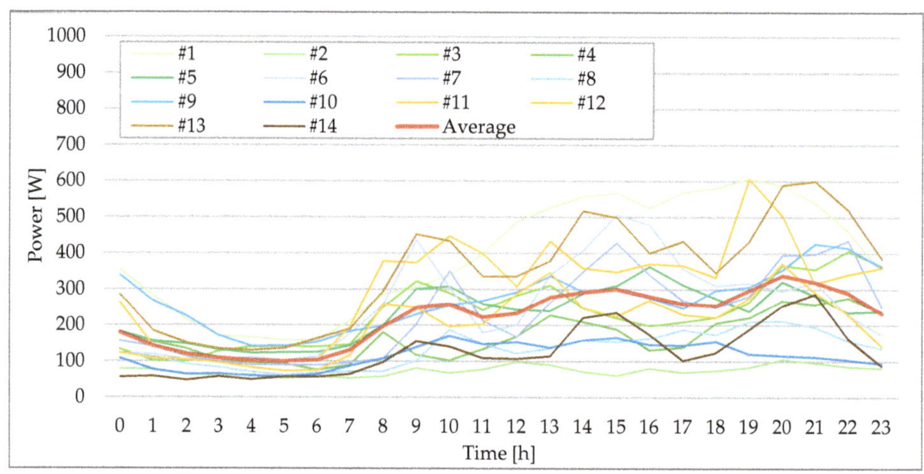

Figure 7. Archetypes average daily profiles over the Saturdays.

In regards to the non-working days profile, from Figure 8 it emerges what follows: (i) on Saturdays, occupant activities in the morning are postponed (i.e., 8:00 to 10:00 a.m.); (ii) since the occupants stay at home longer, the average power remains almost constant from morning up to late afternoon; (iii) the evening peak intensity is greater than on the other days; (iv) there are wider deviations between the archetype profiles referring to the same hours.

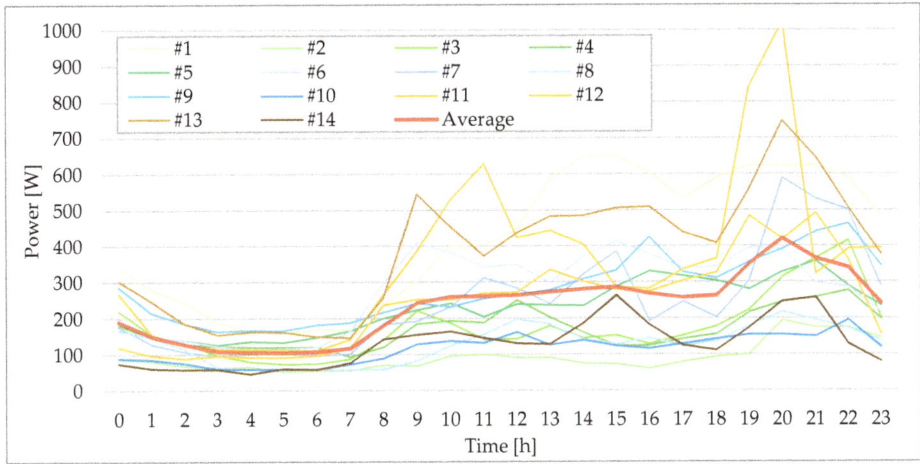

Figure 8. Archetypes average daily profiles over non-working days.

3.2. Users Virtual Aggregation

Once the archetype profiles are known, it is possible to build the consumption aggregate according to Equation (1). The next step is to identify by a selective procedure how many sample dwellings of the database can be considered belonging to each archetype, to compute the overall profile. To do so, the dwellings grade has been calculated following the criteria of Table 1 and compared to the archetype ones, which are based on values reported in Table 5. The best fitting values hailing from each comparison has been fixed equal to the maximum achievable grade. Therefore, the larger the grade value, the lower the sample dwelling deviation is, in comparison with the selected archetype.

Table 5. Archetypes reference parameters.

Parameters	Archetype													
	#1	#2	#3	#4	#5	#6	#7	#8	#9	#10	#11	#12	#13	#14
Storable Loads [kWh]	191	106	111	165	950	213	112	49	181	110	46	92	122	81
Deferrable Loads [kWh]	667	188	549	190	808	714	549	139	915	618	820	1274	835	556
Non-deferrable Loads [kWh]	2648	1024	1085	879	1298	1000	1099	881	2384	1218	1049	1754	1439	959
DHW	0	0	0	0	1	0	0	0	0	0	0	0	0	0
PV array [-]	0	0	0	0	0	0	0	0	0	0	0	0	0	1
Dwelling Floor Surface [m^2]	49	101	66	50	100	50	66	60	94	102	67	134	137	110
Occupants Number [-]	4	2	3	1	4	4	2	2	3	2	3	4	3	2
Occupancy in time span 8–13	1	0	0	0	1	1	0	1	1	0	0	0	0	1
Occupancy in time span 13–19	1	1	1	0	1	1	0	1	1	1	1	1	1	1
Occupancy in time span 19–0	1	1	1	1	1	1	1	1	1	1	1	1	1	1
Occupancy in time span 0–8	1	1	1	1	1	1	1	1	1	1	1	1	1	1

Thereafter, in order to provide further details, a frequency analysis, along with the calculation of other cumulative indicators, have been performed. Indeed, Table 6 summarises the processing outcomes indicating the archetype representativeness in absolute and percentage terms related to the dwellings number, cumulative floor surface, occupants, electrical consumptions, and storable and shiftable loads as well.

Figure 9 shows the results related to the selective procedure, reporting the best fitting for each archetype. It is noteworthy that the best fitting entails, on the average, a grade value equal to 0.81.

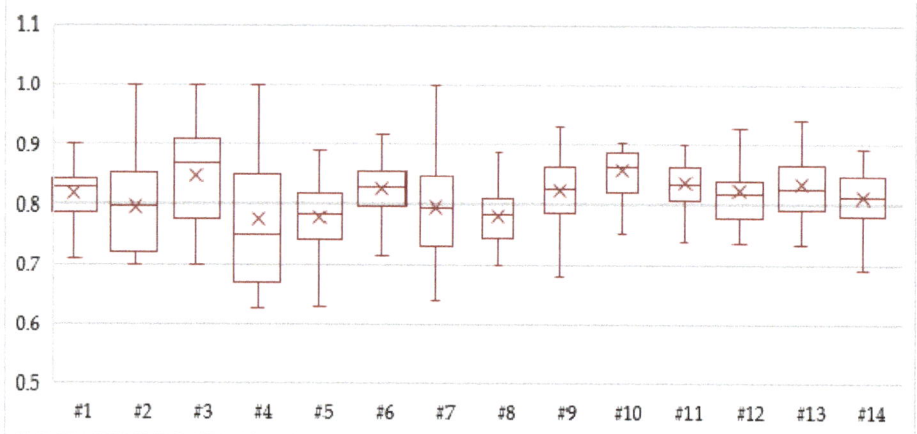

Figure 9. Best fitting statistic distribution.

Table 6. Archetypes representativeness.

Archetype	Dwellings Number	Cumulative Surface [m^2]	Occupants	Electric Consumptions [kWh/year]	Storable Loads [kWh/year]	Shiftable Loads [kWh/year]
#1	31 (4.1%)	4782 (5.2%)	130 (5.1%)	95,963 (6.7%)	5310 (1.6%)	18,081 (8.4%)
#2	16 (2.1%)	1213 (1.3%)	36 (1.4%)	94,444 (6.6%)	5894 (1.8%)	17,602 (8.2%)
#3	18 (2.3%)	1575 (1.7%)	54 (2.1%)	95,718 (6.7%)	10,267 (3.1%)	17,113 (8%)
#4	14 (1.8%)	945 (1.0%)	18 (0.7%)	91,964 (6.4%)	7479 (2.2%)	16,595 (7.7%)
#5	102 (13.5%)	12,419 (13.7%)	369 (14.5%)	262,070 (18.3%)	178,992 (54.8%)	16,305 (7.6%)
#6	138 (18.3%)	18,056 (19.9%)	531 (20.9%)	110,935 (7.7%)	28,897 (8.8%)	16,048 (7.5%)
#7	14 (1.8%)	1186 (1.3%)	31 (1.2%)	83,364 (5.8%)	2507 (0.7%)	15,830 (7.4%)
#8	83 (11%)	5409 (5.9%)	194 (7.6%)	100,642 (7.0%)	21,062 (6.4%)	15,467 (7.2%)
#9	165 (21.9%)	23,820 (26.3%)	630 (24.8%)	108,939 (7.6%)	32,710 (10%)	14,186 (6.6%)
#10	16 (2.1%)	1631 (1.8%)	37 (1.4%)	77,844 (5.4%)	3320 (1.0%)	13,760 (6.4%)
#11	22 (2.9%)	1822 (2%)	69 (2.7%)	71,965 (5%)	1142 (0.3%)	13,468 (6.3%)
#12	33 (4.3%)	4592 (5%)	135 (5.3%)	73,460 (5.1%)	4563 (1.3%)	12,892 (6%)
#13	32 (4.2%)	4975 (5.5%)	115 (4.5%)	76,550 (5.3%)	7890 (2.4%)	12,813 (6%)
#14	67 (8.9%)	7923 (8.7%)	189 (7.4%)	84,345 (5.9%)	16,070 (4.9%)	12,686 (5.9%)
Aggregate	751 (100%)	90,355 (100%)	2538 (100%)	1,428,203 (100%)	326,103 (100%)	212,846 (100%)

Considering the archetype occurrences number, the aggregate load profiles have been built for the average weekdays, Saturdays, and non-working days for each month of the year.

Those load profiles are shown in Figures 10–12, where the numerical values are superimposed on an intuitive colouring code, indicating also the consumptions magnitude in a graphical way. To facilitate readers, the reported numerical values refer to the hourly average uptake, measured in Watt; it is calculated by dividing the aggregate values by the dwellings number (i.e., 751).

In regards to the weekdays (see Figure 9), it can be noticed how the aggregate profile matches partially the average trend line of archetypes, showing a slight morning peak close to 8:00 a.m. and a higher distributed peak in the evening, starting from 8:00 p.m. up to 11:00 p.m. In the hot season (i.e., June, July, and August), higher values of the hourly average power have been registered, due to the air conditioners switching on.

Figure 10 depicts the average load profile over the Saturdays. Here, greater fluctuations in the average hourly uptake have been registered, starting from the last morning hours until the end of the day. That is owing to occupant behaviour variability related to each archetype, as mentioned before. The afternoon and evening peaks occur at different time locations, varying also month by month.

In the end, for non-working days load profile (see Figure 11), the peak region over the evening hours is still the most significant. Additionally, in the hot season, as well as in February and March, the hourly average power gets high values even in the afternoon.

	1	2	3	4	5	6	7	8	9	10	11	12	13	14	15	16	17	18	19	20	21	22	23	24
January	179	126	113	100	101	114	164	270	279	209	208	210	218	229	220	240	213	230	257	364	441	407	412	268
February	201	162	133	140	139	153	298	330	273	208	246	241	225	228	221	308	254	259	307	387	446	454	434	311
March	177	143	124	121	141	126	264	323	232	202	203	193	166	187	214	235	242	205	258	358	458	405	389	271
April	217	169	137	131	142	136	214	268	258	233	228	212	208	224	249	265	249	277	268	373	505	455	427	281
May	172	134	121	112	122	123	203	260	227	200	179	182	204	227	233	239	222	210	232	299	367	376	354	254
June	208	164	138	123	132	131	155	221	203	214	204	210	211	234	286	287	264	234	257	297	382	377	346	288
July	217	182	166	150	147	150	155	198	193	183	172	166	190	223	265	300	288	292	300	304	311	328	318	252
August	257	234	194	187	182	180	189	195	219	216	217	226	254	298	333	347	346	327	312	306	325	330	318	286
September	200	148	137	126	122	122	161	241	203	224	204	208	199	211	261	274	244	217	233	320	399	409	369	296
October	180	150	145	130	134	133	199	309	254	243	219	214	222	210	223	269	271	233	230	310	384	351	332	276
November	223	169	147	126	134	134	226	304	255	229	227	250	232	235	243	309	298	289	286	376	445	463	417	336
December	207	149	134	128	132	121	211	269	255	259	258	234	244	244	261	325	293	332	340	413	479	454	388	317

Figure 10. The aggregate profile of average power: weekdays. (Green: low; White: medium; Red: high).

	1	2	3	4	5	6	7	8	9	10	11	12	13	14	15	16	17	18	19	20	21	22	23	24
January	208	133	110	108	113	104	114	133	252	305	356	264	210	330	371	315	275	290	338	445	427	451	331	256
February	205	158	125	111	104	104	107	156	215	351	347	233	306	404	433	479	490	329	389	335	441	462	390	302
March	187	157	128	119	125	133	134	184	283	401	389	277	300	387	353	398	364	240	286	408	414	477	410	316
April	188	139	110	104	100	96	125	138	207	400	366	246	268	279	303	423	256	196	214	360	474	372	322	330
May	250	211	182	140	114	113	110	167	178	270	253	209	293	385	284	277	268	285	338	285	291	384	298	232
June	197	160	134	122	115	113	150	194	201	262	239	231	251	303	294	274	295	275	313	323	382	399	303	234
July	194	170	166	139	126	135	137	153	175	347	277	198	218	219	319	462	378	365	254	209	183	185	184	229
August	230	208	190	181	154	140	148	154	183	211	228	231	212	269	330	377	306	273	216	233	271	249	286	245
September	225	189	135	123	128	137	125	138	207	264	301	365	307	355	288	274	358	284	303	315	387	366	301	276
October	243	166	135	141	134	119	113	169	258	266	236	238	262	256	280	234	350	270	248	346	300	314	291	250
November	263	228	238	167	130	122	117	158	257	309	281	241	224	317	399	428	373	411	337	310	360	338	331	252
December	215	223	172	144	126	123	142	205	260	321	371	264	239	295	392	437	444	403	345	474	527	469	382	315

Figure 11. The aggregate profile of average power: Saturdays. (Green: low; White: medium; Red: high).

	1	2	3	4	5	6	7	8	9	10	11	12	13	14	15	16	17	18	19	20	21	22	23	24
January	199	189	171	138	125	123	110	131	200	297	313	302	361	340	323	291	356	345	321	448	575	514	478	287
February	255	183	170	114	119	117	115	149	247	255	302	380	362	310	365	475	502	295	299	439	422	434	426	258
March	299	213	157	125	118	122	118	132	244	404	356	253	212	217	303	370	462	223	237	382	382	350	353	273
April	252	168	151	110	102	97	118	131	248	383	345	399	321	256	255	252	272	294	335	533	508	401	362	275
May	166	145	124	106	113	114	119	130	177	236	267	259	277	285	315	291	308	256	228	289	406	376	373	266
June	204	153	149	149	150	151	160	190	207	225	260	247	249	325	321	356	308	252	262	354	419	364	403	314
July	226	200	156	131	133	134	137	149	203	223	185	192	179	240	279	289	300	320	261	203	266	225	219	195
August	201	191	163	138	149	151	152	146	164	186	193	231	228	258	354	376	367	355	283	257	253	284	276	262
September	230	179	150	127	126	132	136	153	214	276	258	253	313	333	369	377	337	288	372	383	463	379	280	226
October	227	160	139	143	107	105	116	152	233	253	274	260	262	278	312	368	252	242	268	360	359	327	278	192
November	256	195	171	177	178	172	188	200	255	346	368	358	353	400	387	398	394	368	422	480	617	520	464	375
December	206	173	157	135	135	134	140	155	275	335	362	341	379	300	309	345	405	398	405	472	520	490	444	319

Figure 12. The aggregate profile of average power: non-working days. (Green: low; White: medium; Red: high).

3.3. *Electricity Price Trend on the Italian Spot Market*

In this section, the PUN Index trend related to 2018 and 2019 has been analysed in order to build the price profile, according to the time-step used for defining the aggregate one. Specifically, by processing data hailing from GME, the PUN Index numerical values in €/MWh (1 € = 1.12 $ based on 2019 yearly average exchange rate [55]) has been calculated by averaging data referred to the aforementioned reference years. With it, the outcomes have been plotted adopting the same graphical approach by superimposing the numerical values onto the coloured cells.

The weekday price profile is shown in Figure 13, where it is possible to find four different time spans. Among those ones, two low-price regions can be easily noticed, occurring over the nighttime from 12:00 p.m. up to 6 a.m. and over the afternoon from 1:00 p.m. to 2:00 p.m. The former is mainly caused by a reduced global energy demand, whereas the latter is due to the PVs production. Peak prices occur in the morning from 8:00 a.m. to 11:00 a.m., where all the working activities usually start; thus, a further peak-price region can be registered in the time span 6:00 p.m.–10:00 p.m., being also affected by the daylight saving time application.

	1	2	3	4	5	6	7	8	9	10	11	12	13	14	15	16	17	18	19	20	21	22	23	24
January	50.3	46.5	43.7	42.8	43.6	47.8	56.9	64.2	71.4	70.1	65.5	62.9	59.4	58.6	60.9	63.9	67.8	74.6	74.3	71.7	65.9	61.6	58.1	53.4
February	50.4	48.1	45.9	45.4	45.8	49.3	59.3	65.4	72.2	68.1	62.0	59.2	55.2	54.2	57.6	59.7	64.5	70.5	77.4	75.5	67.4	62.2	58.1	53.4
March	48.1	45.3	43.2	42.1	42.7	46.9	56.8	63.1	68.8	64.7	59.0	55.7	50.0	49.7	53.3	57.2	60.8	64.0	72.9	79.7	70.7	62.6	56.5	50.9
April	48.6	44.3	41.9	41.1	41.5	46.1	54.2	62.4	70.4	65.1	59.4	55.9	49.2	47.8	51.4	53.9	55.9	55.3	57.7	65.7	70.7	61.7	55.0	49.8
May	49.9	45.6	42.3	41.0	41.3	45.7	53.2	60.9	66.4	62.4	58.5	55.6	50.7	49.8	52.3	54.2	56.1	56.0	57.9	63.0	67.0	63.4	56.5	50.0
June	52.7	49.8	46.1	44.4	44.1	44.6	52.8	58.1	65.1	62.1	58.8	57.0	52.4	52.0	55.5	57.1	59.5	59.8	60.9	65.2	66.0	64.3	58.6	52.0
July	57.6	54.4	51.5	50.0	49.6	50.5	54.9	59.0	63.9	64.0	62.0	60.8	56.9	56.9	60.2	62.2	64.6	65.5	66.6	69.1	68.9	68.7	63.9	58.6
August	60.3	56.0	53.1	51.8	51.4	52.9	55.9	58.2	61.7	61.3	59.0	58.2	56.3	56.2	58.0	59.6	62.4	66.0	69.2	72.3	74.2	70.9	65.6	60.6
September	58.8	56.5	54.8	54.0	54.0	56.7	64.9	69.8	77.1	75.1	70.1	67.1	61.0	61.1	65.9	70.3	73.9	74.2	75.5	85.5	81.2	71.8	65.4	59.7
October	54.9	52.8	50.9	50.4	50.7	54.3	65.6	73.8	78.6	76.0	70.4	67.8	61.5	61.3	65.2	69.0	72.5	73.6	79.0	84.9	75.7	68.3	62.8	56.9
November	50.1	46.5	44.4	42.5	42.7	46.4	56.8	64.5	68.4	67.1	64.3	63.3	60.0	60.6	62.7	65.2	69.9	78.8	78.0	72.1	65.0	59.9	56.6	52.6
December	48.8	44.9	42.6	40.9	41.3	45.7	54.8	61.4	65.4	64.8	61.8	60.4	57.7	57.5	59.9	63.5	67.8	74.0	69.5	66.8	62.7	58.8	55.7	50.9

Figure 13. Average PUN Index trend over the reference months: weekdays. (Green: low; White: medium; Red: high).

In regards to the Saturday profile, it shows a quite similar time distribution compared to the weekdays case. Notwithstanding, the low-price region over the nighttime is less wide, while it is larger in the afternoon (see Figure 14). Furthermore, referring to the profile plot, it can be noticed how the peak-price region over the nighttime is enlarged, since it ranges between 7:00 p.m.–1:00 a.m., especially in the hot season.

	1	2	3	4	5	6	7	8	9	10	11	12	13	14	15	16	17	18	19	20	21	22	23	24
January	55.3	52.0	49.2	47.8	46.8	47.1	53.3	59.5	63.5	63.7	59.7	56.5	54.5	52.1	52.2	55.5	60.0	69.5	72.7	72.0	66.8	63.1	58.3	53.5
February	52.1	49.9	47.5	46.2	45.8	46.9	52.9	58.8	60.8	62.0	57.5	53.8	49.9	46.7	46.7	48.8	52.9	61.6	72.0	72.2	65.4	59.5	53.7	49.2
March	54.9	49.1	46.3	44.9	44.1	46.4	50.4	53.7	55.2	54.8	51.7	47.1	42.9	39.3	38.9	42.4	46.2	53.2	64.2	71.4	65.3	57.7	53.2	47.3
April	52.0	47.7	44.9	44.6	44.1	45.5	48.7	54.1	57.9	56.2	51.4	46.4	40.4	32.3	31.3	36.6	43.7	47.9	52.9	60.2	66.2	59.0	53.4	47.5
May	55.5	50.4	46.4	45.0	44.5	45.7	45.9	52.8	56.1	56.1	55.0	51.6	46.7	41.8	41.1	43.7	48.3	48.4	51.0	57.9	63.1	60.0	53.4	47.3
June	58.2	54.5	49.7	47.2	46.2	43.8	45.1	48.2	51.9	52.3	49.2	43.5	39.2	35.3	34.1	36.3	39.8	46.5	51.9	56.9	59.6	59.3	54.1	50.7
July	61.3	58.2	54.4	53.6	52.2	50.6	50.0	53.0	55.2	54.8	52.1	49.6	47.8	46.2	45.2	46.4	48.9	53.4	58.1	62.4	63.6	63.8	59.2	54.5
August	62.2	58.4	55.5	53.8	53.1	53.1	52.3	53.2	54.5	53.9	51.2	50.1	50.0	48.6	48.2	49.1	51.0	54.1	60.2	66.9	70.2	67.9	61.7	57.4
September	63.2	62.1	58.5	56.1	55.7	56.2	60.5	62.3	61.8	61.1	57.1	54.1	51.3	48.7	47.9	49.4	52.5	58.2	63.4	73.2	72.4	64.7	60.7	55.6
October	61.7	59.7	56.2	54.6	53.7	55.4	62.9	68.2	67.5	65.7	59.8	55.2	52.5	49.1	48.2	49.6	53.1	59.2	70.8	78.4	73.3	65.0	57.3	55.2
November	53.3	49.4	46.8	45.0	43.6	46.8	51.3	54.7	56.7	57.4	56.9	56.5	54.3	52.3	51.6	53.9	57.6	64.8	66.2	66.0	60.6	56.1	53.5	51.0
December	51.4	46.6	42.8	41.4	40.8	41.7	47.2	52.7	55.7	55.4	54.0	52.3	51.5	49.5	50.5	52.8	57.0	62.6	62.8	62.7	58.9	53.7	51.5	47.9

Figure 14. Average PUN Index trend over the reference months: Saturdays. (Green: low; White: medium; Red: high).

Finally, Figure 15 outlines what happens during the non-working days, highlighting that the peak-price region in the morning hours is basically eliminated. Regarding the peak-price region, it remains wide over the nighttime anyhow.

	1	2	3	4	5	6	7	8	9	10	11	12	13	14	15	16	17	18	19	20	21	22	23	24
January	52.6	48.9	46.1	42.6	41.2	39.8	43.8	48.0	49.2	51.0	53.2	53.7	53.8	50.6	50.6	53.8	57.6	63.3	65.4	65.9	63.7	60.7	55.4	50.2
February	48.2	45.5	42.4	39.7	39.3	39.7	43.8	46.1	47.4	48.8	48.3	46.6	45.7	41.9	42.4	44.8	50.1	56.3	64.8	68.3	66.1	62.0	56.3	50.0
March	49.0	45.2	41.2	40.0	41.5	43.8	45.2	44.6	45.6	48.3	51.4	52.0	50.5	41.9	39.6	44.8	50.7	55.2	67.5	69.2	64.9	59.1	52.1	46.8
April	48.8	43.2	40.3	38.5	40.1	44.6	44.5	45.0	45.5	44.9	43.2	38.2	34.9	27.0	24.2	27.6	33.1	40.9	49.8	58.9	69.1	65.1	57.7	51.0
May	49.7	46.2	40.1	36.6	39.7	43.7	40.5	42.6	43.8	43.2	44.1	41.3	37.6	31.1	29.0	33.8	38.7	42.6	49.4	56.4	62.8	62.8	58.1	50.7
June	50.0	47.5	43.8	41.0	41.5	40.8	39.2	39.4	39.3	42.4	44.5	41.1	38.9	32.9	31.9	36.4	44.0	44.7	50.3	56.4	63.1	65.7	62.0	53.7
July	55.8	52.3	49.6	47.9	46.6	44.7	41.9	42.5	41.3	42.6	44.0	42.9	41.4	39.2	37.8	39.4	43.7	47.4	51.6	56.9	61.6	63.4	60.9	55.5
August	57.5	55.7	53.6	52.0	50.8	52.9	50.3	48.5	47.1	47.2	48.3	47.8	47.3	44.5	44.3	46.8	49.9	52.1	55.5	63.1	70.4	69.7	64.7	59.6
September	58.5	54.8	53.2	52.4	52.1	52.3	54.7	53.4	52.2	54.4	54.6	53.5	52.0	48.4	48.1	50.3	52.9	55.5	59.8	68.9	71.0	65.3	62.2	56.0
October	54.5	51.6	49.9	48.9	47.6	48.0	49.5	51.2	51.8	52.6	52.5	51.7	50.3	46.5	46.3	48.8	53.0	55.7	61.9	72.4	71.1	65.5	58.6	53.5
November	48.1	44.8	39.8	38.7	38.0	40.2	42.7	45.5	46.5	49.0	50.5	50.4	51.4	49.8	50.0	52.0	53.8	59.3	61.3	61.5	58.7	54.6	52.4	48.3
December	46.5	40.0	37.1	33.2	32.2	35.2	40.2	45.0	45.8	47.5	48.5	47.2	47.4	44.3	45.7	50.1	54.9	60.2	60.3	61.0	59.1	56.2	51.7	45.7

Figure 15. Average PUN Index trend over the reference months: non-working days. (Green: low; White: medium; Red: high).

3.4. Loads Time-Shifting Strategy Identification

In this section, the implications associated to such a load time-shifting strategy implementation has been addressed and discussed. It is important to point out that the shifting command is delivered to end-users once both threshold conditions on price and power (i.e., on PUN Index value and on average uptake value) are verified. Those thresholds conditions can change dynamically, since their values are calculated in terms of percentiles, according to the Equation (2), and they are strongly dependent on the database content. Indeed, the generic index I_k associated to the desired percentile (e.g., fixing 35th percentile it entails I_k equal to I_{35}) has been calculated as follows:

$$I_k = \left[0.5 + \left(n \cdot \frac{k}{100} \right) \right] \tag{2}$$

where n indicates the number of ordinated sample data.

Having said, the boundary conditions to perform these simulations are summarised in a systemic overview in Appendix B.

Therefore, the Load Shifting Command Function ($LSCF = f(PUN, P)$) reads as:

$$
\begin{aligned}
&If\ (PUN > Limit\ 1) \wedge (P > Limit\ 1)\ then\ LSCF = -2 \\
&\quad Else\ If\ (PUN > Limit\ 2) \wedge (P > Limit\ 2)\ then\ LSCF = -1 \\
&\quad\quad Else\ If\ (PUN < Limit\ 4) \wedge (P < Limit\ 4)\ then\ LSCF = 2 \\
&\quad\quad\quad Else\ If\ (PUN < Limit\ 3) \wedge (P < Limit\ 3)\ then\ LSCF = 1 \\
&\quad\quad\quad\quad Else\ LSCF = 0 \\
&\quad\quad\quad End\ If \\
&\quad\quad End\ If \\
&\quad End\ If \\
&End\ If
\end{aligned}
\tag{3}
$$

Numerical indicators ranging between −2 and 2 have been used to encode the load shifting commands when the limit thresholds are overcome, according to Equation (3). In detail, once PUN Index and Power are higher than the Limit 1, together with the other conditions, the command function provides −2 as output, suggesting the strong load reduction; when the Limit 2 is overcome, the output value is equal to −1, entailing a weak load reduction. On the contrary, weak load increase and strong load increase are suggested for all those values lower than the Limit 3 and laying in the band between Limit 3 and Limit 4, corresponding to 1 and 2, respectively. Finally, according to the flow chart reported in Figure 2 the last used value is 0, corresponding to no-load variation. As a consequence, the load shifting commands distribution have been plotted in Figures 16–18. Applying that methodology it emerges that in the weekdays loads shifting from the evening hours (in the time span 8:00 p.m.–10:00 p.m.) towards the night hours (between 2:00 a.m. and 6:00 a.m.) as well as from the morning hours (8:00 a.m.–9:00 a.m.) towards the afternoon (1:00 p.m.–2:00 p.m.), are recommended (see Figure 16).

	1	2	3	4	5	6	7	8	9	10	11	12	13	14	15	16	17	18	19	20	21	22	23	24
January	1	2	2	2	2	2	1	-1	-2	0	0	0	1	0	0	-1	0	-1	-1	-2	-1	-1	0	0
February	2	2	2	2	2	2	-1	-1	-1	0	0	1	1	1	1	-1	-1	-1	-1	-2	-2	-1	0	0
March	1	2	2	2	2	2	-1	-1	-1	0	0	1	1	1	0	-1	-1	0	-1	-2	-2	-1	0	0
April	1	2	2	2	2	2	1	-1	-1	0	0	0	1	1	0	0	-1	-1	-1	-2	-2	-2	-1	0
May	1	2	2	2	2	2	1	-2	-1	0	0	0	1	0	0	0	-1	0	-1	-2	-2	-2	-1	0
June	1	2	2	2	2	2	1	-1	0	0	0	1	1	0	0	-1	-1	-1	-1	-2	-2	-2	-1	0
July	0	1	2	2	2	2	2	1	0	0	0	0	1	0	0	-1	-1	-1	-2	-2	-2	-2	-1	0
August	-1	1	2	2	2	2	2	1	0	0	1	1	1	0	0	-1	-1	-2	-1	-1	-2	-2	-1	-1
September	1	2	2	2	2	2	1	-1	0	-1	0	0	1	1	0	-1	-1	-1	-1	-2	-2	-1	0	0
October	2	2	2	2	2	2	1	-1	-1	-1	0	0	1	1	1	-1	-1	-1	-1	-2	-2	-1	0	0
November	2	2	2	2	2	2	1	-1	-1	0	0	-1	1	1	0	-1	-1	-1	-1	-2	-1	0	0	0
December	2	2	2	2	2	2	1	-1	0	-1	0	0	1	1	-1	-1	-1	-2	-2	-2	-1	0	0	0

Figure 16. Strategy to optimise the load shifting: weekdays; (−2, Green) Strong Load reduction; (−1, Light Green) Weak Load reduction; (0, White) No Load variation; (1, Light Red) Weak Load increase; (2, Red) Strong Load increase.

	1	2	3	4	5	6	7	8	9	10	11	12	13	14	15	16	17	18	19	20	21	22	23	24
January	1	2	2	2	2	2	1	0	0	-1	-1	0	1	0	0	0	-1	-1	-2	-2	-2	-1	-1	1
February	1	1	2	2	2	2	1	0	0	-1	-1	0	1	0	0	0	-1	-1	-1	-1	-2	-1	-1	1
March	0	1	1	2	2	1	0	0	0	-1	-1	1	0	0	0	0	0	0	0	-2	-2	-2	-1	0
April	0	1	1	1	2	1	0	0	0	-2	-1	1	0	0	0	0	0	0	0	-2	-2	-2	-1	0
May	0	0	1	2	2	2	1	0	0	-1	0	0	0	0	0	0	0	0	-1	-1	-2	-2	-1	1
June	0	0	0	1	1	1	1	1	0	-1	0	1	0	0	0	0	0	0	-1	-2	-2	-2	-2	0
July	0	0	0	0	1	1	1	1	0	-1	0	0	0	0	0	0	0	0	-1	-1	0	0	0	-1
August	-1	0	0	0	1	1	1	1	0	1	1	0	1	0	0	0	0	-1	0	-1	-2	-1	-2	-1
September	0	0	1	1	1	1	0	0	0	0	0	0	0	0	0	0	1	0	-1	-1	-2	-2	-1	1
October	0	0	1	1	2	1	0	0	-1	-1	0	1	0	0	0	1	0	-1	0	-2	-2	-1	0	0
November	1	1	1	2	2	2	1	0	0	-1	-1	0	0	0	0	0	-2	-2	-1	-1	-2	-1	0	1
December	1	1	2	2	2	2	2	0	0	-1	-1	0	1	1	0	-1	-2	-2	-1	-2	-2	-1	0	0

Figure 17. Strategy to optimise the load shifting: Saturdays. (−2, Green) Strong Load reduction; (−1, Light Green) Weak Load reduction; (0, White) No Load variation; (1, Light Red) Weak Load increase; (2, Red) Strong Load increase.

	1	2	3	4	5	6	7	8	9	10	11	12	13	14	15	16	17	18	19	20	21	22	23	24
January	0	1	2	2	2	2	2	2	1	1	-1	-1	-1	0	0	0	-2	-1	-1	-2	-2	-2	-1	1
February	0	1	2	2	2	2	1	1	0	0	-1	0	0	0	0	0	-1	0	-1	-2	-2	-2	-2	0
March	-1	1	2	2	2	2	1	1	1	-1	-1	-1	0	1	0	0	-1	0	0	-2	-2	-1	-1	0
April	0	1	1	1	1	0	0	0	0	-1	0	0	0	1	1	1	0	0	-1	-2	-2	-2	-2	-1
May	0	0	1	2	1	0	1	1	0	0	-1	0	0	0	0	0	0	1	0	-1	-2	-2	-2	-1
June	0	0	0	1	1	1	2	1	1	1	-1	1	1	0	0	0	-1	-1	-1	-2	-2	-2	-2	-1
July	-1	0	0	0	0	1	2	1	1	0	1	1	1	0	0	0	0	-1	-1	-1	-2	-1	-1	0
August	0	0	0	0	1	0	1	1	1	1	1	0	1	0	0	0	0	-1	-1	-1	-1	-2	-1	-1
September	0	0	1	2	2	2	0	1	1	-1	0	1	0	0	0	0	0	-1	-2	-2	-2	-2	-1	0
October	0	1	1	2	2	2	1	1	0	-1	-1	0	0	0	0	0	0	0	-1	-2	-2	-2	-1	0
November	1	2	2	2	2	2	2	1	1	1	-1	0	0	0	-1	-1	-1	-1	-2	-2	-2	-2	-1	0
December	1	1	2	2	2	2	2	1	1	-1	-1	-1	-1	1	1	-1	-2	-1	-2	-2	-2	-2	-1	1

Figure 18. Strategy to optimise the load shifting: non-working days. (−2, Green) Strong Load reduction; (−1, Light Green) Weak Load reduction; (0, White) No Load variation; (1, Light Red) Weak Load increase; (2, Red) Strong Load increase.

In the Saturdays case, the applied strategy recommends loads shifting from the evening hours (from 8:00 p.m. to 10:00 p.m.) towards the night hours (from 3:00 a.m. to 6:00 a.m.) and from the morning hours (10:00 a.m.–11:00 a.m.) towards the central hours of the day (12:00 a.m.–1:00 p.m.). Compared to the weekdays, a greater uncertainty in defining the strategy has been identified. Furthermore, a non-homogeneous commands' distribution between cold and hot season, can be noticed according to Figure 17.

Finally, in the non-working days case, the electric loads occurring in the evening hours (from 7:00 p.m. to 11:00 p.m.) can be moved forward in the night-time (from 3:00 a.m. to 7:00 a.m.). Compared to the previous cases, the load shifting from the morning hours towards the afternoon it is not always required (see Figure 18).

3.5. Flexibility Indicators Calculation

In order to verify the load shifting strategy suitability, several monthly and yearly indicators have been calculated in this section.

Firstly, the average shiftable loads amount for the Residential Cluster (RC) emerging from the applied flexibility strategy has been evaluated.

Referring to each day typology, the suggested loads to move backward or forward have been calculated by the Equation (4), considering only the positive differences in the round brackets.

$$LS = \sum_{i=1}^{24} (P_i - \overline{P}) \cdot \tau \tag{4}$$

Here, P_i represents the required power at the *i-th* hour, while the second term corresponds to the average power of RC profile, and τ is the time span (in this case equal to 1 h).

Secondly, the amount of potential available loads has been deduced from simulations results reported in Table 6. Thus, having supposed in a first approximation that these loads are evenly distributed over the day, the average amount related to each component constituting the RC, is equal to 1966 Wh/day. That value has been, substantially, deduced by dividing the whole flexibility potential of dwelling cluster (i.e., storable loads and shiftable loads) by its sample units (i.e., 751).

The comparison between the flexibility strategy suggestions and the available flexible loads is shown in Figure 19. It can be noticed that the calculated flexible loads are always lower than the available ones. As a consequence, the implemented strategy is suitable in any cases, showing the greater potential in the winter season over the non-working days.

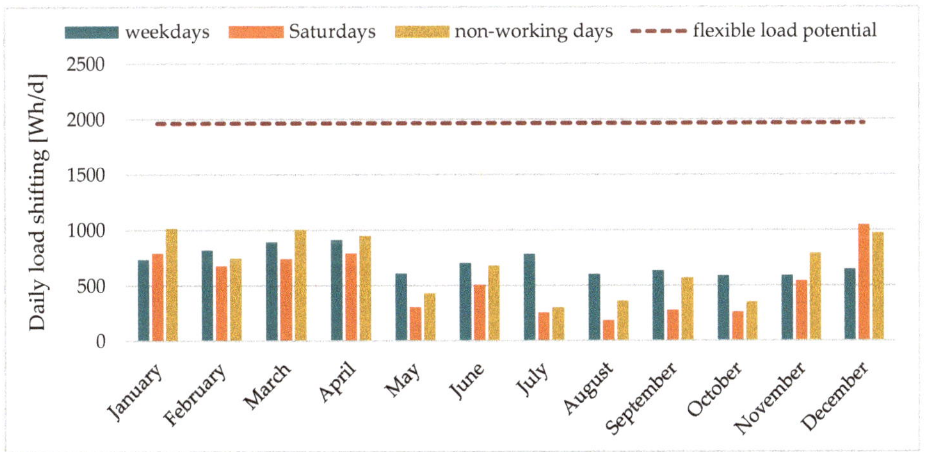

Figure 19. Daily load shifting deriving from the strategy application by month and day typologies. The dot line represents the available daily shiftable loads to participate at the flexibility mechanism.

Finally, once the Load Shifting (*LS*) function is calculated, it is possible to evaluate which is the Flexibility Index (FI_m) referred to the average day trend of each month. That parameter can be evaluated according to the Equation (5), where EN_m is the energy need over 24 h related the *m-th* month.

$$FI_m = \frac{LS_m}{EN_m} \tag{5}$$

Using the same definition, the Flexibility Index (FI_y) over the whole year is computable as a weighted average of Flexibility Index by month, in accordance with the Equation (6).

$$FI_y = \frac{\sum_{m=1}^{12} LS_m}{\sum_{m=1}^{12} EN_m} = \sum_{m=1}^{12} FI_m \cdot w_m \tag{6}$$

$$w_m = \frac{EN_m}{\sum_{m=1}^{12} EN_m} \tag{7}$$

Additionally, since the Available Shiftable Loads (*ASL*) is known, it is possible to define the Available Flexibility Index by month and by year as follows:

$$AFI_m = \frac{ASL_m}{EN_m} \tag{8}$$

$$AFI_y = \frac{ASL_{tot}}{\sum_{m=1}^{12} EN_m} \tag{9}$$

This latter parameter has been computed for the present RC and it is equal to 37.7%. Then, in order to evaluate the effectiveness of the adopted strategy, the Flexibility Index and the Available Flexibility Index can be correlated by Equations (10) and (11).

$$\varepsilon_m = \frac{FI_m}{AFI_m} \tag{10}$$

$$\varepsilon_y = \frac{FI_y}{AFI_y} \tag{11}$$

Figure 20 depicts the strategy effectiveness values ε_m, sorting the results by month and day typologies as usual. Thereafter, all those performance values can be outlined by calculating the effectiveness over the year ε_y, which is equal to 0.34 for this case.

However, once the actual hourly distribution of the available shiftable loads is known, it is possible to assess also the intraday strategy effectiveness. It is important to point out that the results of these comparisons are evidently dependent on the specific choice of threshold limits in the strategy definition. In this case, their values have been deduced from the statistical analysis carried out on both the cluster load and PUN Index, assuming also an identical definition for the variables.

Since it has been assumed Limit 2 = Limit 3 = 50th percentile, it implies that, in all those cases where those thresholds are overcome, a load reduction is recommended; on the contrary, beneath the limits the load increase is required. More generally, this is a conservative strategy, given that any other choice would lead to higher differences between the available and the effective shiftable loads, penalising the strategy effectiveness.

However, it is necessary to verify the precise hourly distribution of the available flexible loads, instead of considering them equally distributed over the 24 h. Furthermore, it is important to take into account where the household appliances operational constraints occur to limit the lowering of flexible loads amount. Indeed, the adoption of a flexibility strategy should not negatively affect the end user's wellbeing, and it must consider the correlations between appliance operation and the occupant presence (i.e., vacuum cleaner, iron, etc.). Those issues have not been extensively addressed in this work, but they will be deeply discussed in the further development of the research project. Anyway, it is noteworthy that accounting for additional operating constraints on household appliances will reduce the flexibility indexes as well as the strategy effectiveness.

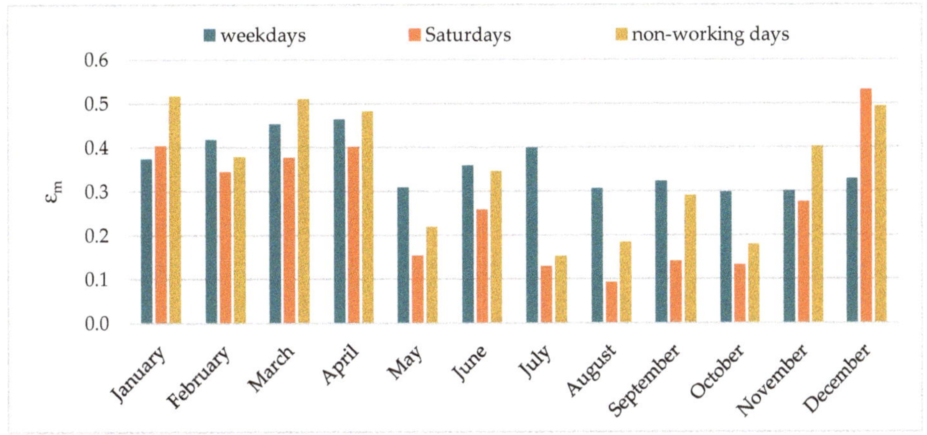

Figure 20. Strategy effectiveness sorted by month and day typologies.

4. Conclusions

The DR activities application to the residential sector represents a viable option to get a higher cost effectiveness for both utilities and end-users. By creating a dwelling cluster, it is possible to gather huge amounts of flexible loads to be shifted over the daytime, in order to participate actively to the market outcomes.

In this work, a procedure to build a dwelling cluster load profile has been presented and discussed, on the basis of a combined approach: the experimental measurements have been coupled with the statistical analysis. Referring to the Italian context, that approach could represent a good opportunity, on one hand to update available data, on the other hand to develop new models for the low voltage grid management.

Finally, a flexibility strategy has been implemented along with the definition of several performance indicators. Using the Italian residential sector as a reference case, and the Italian electricity price trend over 2018 and 2019, it has been possible to evaluate the dwellings contribution to a more flexible system. The most remarkable findings can be listed as follows:

- 14 dwelling archetypes have been defined by the use of a numerical approach based on a grade scale ranging between 0 and 1; each sample household (i.e., 751) has been compared to the archetypes in order to identify its category; this method leads to a good fitting since, on average, the best grade is equal to 0.81;
- the most representative archetypes, in terms of the highest number of dwellings belonging to them, are the #9, #6, and #5 corresponding to 165, 138, and 102 sample households, respectively;
- from data collected by a survey, the available potential of flexibility related to the dwellings cluster has been calculated and it is equal to 538.95 MWh/year; therefore, the average daily value of flexible loads per dwelling is equal to 1966 Wh/d;
- by simulating a flexible strategy on an RC of Italian residential sector, which is based on the hourly pricing mechanism following the day-ahead market outcomes, and on limitations of power uptakes, monthly and annual indicators have been defined; so doing, the flexible strategy effectiveness can be computed to assess its actual suitability;

- the highest monthly effectiveness values have been registered in the cold season over the non-working days ranging between 0.49 and 0.53. Conversely, in the hot season, the maximum effectiveness values are generally lower compared to the winter ones (i.e., 0.3–0.4) and they occur over the weekdays. In the end, all those results can be outlined by means of a single indicator (annual effectiveness), which, in this case, is 0.34.

The calculated values relative to the management effectiveness indicate (at the outset) that the proposed management strategy for the Italian residential sector can be applied. Further developments of present work will be focused on identifying the realistic time distribution of the available flexible loads, and matching the user wellbeing and the appliances technical constraints, due to their contemporary use. In so doing, it will be possible to evaluate more precisely the flexible loads magnitude (which is expected to be lower than the one in the present case study) and to recalculate strategy effectiveness for evaluating actual suitability.

Moreover, the algorithm to identify the electricity peak price variation deriving from the DR strategies adoption has to be implemented. Similarly, a workflow definition associated to the Information and Communication Technologies (ICT) infrastructure has to be built in order to identify and select how many users can effectively apply the load time shifting commands. Finally, evolutive scenarios involving the heating system and DHW electrification, by means of heat pump wide deployment in the Italian residential sector, will be performed and analysed.

Author Contributions: F.M. contributed to this paper by the conceptualization, software implementation and validation; G.L.B. provided the formal analysis and was the writer of the final draft, along with the reviewing and editing process; J.C. was responsible for data collection and post-processing; L.d.S. and S.R. were the project scientific coordinators taking care of funding acquisition and project administration, and the general project supervisors, respectively. All authors have read and agreed to the published version of the manuscript.

Funding: This research received no external funding.

Acknowledgments: This work is part of wider research activity dealing with: "Technologies for efficient use and deployment of electric energy carrier". The project has been carried out in cooperation with ENEA—DTE (Italian National Agency for New Technologies, Energy and Sustainable Economic Development—Department of Energy Technologies) and CITERA (Sapienza University of Rome—Interdepartmental Research Centre for Territory, Construction, Restoration and Environment). The aforementioned institutions are gratefully acknowledged by the authors for their support and funding.

Conflicts of Interest: The authors declare no conflict of interest.

Appendix A

Table A1. Questionnaire structure.

Building location	**Kitchen**
• Province; Municipality	• Cooking plane; Oven; Microwaves oven (type, minutes per day switched-on)
Number of occupants in the dwelling over the day	• Grill; Steak grill pan/electric stove; Toaster Electric coffee maker for espresso; Electric coffee maker mocha; Blender; Food processor (minutes per day switched-on)
• From 8 a.m. to 1 p.m.; from 1 p.m. to 7 p.m.; from 7 p.m. to 12 p.m.; from 12 p.m. to 8 a.m.	**Refrigeration**
Architectural characteristics	• Refrigerator (type, capacity, energy class)
• Building Construction year	**Washing**
• Apartment dimensions and boundary surfaces Vertical Walls and roof external colour	• Washing machine; Tumble dryer; Dishwasher (capacity, weekly cycles, energy class)
• Shading	**Cleaning and ironing**
• Refurbishment actions on building (Walls; Roof; Ground; Windows)	• Vacuum cleaner; Electric broom (minutes per day switched-on)
Heating system	• Iron without water boiler; Iron with built-in water boiler (minutes per day switched-on)
• Centralised; autonomous	**Lighting**
• Heat source type (non-condensing boiler, condensing boiler, heat pump)	• Filament Lamps; Halogen Lamps; Fluorescent Lamps; LED Lamps (number)
• Control (on/of, climatic thermoregulation, chrono-thermostat)	**Audio/Video**
• Emission system (radiators, fan-coils, radiant floor)	• TV, monitor (size, quantity, energy class, hours per day switched-on)
Cooling system	• Decoder; Videorecorder; DVD reader; Radio, stereo; Hi-fi/home theatre (quantity, daily use in hours)
• Electric air conditioner (Energy Class; Number of served rooms)	• **Computer/Internet**
• Fans; Dehumidifiers (number of hours switched-on)	• Desktop PC; Notebook; Modem (quantity, daily use in hours)
Domestic Hot Water (DHW) plant	• Inkjet printer; Laser printer (quantity, copies per day)
• Non-condensing boiler; condensing boiler; heat pump water heater; electric water heater; storage device (yes/no)	**Personal Care**
Solar collectors	• Hairdryer; Hair straightener (daily use in hours)
• Flat solar collectors/Vacuum solar collectors (Number of modules; Slope; Orientation)	**Other equipment**
PV array	• Other equipment (quantity, electric power, daily use in minutes)
• Plant peak power (Slope; Orientation; Self-consumption) **Energy Bills**	
• Natural Gas; Electricity (Monthly consumption; Annual costs)	

Appendix B

Table A2. PUN Index thresholds limit for calculations over the weekdays.

Limit	Percentile	PUN [€/MWh]											
		Jan	Feb	Mar	Apr	May	June	Jul	Aug	Sept	Oct	Nov	Dec
1	75th	66.4	65.9	63.3	60.0	59.1	60.1	64.2	63.2	73.9	73.7	65.7	63.8
2	50th	61.2	59.3	56.6	54.6	54.9	57.0	60.5	59.3	66.5	66.7	61.7	59.4
3	50th	61.2	59.3	56.6	54.6	54.9	57.0	60.5	59.3	66.5	66.7	61.7	59.4
4	25th	52.6	52.6	49.3	48.4	49.8	52.0	56.4	56.1	59.5	56.4	52.0	50.4

Table A3. PUN Index thresholds limit for calculations over the Saturdays.

Limit	Percentile	PUN [€/MWh]											
		Jan	Feb	Mar	Apr	May	June	Jul	Aug	Sept	Oct	Nov	Dec
1	75th	63.2	59.8	54.8	53.6	55.1	52.7	58.1	58.8	62.2	65.2	57.0	55.5
2	50th	56.0	52.9	49.8	47.8	49.4	48.7	53.5	53.8	58.4	58.3	54.1	51.9
3	50th	56.0	52.9	49.8	47.8	49.4	48.7	53.5	53.8	58.4	58.3	54.1	51.9
4	25th	52.2	48.4	45.8	44.5	45.8	43.8	49.9	51.1	55.2	54.4	51.3	47.7

Table A4. PUN Index thresholds limit for calculations over the non-working days.

Limit	Percentile	PUN [€/MWh]											
		Jan	Feb	Mar	Apr	May	June	Jul	Aug	Sept	Oct	Nov	Dec
1	75th	55.9	51.7	52.0	49.1	49.5	50.0	53.1	56.2	56.7	54.8	52.7	52.5
2	50th	51.8	47.0	47.5	43.9	42.9	43.1	45.6	51.4	54.0	51.8	49.9	46.9
3	50th	51.8	47.0	47.5	43.9	42.9	43.1	45.6	51.4	54.0	51.8	49.9	46.9
4	25th	48.7	43.4	44.4	38.4	39.5	39.4	42.3	47.7	52.3	49.4	45.3	43.3

Table A5. Power thresholds limit for calculations over the weekdays.

Limit	Percentile	POWER [W]											
		Jan	Feb	Mar	Apr	May	June	Jul	Aug	Sept	Oct	Nov	Dec
1	75th	268	309	266	271	243	286	294	320	265	272	305	327
2	50th	219	252	212	245	219	220	202	254	216	228	248	259
3	50th	219	252	212	245	219	220	202	254	216	228	248	259
4	25th	175	223	190	213	191	204	169	206	201	212	228	239

Table A6. Power thresholds limit for calculations over the Saturdays.

Limit	Percentile	POWER [W]											
		Jan	Feb	Mar	Apr	May	June	Jul	Aug	Sept	Oct	Nov	Dec
1	75th	333	393	391	338	287	297	260	269	309	272	337	395
2	50th	269	323	297	253	265	248	197	229	282	249	276	310
3	50th	269	323	297	253	265	248	197	229	282	249	276	310
4	25th	133	186	212	167	180	198	171	187	173	201	231	222

Table A7. Power thresholds limit for calculations over the non-working days.

Limit	Percentile	POWER [W]											
		Jan	Feb	Mar	Apr	May	June	Jul	Aug	Sept	Oct	Nov	Dec
1	75th	348	391	354	349	290	322	245	278	345	278	399	400
2	50th	300	298	246	268	258	251	203	230	272	253	365	331
3	50th	300	298	246	268	258	251	203	230	272	253	365	331
4	25th	185	209	184	200	153	198	167	164	183	172	227	216

References

1. European Commission. A Clean Planet for all A European Strategic Long-Term Vision for a Prosperous, Modern, Competitive and Climate Neutral Economy. 2018. Available online: https://ec.europa.eu/clima/sites/clima/files/docs/pages/com_2018_733_en.pdf (accessed on 13 April 2019).
2. Brouwer, A.S.; Van Den Broek, M.; Seebregts, A.; Faaij, A. Impacts of large-scale Intermittent Renewable Energy Sources on electricity systems, and how these can be modeled. *Renew. Sustain. Energy Rev.* **2014**, *33*, 443–466. [CrossRef]
3. Palensky, P.; Dietrich, D. Demand side management: Demand response, intelligent energy systems, and smart loads. *IEEE Trans. Ind. Informatics* **2011**, *7*, 381–388. [CrossRef]
4. Strbac, G. Demand side management: Benefits and challenges. *Energy Policy* **2008**, *36*, 4419–4426. [CrossRef]
5. Behrangrad, M. A review of demand side management business models in the electricity market. *Renew. Sustain. Energy Rev.* **2015**, *47*, 270–283. [CrossRef]
6. Schibuola, L.; Scarpa, M.; Tambani, C. Demand response management by means of heat pumps controlled via real time pricing. *Energy Build.* **2015**, *90*, 15–28. [CrossRef]

7. Diesendorf, M.; Elliston, B. The feasibility of 100% renewable electricity systems: A response to critics. *Renew. Sustain. Energy Rev.* **2018**, *93*, 318–330. [CrossRef]

8. Zappa, W.; Junginger, M.; van den Broek, M. Is a 100% renewable European power system feasible by 2050? *Appl. Energy.* **2019**, *233*, 1027–1050. [CrossRef]

9. Eurostat. Statistics | Eurostat, (n.d.). Available online: https://ec.europa.eu/eurostat/databrowser/view/ten00124/default/table?lang=en (accessed on 31 March 2020).

10. Energy Efficiency Trends & Policies | ODYSSEE-MURE, (n.d.). Available online: https://www.odyssee-mure.eu/ (accessed on 20 May 2020).

11. Jensen, S.Ø.; Marszal-Pomianowska, A.; Lollini, R.; Pasut, W.; Knotzer, A.; Engelmann, P.; Stafford, A.; Reynders, G. IEA EBC Annex 67 Energy Flexible Buildings. *Energy Build.* **2017**, *155*, 25–34. [CrossRef]

12. Rahmani-Andebili, M. Scheduling deferrable appliances and energy resources of a smart home applying multi-time scale stochastic model predictive control. *Sustain. Cities Soc.* **2017**, *32*, 338–347. [CrossRef]

13. Chen, Y.; Xu, P.; Gu, J.; Schmidt, F.; Li, W. Measures to improve energy demand flexibility in buildings for demand response (DR): A review. *Energy Build.* **2018**, *177*, 125–139. [CrossRef]

14. Cumo, F.; Curreli, F.R.; Pennacchia, E.; Piras, G.; Roversi, R. Enhancing the urban quality of life: A case study of a coastal city in the metropolitan area of Rome. *WIT Trans. Built Environ.* **2017**, *170*, 127–137. [CrossRef]

15. Péan, T.; Costa-Castelló, R.; Salom, J. Price and carbon-based energy flexibility of residential heating and cooling loads using model predictive control. *Sustain. Cities Soc.* **2019**, *50*, 101579. [CrossRef]

16. Vázquez, F.V.; Koponen, J.; Ruuskanen, V.; Bajamundi, C.; Kosonen, A.; Simell, P.; Ahola, J.; Frilund, C.; Elfving, J.; Reinikainen, M.; et al. Power-to-X technology using renewable electricity and carbon dioxide from ambient air: SOLETAIR proof-of-concept and improved process concept. *J. CO2 Util.* **2018**, *28*, 235–246. [CrossRef]

17. Irena. Innovation Landscape for a Renewable-Powered Future: Solutions to Integrate Variable Renewables. 2019. Available online: www.irena.org/publications (accessed on 25 June 2020).

18. The European Parliament and the Council of the European Union, Directive 2014/94/eu of the European Parliament and of the Council—of 22 October 2014—on the Deployment of Alternative Fuels Infrastructure. Available online: https://eur-lex.europa.eu/legal-content/EN/TXT/PDF/?uri=CELEX:32014L0094&from=en (accessed on 30 June 2020).

19. De Santoli, L.; Basso, G.L.; Garcia, D.A.; Piras, G.; Spiridigliozzi, G. Dynamic simulation model of trans-critical carbon dioxide heat pump application for boosting low temperature distribution networks in dwellings. *Energies* **2019**, *12*, 484. [CrossRef]

20. Mazzoni, S.; Ooi, S.; Nastasi, B.; Romagnoli, A. Energy storage technologies as techno-economic parameters for master-planning and optimal dispatch in smart multi energy systems. *Appl. Energy.* **2019**, *254*, 113682. [CrossRef]

21. Nastasi, B. Hydrogen policy, market, and R&D projects. In *Solar Hydrogen Production: Processes, Systems and Technologies*; Calise, F., D'Accadia, M.D., Santarelli, M., Lanzini, A., Ferrero, D., Eds.; Academic Press: Cambridge, MA, USA, 2019; pp. 31–44. [CrossRef]

22. Nastasi, B.; Lo Basso, G.; Astiaso Garcia, D.; Cumo, F.; de Santoli, L. Power-to-gas leverage effect on power-to-heat application for urban renewable thermal energy systems. *Int. J. Hydrog. Energy.* **2018**, *43*, 23076–23090. [CrossRef]

23. Roversi, R.; Cumo, F.; D'Angelo, A.; Pennacchia, E.; Piras, G. Feasibility of municipal waste reuse for building envelopes for near zero-energy buildings. *WIT Trans. Ecol. Environ.* **2017**, *224*, 115–125. [CrossRef]

24. Lezama, F.; Faia, R.; Faria, P.; Vale, Z. Demand Response of Residential Houses Equipped with PV-Battery Systems: An Application Study Using Evolutionary Algorithms. *Energies* **2020**, *13*, 2466. [CrossRef]

25. D'Ettorre, F.; De Rosa, M.; Conti, P.; Testi, D.; Finn, D. Mapping the energy flexibility potential of single buildings equipped with optimally-controlled heat pump, gas boilers and thermal storage. *Sustain. Cities Soc.* **2019**, *50*, 101689. [CrossRef]

26. Goy, S.; Finn, D. Estimating demand response potential in building clusters. *Energy Procedia* **2015**, *78*, 3391–3396. [CrossRef]

27. Adhikari, R.; Pipattanasomporn, M.; Rahman, S. An algorithm for optimal management of aggregated HVAC power demand using smart thermostats. *Appl. Energy.* **2018**, *217*, 166–177. [CrossRef]

28. Cai, H.; Ziras, C.; You, S.; Li, R.; Honoré, K.; Bindner, H.W. Demand side management in urban district heating networks. *Appl. Energy.* **2018**, *230*, 506–518. [CrossRef]

Article

Building-Integrated Photovoltaics (BIPV) in Historical Buildings: Opportunities and Constraints

Flavio Rosa 

CITERA, Interdepartmental Centre for Territory, Building, Conservation and Environment, Sapienza University of Rome, 00137 Rome, Italy; flavio.rosa@uniroma1.it

Received: 13 June 2020; Accepted: 4 July 2020; Published: 14 July 2020

Abstract: In this work, we investigate the potential of using last generation photovoltaic systems in traditional building components of historical buildings. The multifunctional photovoltaic components also open new application and implementation horizons in the field of energy retrofitting in historical buildings. Some of the Building-Integrated Photovoltaics (BIPV) solutions lend themselves optimally to solving the problems of energy efficiency in historical buildings. For the next few years, Italian legislation foresees increasing percentages of energy production from renewable sources, including historical buildings. The opportunities and constraints analysed are presented through a specific approach, typical of building processes for innovative technological BIPV solutions on historical buildings.

Keywords: building-integrated photovoltaics—BIPV; building heritage; energy efficiency; traditional materials

1. Introduction

The heritage building stock of which Italy is particularly rich must be studied with an innovative approach for acceptable energy retrofitting solutions. In historical buildings, energy efficiency requirements, in synergy with conservation and protection, show methodological and operational limits that cannot be generalized.

Decarbonization policies framed by the European Union (EU) in the emissions reduction roadmap involve all the economic sectors, such as the civil and construction sector that includes heritage building stock [1]. The need to reduce the incidence of emissions in this sector has stimulated European legislators and international research centres to reduce energy requirements using nearly Zero Energy Building (nZEB) solutions [2–7].

The International Energy Agency (IEA) attributes over 40% savings in expected heating and cooling energy demands under a low-carbon scenario directly to improvements in the building envelope [8].

Driven by policies towards Zero-Energy Buildings and subsequently Plus Energy Buildings (PEB), design and innovation with new Building-Integrated Photovoltaic (BIPV) materials, concepts and combinations of energy-efficient building materials, with BIPV have become essential parts of the development strategies of both, the photovoltaic (PV) sector and the building sector [9–11].

The Strategic Energy Technology Plan (SET Plan) defines renewable technologies as being at the heart of the new energy system, with photovoltaic solar energy as the main pillar [12].

The heritage building stock can provide a contribution to decarbonization policies using Renewable Energy solutions (RES) installed on the building envelope with BIPV appropriate solutions. Most of the Italian heritage building stock does not present stringent protection restrictions as required by current regulations [13]. They can be classified as Traditional Historical Buildings (THB) that are not directly included in the Maintenance and Restoration category of the Code of Cultural and Landscape

Heritage [14]. The historical importance of the building is herein regarded in the function of the specific Italian post-unification period (1871–1942). Urban design, building types, construction techniques and the technology of the material used are not included in any category of particular interest or value to guarantee its safeguarding as foreseen by Cultural Heritage codes. In Rome, over 50% of the existing building stock falls under THB. In Europe, architectural and planning regulations for protected historical buildings lead to major technical constraints in integrating renewable energy, such as photovoltaics. These problems call for innovative and creative solutions for BIPV that must apply both aesthetic and photovoltaic technology to historical buildings that represent the artistic and cultural heritage of a city.

The Energy Performance of Buildings Directive (EPBD) prescribes the Member States to increase the number of buildings that not only fulfil current minimum energy performance requirements, but are also more energy-efficient, thereby, reducing both energy consumption and carbon dioxide emissions [15,16].

In Italy, Ministerial Decree 26 June 2015 completes the transposition of the European EPBD 2002/91/EC. This legislative measure comprises: (a) the application of calculation methodologies for energy performance and the definition of the minimum requirements for buildings; (b) requirements of nearly zero energy buildings; (c) sets the new minimum Energy Performance EP; (d) defines single component requirements to enter into force starting October 2015 [17].

Historical buildings represent a good reference paradigm by which we can investigate energy retrofit approaches and methodologies using BIPV solutions [18].

It is more difficult to apply an adequate retrofitting scheme on historic and/or listed public buildings, particularly in Italy, because there are strict regulations on the changes of this type of building. In the case of Rome, solar energy harvesting in Italian urban scenarios were analysed, by considering the geographical and morphological constraints with respect to the Sustainable Energy Action Plans (SEAP) [19].

The production of energy with solar systems as well as depending on local radiation is highly conditioned by the solar conversion technology adopted and the orientation of the modules. These factors are even more stringent on THB, as not all optimal exposures can be used for mounting modules due to the magnitude of impact. The THBs considered in this work can be assimilated to the provisions of the Annex 3A-B for the Built Heritage or Historic Urban Landscape of the International Council on Monuments and Sites ICOMOS Guidelines, where a grading scale for assessing the value and magnitude of the impact of Heritage assets is proposed. ICOMOS Guidance on Heritage Impact Assessments for Cultural World Heritage Properties [20,21].

In relation to the public heritage building stocks, it is possible to identify thresholds beyond which nZEB targeted non-reachability allows for a more defined vision of the issues, and to implement specific retrofitting strategies [22].

The requirements in the improvement of energy efficiency in cultural heritage buildings using active solar systems in historical constructions establishes the use of the general principles of restoration, including the reversibility and non-invasiveness of interventions on historical structures [23]. This approach can also be reworked for interventions on THBs, as illustrated below, starting from a holistic point of view and by examining the relationship between energy efficiency and preservation of the THB. This approach can also be reworked for interventions on THBs as illustrated below. PV technology in architecture has two types of solutions: Building Attached Photovoltaics (BAPV), in which the element is mounted on the casing using various techniques; and BIPV modules that form a building component and provide a function as described in EU regulation 305/2011, that defines the seven basic requirements for construction that have to be fulfilled, in addition to the electrical requirements [24]. PV Active Solar components are also defined as multifunctional as they must simultaneously perform functions so that the active architectural element of the building system produces and distributes energy [25–27]. In addition to being a source of electricity, several other purposes can be achieved, such as weather protection, thermal insulation, noise reduction and daylight modulation.

In recent years, BIPV applications in Europe have been applied to one-third of all renovation projects and two-thirds of new building construction [28]. The photovoltaic productivity on roofs and façades in Europe was estimated at around 1 TWp by 2030 [29]. For integrated solutions like BIPV, it has been calculated that Italy can reach 40% of the national electricity needs [30–32]. BIPV is a solution that can be adapted to the historical building stock to satisfy solutions calibrated on these particular buildings. When changes and adaptations to sustainability standards are proposed, even the slightest alterations, particularly external, can be damaging [33]. BIPV Solutions should be assessed as opportunities in the energy efficiency processes of historical buildings. At the same time, constraints must be assessed to mitigate the risks of impact on architectural and landscape heritage of merit and value.

The major problem of BIPV on THB solutions is the visual impact determined by the color of the photovoltaic cells that affect the level of the overall insertion on valuable casings and traditional materials. The color of a solar module is determined by the color of the cells in the module. [34]. Layering techniques to camouflage the PV element and solar damage to traditional materials, such as brick, tile and plaster developed in nanoscale, are in an advanced phase of the study and solutions are already on the market [35].

BIPV applications with high integration and almost no visual impact are already on the market [36]. Solutions for the treatment of glazed surfaces, Window-Integrated Photovoltaics (WIPV), with nanotechnologies are also very interesting [37]. The colored glass is selective, designed to reflect a narrow spectral band of visible light to provide color. The rest of the solar spectrum is transmitted to the solar device and converted into energy.

This duality in a historical building clearly influences the evaluation of materials and technologies, restricting solutions first and foremost to the aesthetic factor. The visual impact component conditions the materiality of the element that is firmly correlated to the materials with which they are made.

In this work, we want to investigate the opportunities and constraints of BIPV solutions in historical buildings, evaluate the potential applications in the regeneration of historical and/or traditional building heritage, without detriment to conservation and protection. Three macro-factors, technologies, market, and innovative solar-integrated solutions that describe state-of-the-art BIPV are analyzed beforehand. A final evaluation scheme of the opportunities and constraints of BIPV solutions on historical building enclosures is then proposed.

Then, a synthesis approach is defined to meet the needs of the energy retrofit, the protection and preservation of the historical building stock and the BIPV solutions, named Heritage Building Energy Solar Solution Technologies (hBESST).

The typical structuring modalities of the BIM-based building design process [38–40] and the correlations with BIPV solutions on historical buildings will be briefly analyzed.

2. Materials and Methods

2.1. Historical Buildings and Traditional Heritage Buildings (THB) Retrofitting

The relation between sustainability and heritage cannot be reduced to the mere energy efficiency of the buildings, simplifying a complex problem into an exclusive element of energy savings [41].

In Italy, over 7 million buildings are over 50 years old, which is equal to 61% of the building stock. From these, over 2 million, or 18%, are in a state of conservation ranging from mediocre to bad. The planned objectives of greenhouse gas (GHG) reduction in the construction sector, introduced by international and national regulations, must be compared to find economically sustainable solutions for each type of user.

The guidelines for improving energy efficiency in cultural heritage indicate that no solution can be considered in itself decisive and that the perspective, in which to move, can only be that which pragmatically proceeds on a case-by-case basis. The current approach is to reduce visual impact, using shields which obstruct the view of the modules to obtain a low magnitude of impact [21,42]. However, the approach cannot be simply one of concealment. Layering technologies allow overlapping layers, of which the last one is visible, made with finishes similar to historical and traditional building materials that hide the underlying photovoltaic element [43].

In principle, multi-layering techniques modify the appearance of a solar cell through the variation of the Anti-Reflection Coatings ARC [44]. The placement of stacks with large numbers of layers is a challenge, especially in a high-throughput, low-cost production environment [34].

The processes used in manufacturing the new transparent PVs are environmentally friendly and not energy-intensive. The coatings are placed at nearly room temperature so the transparent PV can be laid on essentially any type of surface [45]. There is no need to use glass, which is costly in the manufacturing of conventional systems [46].

All Italian historical building stock, bound and not, will be increasingly subject to interventions of energy retrofitting in the coming years. In Italy, Legislative amendment 28 of 2011 in the implementation of Directive 20-20-20 (2009/28/EC) introduced the requirements for renewables in buildings (RES) in the event of major renovations. Regarding buildings that fall into the category of historical centres, as defined by the legislation [13], the percentage of RES coverage is expected to be 25% starting on 1 January 2018. It can also be obtained through technological combinations.

In the building sector, photovoltaic technology has taken over as a low-level integration technology element, with installations, usually adhering to the enclosure, known as Building Attached Photovoltaic (BAPV) [47,48].

The problems of technology and protection integration are addressed as follows: "doing something about historical buildings with energy efficiency measures (obviously compatible with the cultural characteristics of the artefacts) is the first significant step for the real conservation of said heritage, so widespread, so fragile, so difficult and expensive to preserve [49]." In historical buildings, listed and not, the installation of photovoltaic systems and components has always had to deal with the visual impact on traditional materials and urban and landscape contexts.

The production of electricity from photovoltaic sources is one of the technological solutions that can be integrated into the building with plug and play solutions. In geographical areas, such as Southern Europe, the annual productivity in the face of well-exposed available surfaces allows for the coverage of high levels of electricity requirements: see Figure 1.

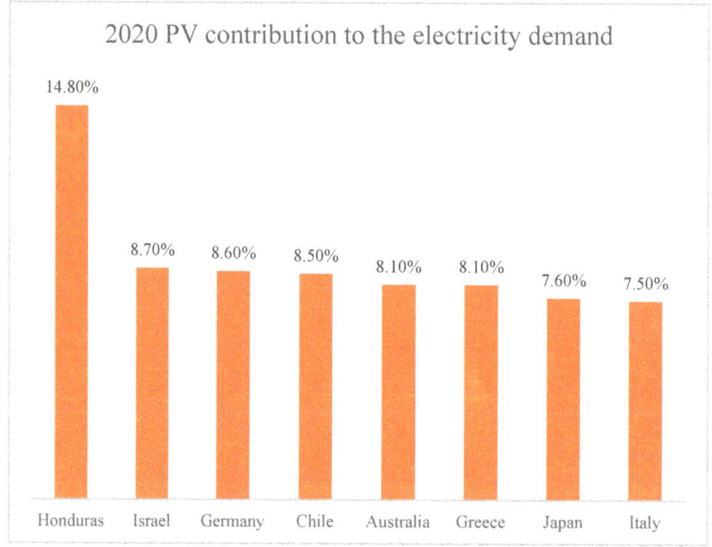

Figure 1. Theoretical PV production by nation (Source: 2018 snapshot of Global Markets—IEA PVPS).

Energy THB retrofitting with BIPV solutions has been an important subject in an R and D laboratory of ideas, which are developed to create solutions for the support and guardianship of the architectural property [50–53].

The increase in the production of electricity from solar sources is constantly on the rise thanks to progressively improved high-performance materials and technologies [54,55]. The active surfaces on the building envelopes, using efficient materials and technologies, will allow for an increase in the density of electricity from solar sources to single buildings, and in the single exposed surface of the casing. The energy density factor in BIPV is decisive in the choice of photovoltaic technologies to optimize the available surfaces concerning the production of electricity, as shown in Figure 2 [56–58].

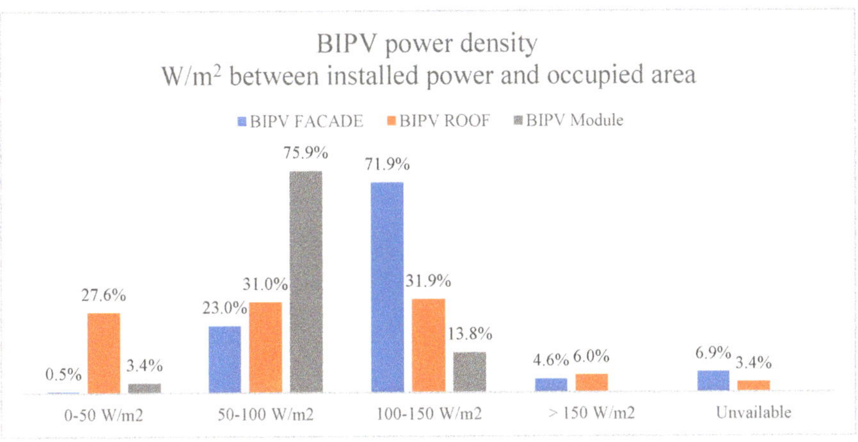

Figure 2. Power density (W/m²) BIPV modules by the installation.

The use of integrated low visual impact solutions and the increased availability of electricity from RES will permit the use of the primary energy from the sun for the production of electricity with H2G (Hydrogen to Gas) hybrid solutions [59,60] and the possibility of achieving the objectives set for nZEBs.

The energy response (using a network of active and passive systems and technologies) from a historical building (or a "cultural landscape", using due caution in terminology), can be improved through appropriate and well-balanced solutions [61].

Plant components have become an increasingly invasive part of the building-plant system since the Second World War [62,63]. Moreover, historical buildings have undergone the highest number of (sometimes damaging) alterations due to the inclusion of new plant solutions over the years as per Figure 3.

Technological infrastructures require technical spaces such as cavedium, traces, and niches that are often obtained without preliminary studies on stone and/or masonry bearing structures. Historical heritage in real estate, as buildings, must always be able to testify to their historical importance through all the material components that constitute it. Conservation can be implemented to all the activities and work carried out to control the conditions of the cultural property and maintain the integrity, functional efficiency and identity of the property and its parts [13] through appropriate maintenance methods.

With a view to an integrated process between HB retrofitting and solar energy solutions like BIPV, Heritage Building Information Modeling (HBIM) can offer solutions that allow framing information flows and critical factors. The application of BIM design process to existing buildings it will become mandatory in the coming years for all types of design especially in regards to maintenance and large refurbishments [40,64,65], but these applications do not contemplate the historical and cultural legacy of the buildings and sites [66].

The BIM approach is a 'workflow', in which all of the building objects that combine to make up the building design coexist in a single database. This concept is important because it allows us to understand the entire building lifecycle (encompassing design, build and operation) from a single, central data storage unit. Therefore, in theory, a BIM implementation should facilitate a single, logical, consistent source of information associated with the building.

The definition of Level of Definition (LOD) within the workflow aimed at the HBIM, and specifically the PV components, is the first methodological step to be implemented. One of the first objectives in developing a BIM model of infrastructure is to define its LOD. Design phase outputs require a very detailed definition of project components to enable the user to extract automatic configurations of element quantities and related costs, and also to generate two-dimensional (2D) technical drawings [67].

Figure 3. Implementation of somewhat controversial photovoltaic systems on heritage buildings.

There are still gaps in the know-how that forestall insights into the future development of methods and tools of Historical Building Information Modelling for refurbishment projects. Therefore, it will prevent a complete automated diagnosis of the residual performances and designs of energy retrofitting using BIPV solutions.

In this work, we present a series of fundamental elements for energy retrofitting interventions on THBs that are strongly integrated with each other. The process presented requires active involvement of the stakeholders. The innovations in the design process of the BIM-HBIM is transformation in all this phases throughout the entire process: Design, management, maintenance, decommissioning. The multi- or trans-disciplinarily that governs energy retrofitting is even wider in the case of Built heritage or Historic Urban Landscape.

For the above mentioned, and, this paper proposes a holistic approach to the entire design process starting from the assessment of the conservative impact in relation to BIPV energy efficiency solutions on Built Heritage. This approach is presented with a preliminary integrated assessment of the benefits for energy efficiency, heritage impact and intervention costs of, highlighting opportunities and constraints.

2.2. BIPV Standardization

The concept of a multi-functionality for a photovoltaic module poses a problem shared definition between technological or construction component.

It is important to remember that any use of PV modules must adhere to the specific standards in force in the country of use. The European standard EN 50583 for BIPV that applies to photovoltaic modules used as construction products, was published in 2016. This new standard consists mainly of the compilation and modification of existing standards related to BIPV. To be suitable for building integration, PV products must fulfil both the standards of the PV sector and the construction sector as presented in Table 1 [11,68].

Table 1. Electrical and building reference standards for PV modules [69].

PV Module Standards	Building Standards
IEC International Electrotechnical Commission	**ISO** International Organization for Standardization
CENELEC European Commission for Electrotechnical Standardization	**CEN** European Committee for Standardization
CES Comité Electrotechnique Suisse	**SIA** Schweizerische Ingenieur- und Architekten-Verein

The scope of EN 50583 is to implement technical requirements for photovoltaic modules used as building components that are subjected to both electrical standards; the Low Voltage Directives 2006/95/EC and IEC/CENELEC, and the European Construction Product Regulation 305/2011 [70].

The standard is divided into two parts: (1) Photovoltaics in buildings: Panels, two-part systems, based on three levels of differentiation as reported in Table 2:

Table 2. Levels of differentiations EN 50583.

Level 1	General requirements for all BIPV resulting from requirements of the Low Voltage Directives and the Construction Products Directive of the European Union
Level 2	Requirements resulting from panel material (e.g., glass)
Level 3	Requirements resulting from panel mounting location within the building (5 mounting categories are differentiated)

Until the norm EN 50583-1/2 "Photovoltaics in Buildings" was published, BIPV modules and systems were not regulated through any unified European standard. PV panels mounted on the building envelope in either, one of the five mounting categories, stated in the EN 50583 standard, were thus treated both as an electrical component and a building product. However, the EN 50583 standard does not focus largely on BIPV modules. They are mentioned briefly in part 1 with regards to categorization dependent on the material (glass/polymer/metal sheet/other), and in part 2, with regards to categorization dependent on the installation method.

To be suitable for building integration, PV products have to fulfil both the PV sector and construction sector standards. Most of the time, the requirements for these standards are related to a country or a region [68].

In Italy, a definition of BIPV is given by the GSE Gestore Servizi Energetici (Energy Service Management). The photovoltaic model that is possible and effective for applications of an architectural type alone, is the building element itself [71]. This document extends particular relevance to the concept of integration for which the photovoltaic surface, together with the assembly system (in the case of a special component), replaces traditional building elements, and in addition to the production of electricity, guarantees the following functions typical of a building envelope: Water tightness, mechanical seal comparable with that of the replaced building element, and thermal resistance that does not compromise the performance of the building casing.

Building-Integrated Photovoltaics is classified as a new building system technology drive as established in the UNI 8290 standard [25]. The problem of the classification of photovoltaic models to be supplemented in the covering of a building has also been described from a fire protection perspective. This has highlighted a normative gap in all PV systems not recognized as construction products. This was dealt with in EN 15583.

A shared definition of the solutions for the integration of photovoltaic technologies on the building envelope are the following:

- BAPV: Photovoltaic modules are considered to be building attached if the PV modules are mounted on a building envelope and do not fulfil the above criteria for building integration.
- BIPV: Building-Integrated Photovoltaics modules are considered to be building-integrated if the PV modules from a construction product providing a function as defined in the European Construction Product Regulation CPR 305/2011. Thus, the BIPV module is a prerequisite for the integrity of building functionality. If the integrated PV module is dismounted (in the case of structurally bonded modules, dismounting includes the adjacent construction product), the PV modules would have to be replaced by an appropriate construction product.

The action of harmonizing CENELEC and ISO regulations is the subject of a study conducted by the IEA Pvps Task 15 to subtask C to achieve:

- The international definition of BIPV
- BIPV needs and functions analysis
- BIPV requirements overview
- Multifunctional BIPV evaluation
- Suggested topics for exchange between different standardization activities on an international level.

The work on BIPV standardization has gained new impetus due to the decision by IEC/TC 82 to create a new Project Team (PT 63092) to prepare an international BIPV standard [72].

Building-Integrated Photovoltaics on historical buildings is an approach that must take into account the peculiarities of traditional materials and visual impact in a logic of high integration and reversibility.

2.3. Three Macro Factors for Building-Integrated Photovoltaics

The reduction of consumption and the consequent lowering of the energy requirement is chosen based on high-efficiency plant elements.

The adoption of multifunctional BIPV innovative solutions and colored active PV components with high energy efficiency will allow remedying, or at least mitigate, the historical conflict between technological systems and the safeguarding and protection of the Architectural historical building.

Three factors determine favorable conditions for the development of BIPV as illustrated in Table 3: Technology, market and innovative solar-integrated technologies. The first two can be measured and evaluated directly. The third factor must be assessed by less deterministic formalities through a Building Management Approach for Heritage (BMAH). The first two are exogenous factors independent of the historical importance or otherwise the building and materials with which it was built. The third one must be analyzed with all its components. This analysis can be conducted through a case-by-case approach, but with a holistic methodology that respects the valorization and historical building energy retrofit.

Table 3. Macro factors of BIPV on historical buildings.

PV Technologies	Market	Innovative Solar-Integrated Technologies
PV efficiency	LCOE	Multi-functionality
Load/generation management	Grid Parity	Energy Efficiency
Smart Buildings synergies	ROI/EROI	Colored and Glazed cells
Smart Grid focused	Energy Pay Back Time EPBT	Light selective compound for traditional materials (multi-layering)
Low visual impact		

In constrained or THB buildings, BIPV is the bottom-up approach. The building process is based on inductive reasoning: from specific to universal, through analyses and solutions that must be adopted using an innovative methodology.

The three factors listed above are analyzed individually, highlighting their significance in the case of BIPV on heritage buildings.

2.3.1. Photovoltaic Technologies

Research in the PV field still focuses greatly on silicon-based materials. Since no single technology, either established or in development, offers benefits on all fronts, researchers recommend scaling up current silicon-based systems quickly. While, continuing to work on other technologies to increase efficiency, decrease materials used, and reduce the manufacturing complexity and cost [45]. PV classification is arranged by generations on the grounds of the evolution of the technologies [73]:

- I Generation: Silicon-based solar cells (mono- and poly-crystalline silicon) constituted the first PV sector to emerge. Currently, crystalline silicon technologies account for more than 97% of the overall cell production and more than 94% in the IEA PVPS countries [8].
- II Generation: Thin-film solar cells based on CdTe, copper indium gallium selenide (CIGS), or amorphous silicon was developed as a cheaper alternative to crystalline silicon cells.
- III generation' solar cells (tandem, perovskite, dye-sensitized, organic, new concepts, …) account for a broad spectrum of concepts, ranging from low-cost low-efficiency systems (dye-sensitized, organic solar cells) to high-cost high-efficiency systems (III–V multifunction), with various purposes from building integration to space applications [8,59,74–76].

Emerging technologies encompass advanced thin films and organic solar cells (OSC). The latter is about to enter the market through niche applications.

- Concentrator technologies (CPV) use an optical concentrator system, which focuses solar radiation onto a small high-efficiency cell.
- Multi-junction cells design involves superposing several cells in a stack [75].
- Novel PV concepts aim at achieving ultra-high efficiency solar cells using advanced materials and new conversion concepts and processes [8]. The flexibility and light weight make them useful for nomad applications, while the possibility to tune the color, the shape, and the transparency opens the route of the integration in modern and esthetic features [73,76].
- Glazing integrated photovoltaics based on dye-sensitized solar cells are in a relatively early stage of development. These solutions are very interesting for installations on glazed surfaces for any type of building, historical or otherwise, as they preserve the aesthetic appearance of the enclosure. The colour scale available offers designers greater choice [77].

Table 4 shows the targets foreseen by the Technology Roadmap for photovoltaics up to 2050.

Table 4. General technology target (Source: Technology Roadmap, solar photovoltaic energy, IEA 2010).

Targets	2008	2020	2030	2050
Typical flat-plate module efficiencies	<16%	<23%	<25%	<40%
Pay-back time in 1500 kWh/kWp regime	2 years	1 year	0.75 year	0.5 year
Durability	25 years	30 years	35 years	40 years

For the next few years, Crystalline Technologies will still represent a reference point, given their significant presence of maturity and technology on the market.

The main objectives of research on photovoltaic technologies are:

- Increase module efficiencies
- Reduce Si consumption
- Develop specific PV materials/solutions for building integration.
- Low energy-intensive production processes
- Readily available raw materials

Many studies are addressing the need to synthesize BIPV solutions with nZEB references, highlighting that each new intervention, including energy efficiency, has to maintain the values mentioned, and further achieve suitable landscape integration within the urban context [7,78–81].

The availability of primary energy from PV provides a fundamental contribution to the net-zero energy balance that can be determined, either from the balance between delivered and exported energy or between load and generation. Import and export balance and the latter load/generation balance [82,83].

National strategies towards climate-neutral buildings have to reflect the (future) climate, the building standard and energy system as well as the associated (future) energy grid infrastructure. Individual strategies differ depending on the climate, the resources for renewable power in the grid and the heating and cooling grid infrastructures but they must keep in mind development towards a more networked European energy grid in the future (smart European grid) [4].

The visual impact of BIPV is a problem that encounters more difficulties in the solution due to the synthesis of issues to be addressed within the photovoltaic architecture of valuable envelopes. In the present work we provide a multidisciplinary approach that is considered unavoidable as a result of the convergence of issues regarding antithetical appearance.

2.3.2. The Market

The cost of photovoltaic technologies is constantly changing. By 2025, the global weighted average levelized cost of electricity (LCOE) of solar photovoltaics (PV) could fall as much as 59% [55]. Moreover, together with the achievement of Grid Parity in Italy [78], they are a valid driver for the energy efficiency of the historical building stock [78,79,84,85].

Building-Integrated Photovoltaics is seen as one of the five major tracks for large market penetration of PV, in addition to price decrease, efficiency improvement, lifespan, and electricity storage [86].

The increase is justified through the rapidly plunging installation cost per watt; enhanced aesthetics of BIPV; improving the efficiency of c-Si modules as well as flexible thin-film panels, and the unabated desire among residential and commercial owners to "go green" and to reach the national energy efficiency target [87].

Building-Integrated Photovoltaics is still too costly, especially when compared to its rival technology, building-added PV (BAPV) since its added value as a multifunctional building element is only now beginning to be recognized [88].

It is expected that the BIPV/BAPV prize will induce the integration of new photovoltaic systems in protected historical urban districts by accelerating innovations in photovoltaic technologies and improving the architectural enhancement needed for sustainable use in European protected historical urban districts [89].

The growth opportunities for the BIPV market analyzed by the NREL [90–92] for the residential sector, are among the factors to be enhanced in the future through government support in maintaining historical/cultural building designs and through incentives.

The levelized cost of energy (LCOE) from PV systems is already below retail electricity prices (per-kWh charge) in several countries [93]. Innovations in the solar technologies field, with more and more politically efficient and appearance incentives, have introduced the Grid Parity including an LCOE between 200-300 USD/MWh in Italy.

LCOE of PV and BIPV are different. From 2010 to 2018 the BAPV LCOE in Italy suffered an 80% drop [94].

At the basis of this reduction is obviously 26% to 32% the drop-in prices of crystalline silicon modules. BIPV lowering of the LCOE is related to the reduction of costs in project management, due to the strong integration of BIPV in the building project, design and construction and performance cost reductions of technological components. For example, some savings derive from the elimination of the cost of BIPV mounting hardware.

The Energy Pay Back Time (EPBT) is the time it takes for the PV system to generate as much energy as has been used to produce it. High energy return on energy investment (EROI) corresponds to a short EPBT as reported in Table 5.

Table 5. Mean harmonized EPBT and EROI for multiple insulations in the Italian area.

Insolation (kWh/m^2/yr)	Italy Average 1700	Centre Italy 1436	South Italy 2032
Mono-Si			
Mean EPBT (years)	4.11	**5.22**	2.90
Mean EROI	8.73	11.2	**4.9**
Poly-Si			
Mean EPBT (years)	3.06	3.89	2.16
Mean EROI	11.62	15.0	6.5
a:Si			
Mean EPBT (years)	2.28	2.90	1.61
Mean EROI	14.45	18.6	8.0
CdTe			
Mean EPBT (years)	1.02	1.29	**0.72**
Mean EROI	34.18	**44.0**	19.0
CIGS			
Mean EPBT (years)	1.73	2.20	1.23
Mean EROI	19.94	25.7	11.1

An EPBT of one year and a life expectancy of 30 years corresponds to EROI of 30:1 [8]. EROI is simply a measure of the marginal amount of additional energy that certain technologies can provide in society for a given energy investment [95,96]. In the following table, the results of research, conducted in the systematic review, are re-established and meta-analysis of embedded energy, energy payback time and energy return on energy invested for the crystalline silicon and thin-film photovoltaic systems [97]. The results highlight, depends primarily on their embedded energy and not their efficiency.

The following table, elaborated by the author on data from Defne et al. [97,98], shows the relative results given the mean harmonized EPBT values for the following insolation (kWh/m^2/yr): 1700 (Italian average), 1436 (Central Italy), and 2032 (Sicily in Southern Italy) for roof mounted PV.

A PV system with a multicrystalline module in Sicily has an EPBT of around one year. Considering a lifespan of 20 years this system can produce twenty times the energy needed to produce it [99,100].

2.3.3. Innovative Solar-Integrated Technologies

Low impact integration on traditional elements of historical buildings, following the innovations on the single photovoltaic cells, opens new perspectives with important repercussions for the valorization and protection of the historical building heritage.

The innovative BIPV solutions rely on the multifunctional qualities that determine the level of integration on building envelopes, be they modern or historical [86,91,92,99,100]. The two factors define a specific "Photovoltaic architecture" approach as proposed by Antec Solar in Figure 4.

Photovoltaic architecture is a discipline called to find innovative, energy sustainable solutions with a low magnitude impact on sensitive construction like THB. They demand more and more about the BIPV components to enable designers to develop appropriate design and effective integration [8]. The multi-functionality of the design must provide BIPV and THB solutions considering (1) Heritage Building envelope characteristics; (2) function and performance; (3) product customization.

The various retrofitting levels on buildings testify to the cultural, historical and aesthetic values that need to be implemented through an appropriate workflow of knowledge, information and solutions. In this case, the BIM/BIPV approach on historical buildings and/or THB represents an innovative design and management solution for the entire workflow [101–106]. This analysis is not detailed in this paper and will be addressed in the second phase of research. The main components that determine the innovations and integration of BIPV solutions on historical buildings are shown below.

Figure 4. Multi-color solar wall Antec Solar headquarters in Arnstad Germany (source: [37]).

2.3.4. Multifunctional PV Solutions

If BIPV solutions are designed and installed to provide multifunctional solutions like solar components with low impact on traditional materials on THB envelopes. They must also guarantee results and performances in the area of valorization and conservation of cultural architectural heritage.

The steep increase in the levels of production of innovative materials opens new avenues to materials, technologies and solutions in the Construction Management Process (CMP) even on historical buildings. The field of nanotechnologies seems to be the most promising, since it is expected that active photovoltaic supports will reach the thickness of paint [45,107–109].

However, the BIPV concept raises the likelihood that a thin layer of PV-active material, possibly lain as a paint, could become a standard feature of building elements, such as roofing tiles, façade materials, glass, and windows, just as double-glazed windows have become standard in most countries [8].

There are photovoltaic films already on the market, which allow the glazed components of the enclosures to be active solar surfaces [110]. Power generation, through window coatings, is a relatively new idea, and is based on the use of semi-transparent solar cells as windows [77]. Transparent PV additive technology has a limited mechanical flexibility is a high cost of the modules [111]. Flexible or light-weight PV modules are also part of R and D efforts. BIPV applications are expected to become a major market for flexible PV modules [55]. In the area of walkable materials, solutions are available on the market, which includes the solarization of roads, floors and terraces [112]. Some solutions, like those by Onyx Solar in the Figure 5, still represent a solution for THB because the invisibility of the intervention also allows chromatic solutions not typical of historical materials. Solvable problems with external finishes and colors more similar to those of historical buildings.

Figure 5. Walkable photovoltaics solution for the flat roof (source: Onyx Solar).

Interestingly, the PVT prototype developed by Greppi et al. [113] for a new hybrid solar panel, as per Figure 6, can be used as a tile to pave driveways, areas, and terraces and to cover roofs due to its particular robustness and compactness. The main feature, that characterizes the new hybrid tile, is its walkability and the simple laying procedure.

The problem of laying in historical buildings subject to difficult retrofitting must be fully assessed also with regards to the additional loads that are applied to the underlying structures and the technical bulky structures.

Figure 6. Photovoltaic cells, heat sink enclosed in transparent and opaque resins with on left hydraulic and electrical connectors [113].

Organic solar technologies seem to be a very promising cell-based solution. The 6% efficiency is, of course, a big minus, however, through the continued development in technology and research, its efficiency is gradually increasing. One of the advantages of organic cells is that, when compared to CIGS cells (copper indium gallium selenide solar cell), organic cells don't require rare elements such as Indium.

The lifespan of these solar modules is approximately 15 years. This technology is expected to be a bit less durable than glass modules, but a 15-year lifespan is not bad at all [114].

2.4. Energy Efficiency

The main objective of the research in the photovoltaic sector is to achieve a high solar energy conversion efficiencies using widely available raw materials, and sustainability both, from an economic and environmental point of view.

As reported in Table 6, the growth and development forecasts for the next few years of photovoltaic technologies are still very much linked to Silicon-based components.

As already mentioned above, the main limitation of BIPV on the listed and THB is the visual impact; mainly the PV cell.

Table 6. Technology goals and key R and D issues for crystalline silicon technologies (Source: Technology Roadmap, solar photovoltaic energy, IEA 2010).

Crystalline Silicon Technologies	2010–2015	2015–2020	2020–2030/2050
Efficiency targets in %	• Monocrystalline: 21% • Multicrystalline: 17%	• Monocrystalline: 23% • Multicrystalline: 19%	• Monocrystalline: 25% • Multicrystalline: 21%
Manufacturing aspects	• Si consumption < 5 g/Watt (g/W)	• Si consumption < 3 g/W	• Si consumption < 2 g/W
Research &Development areas	• New silicon materials and processing • Cell contacts, emitters and passivation	• Improved device structures • Productivity and cost optimization in production	• Wafer equivalent technologies • New device structures

2.5. Colored Cells and Glazing Modules

One of the main problems in BIPV solutions on traditional materials are factors, related to module/cell color, pattern, texture, and visible materials [115].

The concept of a low impact solution for photovoltaic components on historical materials is based on preserving the history and value of the product, and the final result obtained. Reversibility to the original state is fundamental if the BIPV solutions are replaced or removed.

The first and most important element pointed out by market surveys was the need to change the color of the PV module from the blue-purple tones of standard technologies to a more terra-cotta like color matching traditional roofing materials [116].

The need to make photovoltaic components "invisible" on building envelopes is a very advanced research objective. The color of the external finish is one of the main obstacles of BIPV solutions.

Reflection losses limit all types of photovoltaic devices. The first reflection loss occurs at the glass-air interface of the photovoltaic module. If no light trapping mechanism is used, about 4% of the solar energy is lost on this surface [117]. The most commonly used techniques in reducing reflection, include texturing of the glass surface and the application of an anti-reflective coating (ARC). The simplest ARC consists of a single layer of refractive index matching material [118]. High quality anti-reflection (AR) coatings have become a vital feature of high-efficiency silicon solar cells. Solar cell efficiency can be improved by antireflection gratings. The silicon material has a high refractive index. Understanding how to develop broadband and omnidirectional antireflection is a key technology for increasing solar energy efficiency [119].

Most of these ARCs are manufactured by film deposition techniques such as chemical vapour deposition (CVD), sputtering, or evaporation, as well as possible lithography steps on perovskite solar cells [120].

Therefore, RCs are of great importance in improving the efficiency of the solar cell by reducing loss due to reflection. ARCs containing a single layer can be non-reflective only at a single wavelength, generally up to the middle of the visible spectrum. Whereas, ARCs containing double layers are effective over the entire visible spectrum [121].

The predominantly dark color of the photovoltaic cells is designed to reflect as little light as possible. This way the solar cell will produce the maximum power output. The color of the solar cells can be changed by varying the thickness of the anti-reflection coating. By reducing the thickness of the anti-reflection layer, the overall reflection will increase and the efficiency decrease by 15–30%, depending on the color.

Using antireflection coatings (ARCs) processes suitable for large-scale manufacturing, many authors have explored new materials and process modifications to systematically reduce conversion losses in devices manufactured on low-cost glass substrates [122].

More precisely, colored modules with high saturation, angle independent color appearance, and a minimized color, induced targeted efficiency loss. Homogeneous colored layers have a high transmittance in wavelength range with non-negligible spectral responsivity of the PV-cells [123].

BIPV latest generation solutions allow access to solutions with high integration and low visual impact even on historical buildings enclosures.

The technological solutions in the field of colored and transparent solar technologies represent one of the most promising for application on historical buildings. The colored solutions, including different crystalline Si cells and Thin-film technologies have been available on the market for several years.

The coloring of the active PV component and a specific design process offer opportunities and innovative BIPV solutions that can also be implemented on valuable enclosures.

There are two very promising and innovative solutions for heritage building envelopes: transparent solar technologies and selective materials, with low molecular density solar radiation that reproduce the components and colors of traditional material.

Further advances have been made in applied research on the color of materials used in the production of photovoltaic cells. Market projections predict that in 2020 the price of these modules will decrease by 20% concerning 2013.

The adoption of innovative technologies for colored photovoltaic cells with a high energy yield will make it possible to correct, or at least mitigate, the historical conflict of technological systems versus safeguarding and preserving the historical and architectural heritage.

2.6. Glazing

The glazing application on a THB envelope represents one of the more interesting opportunities in energy retrofitting solutions. The minimum glazing available surface on a building is approximately 30%. The potential of these solutions on historical buildings from an nZEB perspective offers a vast scope of very interesting opportunities in future years when the current more promising solutions will have acceptable EPBT. Window-integrated Photovoltaic (WIPVs) is a system that generates electrical energy from glass and windows [124].

2.7. Light Selective Compound for Traditional Materials (multilayering)

Any alternative energy technology that is supposed to address the problem of energy sufficiency and security and climate change must be: Simple, easily scalable and inexpensive.

This results from the controlled and coherent integration of the solar collectors simultaneously from all functional, constructive, and formal (aesthetic) points of view [115].

These so-called "invisible" solutions are available on the market [23]. The silicon cells are completely embedded in the body of the same as these are made of polymeric compounds loaded with natural powders. The result is that they are similar in appearance to traditional materials, and at the same time, allow sunlight to filter through the outer surface (as if it were transparent) and reach the photovoltaic cells.

The principle is stratification (layering) achieved by superimposing low molecular density polymeric materials over the PV cell [35]. In this case, the layering technique does not concern the cell but the external finish, with particular optical properties that allow camouflaging the tile on the components of the building envelope.

The single roof coppo-tile, as per Figure 7, has a power of 4.5 Wp. To obtain a power of 1 kWp, 223 roof tiles are necessary for a surface of 15 m². Technological innovations prove that multifunction optics open up very interesting solutions. It is possible to propose low impact BIPV solutions, not only on roofs, but also on architectural technological units and historical, as well as traditional building materials.

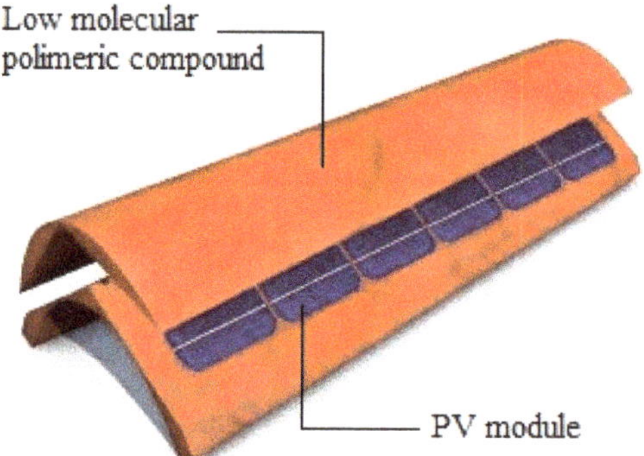

Figure 7. PV coppo with a polymeric compound developed to encourage the photon absorption (source [35]).

A reference model of BIPV solutions is proposed as functional for technologies, visual impact, the level of integration per architectural unit and potential integration on traditional materials.

Solutions that can have win-win a result on the extended historical buildings stock not subject to restrictions and in non-prestigious urban areas. Such an extension of active solar surfaces contributes to an increase in the production of electricity from a primary source for each building beyond certain protection constraints. Some architectural elements of historical buildings, such as cornices, terraces, vertical walls and traditional roofs in Roman tile, are presented in this paper for their photovoltaic solarization.

3. Results

The availability of BIPV multifunctional components opens up the possibility of low-visual impact energy retrofit interventions on the heritage buildings stock.

The limits of integration are not of a technological or aesthetic nature to respect the historical value of the enclosures. Given the current levels of integration achieved, in the present work, an analysis was carried out comparing BAPV and BIPV approaches for heritage buildings to highlight opportunities and constraints. This analysis provides an effective means of 'mapping' the current situation and identifying chances for future developments.

In the present work, practical involvement is explored not only from a holistic point of view but also to examine the relationship between supportability and conservation in the field of the energy efficiency applied to the historical building property as per Figure 8.

The solutions supplied by BAPV to BIPV on historical buildings need an innovative methodological approach that involves guardianship organs, seekers and planners for a reverse photovoltaic design. In this work, we propose the Heritage Building Energy Solar Solution Technologies (hBESST) as a synthesis approach to meet the needs of the energy retrofit, the protection and preservation of the historical building stock and the BIPV solutions.

A solution to the integration of integrated solar technologies on a historical building is proposed in Table 7 through various levels; from a high-impact BAPV to a low-impact BHESST concerning the opportunities and constraints exposed above.

The holistic approach frames define and quantify the opportunities and constraints that emerged using a simple 3-value scale of Low (*), medium (**) and High (***). Subsequently, to provide typical indications of the building process, through BIPV design solutions and practical implementation, it was possible to define an initial example on how to intervene on typical architectural elements of THBs already available on the market.

Figure 8. (Source: www.lofsolar.com) and Perovskite transparent PV cell.

Table 7. Approaches and criteria within the Heritage Building Management Process for different BIPV solutions.

Criteria Approach	BIPV-THB Building Management Process				
	Project Management	BIPV Design	Construction	Visual Impact	Codes and Regulation
BAPV Building Attached (Applied/Added) PV	*	*	*	***	***
BIPV Building Integrated PV	**	**	***	**	***
BIPPV Building Integrated Project PV	***	***	*	*	**
hBESST Building Heritage Energy Solar Solutions Technology	***	***	*	*	?

3-value scale: Low (*), Medium (**), High (***).

The methodology of the building process for heritage buildings in the field of energy retrofit shows clear limits of replicability given the specificity of each case on which to intervene. However, this does not mean that we can generalize the process with the detriment of the principles of protection and safeguarding, by identifying three objectives: BIPV, Heritage Buildings (Hbuil) and nZEB.

Within the hBESST process, three BIPV, nZEB and hBuil targets were defined and four interactions were identified, as shown in Figure 9.

Point 4 represents the maximum number of integration conditions required to establish a structured process that goes from the needs of energy sustainability up to the development phase on the scale of the single architectural element on historical buildings, including THB.

For each interaction, through a holistic approach, the hBESST has made it possible to highlight 23 factors that can be assessed according to opportunities and constraints as reported in Table 8.

The following table provides comparative solutions for laying surfaces by evaluating BIPV and hBESST solutions.

The technologies are presented based on PV technology, the efficiency of electricity production and annual producibility per 1 m2 installed.

Table 9 shows how some hBESST solutions on building envelope components on historical buildings can be developed with technologies already available on the market. For the glazing and WIPV solutions, market solutions with acceptable EROI and EPBT are not yet available. In 2030, the solarization of glazed surfaces is expected to reach levels of economic acceptability.

Table 8. Opportunities and constraints assessment according to the factors determined by the interactions.

Dual Objectivies	Interactions	Factors	Opportunities	Constraints
1. BIPV and Heritage buildings	Integration on the building envelope	Visual impact	***	***
		Component visibility	**	*
		Multifunction BIPV solutions	*	*
		Reversibility	***	*
		Cultural heritage code compliance	*	***
2. Heritage buildings and nZEB	Energy retrofit	Maintenance target	***	*
		Adapting and updating new rules and codes	***	*
		Primary energy reuse (Fuel cells, H2G)	***	*
		New materials	***	*
		Integrated solutions on traditional materials	***	*
3. nZEB and BIPV solutions	Energy sustainability	nZEB compliance on HBuil	***	***
		High solarization of building surface	***	**
		Primary energy maximum producibility	***	*
		Optimized Energy manage demand	***	*
		LCA analysis	***	**
4. BIPV, Heritage buildings and nZEB	hBESST	Respecting historical buildings	***	***
		Crossing barrier on regulations, technologies	***	*
		Foster and manage innovative approaches	***	*
		Guidelines for harmonization	***	*
		BIPV needs & function analysis	***	*
		Dissemination and new professionals training	***	*
		New specific business models	***	*
		PV solution HBIM oriented	***	*

3-value scale: Low (*), Medium (**), High (***).

Table 9. Comparison between traditional BIPV and BHESST solutions on historical building envelopes and technological and visual impact parameters.

	BIPV	hBESST	PV Technology	Efficiency	Producibility kW/sqm/yr	BIPV Modules Solutions	Visual Impact	Turnkey Solution
Roofs			Mono-Poly Si	22%	317	Color Layer	*	*
Terraces			Mono-Poly Si Amorphous	22%	274	Floor Pavement Sun decks	*	*
Cornices			Mono-Poly Si Amorphous	22%	274	Perimetric Cornices buildings surfaces	*	*
Walls			Mono-Poly Si Amorphous ThinFilm (CdTe)	22% 12% 16%	204 0 149	Vertical walls with good exposition. PV surfaces with Opaque finishing	**	**
Glazing WIPV			DSSC OPV	3.4	3.60 kWh/kWp·day	Windows Glazing surfaces	*	*

3-value scale: Low (*), Medium (**).

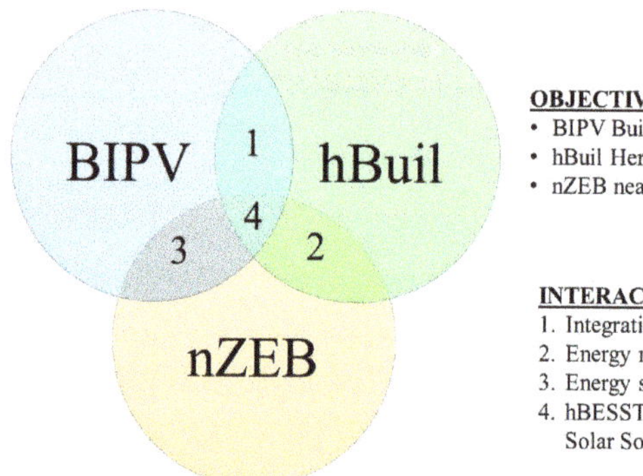

Figure 9. Four interactions between three objectives to obtain the heritage building management system hBESST.

A bottom-up approach was adopted which, starting from the level of integration on the historical building component, has allowed formulating interactions and factors related to energy retrofit on the enclosures of historical buildings [86].

Figure 10 shows the bottom-up approach concerning the hBESST and hBMP objectives.

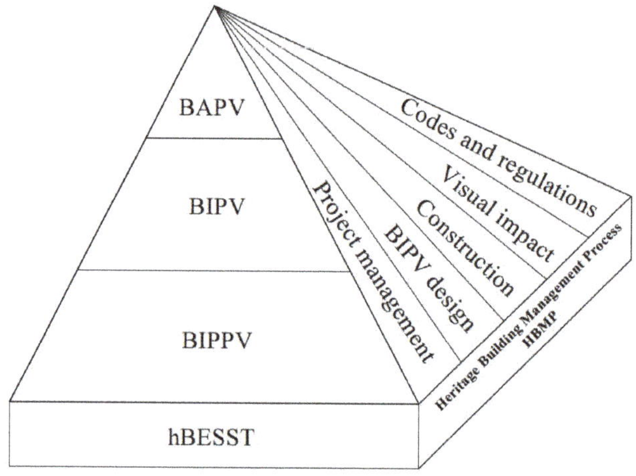

Figure 10. Bottom-up approach schematization of the hBESST procedure.

Related adoptions have been assessed and incorporated in this work.

Table 10 shows the main opportunities and constraints of hBESST regarding know-how, training and economic barriers, and to a lesser extent, technological barriers. The need to carry out case studies with innovative and low impact solutions starting from the unconstrained historical building stock can also accelerate overcoming barriers linked to codes and conservation rules.

The process outlined in this paper can be implemented on that part of a historical building stock, such as THB in urban contexts, without constraints and architectural and landscape protection. An hBESST approach allows you to tackle solutions with a low visual impact on traditional materials and surfaces.

Heritage BIM solutions and applications for energy retrofitting, and in particular for BIPV solutions, represent a huge potential for numerous researches. HBIM applied to the specificity of THB defines a new relationship between BIPV, traditional materials and optimization of a workflow, where BIPV, nZEB and the Heritage/Historical buildings represent an opportunity and not a constraint.

The BIM informative process is classified in seven levels of which the first three are linked to the geometry of the objects while 4D is applied to scheduling, 5D is applied to estimation, 6D is applied to sustainability and 7D is applied to facility management applications. The energy component turns out to be transversal for all seven levels but determining for levels from 4D to 7D.

The BIM and HBIM solutions in a multilevel and interoperable digitization process represent obligatory paths as they allow for the analysis and evaluation of hBESST choices, thereby facilitating the choice at all levels of design for BIPV solutions.

The constraints concern the LOD definition, the parametric digitization of all the BIPV components within the typical workflow of the BIM processes.

This first phase of the research will be integrated later in the second phase with an evaluation procedure on the components of BIPV on historical buildings through votes and opinions formulated by experts in the field.

Table 10. Compare BIPV and hBESST and related opportunities and constraints.

	Opportunities	Constraints
BIPV	<ul><li>Larger collection surface</li><li>High integration</li><li>Smart grids will increase the use of photovoltaics</li></ul>	<ul><li>Additional complexity in building design</li><li>More complex system design</li><li>Visual impact</li><li>Lack of cooperation in the design workflow</li><li>Higher costs compared to market alternatives and technological solutions</li></ul>
hBESST	<ul><li>Fitted solutions</li><li>High respect for traditional materials</li><li>High RES production</li><li>Designing the PV as a building material with high architectural integration that meets technical, functional and aesthetic requirements at acceptable costs</li><li>Enabling mass realization of nZEB by BIPV</li><li>Develop BIPV module and system design concepts that enable fast and highly automated installation,</li><li>Innovation and integrated solutions between the photovoltaic industry and the building industry.</li><li>HBIM process in an innovative workflow process</li></ul>	<ul><li>Crossing barrier on regulations, technologies</li><li>Define a specific business model</li><li>Limited availability on the market of highly integrated BIPV solutions on traditional historical building materials</li><li>Few realizations of BIPV on historical buildings</li><li>Strengthen research, development and demonstration (RD&D) efforts for BIPV to further reduce costs.</li><li>Transform the niche market of BIPV into a mass market</li><li>LOD define for specific Heritage on PV materials in BIM/HBIM design process</li></ul>

4. Discussion

The process outlined in this paper can be implemented on that part of a historical building stock, such as THB in urban contexts, without constraints, architectural and landscape protection. An hBESST approach allows you to tackle solutions with a low visual impact on traditional materials and surfaces.

Heritage BIM solutions and applications for energy retrofitting, and in particular for BIPV solutions, represent a huge potential for numerous researches. HBIM, applied to the specificity of THB, defines a new relationship between BIPV, traditional materials and optimization of a workflow where BIPV, nZEB and the Heritage/Historical buildings represent an opportunity and not a constraint.

The BIM informative process is classified in seven levels of which the first three are linked to the geometry of the objects while 4D is applied to scheduling, 5D is applied to estimation, 6D is applied to sustainability and 7D is applied to facility management applications. The energy component turns out to be transversal for all seven levels but determining for levels from 4D to 7D.

The BIM and HBIM solutions in a multilevel and interoperable digitization process represent obligatory paths as they allow for the analysis and evaluation of hBESST choices, thereby facilitating the choice at all levels of design for BIPV solutions.

The constraints concern the LOD definition, the parametric digitization of all the BIPV components within the typical workflow of the BIM processes.

This first phase of the research will be integrated later in the second phase with an evaluation procedure on the components of BIPV on historical buildings through votes and opinions formulated by experts in the field.

5. Conclusions

The present work aims to investigate the opportunities and to outline the limits and the opportunities that BIPV design approaches offer during architectural and energetic upgrading interventions on historical buildings. The interactions of the three main objectives, BIPV, Hbuil and nZEB, have been analyzed using traditional Heritage Building Solar Solutions Technologies (hBESST). Subsequently, through a bottom-up process for individual materials and PV technologies on enclosures of historical buildings, we defined a Heritage Building Energy Solar Solution Technologies (hBESST) as the first approach of synthesis to the needs of the energy retrofit, BIPV solutions, and protection and preservation of the historical building stock.

The challenge in hBESST design is to find the best combination of design strategies that will face the energy performance problems of a THB and future listed buildings.

The PV's maximization producibility on a historical building property cannot be the only methodological reference. The technical solutions are shown in the present paper. An hBESST approach can be defined if, for example, cornice solarization and roof floors are opportunely inserted into project integration management. Integration interventions on other traditional materials can be designed to keep the natural balance and protect the landscape, the historical patrimony and the energetic sustainability. The latest generation PV components, colored and highly invisible, open very interesting design perspectives.

The results highlighted have been submitted to give indications and cues in evaluating the limits that RES solutions, and in this case BIPV, can present on the historical building stock.

Funding: This research received no external funding

Conflicts of Interest: The authors declare no conflict of interest.

References

1. European Commission. *Communication From the Commission: A Roadmap for Moving to a Competitive Low Carbon Economy in 2050;* COM(2011) 112 Final 34; European Commission: Brussels, Belgium, 2011; pp. 1–34. [CrossRef]
2. D'Agostino, D.; Mazzarella, L. What is a Nearly zero energy building? Overview, implementation and comparison of definitions. *J. Build. Eng.* **2019**, *21*, 200–212. [CrossRef]
3. Kurnitski, J.; Allard, F.; Al, E. How to Define Nearly Net Zero Energy Buildings nZEB—REHVA Proposal for Uniformed National Implementation of EPBD Recast. Available online: https://www.rehva.eu/fileadmin/hvac-dictio/03-2011/How_to_define_nearly_net_zero_energy_buildings_nZEB.pdf (accessed on 9 May 2018).
4. Hermelink, A.; Schimschar, S.; Boer, T. Towards Nearly Zero-Energy Buildings Definition of Common Principles under the EPBD Final Report, Köln. 2012. Available online: https://ec.europa.eu/energy/sites/ener/files/documents/nzeb_full_report.pdf (accessed on 11 May 2018).
5. Romagnoni, P. Edifici a Basso Consumo Energetico: Tra ZEB e NZEB, Ppt. Available online: https://sistemaedificio.files.wordpress.com/2015/03/romagnoni_venezia.pdf (accessed on 8 May 2018).
6. Deng, S.; Wang, R.Z.; Dai, Y.J. How to evaluate performance of net zero energy building—A literature research. *Energy* **2014**, *71*, 1–16. [CrossRef]
7. Mauri, L. Feasibility Analysis of Retrofit Strategies for the Achievement of NZEB Target on a Historic Building for Tertiary Use. *Energy Procedia* **2016**, *101*, 1127–1134. [CrossRef]
8. International Energy Agency (IEA). *Technology Roadmap Solar Photovoltaic Energy—2014 Edition;* Technology Roadmap; IEA: Paris, France, 2014. [CrossRef]

9. *SET-Plan—Declaration on Strategic Targets in the context of an Initiative for Global Leadership in Photovoltaics (PV)*; European Commission: Brussels, Belgium, 2016.

10. Aste, N.; Adhikari, R.S.; Del Pero, C. Photovoltaic Technology for Renewable Electricity Production: Towards Net Zero Energy Buildings. In Proceedings of the 2011 International Conference on Clean Electrical Power (ICCEP), Ischia, Italy, 14–16 June 2011; pp. 446–450. [CrossRef]

11. Ferrara, C.; Wilson, H.R.; Sprenger, W. Building-integrated photovoltaics (BIPV). In *The Performance of Photovoltaic (PV) Systems: Modelling, Measurement and Assessment*; Elsevier Inc.: Freiburg, Germany, 2017; pp. 235–250. [CrossRef]

12. SET-Plan TWP PV Implementation PlanFinal Draft. 2017. Available online: https://setis.ec.europa.eu/system/files/set_plan_pv_implmentation_plan.pdf (accessed on 15 January 2020).

13. MIBACT Ministero per i Beni e le Attività Culturali, Legislative Decree n. 42 of 22 January 2004. Code of Cultural and Landscape Heritage. 2004. Available online: http://www.unesco.org/culture/natlaws/media/pdf/italy/it_cult_landscapeheritge2004_engtof.pdf (accessed on 23 March 2017).

14. Rosa, F.; Carbonara, E. An analysis on technological plant retrofitting on the masonry behaviour structures of 19th century Traditional Historical Buildings (THB) in Rome. *Energy Procedia* **2017**, *133*, 121–134. [CrossRef]

15. European Union (EU). *Directive 2010/31/EU of the European Parliament and of the Council of 19 May 2010 on the Energy Performance of Buildings*; EU: Brussels, Belgium, 2010. [CrossRef]

16. EU. *DIRECTIVE (EU) 2018/844 of the European Parliament and of the Council of 30 May 2018 Amending Directive 2010/31/EU on the Energy Performance of Buildings and Directive 2012/27/EU on Energy Efficiency (Text with EEA Relevance)*; EU: Brussels, Belgium, 2018.

17. Decree of June 26th, 2015 Concerning New Minimum Requirements and Methodology for Calculating Energy Performance of Buildings. 2015. Available online: https://www.mise.gov.it/index.php/it/normativa/decreti-interministeriali/2032966-decreto-interministeriale-26-giugno-2015-applicazione-delle-metodologie-di-calcolo-delle-prestazioni-energetiche-e-definizione-delle-prescrizioni-e-dei-requisiti-minimi-degli-edifici (accessed on 9 January 2020).

18. De Santoli, L. Guidelines on energy efficiency of cultural heritage. *Energy Build.* **2015**, *86*, 534–540. [CrossRef]

19. Nastasi, B.; di Matteo, U. Solar Energy Technologies in Sustainable Energy Action Plans of Italian Big Cities. *Energy Procedia* **2016**, *101*, 1064–1071. [CrossRef]

20. Feilden, B.M. *Conservation of Historic Buildings*; Architectural Press: New York, NY, USA, 2003. Available online: http://www.hoepli.it/libro/conservation-of-historic-buildings/9780750658638.html (accessed on 24 March 2017).

21. Guidance on Heritage Impact Assessments for Cultural World Heritage Properties, 2011, ICOMOS. Available online: https://www.icomos.org/world_heritage/HIA_20110201.pdf (accessed on 10 July 2020).

22. Mancini, F.; Nastasi, B. Energy retrofitting effects on the energy flexibility of dwellings. *Energies* **2019**, *12*, 2788. [CrossRef]

23. MiBACT. *Linee di Indirizzo per il Miglioramento Dell'efficienza Energetica nel Patrimonio Culturale. Architettura, Centri e Nuclei Storici ed Urbani*; MiBACT: Rome, Italy, 2014.

24. WP2: Standardisation and Testing | Construct PV. Available online: http://www.constructpv.eu/wp2-standardisation-and-testing/ (accessed on 15 May 2018).

25. Italian Company for Standardization. *UNI 8290-1:1981+A122:1983: Residential Building. Building Elements. Classification and Terminology*; Italian Company for Standardization: Milan, Italy, 1983.

26. Pagliaro, M.; Ciriminna, R.; Palmisano, G. BIPV: Merging the photovoltaic with the construction industry. *Prog. Photovolt. Res. Appl.* **2010**, *18*, 61–72. [CrossRef]

27. Heinstein, P.; Ballif, C.; Perret-Aebi, L.-E.E. Building integrated photovoltaics (BIPV): Review, potentials, barriers and myths. *Green* **2013**, *3*, 125–156. [CrossRef]

28. Delponte, E.; Marchi, F.; Frontini, F.; Polo, C.; Fath, K.; Batey, M. BIPV in eu28, from niche to mass market: An assessment of current projects and the potential for growth through product innovation. In Proceedings of the 31st European Photovoltaic Solar. Energy Conference and. Exhibition (EU PVSEC 2015), Hamburg, Germany, 14–18 September 2015; pp. 3046–3050.

29. Defaix, P.R.; van Sark, W.G.J.H.M.; Worrell, E.; de Visser, E. Technical potential for photovoltaics on buildings in the EU-27. *Sol. Energy* **2012**, *86*, 2644–2653. Available online: http://linkinghub.elsevier.com/retrieve/pii/S0038092X12002186 (accessed on 14 May 2018). [CrossRef]

30. El Gammal, A.; Mueller, D.; Buerkstuemmer, H.; Vignal, R.; Macé, P. Technical Evaluation of BIPV Power Generation Potential in EU-28. 2016; pp. 2518–2522. Available online: http://becquerelinstitute.org/wp-content/uploads/2014/08/6DO-8-final-BIPV-Technical-Potential.pdf (accessed on 14 May 2018).

31. Osseweijer, F.J.W.; van den Hurk, L.B.P.; Teunissen, E.J.H.M.; van Sark, W.G.J.H.M. A Review of the Dutch Ecosystem for Building Integrated Photovoltaics. *Energy Procedia* **2017**, *111*, 974–981. [CrossRef]

32. Mancini, F.; Nastasi, B. Solar energy data analytics: PV deployment and land use. *Energies* **2020**, *13*, 417. [CrossRef]

33. Godwin, P.J. Building Conservation and Sustainability in the United Kingdom. *Procedia Eng.* **2011**, *20*, 12–21. [CrossRef]

34. Selj, J.H.; Mongstad, T.T.; Søndenå, R.; Marstein, E.S. Reduction of optical losses in colored solar cells with multilayer antireflection coatings. *Sol. Energy Mater. Sol. Cells* **2011**, *95*, 2576–2582. [CrossRef]

35. Invisible Solar | Dyaqua. Available online: http://www.dyaqua.it/invisiblesolar/_it/index.php (accessed on 19 May 2018).

36. Escarre, J.; Li, H.-Y.Y.; Sansonnens, L.; Galliano, F.; Cattaneo, G.; Heinstein, P.; Nicolay, S.; Bailat, J.; Eberhard, S.; Ballif, C.; et al. When PV modules are becoming real building elements: White solar module, a revolution for BIPV. In Proceedings of the 2015 IEEE 42nd Photovoltaic Specialist Conference (PVSC), New Orleans, LA, USA, 14–19 June 2015; pp. 1–2. [CrossRef]

37. Swissinso–Kromatix. Available online: http://www.swissinso.com/company/technology.html (accessed on 17 March 2017).

38. Kuo, H.J.; Hsieh, S.H.; Guo, R.C.; Chan, C.C. A verification study for energy analysis of BIPV buildings with BIM. *Energy Build.* **2016**, *130*, 676–691. [CrossRef]

39. Jakica, N. State-of-the-art review of solar design tools and methods for assessing daylighting and solar potential for building-integrated photovoltaics. *Renew. Sustain. Energy Rev.* **2018**, *81*, 1296–1328. [CrossRef]

40. Piselli, C.; Romanelli, J.; di Grazia, M.; Gavagni, A.; Moretti, E.; Nicolini, A.; Cotana, F.; Strangis, F.; Witte, H.J.L.; Pisello, A.L. An integrated HBIM simulation approach for energy retrofit of historical buildings implemented in a case study of a medieval fortress in Italy. *Energies* **2020**, *13*, 2601. [CrossRef]

41. Magrini, A.; Franco, G.; Guerrini, M. The Impact of the Energy Performance Improvement of Historic Buildings on the Environmental Sustainability. *Energy Procedia* **2015**, *75*, 1399–1405. [CrossRef]

42. Xu, R.; Wittkopf, S.; Roeske, C. Quantitative Evaluation of BIPV Visual Impact in Building Retrofits Using Saliency Models. *Energies* **2017**, *10*, 668. [CrossRef]

43. van Berkel, B.; Minderhoud, T.; Piber, A.; Gijzen, G. Design innovation from PV_module to building envelope (EU PVSEC 2014). In Proceedings of the 29th European Photovoltaic Solar Energy Conference and Exhibition, Amsterdam, The Netherlands, 22–26 September 2014; pp. 3606–3612.

44. MacLeod, H.A. Thin Film Optical Filters. 2001. Available online: https://kashanu.ac.ir/Files/thin%20film%20optical%20filter(macklod).pdf (accessed on 22 May 2020).

45. Stauffer, N.W.M. Solar photovoltaic technologies: Silicon and beyond. *Energy Future* **2015**, 8–12. Available online: http://energy.mit.edu/news/transparent-solar-cells/ (accessed on 20 December 2019).

46. Lim, J.W.; Kim, G.; Shin, M.; Yun, S.J. Colored a-Si:H transparent solar cells employing ultrathin transparent multi-layered electrodes. *Sol. Energy Mater. Sol. Cells* **2017**, *163*, 164–169. [CrossRef]

47. Decree, I. Decreto Interministeriale n.1444. Available online: https://www.gazzettaufficiale.it/eli/id/1968/04/16/1288Q004/sg (accessed on 3 February 2020).

48. Clarke, J.A.; Hand, J.W.; Johnstone, C.M.; Kelly, N.; Strachan, P.A. Photovoltaic-integrated building facades. *Renew. Energy* **1996**, *8*, 475–479. [CrossRef]

49. Yang, T.; Athienitis, A.K. A review of research and developments of building-integrated photovoltaic/thermal (BIPV/T) systems. *Renew. Sustain. Energy Rev.* **2016**, *66*, 886–912. [CrossRef]

50. Recchia, A.P. Problematiche e Opportunità Nell'uso Delle Fonti Energetiche Rinnovabili nel Patrimonio Storico Monumentale, MIBACT Ministero per i Beni e le Attività Culturali, Milan. 2011. Available online: http://www.iborghisrl.it/new/wp-content/uploads/2011/10/Convegno-Milano-5-ottobre-2011.pdf (accessed on 21 February 2017).

51. Frontini, F.; Scognamiglio, A.; Graditi, G.; Lopez, C.P.; Pellegrino, M. From BIPV to Building Component. Available online: http://www.constructpv.eu/wp-content/uploads/2015/06/PVSEC_FF_1.pdf (accessed on 15 May 2018).

52. Pacheco-torgal, F.; Granqvist, C.G.; Jelle, B.P.; Vanoli, G.P. *Cost-Effective Energy Efficient Building Retrofitting: Materials, Technologies, Optimization and Case Studies*; Woodhead Publishing: Duxford, UK.

53. López, C.S.P.; Frontini, F. Energy Efficiency and Renewable Solar Energy Integration in Heritage Historic Buildings. *Energy Procedia* **2014**, *48*, 1493–1502. [CrossRef]

54. Pellegrino, M.; López, C.P.; Graditi, G.; Scognamiglio, A.; Frontini, F. From BIPV to Building Component. In Proceedings of the 28th European Photovoltaic Solar Energy Conference and Exhibition, Paris, France, 30 September–4 October 2013; pp. 3757–3761. [CrossRef]

55. Materials and Devices | Photovoltaic Research | NREL. Available online: https://www.nrel.gov/pv/materials-devices.html (accessed on 24 May 2018).

56. IEA PVPS Trend 2017 in Photovoltaics Applications. 2017. Available online: http://iea-pvps.org/fileadmin/dam/public/report/statistics/IEA-PVPS_Trends_2017_in_Photovoltaic_Applications.pdf (accessed on 22 February 2020).

57. Green, M.A.; Emery, K.; Hishikawa, Y.; Warta, W.; Dunlop, E.D. Solar cell efficiency tables (version 45). *Prog. Photovolt. Res. Appl.* **2015**, *23*, 1–9. [CrossRef]

58. Cerón, I.; Caamaño-Martín, E.; Neila, F.J. 'State-of-the-art' of building integrated photovoltaic products. *Renew. Energy* **2013**, *58*, 127–133. [CrossRef]

59. Green, M.A.; Hishikawa, Y.; Dunlop, E.D.; Levi, D.H.; Hohl-Ebinger, J.; Ho-Baillie, A.W.Y. Solar cell efficiency tables (version 51). *Prog. Photovolt. Res. Appl.* **2017**, *26*, 3–12. [CrossRef]

60. Nastasi, B.; Basso, G.L. Hydrogen to link heat and electricity in the transition towards future Smart Energy Systems. *Energy* **2016**, *110*, 5–22. [CrossRef]

61. de Santoli, L.; Basso, G.L.; Nastasi, B. The Potential of Hydrogen Enriched Natural Gas deriving from Power-to-Gas option in Building Energy Retrofitting. *Energy Build.* **2017**, *149*, 424–436. [CrossRef]

62. Carbonara, G. Energy efficiency as a protection tool. *Energy Build.* **2015**, *95*, 9–12. [CrossRef]

63. Albo, A.; Rosa, F.; Tiberi, M.; Vivio, B. High-efficiency and low-environmental impact systems on a historical building in Rome: An InWall solution. *WIT Trans. Built Environ.* **2014**, *142*, 529–540. [CrossRef]

64. Rosa, F.; Cumo, F.; Calcagnini, L.; Vivio, B. Redevelopment of Historic Buildings through the Implementation of Green Roofs: A Study of a Design Methodology. Available online: http://www.academia.edu/2421781/Redevelopment_of_historic_buildings_through_the_implementation_of_green_roofs_a_study_of_a_design_methodology (accessed on 9 September 2013).

65. Jordan-Palomar, I.; Tzortzopoulos, P.; García-Valldecabres, J.; Pellicer, E. Protocol to Manage Heritage-Building Interventions Using Heritage Building Information Modelling (HBIM). *Sustainability* **2018**, *10*, 908. [CrossRef]

66. Mazzola, E.; Mora, T.D.; Peron, F.; Romagnoni, P. An Integrated Energy and Environmental Audit Process for Historic Buildings. *Energies* **2019**, *12*, 3940. [CrossRef]

67. Bruno, S.; De Fino, M.; Fatiguso, F. Historic Building Information Modelling: Performance assessment for diagnosis-aided information modelling and management. *Autom. Constr.* **2018**, *86*, 256–276. Available online: https://www.sciencedirect.com/science/article/pii/S0926580517301164?via%3Dihub (accessed on 27 March 2018). [CrossRef]

68. Osello, A.; Rapetti, N.; Semeraro, F. BIM Methodology Approach to Infrastructure Design: Case Study of Paniga Tunnel. *IOP Conf. Ser. Mater. Sci. Eng.* **2017**, *245*, 62052. [CrossRef]

69. Standardization Needs for BIPV. 2016. Available online: http://www.pvsites.eu/downloads/download/report-standardization-needs-for-bipv (accessed on 28 January 2020).

70. BIPV Standards. Available online: http://www.bipv.ch/index.php/en/technology-top-en/quality/standards (accessed on 22 May 2018).

71. Pellegrino, M.; Flaminio, G.; Graditi, G. Testing and Standards for new BIPV products. In Proceedings of the IECON 2013—39th Annual Conference of the IEEE Industrial Electronics Society, Vienna, Austria, 10–13 November 2013; pp. 8127–8132. [CrossRef]

72. GSE—Building-Integrated PV (BIPV) Plants with Innovative Features. Available online: http://www.gse.it/en/feedintariff/Photovoltaic/Fourthfeed-intariff/PVintegratedwithinnovativecharacteristics/Pages/default.aspx (accessed on 17 February 2017).

73. *PVPS Annual Report 2017: Photovoltaic Power Systems Technology Collaboration Programme*; IEA: Fribourg, Switzerland, 2017.

74. Almosni, S.; Delamarre, A.; Jehl, Z.; Suchet, D.; Cojocaru, L.; Giteau, M.; Behaghel, B.; Julian, A.; Ibrahim, C.; Tatry, L.; et al. Material challenges for solar cells in the twenty-first century: Directions in emerging technologies. *Sci. Technol. Adv. Mater.* **2018**, *19*, 336–369. [CrossRef] [PubMed]

75. Building Integrated Photovoltaics (BIPV) | WBDG Whole Building Design Guide. Available online: https://www.wbdg.org/resources/building-integrated-photovoltaics-bipv (accessed on 21 May 2018).

76. Vikram, K. Fourth Generation Photovoltaics Research in India. 2016, p. 101. Available online: http://www.mineco.gob.es/stfls/MICINN/Investigacion/FICHEROS/Presentacion_participantes_Sevilla/06_Organic_Photovoltaics_RD_India_V_Kumar.pdf (accessed on 23 November 2019).

77. Kukreti, K.; Rathod, A.P.S.; Kumar, B. *Recent Advancements and Overview of Organic Solar Cell*; Institute of Electrical and Electronics Engineers Inc.: Piscataway, NJ, USA, 2017; pp. 1539–1544. [CrossRef]

78. Berny, S.; Blouin, N.; Distler, A.; Egelhaaf, H.-J.; Krompiec, M.; Lohr, A.; Lozman, O.R.; Morse, G.E.; Nanson, L.; Pron, A.; et al. Solar Trees: First Large-Scale Demonstration of Fully Solution Coated, Semitransparent, Flexible Organic Photovoltaic Modules. *Adv. Sci.* **2015**, *3*, 1500342. [CrossRef]

79. Anderson, A.-L.; Chen, S.; Romero, L.; Top, I.; Binions, R. Thin Films for Advanced Glazing Applications. *Buildings* **2016**, *6*, 37. [CrossRef]

80. Ferrari, S.; Romeo, C. Retrofitting under protection constraints according to the nearly Zero Energy Building (nZEB) target: The case of an Italian cultural heritage's school building. *Energy Procedia* **2017**, *140*, 495–505. [CrossRef]

81. Ascione, F.; de Masi, R.F.; de Rossi, F.; Ruggiero, S.; Vanoli, G.P. NZEB target for existing buildings: Case study of historical educational building in Mediterranean climate. *Energy Procedia* **2017**, *140*, 194–206. [CrossRef]

82. Poggi, F.; Firmino, A.M.V.; Amado, M.P.; Pinho, F.F.S. Natural stone walls in vernacular architecture: What contribution towards rural nZEB concept? *Bulletin de La Société Géographique de Liège* **2015**, *65*, 51–66.

83. Sala, M.; López, C.S.P.; Frontini, F.; Tagliabue, L.C.; de Angelis, E. The Energy Performance Evaluation of Buildings in an Evolving Built Environment: An Operative Methodology. *Energy Procedia* **2016**, *91*, 1005–1011. [CrossRef]

84. Sartori, I.; Napolitano, A.; Voss, K. Net zero energy buildings: A consistent definition framework. *Energy Build.* **2012**, *48*, 220–232. Available online: https://ac.els-cdn.com/S0378778812000497/1-s2.0-S0378778812000497-main.pdf?_tid=4411dcf4-d5f7-46a8-b232-310af439279b&acdnat=1524817651_ae3ea6d3b3e462f45c299317e1ece6f0 (accessed on 27 April 2018). [CrossRef]

85. Klein, K.; Kalz, D.; Herkel, S. Grid impact of a net zero energy building with BiPV using different energy management strategies. In Proceedings of the International Conference CISBAT 2015 Future Buildings and Districts Sustainability from Nano to Urban Scale, Lausanne, Switzerland, 9–11 September 2015; pp. 579–584. [CrossRef]

86. Mancini, F.; Cecconi, M.; de Sanctis, F.; Beltotto, A. Energy Retrofit of a Historic Building Using Simplified Dynamic Energy Modeling. *Energy Procedia* **2016**, *101*, 1119–1126. [CrossRef]

87. Garcia, D.A.; Cumo, F.; Tiberi, M.; Sforzini, V.; Piras, G. Cost-Benefit Analysis for Energy Management in Public Buildings: Four Italian Case Studies. *Energies* **2016**, *9*, 522. [CrossRef]

88. Ritzen, M.; Reijenga, T.; El Gammal, A.; Warneryd, M.; Sprenger, W.; Rose-Wilson, H.; Payet, J.; Morreau, V.; Boddaert, S. IEA-PVPS Task 15. In Proceedings of the 48th IEA PVPS Executive Commitee Meeting, Vienna, Austria, 15–16 November 2016.

89. Transparency Market Research, BIPV Market Report. Available online: http://www.dem4bipv.eu/bipv/bipv-market/ (accessed on 20 February 2017).

90. General Motors Research. Building Integrated Photovoltaics: An Emerging Market. Available online: https://www.solarserver.com/solar-magazine/solar-report/solar-report/building-integrated-photovoltaics-an-emerging-market.html (accessed on 17 May 2018).

91. Horizon Prize for Integrated Photovoltaic System in European Protected Historic Urban Districts. Available online: https://ec.europa.eu/research/participants/portal/desktop/en/opportunities/h2020/topics/lce-prize-photovoltaicshistory-01-2016.html (accessed on 19 May 2018).

92. Adamson, K.-A. Philip Drachman Industry Analyst Executive Summary: Building Integrated Photovoltaics Section 1. 2012. Available online: http://webcache.googleusercontent.com/search?q=cache:ZVsjTWMFpxIJ:www.sustainworldwide.com/uploads/5/0/6/1/5061561/bipv-10-executive-summary.pdf+&cd=1&hl=zh-TW&ct=clnk (accessed on 10 July 2020).

93. Ritzen, M.; Reijenga, T.; El Gammal, A.; Warneryd, M.; Sprenger, W.; Rose-Wilson, H.; Payet, J. Enabling Framework for BIPV Acceleration. Available online: https://webcache.googleusercontent.com/search?q=cache:96NKZsUZDxgJ:https://www.photovoltaic-conference.com/images/2016/2_Programme/parallel_events/AccelerationBIPV/Michiel_RITZEN.pdf+&cd=1&hl=zh-TW&ct=clnk&gl=us (accessed on 10 July 2020).

94. James, T.; Goodrich, A.; Woodhouse, M.; Margolis, R.; Ong, S. *Building-Integrated Photovoltaics (BIPV) in the Residential Sector: An Analysis of Installed Rooftop System Prices*; U.S. Department of Energy: Washington, DC, USA, 2011; p. 50.

95. IRENA. *The Power to Change: Solar and Wind Cost Reduction Potential*; IRENA: Abu Dhabi, UAE, 2016.

96. Fthenakis, V. PV Energy ROI. Track Efficiency Gains. Solar Today. 2012; p. 3. Available online: https://www.bnl.gov/pv/files/pdf/240_SolarTodayJune12_c.pdf (accessed on 20 December 2019).

97. Raugei, M.; Sgouridis, S.; Murphy, D.; Fthenakis, V.; Frischknecht, R.; Breyer, C.; Bardi, U.; Barnhart, C.; Buckley, A.; Carbajales-Dale, M.; et al. Energy Return on Energy Invested (ERoEI) for photovoltaic solar systems in regions of moderate insolation: A comprehensive response. *Energy Policy* **2017**, *102*, 377–384. [CrossRef]

98. Deriche, M.A.; Hafaifab, A.; Mohammedia, K. EPBT and CO_2 emission from solar PV monocrystaline silicon. In Proceedings of the 2018 International Conference on Applied Smart Systems (ICASS), Medea, Algeria, 24–25 November 2018. [CrossRef]

99. Bhandari, K.P.; Collier, J.M.; Ellingson, R.J.; Apul, D.S. Energy payback time (EPBT) and energy return on energy invested (EROI) of solar photovoltaic systems: A systematic review and meta-analysis. *Renew. Sustain. Energy Rev.* **2015**, *47*, 133–141. [CrossRef]

100. Photovoltaics Report, Freiburg, Germany. 2018. Available online: https://www.ise.fraunhofer.de/content/dam/ise/de/documents/publications/studies/Photovoltaics-Report.pdf (accessed on 15 March 2020).

101. Norton, B.; Eames, P.C.; Mallick, T.K.; Huang, M.J.; McCormack, S.J.; Mondol, J.D.; Yohanis, Y.G. Enhancing the performance of building integrated photovoltaics. *Sol. Energy* **2011**, *85*, 1629–1664. [CrossRef]

102. Bonomo, P.; Frontini, F.; Chatzipanagi, A. Overview and analysis of current BIPV products: New criteria for supporting the technological transfer in the building sector. *Vitr. Int. J. Arch. Technol. Sustain.* **2015**, *1*, 67–85. [CrossRef]

103. Gupta, A.; Cemesova, A.; Hopfe, C.J.; Rezgui, Y.; Sweet, T. A conceptual framework to support solar PV simulation using an open-BIM data exchange standard. *Autom. Constr.* **2014**, *37*, 166–181. [CrossRef]

104. Kiviniemi, A. The Effects of Integrated BIM in Processes and Business Models. In *Distributed Intelligence in Design*; Wiley-Blackwell: Oxford, UK, 2011. [CrossRef]

105. Arayici, Y.; Coates, P.; Koskela, L.; Kagioglou, M.; Usher, C.; O'Reilly, K. Technology adoption in the BIM implementation for lean architectural practice. *Autom. Constr.* **2011**, *20*, 189–195. [CrossRef]

106. Saygi, G.; Remondino, F. Management of Architectural Heritage Information in BIM and GIS: State-of-the-Art and Future Perspectives. *Int. J. Herit. Digit. Era* **2013**, *2*, 695–713. [CrossRef]

107. Liu, X.; Wang, X.; Wright, G.; Cheng, J.; Li, X.; Liu, R. A State-of-the-Art Review on the Integration of Building Information Modeling (BIM) and Geographic Information System (GIS). *ISPRS Int. J. Geo-Inf.* **2017**, *6*, 53. [CrossRef]

108. Ning, G.; Kan, H.; Qiu, Z.; Weihua, G.; Geert, D. e-BIM: A BIM-centric design and analysis software for Building Integrated Photovoltaics. *Autom. Constr.* **2018**, *87*, 127–137. [CrossRef]

109. IEA. Publication: Technology Roadmap: Solar Photovoltaic Energy–Foldout—Chinese Version. Available online: http://www.iea.org/publications/freepublications/publication/name,34345,en.html (accessed on 18 December 2013).

110. Fantechi, S.; Weir, I.; Fillon, B. Photovoltaics and Nanotechnology: From Innovation to Industry. In *Nanotechnology for Sustainable Manufacturing*; CRC Press: Boca Raton, FL, USA, 2014; pp. 37–58. [CrossRef]

111. Kramer, I.J.; Minor, J.C.; Moreno-Bautista, G.; Rollny, L.; Kanjanaboos, P.; Kopilovic, D.; Thon, S.M.; Carey, G.H.; Chou, K.W.E.; Zhitomirsky, D.; et al. Efficient spray-coated colloidal quantum dot solar cells. *Adv. Mater.* **2014**, *27*, 116–121. [CrossRef]

112. Energy Glass™. Available online: http://www.energyglass.com/egl/about-us.php (accessed on 28 February 2017).

113. Lunt, R.R.; Bulovic, V. Transparent, near-infrared organic photovoltaic solar cells for window and energy-scavenging applications. *Appl. Phys. Lett.* **2011**, *98*, 113305. [CrossRef]

114. Onyxsolar. Walkable Photovoltaic Floor (For BIPV)—Onyx Solar—PV Floor. Available online: http://www.onyxsolar.com/walkable-photovoltaic-roof.html (accessed on 20 February 2017).

115. Greppi, M.; Fabbri, G. Experimental characterization of a hybrid industrial solar tile. *Energy Procedia* **2017**, *126*, 621–627. [CrossRef]

116. Intersolar Munich 2015—Technology Highlights. Available online: http://sinovoltaics.com/events/intersolar-munich-2015-technology-highlights/ (accessed on 20 May 2018).

117. Farkas, K. Designing Photovoltaic Systems for Architectural Integration. Criteria and Guidelines for Product and System Developers. 2013. Available online: http://task41.iea-shc.org/data/sites/1/publications/task41A3-2-Designing-Photovoltaic-Systems-for-Architectural-Integration.pdf (accessed on 27 April 2020).

118. Perret-Aebi, L.-E.; Ballif, C.; Leterrier, Y.; Roecker, C.; Schüler, A.; Scartezzini, J.-L.; Leibundgut, H.; Carmeliet, J. ARCHINSOLAR Innovative PV Products for Better Aesthetical Integration of Green Energies in the Built Environment, Bern. 2014. Available online: https://webcache.googleusercontent.com/search?q=cache:2GL59LKAbkwJ:https://isfh.de/wp-content/uploads/2017/09/Giovannetti_EuroSun_2016.pdf+&cd=9&hl=zh-TW&ct=clnk&gl=us (accessed on 10 July 2020).

119. Kaminski, P.M.; Lisco, F.; Walls, J.M. Multilayer Broadband Antireflective Coatings for More Efficient Thin Film CdTe Solar Cells. *IEEE J. Photovolt.* **2013**, *4*, 452–456. [CrossRef]

120. Nubile, P. Analytical design of antireflection coatings for silicon photovoltaic devices. *Thin Solid Films* **1999**, *342*, 257–261. [CrossRef]

121. Li, X.; Tan, Q.; Jin, G. Surface profile optimization of antireflection gratings for solar cells. *Optik* **2011**, *122*, 2078–2082. Available online: http://linkinghub.elsevier.com/retrieve/pii/S0030402611000441 (accessed on 27 June 2018). [CrossRef]

122. Luo, Q.; Deng, X.; Zhang, C.; Yu, M.; Zhou, X.; Wang, Z.; Chen, X.; Huang, S. Enhancing photovoltaic performance of perovskite solar cells with silica nanosphere antireflection coatings. *Sol. Energy* **2018**, *169*, 128–135. [CrossRef]

123. Sharma, R.; Gupta, A.; Virdi, A. Effect of single and double layer antireflection coating to enhance photovoltaic efficiency of silicon solar. *J. Nano-Electron. Phys.* **2017**, *9*, 02001-1–02001-4. [CrossRef]

124. Munshi, A.H.; Kephart, J.M.; Abbas, A.; Shimpi, T.M.; Barth, K.L.; Walls, J.M.; Sampath, W.S. Polycrystalline CdTe photovoltaics with efficiency over 18% through improved absorber passivation and current collection. *Sol. Energy Mater. Sol. Cells* **2018**, *176*, 9–18. [CrossRef]

Article

Bioclimatic Architecture and Urban Morphology. Studies on Intermediate Urban Open Spaces

Alessandra Battisti

Department of Planning, Design and Technology of Architecture, Sapienza University of Rome, Via Flaminia 72, 00196 Rome, Italy; alessandra.battisti@uniroma1.it

Received: 7 October 2020; Accepted: 4 November 2020; Published: 6 November 2020

Abstract: This paper deals with the interactions between biophysical and microclimatic factors on the one hand with, on the other, the urban morphology of intermediate urban open spaces, the relationship between environmental and bioclimatic thermal comfort, and the implementation of innovative materials and the use of greenery, aimed at the users' well-being. In particular, the thermal comfort of the open spaces of the consolidated fabrics of the city of Rome is studied, by carrying out simulations of cooling strategies relating to two scenarios applied to Piazza Bainsizza. The first scenario involves the use of cool materials for roofs, cladding surfaces, and pavement, while the second scenario, in addition to the cool materials employed in the first scenario, also includes the use of greenery and permeable green surfaces. The research was performed using summer and winter microclimatic simulations of the CFD (ENVI-met v. 3.1) type, in order to determine the different influences of the materials with cold colors, trees, and vegetated surfaces on the thermal comfort of the urban morphology itself. Meanwhile, the comfort assessment was determined through the physiological equivalent temperature (PET) calculated with the RayMan program. The first scenario, with the use of cool materials, improves summer conditions and reduces the urban heat island effect but does not eliminate thermal discomfort due to the lack of shaded surfaces and vegetation. The second scenario, where material renovations is matched with vegetation improvements, has a slightly bad effect on winter conditions but drastically ameliorates the summer situation, both for direct users and, thanks to the strong reduction of the urban heat island effect, to urban inhabitants as a whole.

Keywords: resilience; urban regeneration; adapting to change; climate performance; innovative technologies

---

1. Introduction

The 2015 Paris COP 21 and those that followed (until the last one in Madrid), forcefully underscored that the struggle against climate change is one of the environmental issues most debated on the global level in recent years and that an innovative vision for cities has never been so important as in our century. In fact, currently more than half of the world's population lives in cities, and this migratory trend should continue in the years to come: by 2050, more than two thirds of the world's population will be concentrated in urban settlements. The IPCC's 2018 report indicated that the risks—for the human race, the biosphere, and economies—were higher than previously stated and that encouraged actions aimed at reducing global temperatures [1], as well as reducing greenhouse gas concentrations [2]. These actions were polarized on two levels: the first regarding the planet's major ecosystems (rain forests, major river systems, etc.) and the second focusing on urban areas, since these are the ones most seriously stricken by the effects of climate change, showing a high level of vulnerability and of environmental risk.

It is thus brought to light that with the gradual phenomenon of urbanization taking place, it will in the future be cities determining the possible achievement of a radical turning point, capable of dealing with climate change and inclusive economic growth able to create jobs, without making foolish use of nonrenewable resources, on the path shown by environmental sustainability and resilience. This is why the Sustainable Development Goal (SDG) 11, "Make cities and human settlements inclusive, safe, resilient and sustainable" (https://www.un.org/sustainabledevelopment) holds such importance at the moment. In particular, objective 11.B states that by 2020 we might see substantial growth in the number of cities and urban settlements that have adopted and implemented integrated policies and plans towards inclusion, the efficiency of resources, mitigation of and adaptation to climate change, and resilience to disasters, and that have developed and implemented, in line with the Sendai framework for disaster risk reduction 2015–2030, a disaster risk management that is holistic on all levels. There are, however, still many challenges to be faced in order to achieve SDG 11. It will be necessary to conceive necessary innovative solutions to help inhabitants behave in a more aware fashion (consider Switzerland's 2000 W citizen) and to design models of city life in which the inhabitants are involved in positively influencing their lifestyle and can interact actively with their community.

The increasingly frequent heat waves triggered by climate change focus attention on the important relationship between land cover and temperature, particularly in urban areas with a greater presence of artificial surfaces, often asphalted and built in concrete or with fewer planted areas. Accompanied by the reduction in urban green areas, these are all characteristics that cause open spaces to absorb heat in the summer and do not permit the land's adequate transpiration and evaporation, thus resulting in the phenomenon of urban heat islands. In this perspective, the strategies of mitigation and adaptation to resilience of public spaces are seen as the most important ones to realize in the short term and, as in many other cases, the identified objectives must be tangible, concrete, and measurable so as to be able to make a significant impact on changing course and having the power to make a difference.

According to Landsberg [3], the first scholar to use the term "heat island" was Manley in 1958, in his discussion on the effect of London's artificial heat [4]. The term has now taken on the meaning of a phenomenon taking place in urban environments, characterized by higher summer temperatures—varying from 2 °C to 12 °C—than the surrounding rural areas [5–7], attributable to the built urban environment and presenting particular risks for the urban population [8]. Moreover, experts often divide heat islands into three different types:

1. Surface heat islands: the infrared radiation emitted and reflected by surfaces, which allow the portions of building where the surfaces are hottest to be identified, must be measured.
2. Canopy-level heat islands: which exist in the air stratum where human activities are performed, from the ground to beneath the tree canopy and building roofs.
3. Urban and boundary layer heat islands: which start from the level of building roofs and tree canopies, extending to the point where the urban landscapes no longer influence the atmosphere. This region generally extends no more than a mile from the surface [5].

Urban heat islands at the surface and canopy level, more commonly studied in architecture as cited in this document, are those in which the built environment and its artificialization has the most pronounced effect because air flow and energy exchanges are controlled for micro-scale, and in which the site's specific characteristics are extremely important [9,10].

In Europe, massive urbanization today sees the intensification of the phenomenon of the urban heat island, which contributes to the increased thermal discomfort of the inhabitants of the city, and therefore to the increase in their behaviours and energy-intensive consumption to counteract the phenomenon of summer overheating (e.g., through the intense use of artificial ventilation systems). Some cities like Paris and London, after intense heat waves, recorded exponentially increased mortality [11] and adopted strategies to mitigate urban heat islands, seeking to react with initiatives making it possible to protect the population by increasing its capacity to adapt to these extreme levels of the phenomenon. Therefore, based on existing knowledge, several questions were raised from the standpoint of urban planning, relating to the mitigation and adaptation strategies to combat the risks of UHIs (Urban Heat Islands) and of climate change. In particular, in line with the growing importance of mitigation and adaptation measures, and in light of the concept of energy retrofitting, especially of consolidated and historic fabrics, the identification, study, and design of high-performance, innovative systems is one of the main fields of investigation and a significant contribution that technology can provide to combat the negative effects of the UHI and the risks due to climate change in highly manmade settings. In this sense, technological research never ceases to suggest alternative, innovative, and creative paths and to offer new components and materials for designing; it is the architect's task to grasp their possibilities in power, in order to trigger that process of acting, operating, and working, in which to express one's will to change course towards a resilient planning of public space [12,13].

2. Mitigation and Adaptation Strategies toward Urban Resilience

In recent decades there has been an increasing interest in the concept of "urban resilience" [14]. The theory of resilience was introduced for the first time in the 1960s–1970s in the discipline of ecology [15], thanks to the work of the American C.S. Holling, considered the founder of modern thinking in ecological resilience [16]. Built on the American ruins of the thinking linked to the Keynesian Fordist regime of unlimited, unrestricted growth, this concept essentially represents an adaptive resource management strategy. Thereafter, the concept was notably used to describe the behaviour of natural systems when confronted with external disturbances [17], and it is in this area that, in the 1990s, "engineering resilience" was identified. Contrary to ecological resilience, which regarded the adaptive capacities of ecological systems in conditions of irreducible uncertainty [18], engineering resilience admitted multiple states of equilibrium and granted the system the possibility of absorbing disturbances under a given threshold. Resilient systems thus took shape, which, when confronted with a stress, were able to react by renewing themselves, while maintaining the functionality and recognizability of the initial systems [19]. In this perspective, urban resilience refers to an urban system's ability to maintain or to rapidly return to its vital functions when confronted with a disturbance and to adapt to change [20]. As the OECD (Organization for Economic Co-operation and Development) has stressed, a resilient city has the ability to absorb, rebound, and prepare for future shocks, of whatever nature they are [21]. In this line of research, the concept of the resilient city is articulated with respect to the challenges and risks that climate changes are heralding, by including two main strategies: those of mitigation and those of adaptation, strategies in which mitigation aims to reduce the impact of climate change [22], while adaptation aims to diminish its effects [23]. In the sector of constructions in simplified fashion, three prevalent strains of action aimed at putting these two strategies into practice may be identified: (i) those dedicated to the use of smart materials (nanomaterials, smart materials, cool materials); (ii) those introducing the implementation of green infrastructures (green wall, green roof, trees); and (iii) those relating to the development of blue infrastructures (nebulisation systems and use of water for cooling air or surfaces) [24].

2.1. Cool Materials

Among the mitigation techniques that aim to balance the heat load of the cities by increasing the heat losses and decreasing the relative thermal increases are those based on geoengineering. Since the 1970s, there have been many studies that have investigated the optical characteristics of materials, leading to the application of highly reflective materials (cool materials) capable of maintaining low surface temperatures, and other research that has instead investigated the thermal characteristics of materials leading to the applications in construction of phase change (PCM) and thermochromic materials. As pointed out by Pisello, the class of cool materials is the broadest and includes natural materials with high reflectivity, artificial white materials, cold coloured materials (traditional dark materials with a low surface temperature), and nanomaterials [25].

As observed by Yang et al. [26] previous studies on high reflective materials were focused primarily on their effects on the outdoor temperatures, thus focusing only on microclimate parameters such as air temperature and surface temperature variations and in that regard a consistent review has been made by Santamouris and Hakbari et al. [22,27] and recently by Pisello [25]; nevertheless, recently studies have highlighted the importance to comprise also other parameters such as mean radiant temperature (Tmrt) and above all comfort analysis (in terms of physiological equivalent temperature—PET or Universal Thermal Comfort Index—UTCI), since it appears that "cool materials" have different side effects on outdoor thermal comfort depending on the surface of application, on their exposure to solar radiation, and on the combination with other elements [26].

During the day, a greater reflectivity of surface materials allows a considerable reduction of daytime surface temperature to be achieved, while during the night the materials' albedo has no influence as concerns lower nocturnal surface temperatures. Cool materials can be applied to roofs (e.g., Cool roof), walls (e.g., Cool wall), and sidewalks (e.g., cool pavement). A field study conducted in Athens, regarding pavement, in an urban park, made with cool materials (Flisvos project) [28], showed that using pavement made with cool materials can reduce surface temperatures by up to 7.6 °C in unshaded conditions. However, it is essential to assess what part of the reflected radiation is absorbed by the surrounding surfaces of the built environment, like walls or pedestrians themselves, as several studies [26,29–31] have shown. A study by Chatzidimitriou and Yannas [31] shows that the lower surface temperatures obtained using reflecting materials, at the level of the users' comfort, do not offset the greater quantity of reflected radiation. On the other hand, the combination of cool materials and vegetation, particularly with trees, has shown positive effects both on reducing outdoor temperature and on the users' comfort [32,33].

Recently, some authors have shown that the use of the cool roof was able to bring about a 2–44% reduction in the cooling requirement, with an average of 20% [34,35]. It was pointed out that cool roofs can obtain a greater potential of mitigation of the urban heat island when they have an albedo coefficient greater than or equal to 0.7. In spite of the excellent performance, cool roofs, in a matter of a few years, lose much of their initial reflection due to constant exposure to atmospheric agents, which accelerates its ageing process.

Thermochromic materials are particularly used on construction envelope surfaces for their ability to dynamically modify their optical and thermal properties, changing colour reversibly in response to the thermal stresses of the outdoor environment—passing from darker hues to lighter ones as the temperature rises and returning to the original colour when it falls. Karlessi et al. [36] have perfected 11 thermodynamic claddings for building envelopes, studying their performance in comparison with similarly coloured claddings, both of the highly reflective type (cool) and traditional. The study casts light on how the surface temperatures of thermochromic claddings were lower than the temperatures of cold claddings and of traditional ones of the same tone of colour, by 7 °C and 11 °C, respectively. In addition, in the case of thermochromic materials, there are still factors limiting widespread use, due to problems of durability of performance and stability over time [37].

2.2. Green Infrastructures

The second type of mitigation and adaptation measures regarding open urban space is that of "green infrastructures", considered by many as one of the best and most effective strategies for cooling, combating, and mitigating the urban heat island, thanks to the shade, with the consequent reduction of ground surface temperatures and evapotranspiration [38]. They are defined as a "strategically planned network of natural and semi-natural areas with other environmental features designed and managed to deliver a wide range of ecosystem services" [39] thanks to their cooling potential and their mitigation effects in and out of doors [38]. This type comprises all the vegetated surfaces, from parks to trees, including the more artificial ones like green roofs and walls. Many studies have shown that green infrastructures make a significant contribution to reducing the impacts of climate change [40], particularly by acting in a significant way improving air quality, reducing heat stress, and containing floods. Parks with their large, grassy surfaces are important in night-time cooling action [31], but during the daytime can at times, especially if not well irrigated, be hotter than the surrounding environment [41]. This temperature difference between urban green spaces and the surrounding built terrain, called the "park cool island" (PCI) [42], shows a considerable diversity in the effectiveness of the surrounding areas and over the course of the day, and thus raises some disputed observations. Bowler et al. [43] have shown that most of the research that has been done has dealt with the study of PCIs at the "micro-scale" level, localizing their effect only in the immediate vicinity of the vegetated area and not considering the effect of the cooling contributed by the vegetation beyond the green area. In a recent study, Ketterer and Matzarakis, for a consolidated urban area in Stuttgart, proposed three intervention scenarios with different solutions of implementation of green infrastructures [44]. The results obtained by the two scholars have shown how air temperature is higher in the scenario with the asphalt paving and lower in the same area that has been planted with trees, while the scenario with only substitution of the asphalt with grass and low vegetation was revealed as without influence, presenting the same temperatures as obtained in the scenario with asphalt alone. Considering the thermal comfort these results are in line with a recent study conducted by Lee et al. in Friburg [11], where trees display a mean PET variation of 3 K and grassland of 1 K. Thus, it appears that the shading factor is the most dominant in the comfort perception, and it may vary considerably if it is from a building, a vegetated element (e.g., tree), or an artificial element (e.g., umbrella).

A recent Canadian study has shown how the shadow cast by buildings is one of the most effective cooling strategies in an urban environment, followed by the planting of trees and the construction of canopies [45]. If in a park area we calculate only the performance for comfort and that for mitigating heat, trees taken individually have clearly better results than grassy surfaces—results due to the geometric shape and the shady effect of tree canopies, as well as to the phenomenon of evapotranspiration and the ability of the foliage to modify the direction of the natural ventilation and to interact with it [43]. Of all the green infrastructures, trees are the best mitigating agents during heat waves, able to dissipate heat by long-wave radiation, convection of heat into the air, and transpiration. In particular, as pointed out by Upreti et al. [46]: "the effect of the shading of urban trees reduces net energy absorption, thereby modifying the urban energy balance and cooling the urban canopy and limit strata while reducing the sensible heat [47,48]; trees also influence indoor energy consumption for cooling [49]". The study's results also [46] clearly show, through calculations, how the shade from trees is able, during the daytime, to reduce the quantity of heat absorbed by roads and building envelopes by approximately 6 °C in their immediate vicinity, and by 1–4 °C for the expanded surrounding area, while during the night-time a reduction in heat due to the UHI phenomenon, by about 1.5 °C, was calculated.

In all the analysed studies, trees exert a greater influence on surface temperature than on air temperature, given that the surface heating caused by the sensible heat of construction materials is more significant in comparison with atmospheric heating. Moreover, trees are a good solution for the absorption and evaporation of water, creating a water evaporation path also through impermeable layers like asphalt [50].

However, there is a recent controversial side-effect due to climate change and air pollution on the plant behaviour and thus on parks efficiency. A recent collaborative study with plant biologists and climate scientists [51] has pointed out how under certain extreme outdoor conditions, such as summer heatwaves and high air pollution rate, parks do not always "behave" as they should, in fact they decrease their cooling capacity and may not improve outdoor thermal comfort.

In an urban setting, the extensive use of trees also presents other risks, and in particular: excessive shading of certain areas, wind shadow, maintenance problems, and pollen allergies [47].

Thus, in evaluating and designing green spaces it is fundamental to consider all the risks derived from the initial environmental conditions and consequentially to design according to local characteristics. Finally, regarding systems of "superficial greenery", green roofs represent an interesting measure because they proved to have positive mitigation effects both on indoor as well on the outdoor [52], they can reduce the risks of heat waves, catch run-off water, and reduce the need for indoor cooling. By preventing incoming solar radiation from reaching the building structure below and by the evapotranspiration process of plants, green roofs guarantee a mitigation of the ambient outdoor temperatures.

In these cases, thermal performance is determined by: roof shading; evaporative cooling due to the presence of plants and of substratum of cultivation; and additional insulation levels of plants, caused by their substratum of cultivation and by the thermal mass of the soil [53]. Moreover, the energy benefits that can be obtained through the implementation of green roofs are also influenced by the local climate [22], by the morphology of the green roof, and by the technological components constituting the building. It has been found that the average solar radiation absorbed by vegetation is nearly 23% [54], and research conducted on the interiors of buildings in Southern Italy have found, during the summer, average surface temperatures inside buildings with green roofs are approximately 12 °C lower than those measured in buildings with traditional roofs [30]. Many recent studies have focused on the variations in air temperature outside the buildings, by comparing the values that can be obtained via mitigation of the UHI through the use of reflecting roofs on the one hand and green roofs on the other [22]. The results on an urban scale obtained on the applied green roofs have provided evidence of a reduction in urban room temperatures between 0.3 and 3 °C. In a recent study conducted by Berardi [55], the results show how the increase in the leaf area index (LAI) leads, during the daytime, to a cooling of the air temperature of to up to 0.4 °C more at the pedestrian level. In New York City, the use of numerous green roofs has made it possible to reduce summer afternoon temperatures by up to 0.6 °C [56], while in Hong Kong air temperature reductions due to the implementation of vegetation on the roofs of up to 0.7 °C at 2:00 p.m. have been measured [57].

Another interesting strategy belonging to GI and to the adaptation measures are rain gardens, also referred to as Sustainable Drainage Systems (SuDS).

SuDSs are able to reduce the impacts of the actions of the artificialization and in particular of the sealing of the soil, seeking to re-establish the natural water cycles and managing the rainwater system through processes of infiltration and evapotranspiration in situ. Sustainable Drainage Systems (SuDS), as rain gardens, bring back water to the surface, encouraging an interactive relationship with local residents and helping to improve urban landscape; they focus on the use of new paradigms, acting at the root point of the problem, that is to say, at the points of contact between rain water and the city: rooftops, private plots, gardens, squares, streets, etc.

Thanks to their ability to reduce air pollution, prevent flooding disasters, and improve urban open spaces, these systems are no longer considered as adaptation measures but as mitigation measures as well. SuDSs contain within them soil desealing strategies obtained through the use of permeable and/or porous pavements that bring beneficial effects to the soil, the air, and human comfort; porous pavement, for example, allows rainwater to infiltrate and evaporate in a significant quality in comparison with what takes place on asphalted surfaces, helping to mitigate the phenomenon of the urban heat island. In a recent study in Como [50], experiments on four different types of pavement were conducted and considerable reductions of evaporative cooling were measured from land with permeable pavement, in comparison with impermeable solutions: at a depth of 20 cm, the soil temperatures recorded beneath concrete and asphalt pavement were respectively 4 and 5 °C warmer than those recorded at the same depth beneath the soil with porous pavement or from unpaved surfaces [58].

3. Case Study Piazza Bainsizza, Rome

Piazza Bainsizza (Figure 1) is an area around 4500 square meters (90 m × 50 m), located in Rome in Prati|della Vittoria district: latitude 41° 53′ Q″ N, longitude 12° 30′ 67″ E, elevation 18 m and its climate is dry-subtropic Csa, according to Köppen-Geigen classification. The fronts of the buildings around the square arose in the post-war period, including a bus depot and three tree-lined roads that cross it and form the backdrop to the square. These buildings are approximately 5–7 floors high, with a height/width (H/W) ratio between 0.30 and 0.36. Over time, the oval-shaped centre of the square has taken on the use of a parking lot, forming a real roundabout for vehicles crossing it. Activities and shops are articulated along the wings outlined by the buildings facing the square, while inside, in the area not used as a parking lot, there are examples of black poplar over 20 m high. During the winter, the area is shaded all day long due to its compactness; on the contrary, Piazza Bainsizza receives direct radiation because of its morphology and dimension. During the summer, the piazza is almost completely unshaded all day long (Figure 2).

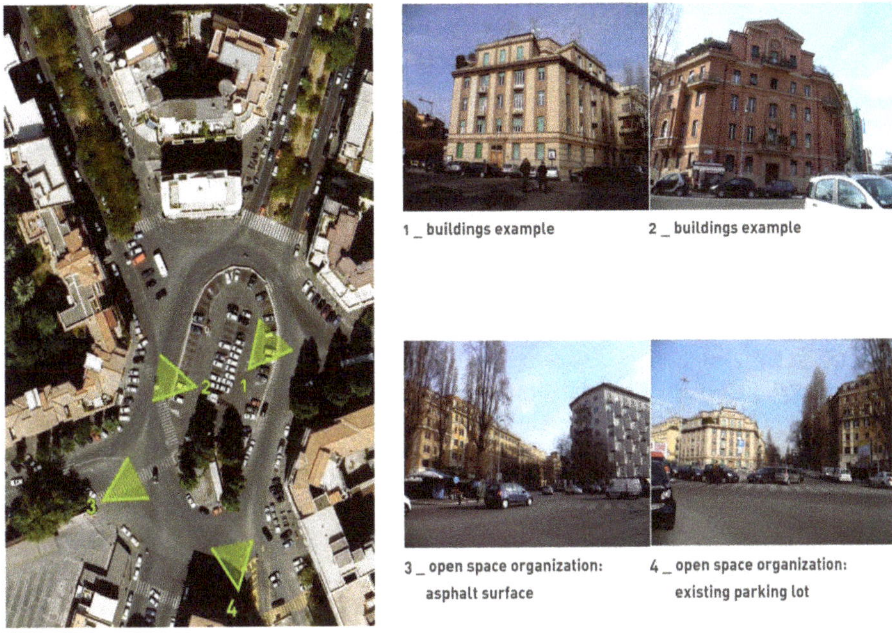

Figure 1. Site analysis and urban morphology.

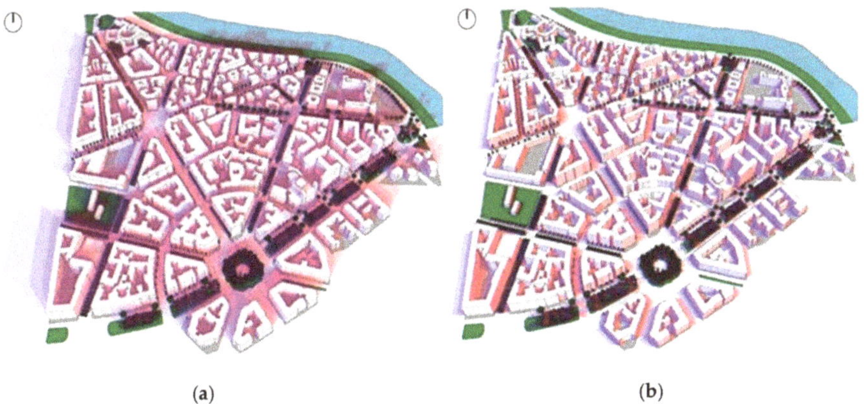

(a)　　　　　　　　　　(b)

Figure 2. Prati area shadow factor: (**a**) Winter shading analysis based on 21 December, with analysis range 9–12 (blue) 12–15 (red) hours and step 120 min; (**b**) Summer shading analysis based on 21 June, with analysis range 9–12 (blue) 12–15 (red) hours and step 120 min (analysis software: ECOTECT ANALYSIS 2010, software preset grid cell dimension).

The research methodology included a site analysis preliminary phase. The main climatic factors, represented by the average air temperature, solar radiation, wind speed and humidity, and the related seasonal and daily fluctuations were determined [59]; this data was determined looking at the related meteorological stations (in Prati|della Vittoria District). Then the morphological characteristics that best show the relationship with the microclimate were identified: form and size of open spaces and surrounding buildings; building density; SVF, Sky View Factor (defined by Nikolopoulou et al. [60], as the three dimensional measurement of the solid angle of sky view from an urban space, which determines the radiating heat exchange between city and sky); shading; the reflection and emissivity characteristics of materials and surfaces present in outdoor spaces (facades, roofs, paving systems); vegetation quality through an accurate survey of the existing vegetation.

Afterwards, different possible indicators were considered that could describe the microclimatic conditions and the thermo hygrometric comfort of the site, operating through Mean Radiant Temperature (MRT), and Physiological Equivalent Temperature (PET). The first parameter measures the average of temperatures for surfaces in an environment that are involved in a radiating thermal exchange and it represents the influence of the urban reservoir, the morphological system, and the materials on microclimate [61]. The second indicator, specifically developed for exterior spaces, defines the temperature as users perceive it, taking into account overall environmental meteorological factors, thermal balance, and thermoregulation processes of the human body, and therefore delivering a comfort value estimate that can easily be understood. Its accuracy, together with the other outdoor comfort indexes, has recently been questioned since it underestimates the strong adaptation capability of users to variable outdoor conditions and therefore it may result in an excessively negative evaluation compared to the actual perceived sensation; despite this, the PET allows us to define at least the trend of comfort with a fair amount of accurateness. The next step was to analyse the possible space morphological transformation strategies that can be applied to the specific context; since it was impossible to perform invasive actions variations on surfaces, integration of vegetation was considered. The possible actions operable on surfaces include: exterior paint or substitution of plaster coating on facades and roofs and the introduction of new materials on roofs and paving systems; allowing both colour variation and innovative systems such as "cool pavements" and "cool pigments" that present a high reflecting coefficient and high emissivity which lower heat gain. Moreover, it is possible to hypothesize the implementation of green areas in punctual, linear, and spread configurations of ground cover or trailing vegetal elements.

After this preliminary phase, characterized by information gathering procedures, a computer model was assembled within suitable simulation software (ENVI-met 3.1 by ENVI-met GmbH, Essen, Germany, RayMan by Albert-Ludwigs-Universität Freiburg, Freiburg, Germany), where previously collected climatic and morphological data concerning the research field could be merged. Such a model allows one to calculate the indicators related to microclimate and comfort (MRT e PET), selected both during the system's initial stage, highlighting problems and faults, and following the implementation of the design strategies analysed, in order to evaluate their effectiveness and highlight the relationship between microclimate and morphology.

The strategies can be considered both singularly and within a system, in order to define implementation of two different scenarios and classify them based on invasiveness of the actions, on financial feasibility and environmental compatibility. In this case, defining the simulation of the ante-operam benchmark state means establishing a standard model to refer to and with respect to which propose in the final phase a scale for evaluating the performance of the outdoor comfort index (PET) scenario; it will be put into evidence, from time to time, the most relevant factors and the most interesting time slots to be able to extrapolate considerations regarding the single material (thermal and physical characteristics, distribution), the single surface (exposure, configuration), and between different materials (e.g., innovative material-trees, innovative material-vegetation, etc.) in order to be able to investigate the relationship between the technical characteristics (mainly related to the material) and morphological (related to the spatial configuration and distribution) of the urban outdoor spaces. Within this treatment structure, the simulation of the starting scenario represents the basic reference on which to build connections and comparisons between the various scenarios of urban regeneration, constituting in effect the benchmark case of reference.

3.1. Analysis of Initial Meteorological Conditions and Simulated Models

Piazza Bainsizza was chosen as main reference area due its morphological, materials vegetation condition with its close fronts (70%), high of buildings around 25–30 m, little vegetation, pavements in asphalt and low buildings albedo (0.2–0.3).

The study was conducted during the summer and winter season on the outdoor spaces, and in particular the 21st of June and the 21st of December were selected in that they were considered two appropriate periods of time to analyse the effects of cool materials and vegetation on the outdoor microclimate and thermal comfort conditions in Rome.

The ENVI-met v.3.1 model, which is a Computational Fluid Dynamic (CFD) modelling, was used to calculate the human biometeorological conditions in the two urban regeneration scenarios proposed through the modelling of the four systems: soil, vegetation, atmosphere, and buildings.

The scenario was approximated in terms of spatial configurations within the specific constraints of the software, thus:

1. The grid cell dimension used for the ENVI-met models measures 2 m × 2 m.
2. Walls and roofs were modelled in the Database Manager specifying their three characteristic layers.
3. In the exterior layer of walls and roofs it was chosen to apply the prevalent material, that is concrete tiles for the roofs and brick block covered with lime plaster for the walls.

The ENVI-met simulations were carried out from 8:00 to 20:00 on 21st June 2015 and 21st December 2015. The input data used in ENVI-met include air temperature (Ta) and relative humidity, air velocity (v), and cloud cover (cc) from the Prati|della Vittoria District meteorological station (Table 1) and the others related to materials and vegetation setting in the different scenarios (S0, S1, S2) represented in Table 2.

Table 1. Initial settings for winter and summer simulation used in ENVI-met 3.1.

Input Data	Winter Simulation	Summer Simulation
Starting date and time	21 December 2015, 8:00 a.m.	21 June 2015, 8:00 a.m.
Total simulation time	12 h	12 h
Wind speed in 10 m (v)	0.9 m/s	1.3 m/s
Wind direction	50°	230°
Initial air temperature (T_{air})	9 °C	25 °C
Relative humidity (RH)	81%	69%
Roughness length (z_0)	0.1	0.1
Number x grid cells	44	44
Number y grid cells	44	44
Number z grid cells	30	30
Dimension of the grid in dx	2 m	2 m
Dimension of the grid in dy	2 m	2 m
Dimension of the grid in dz	2 m	2 m
Cloud cover (cc)	3	0
Albedo ground	0.15	0.15
Albedo roof	0.3	0.3
Albedo wall	0.2	0.2

The evaluation of outdoor thermal comfort was calculated (with RayMan) through the PET index (physiological equivalent temperature) defined as the air temperature at which, in an indoor environment (without wind and solar radiation), the heat balance of the human body between the internal temperature and that of the skin is balanced with the outdoor conditions to be assessed.

The input value in summer and winter are the same used for ENVI-met. The SVF, one of the main features of RayMan, was determined with ENVI-met. At the same time, the Leaf Area Density (LAD) was calculated with ENVI-met (Figure 3).

Table 2. Materials and vegetation settings for the different scenarios: S0, S1, S2.

Surface	S0		S1		S2		
	Name	Albedo (a)	Name	Albedo (a)	Name	Albedo (a)	LAD (m²/m³)
Roofs	concrete tile	0.30	cool pigment concrete tile	0.65	grass	0.20	-
Walls	concrete wall + cement plaster	0.40	concrete wall + cement plaster + cool pigmented	0.60	concrete wall + cement plaster + cool pigmented	0.60	-
Pavements	asphalt	0.20	cool pigment concrete tile	0.65	asphalt + grass	0.20	-
Vegetation	-	-	-	-	*Cedrus Atlantica* evergreen (Cat)	0.18	0.70–0.80
	-	-	-	-	*Pinus Pinea* evergreen (Pp)	0.18	0.70–0.80
	-	-	-	-	*Quercus ilex* deciduous (Qi)	0.20	0.80-0.90

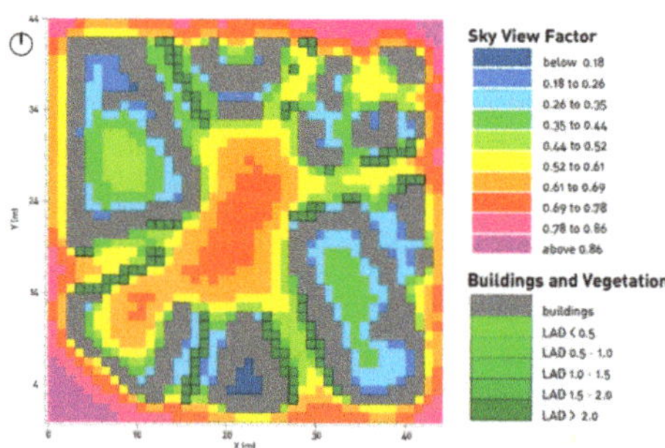

Figure 3. Piazza Bainsizza reference area: Sky view factor distribution and Leaf Area Density (analysis software: ENVI-met 3.1).

The Global radiation G (W/m²) was part of a meteorological datafile calculated from time, date, geographic position, and a cloud cover observation. For the thermophysiological parameter of the human body, three different users were simulated, based on the Munich Energy-balance Model for Individuals (MEMI): a standard European male (35 years old, 1.75 m tall, weight 75 kg), a standard European woman (35 years old, 1.60 m tall, 50 kg weight), and a standard European child (5 years old, 1.10 m tall, 18 kg weighted), both with a clothing index of 0, 6 clo in summer, 0.9 clo in winter, and an average activity rate of 80 W for adults and an activity rate of 120 W for the child (Figure 4).

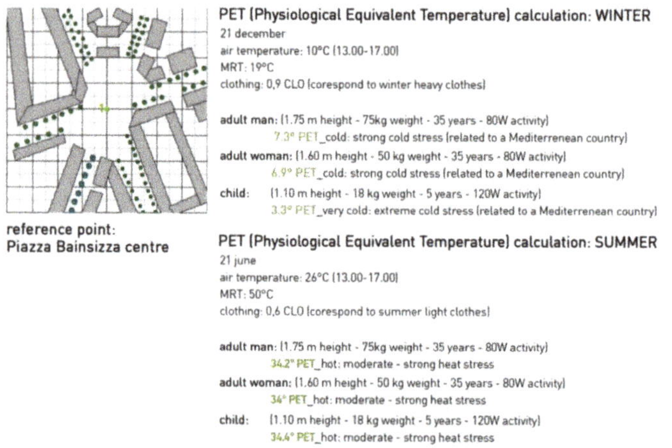

Figure 4. Microclimatic comfort analysis of initial scenario (S0): PET during winter and summer.

3.2. Definition of Scenario Simulation

The second study phase focused on simulations and the comparison of two different scenarios of regeneration of urban exterior spaces, in order to reduce the urban heat island effect and improve microclimatic comfort. The first scenario involves the use of cool materials for pavements, façades, and roofs, while the second scenario, beyond using cool materials, in addition to what was done in the first scenario inserts the use of vegetation (green roof, green pavements, and trellis). The yardstick of comparison of both scenarios was the initial condition, scenario 0, used as a reference model for assessing the validity and the improvement from time to time due to the implementation of the various innovative materials and of vegetation. In scenario 0, we may notice that most of the street presents a high value of SVF; in summer, they are shaded during the day and buildings store heat and release during the night, increasing the UHI (Figure 3).

In Figure 5a we can notice that during winter, wind from the north can pass through Piazza Bainsizza, which is unprotected. Surface temperature is medium because asphalt pavement is the common building material with low albedo value (0.15–0.3). Meanwhile during summer, as it is shown in Figure 5b, wind from the south-west can pass through Piazza Bainsizza due to urban streets configuration. Although, surface temperature is very high due to the low albedo of building and pavements material and the absence of vegetation.

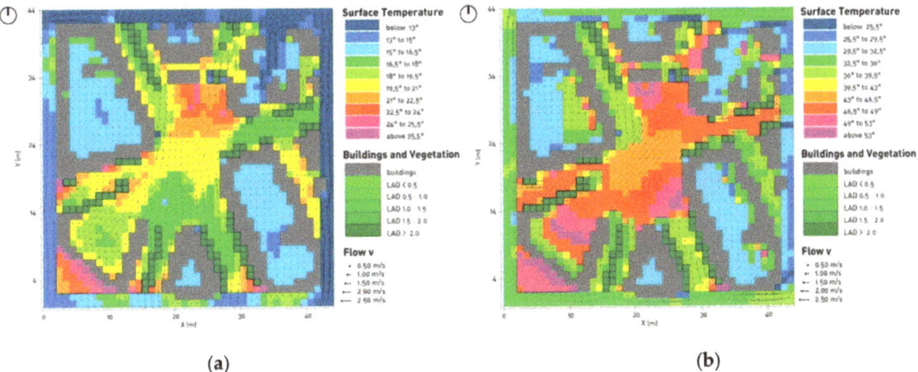

(a)   (b)

Figure 5. Piazza Bainsizza surface temperature and wind analysis: (**a**) Winter simulation based on 21st December at 3:00 p.m.; (**b**) Summer simulation based on 21st June at 3:00 p.m. (analysis software: ENVI-met 3.1).

The cool materials used in the urban regeneration process were selected and applied to meet the following criteria:

1. Optimization of possible nonmirrored reflectivity to solar radiation on the surfaces involved in the intervention.
2. Optimization of the emissivity factor.
3. Aesthetic value and integration due to the consolidated urban context.

The natural materials and elements like grass and trees were selected to meet the following criteria:

1. Planting with trees having medium-high leaf area density (LAD).
2. Planting implemented in keeping with the consolidated urban and Mediterranean context.

The first restructuring scenario (S1) is configured as a minimum intervention scenario that acts upon the surfaces at pedestrian level, in the replacement of the urban paving asphalt and the cement tiles of roofs with tiles made using cool materials and upon the building envelope through the application of cool pigments (Figure 6).

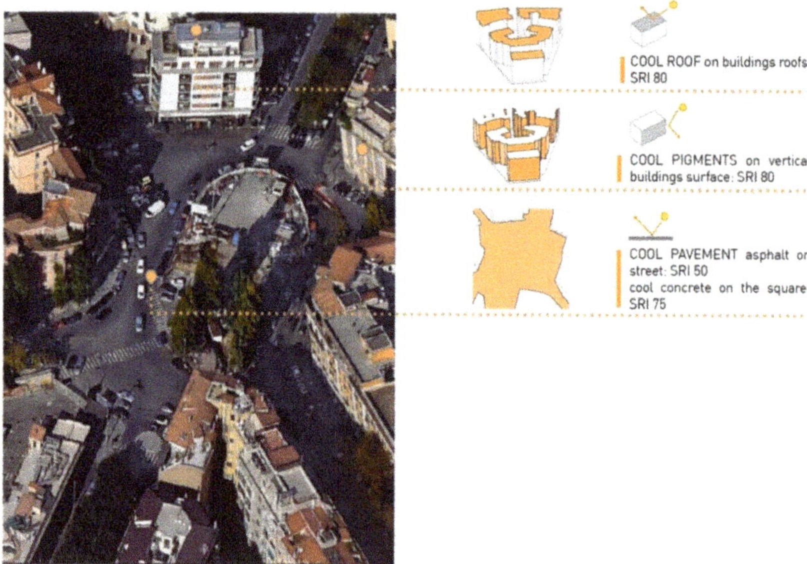

Figure 6. Scenario of intervention 1: surface renovation.

The second proposed requalification scenario (S2) applies the cool materials of scenario 1 in the same way, adding trees (28 new trees on the South West side in total) and vegetated surfaces among the urban pavements (Figure 7).

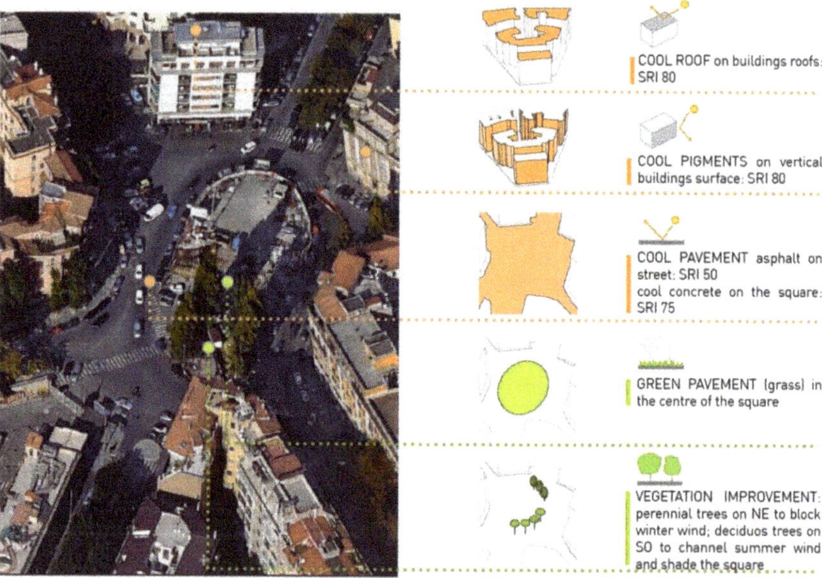

Figure 7. Scenario of intervention 2: surface renovation and vegetation improvement.

Figure 8 represents the initial condition scenario S0 (initial condition) and the two restructurings, scenario S1 (cool materials) and scenario S2 (cool materials + trees + grass).

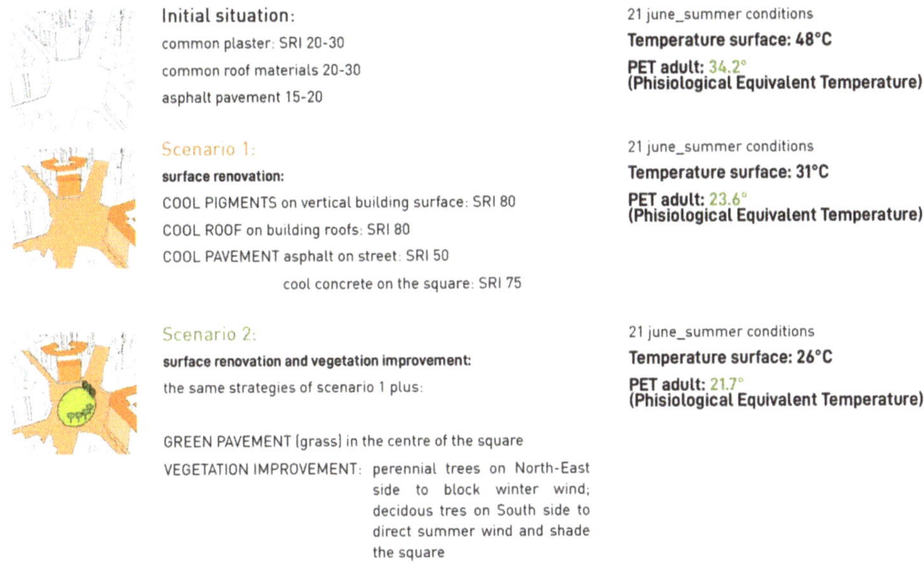

Figure 8. Summary of the comparison between the initial situation (S0) and the two different scenarios (S1 and S2) during summer (21 June).

4. Discussion of Results

In scenario 1, during winter, using a cool pavement and cool pigments and building materials, there is a slight decrease of surface temperature, as heat storage in urban fabric is reduced (Figure 9a). Meanwhile during summer (Figure 9b), cool materials use really improves microclimatic comfort by decreasing surface temperature up to almost 10 °C; however, direct radiation is still very high due to the complete lack of shading surfaces.

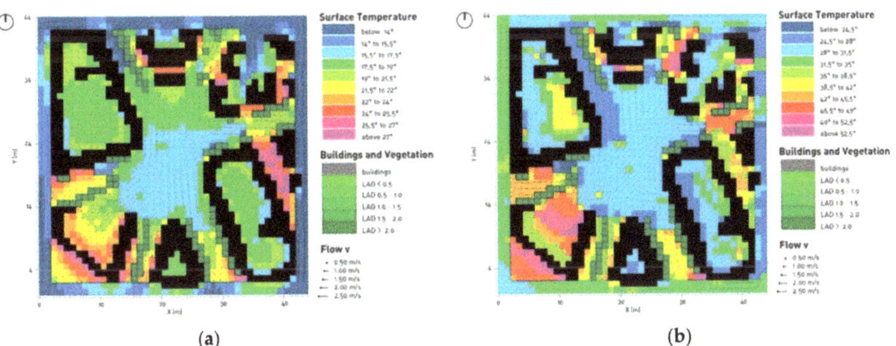

(a)   (b)

Figure 9. Scenario 1 surface temperature and wind speed analysis: (**a**) Winter analysis based on 21st December at 3:00 p.m.; (**b**) Summer analysis based on 21st June at 3:00 p.m. (analysis software: ENVI-met 3.1).

In scenario 2, during winter, vegetated pavement strongly reduces surface temperature, but coniferous trees on wind direction decrease cold winter wind speed (Figure 10a). During summer, cool materials, vegetation, and shading plants drastically decrease surface temperature, thus reducing heat storage in urban fabric and so lowering urban heat island effect (Figure 10b).

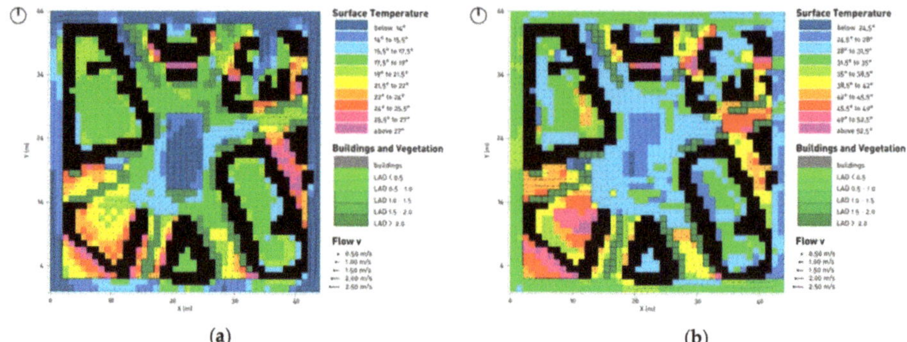

Figure 10. Scenario 2 surface temperature and wind speed analysis: (**a**) Winter analysis based on 21st December at 3:00 p.m.; (**b**) Summer analysis based on 21st June at 3:00 p.m. (analysis software: ENVI-met 3.1).

Assessing the PET variations found in the two urban regeneration scenarios, during the summer both scenario S1 and scenario S2 show a broader distribution of PET levels that are lower during the day than the initial condition evidenced in scenario S0, demonstrating their potential for thermal comfort, attenuating summer overheating.

As regards scenario S1 done with the use of cool materials, it is important to emphasize that regardless of whether they are used for pavement or on façades, they show a significant cooling effect on the surface temperature.

As to the scenario S2 results, the study shows the significant mitigation potential of trees, due to the shading determined by the geometry and characteristics of the canopy and the leaf area density (LAD).

The first scenario, with the use of cool materials, improves summer conditions and reduces the urban heat island effect but does not eliminate thermal discomfort due to the lack of shaded surfaces or vegetation. The second scenario, where material renovations are matched with vegetation improvements, has a small bad effect on winter conditions but drastically ameliorates the summer situation, both for direct users and, thanks to the strong reduction of the urban heat island effect, to urban inhabitants as a whole.

The air temperature (Tair) is the parameter that registers the least changes between the base scenario and the two proposed upgrading scenarios. In scenario 1 (cool materials) the delta is equal to $-0.5\,°C$ and $-1.0\,°C$ at 12.00 while there is a greater extension of the intermediate values, i.e., what is interesting to note is that it is not so much the delta that changes and the extension in the distribution of intermediate temperature values, which involve a greater area of the pedestrian square. Scenario 2 (cool pavements + trees) records a sharper lowering of the Tair value equal to $-1\,°C$, at 16.00 there is a delta of $-0.5\,°C$ compared to 16.00 of the starting scenario (Figure 11).

It is possible to see how the albedo factor affects convective exchanges more decisively, since by decreasing the surface temperature of the materials, the convective exchanges between surface and air and therefore the temperature of the air itself decrease. The parameter of the surface temperature (T surfaces) is perhaps one of the most easily read parameters of microclimatic simulations, since it is in direct correspondence with the physical and thermal characteristics of the material used in particular its reflection coefficient, as well as in direct relation to the shading factor. The most performing scenario is scenario 2, where the combination of trees and cool coloured flooring results in a delta equal to $-2\,°C$ (Figure 11).

The scenario that presents the highest values of Tmrt and therefore pejorative compared to the basic scenario turns out to be scenario 2, combination-application of cool coloured materials to all the architectural surfaces of the urban area, which records an increase in Tmrt equal to $+4.5\,°C$ at

8.00 and +7.3 °C at 12.00. During the day, therefore, the increase in the solar reflection of the surfaces leads to a lowering of the air temperature and a significant lowering of the temperatures of the surfaces themselves; however, the effect on the average radiant temperature is worse, with consequent modification of the PET (Figure 12).

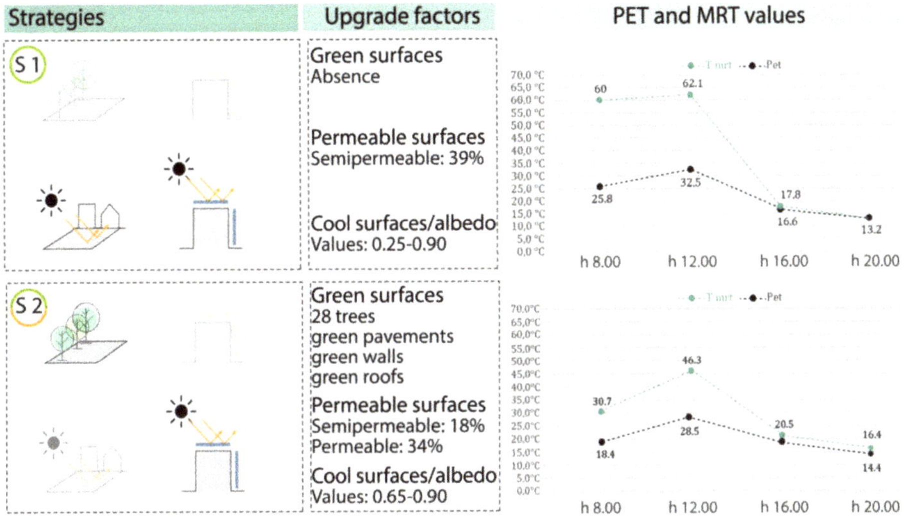

Figure 11. Variation of Tair, surface T and mrt T between S0, S1 and S2 during summer (21 June).

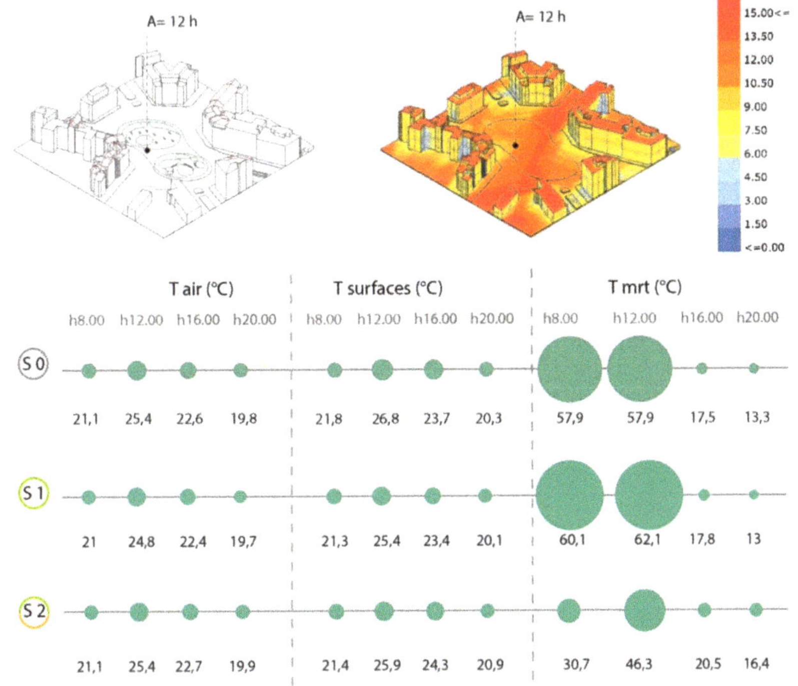

Figure 12. Final comparison between S1, S2 during summer (21 June).

5. Conclusions

The results of this study reveal certain aspects of importance for the regeneration of open spaces in the setting of an established Mediterranean city. In the first place, the study casts light on how, during the summer, significant results on attenuating the urban heat island can be obtained thanks to the use of innovative materials, particularly if accompanied by the use of shading vegetation.

Moreover, the use of vegetation has a better effect on outdoor thermal comfort, as shown by the scenario S2 results (Figures 11 and 12).

The open spaces with trees in combination with pavement made using cool materials, and cool roofs, are demonstrated to be an effective strategy for mitigating the phenomenon of summer overheating and for improving outdoor thermal comfort. As regards the use of cool material for the renewal of pavement, roofs, and façades in a consolidated urban environment, they improve the conditions of summer outdoor with a mitigation effect as concerns surface temperatures.

Additional investigations should be done on their behaviour in different orientations and design configurations with patterns of other materials, leading research to analyse additional combinations of hybrid technologies of cool materials, the use of vegetation like the green wall, and the use of water to promote evaporative cooling on the pedestrian level. Lastly, as to the use of vegetation, additional simulations might permit a more targeted choice of newly planted trees, assessing the performance of different trees and the cooling effect on the outdoor environment depending on the type, position, and orientation of the foliage.

In particular, more targeted studies could be conducted on the design of more comfortable open spaces, especially during summer overheating, for the fragile segment of the population such as children and even the elderly.

In fact, as shown in Figures 13 and 14, which compares the hottest hours between 13 and 17, the children in scenario 1 and 2 are in more disadvantaged conditions than a 35 year old man and woman (perceiving a PET of 3.5° lower in winter and only 0.7° lower in summer in scenario 1 and 2.8° lower in winter and 1.2° lower in summer in scenario 2 than that of an adult man and 3.1° lower in winter and only 0.3° lower in summer than that of an adult woman in scenario 1, while in scenario 2 it is 2.5° lower in winter and 0.8° lower in summer).

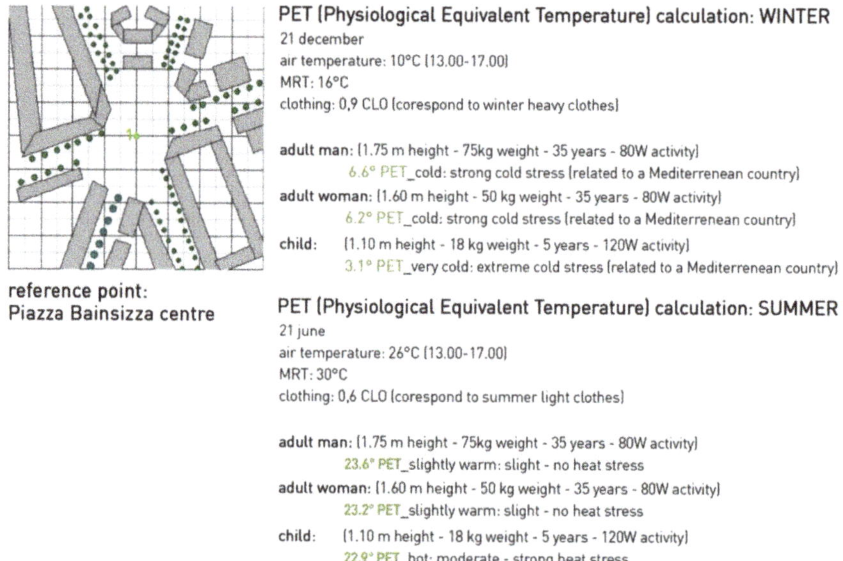

Figure 13. Microclimatic comfort analysis of scenario 1 (S1): PET during winter and summer.

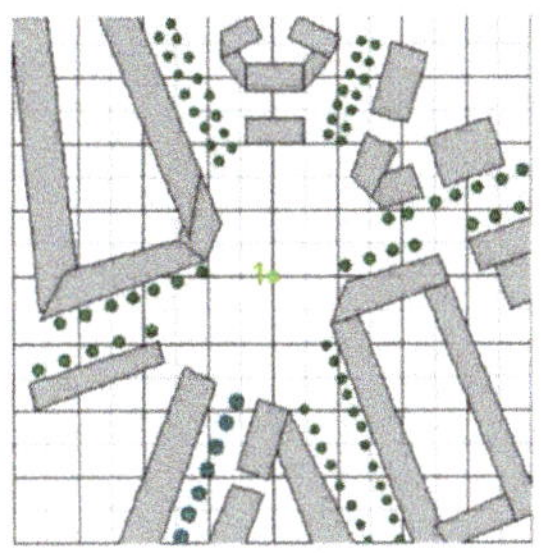

PET (Physiological Equivalent Temperature) calculation: WINTER

21 december
air temperature: 10°C (13.00-17.00)
MRT: 16°C
clothing: 0,9 CLO (corespond to winter heavy clothes)

adult man: (1.75 m height - 75kg weight - 35 years - 80W activity)
    5.8° PET_cold: strong cold stress (related to a Mediterrenean country)
adult woman: (1.60 m height - 50 kg weight - 35 years - 80W activity)
    5.5° PET_cold: strong cold stress (related to a Mediterrenean country)
child: (1.10 m height - 18 kg weight - 5 years - 120W activity)
    3.0° PET_very cold: extreme cold stress (related to a Mediterrenean country)

PET (Physiological Equivalent Temperature) calculation: SUMMER

21 june
air temperature: 26°C (13.00-17.00)
MRT: 30°C
clothing: 0,6 CLO (corespond to summer light clothes)

adult man: (1.75 m height - 75kg weight - 35 years - 80W activity)
    21.7° PET_comfortable: no thermal stress
adult woman: (1.60 m height - 50 kg weight - 35 years - 80W activity)
    21.3° PET_comfortable: no thermal stress
child: (1.10 m height - 18 kg weight - 5 years - 120W activity)
    20.5° PET_comfortable: no thermal stress

Figure 14. Microclimatic comfort analysis of scenario 2 (S2): PET during winter and summer.

Funding: This research received no external funding.

Acknowledgments: A special thanks to L. Martinelli for the simulations and calculations with Ecotect Analysis, ENVI-met, and RayMan for the data graphic visualization; thanks to F. Laureti for the graphic elaboration of Figures 11 and 12.

Conflicts of Interest: The author declares no conflict of interest.

References

1. Intergovernmental Panel for Climate Change. *IPCC Global Warming of 1.5 °C An IPCC Special Report on the Impacts of Global Warming of 1.5 °C Above Pre-Industrial Levels and Related Global Greenhouse Gas Emission Pathways, in the Context of Strengthening the Global Response to the Threat of Climate Change, Sustainable Development, and Efforts to Eradicate Poverty*, 1st ed.; IPCC: Geneva, Switzerland, 2018; ISBN 978-92-9169-151-7.
2. UNFCCC. *Yearbook of Global Climate Action 2019*, 1st ed.; UNFCCC: Bonn, Germany, 2019; ISBN 978-92-9219-186-3.
3. Landsberg, H.E. *The Urban Climate*; Academic Press: New York, NY, USA, 1983; ISBN 978-0-12-435960-4.
4. Manley, G. On the frequency of snowfall in metropolitan England. *Q. J. R. Meteorol. Soc.* **1958**, *84*, 70–72. [CrossRef]
5. US EPA. Reducing Urban Heat Islands: Compendium of Strategies. 2009. Available online: http://www.epa.gov/heatisld/resources/compendium.html (accessed on 12 May 2019).
6. Pearlmutter, D. Patterns of sustainability in desert architecture. *Arid Lands Newsl.* **2000**, *47*, 1–12. Available online: http://ag.arizona.edu/oals/ALN/aln47/pearlmutter.html (accessed on 12 May 2019).
7. US EPA. Smart Growth and Urban Heat Islands. 2009. Available online: http://www.epa.gov/heatisld/resources/pdf/smartgrowthheatislands.pdf (accessed on 12 May 2019).
8. Oke, T.R. The energetic basis of the urban heat island. *Q. J. R. Meteorol. Soc.* **1982**, *108*, 1–24. [CrossRef]
9. Golden, J. The built environment induced urban heat island effect in rapidly urbanizing arid regions—A sustainable urban engineering complexity. *Environ. Sci.* **2004**, *1*, 249–321. [CrossRef]
10. Santamouris, M. Heat island research in Europe—The state of the art. *ABER* **2007**, *1*, 123–150. [CrossRef]
11. Lee, H.; Mayer, H.; Chen, L. Contribution of trees and grasslands to the mitigation of human heat stress in a residential district of Freiburg, Southwest Germany. *Landsc. Urban Plan.* **2016**, *148*, 37–50. [CrossRef]

12. Piselli, C.; Castaldo, V.; Pigliautile, I.; Pisello, A.; Cotana, F. Outdoor comfort conditions in urban areas: On citizens' perspective about microclimate mitigation of urban transit areas. *Sustain. Cities Soc.* **2018**, *39*, 16–36. [CrossRef]

13. Nouri, A.S.; Costa, J.P. Addressing thermophysiological thresholds and psychological aspects during hot and dry mediterranean summers through public spacedesign: The case of Rossio. *Build. Environ.* **2017**, *118*, 67–90. [CrossRef]

14. Hassler, U.; Kohler, N. Resilience in the built environment. *Build. Res. Inf.* **2014**, *42*, 119–129. [CrossRef]

15. Holling, C.S. Myths of ecology and energy. In *Perspectives on Energy: Issues, Ideas, and Environmental Dilemmas*, 2nd ed.; Ruedisili, L.C., Firebaugh, M.W., Eds.; Oxford University Press: New York, NY, USA, 1982; Volume 3, pp. 8–15.

16. Cretney, R. Resilience for whom? Emerging critical geographies of socio-ecological resilience. *Geogr. Compass* **2014**, *8*, 627–640. [CrossRef]

17. Holling, C.S. Resilience and stability of ecological systems. *Annu. Rev. Ecol. Syst.* **1973**, *4*, 1–23. [CrossRef]

18. Nelson, S.H. Resilience and the neoliberal counter-revolution: From ecologies of control to production of the common. *Resilience* **2014**, *2*, 1–17. [CrossRef]

19. Gunderson, H.; Pritchard, L. *Resilience and the Behavior of Large-Scale Systems*, 1st ed.; Island Press: Covelo, CA, USA, 2002; ISBN 1-55963-970-9.

20. Meerow, S.; Newell, J.P.; Stults, M. Defining urban resilience: A review. *Landsc. Urban Plan.* **2016**, *147*, 38–49. [CrossRef]

21. OECD. Resilient Cities. Available online: http://www.oecd.org/cfe/regional-policy/resilient-cities.html (accessed on 21 July 2018).

22. Santamouris, M. Cooling the cities—A review of reflective and green roof mitigation technologies to fight heat island and improve comfort in urban environments. *Sol. Energy* **2014**, *103*, 682–703. [CrossRef]

23. European Environment Agency. *Urban Adaptation to Climate Change in Europe 2016: Transforming Cities in a Changing Climate*, 1st ed.; EEA: Copenhagen, Denmark, 2016; ISBN 978-92-9213-741-0.

24. Shashua-Bar, L.; Hoffman, M.E. Vegetation as a climatic component in the design of an urban street: An empirical model for predicting the cooling effect of urban green areas with trees. *Energy Build.* **2000**, *31*, 221–235. [CrossRef]

25. Pisello, A.L. State of the art on the development of cool coatings for buildings and cities. *Sol. Energy* **2017**, *144*, 660–680. [CrossRef]

26. Yang, J.; Wang, Z.H.; Kaloush, K.E. Environmental impacts of reflective materials: Is high albedo a 'silver bullet' for mitigating urban heat island? *Renew. Sustain. Energy Rev.* **2015**, *47*, 830–843. [CrossRef]

27. Akbari, H.; Cartalis, C.; Santamouris, M.; Synnefa, A.; Kolokotsa, D.; Muscio, A.; Pisello, A.L.; Rossi, F.; Wong, N.H.; Zinzi, M. Local climate change and urban heat island mitigation techniques—The state of the art. *J. Civ. Eng. Manag.* **2016**, *22*, 1–16. [CrossRef]

28. Santamouris, M.; Gaitani, N.; Spanou, A.; Saliari, M.; Giannopoulou, K.; Vasilakopoulou, K.; Kardomateas, T. Using cool paving materials to improve microclimate of urban areas—Design realization and results of the flisvos project. *Build. Environ.* **2012**, *53*, 128–136. [CrossRef]

29. Erell, E.; Pearlmutter, D.; Boneh, D.; Kutiel, P.B. Effect of high-albedo materials on pedestrian heat stress in urban street canyons. *Urban Clim.* **2014**, *10*, 367–386. [CrossRef]

30. Taleghani, M.; Berardi, U. The effect of pavement characteristics on pedestrians' thermal comfort in Toronto. *Urban Clim.* **2018**, *24*, 449–459. [CrossRef]

31. Chatzidimitriou, A.; Yannas, S. Microclimate development in open urban spaces: The influence of form and materials. *Energy Build.* **2015**, *108*, 156–174. [CrossRef]

32. Shahidan, M.F.; Jones, P.J.; Gwilliam, J.; Salleh, E. An evaluation of outdoor and building environment cooling achieved through combination modification of trees with ground materials. *Build. Environ.* **2012**, *58*, 245–257. [CrossRef]

33. Laureti, F.; Martinelli, L.; Battisti, A. Assessment and mitigation strategies to counteract overheating in urban historical areas in Rome. *Climate* **2018**, *6*, 18. [CrossRef]

34. Kolokotsa, D.; Maravelaki-Kalaitzaki, P.; Papantoniou, S.; Vangeloglou, E.; Saliari, M.; Karlessi, T.; Santamouris, M. Development and analysis of mineral based coatings for buildings and urban structures. *Sol. Energy* **2012**, *86*, 1648–1659. [CrossRef]

35. Synnefa, A.; Santamouris, M.; Livada, I. A study of the thermal performance of reflective coatings for the urban environment. *Sol. Energy* **2006**, *80*, 968–981. [CrossRef]

36. Karlessi, T.; Santamouris, M.; Apostolakis, K.; Synnefa, A.; Livada, I. Development and testing of thermochromic coatings for buildings and urban structures. *Sol. Energy* **2009**, *83*, 538–551. [CrossRef]

37. Bretz, S.E.; Akbari, H. Long-term performance of high albedo roof coatings. *Energy Build.* **1997**, *25*, 159–167. [CrossRef]

38. Saaroni, H.; Amorim, J.; Hiemstra, J.; Pearlmutter, D. Urban Green Infrastructure as a tool for urban heat mitigation: Survey of research methodologies and findings across different climatic regions. *Urban Clim.* **2018**, *24*, 94–110. [CrossRef]

39. Building a Green Infrastructure for Europe. Available online: https://www.worldcat.org/title/building-a-green-infrastructure-for-europe/oclc/875969246&referer=brief_results (accessed on 4 August 2018).

40. United Nations Human Settlements Program (UN-Habitat). *The State of European Cities 2016 Cities Leading the Way to a Better Future*, 1st ed.; Publications Office: Luxembourg, 2016; ISBN 978-92-79-64259-3.

41. Human-Biometeorological Conditions and Thermal Perception in a Mediterranean Coastal Park. Available online: https://link.springer.com/article/10.1007%2Fs00484-014-0944-z (accessed on 22 July 2018).

42. Spronken-Smith, R.A.; Oke, T.R. The thermal regime of urban parks in two cities with different summer climates. *Int. J. Remote Sens.* **1998**, *19*, 2085–2104. [CrossRef]

43. Bowler, D.E.; Buyung-Ali, L.; Knight, T.M.; Pullin, A.S. Urban greening to cool towns and cities: A systematic review of the empirical evidence. *Landsc. Urban Plan.* **2010**, *97*, 147–155. [CrossRef]

44. Ketterer, C.; Matzarakis, A. Human-biometeorological assessment of heat stress reduction by replanning measures in Stuttgart, Germany. *Landsc. Urban Plan.* **2014**, *122*, 78–88. [CrossRef]

45. Analysis and Comparison of Shading Strategies to Increase Human Thermal Comfort in Urban Areas. Available online: http://www.mdpi.com/2073-4433/9/3/91 (accessed on 22 July 2018).

46. Upreti, R.; Wang, Z.H.; Yang, J. Radiative shading effect of urban trees on cooling the regional built environment. *Urban For. Urban Green.* **2017**, *26*, 18–24. [CrossRef]

47. Roy, S.; Byrne, J.; Pickering, C. A systematic quantitative review of urban tree benefits, costs, and assessment methods across cities in different climatic zones. *Urban For. Urban Green.* **2012**, *11*, 351–363. [CrossRef]

48. Rahman, M.A.; Armson, D.; Ennos, A.R. A comparison of the growth and cooling effectiveness of five commonly planted urban tree species. *Urban Ecosyst.* **2015**, *18*, 371–389. [CrossRef]

49. Akbari, H.; Pomerantz, M.; Taha, H. Cool surfaces and shade trees to reduce energy use and improve air quality in urban areas. *Sol. Energy* **2001**, *70*, 295–310. [CrossRef]

50. Fini, A.; Frangi, P.; Mori, J.; Donzelli, D.; Ferrini, F. Nature based solutions to mitigate soil sealing in urban areas: Results from a 4-year study comparing permeable, porous, and impermeable pavements. *Environ. Res.* **2017**, *156*, 443–454. [CrossRef] [PubMed]

51. Kala, J.; De Kauwe, M.G.; Pitman, A.J.; Medlyn, B.E.; Wang, Y.P.; Lorenz, R.; Perkins-Kirkpatrick, S.E. Impact of the representation of stomatal conductance on model projections of heatwave intensity. *Sci. Rep.* **2016**, *6*, 23418. [CrossRef]

52. A Practical Guide to Cool Roofs and Cool Pavements. Available online: https://www.coolrooftoolkit.org/read-the-guide/ (accessed on 22 July 2018).

53. Liu, K.; Baskaran, B. Thermal performance of green roofs through field evaluation. In Proceedings of the first North American Green Roof Infrastructure Conference, Awards and Trade Show: Greening Rooftops for Sustainable Communities, Chicago, IL, USA, 29–30 May 2003.

54. Lazzarin, R.M.; Castellotti, F.; Busato, F. Experimental measurements and numerical modelling of a green roof. *Energy Build.* **2005**, *37*, 1260–1267. [CrossRef]

55. Berardi, U. The outdoor microclimate benefits and energy saving resulting from green roofs retrofits. *Energy Build.* **2016**, *121*, 217–229. [CrossRef]

56. Smith, K.R.; Roebber, P.J. Green roof mitigation potential for a proxy future climate scenario in Chicago, Illinois. *J. Appl. Meteorol. Climatol.* **2011**, *50*, 507–522. [CrossRef]

57. Rosenzweig, C.; Solecki, W.; Parshall, L.; Gaffin, S. *Mitigating New York City's Heat Island with Urban Forestry, Living Roofs and Light Surfaces*, 1st ed.; Columbia University: New York, NY, USA, 2006.

58. Asaeda, T.; Ca, V.T.; Wake, A. Heat storage of pavement and its effect on the lower atmosphere. *Atmos. Environ.* **1996**, *30*, 413–427. [CrossRef]

59. Mancini, F.; Nastasi, B. Solar energy data analytics: PV deployment and land use. *Energies* **2020**, *13*, 417. [CrossRef]
60. Nikolopoulou, M.; Steemers, K. Thermal comfort and psychological adaptation as a guide for designing urban spaces. *Energy Build.* **2003**, *35*, 95–101. [CrossRef]
61. Dessì, V.; Rogora, R. *Il Comfort Ambientale Degli Spazi Aperti*; EdicomEdizioni: Gorizia, Italy, 2005.

Publisher's Note: MDPI stays neutral with regard to jurisdictional claims in published maps and institutional affiliations.

Article

Thermo-Fluid Dynamics Analysis of Fire Smoke Dispersion and Control Strategy in Buildings

Ricardo S. Gomez [1,*], **Túlio R. N. Porto [2]**, **Hortência L. F. Magalhães [3]**, **Antonio C. Q. Santos [2]**, **Victor H. V. Viana [4]**, **Kelly C. Gomes [1,5]** and **Antonio G. B. Lima [2]**

[1] Postgraduate Program in Mechanical Engineering, Federal University of Paraiba, João Pessoa 58051-900, PB, Brazil; gomes@cear.ufpb.br

[2] Department of Mechanical Engineering, Federal University of Campina Grande, Campina Grande 58429-900, PB, Brazil; trnporto@gmail.com (T.R.N.P.); antoniocarlos_queiroz@hotmail.com (A.C.Q.S.); antonio.gilson@ufcg.edu.br (A.G.B.L.)

[3] Department of Chemical Engineering, Federal University of Campina Grande, Campina Grande 58429-900, PB, Brazil; hortencia.luma@gmail.com

[4] Postgraduate Program in Security Engineering, University Center of Patos (UNIFIP), Campina Grande 58416-440, PB, Brazil; vhviana13@gmail.com

[5] Department of Renewable Energy Engineering, Federal University of Paraiba, João Pessoa 58051-900, PB, Brazil

* Correspondence: ricardosoaresgomez@gmail.com; Tel.: +55-83-98893-7991

Received: 15 October 2020; Accepted: 12 November 2020; Published: 17 November 2020

Abstract: Smoke is the main threat of death in fires. For this reason, it becomes extremely important to understand the dispersion of this pollutant and to verify the influence of different control systems on its spread through buildings, in order to avoid or minimize its effects on living beings. Thus, this work aims to perform thermo-fluid dynamic study of smoke dispersion in a closed environment. All numerical analysis was performed using the Fire Dynamics Simulator (FDS) software. Different simulations were carried out to evaluate the influence of the exhaust system (natural or mechanical), the heat release rate (HRR), ventilation and the smoke curtain in the pollutant dispersion. Results of the smoke layer interface height, temperature profile, average exhaust volumetric flow rate, pressure and velocity distribution are presented and discussed. The results indicate that an increase in the natural exhaust area increases the smoke layer interface height, only for the well-ventilated compartment (open windows); an increase in the HRR accelerates the downward vertical displacement of the smoke layer and that the 3 m smoke curtain is efficient in exhausting smoke, only in the case of poorly ventilated compartments (i.e., with closed windows).

Keywords: smoke; natural exhaust; mechanical exhaust; smoke curtain; fire dynamics simulator

---

1. Introduction

Fire is an irreversible process that involves the production of flame, heat, smoke and toxic gases, which can cause material losses, physical trauma, severe burns, respiratory and cardiovascular diseases, and death [1,2]. The main causes of fire are electrical failures, activities related to cooking, friction that occurs in machines and equipment, cutting and welding processes, improper handling of materials and equipment, leakage or release of flammable liquids and/or gases, human errors or unsafe human behavior, such as smoking in inappropriate places, and arson [1,3–9].

Researchers are unanimous in asserting that the main threat of death from fires is smoke [2,10–17]. According to Anseeuw et al. [14], 60% to 80% of deaths at the fire scene are attributed to smoke inhalation. This is because, during a fire, the burning of solid fuels causes a reduction in the oxygen concentration in the environment and an incomplete combustion of gases, generating highly toxic products such as CO and HCN, contributing to the occurrence of death by asphyxiation at the fire scene [10]. In addition to the problem of inhalation, smoke can cause fear, panic, tearing and irritation of the eyes, and reduced visibility, factors in turn make it difficult to safely exit the building.

Stefanidou et al. [2] presented the main toxic and irritating chemicals generated by the combustion of common building materials, as well as the main factors that contribute to the development of smoke inhalation injuries. In general, the smoke initially affects the upper airways (upper respiratory tract) of the fire victim and may, in a short time, becoming a complex life-threatening systemic disease, affecting all organs in the body [18].

In a fire, as the materials undergo combustion, they release hot smoke that, being lighter than ambient air, moves vertically upwards faster than horizontally, developing an inverted cone shape, well known as a plume. When the plume reaches the top of the building, with a certain speed, the smoke spreads radially across the ceiling, forming a layer of smoke (or hot gases). After the smoke covers the entire ceiling, it tends to move vertically downwards until the entire environment is filled with smoke or until mass flow rate entering the hot gas layer is balanced by the mass flow rate of exhaustion [12].

The smoke layer interface height, an extremely important parameter in the design of smoke exhaust systems, is defined as the distance, vertically, between the building floor and the smoke layer interface. Thus, it is of great importance, for people's survival, that the smoke layer interface height is above the height of their heads, for a sufficient time, so that an efficient evacuation from the fire scene is possible, that is, that this procedure occurs with minimal toxic gases inhalation.

In Brazil, the subject of fire safety started to be widely discussed after the tragedy that occurred at a nightclub located in the city of Santa Maria-RS, in January 2013. At that time, where 242 people died and almost 700 were injured. Even so, until the present moment, there is no national standard in the country that defines the parameters to be adopted in smoke control design. In the absence of these standards, the Technical Instructions established by the Fire Department of the Military Police of the State of São Paulo, 2019, which have as reference several specifications contained in international regulations, such as the NFPA and DIN standards, provide minimum fire safety requirements and, in some cases, specifying design parameters and fire protection systems installation.

Given the above, scientific studies related to fires in closed compartments are crucial for the design of smoke control systems, allowing the development and improvement of engineering strategies and techniques aimed at protecting the lives of its occupants, the facilities and equipment.

The use of computational fluid dynamics (CFD) to solve problems related to confined fires has become quite popular due to advances in computational power and numerical methods [19]. This technique has provided a better understanding of the behavior of this phenomenon and has made it possible to reduce costs, time and risks, when compared to experimental analyzes, especially in hypothetical scenarios of fires that are difficult to be implemented through experiments [17]. However, it is important to emphasize that experimental tests play important role in the development and validation of mathematical models to be used as the CFD tools.

One of the main CFD software packages used to study the behavior of fires in buildings is the Fire Dynamics Simulator (FDS), developed and available at no cost from the National Institute for Standards and Technology (NIST, Gaithersburg, MD, USA). In order to reduce or even eliminate the uncertainties of the numerical results, several studies proposed to verify [20,21] and validate [12,16,17,22–26] the models used by the FDS software.

In addition to the FDS software, other CFD tools have already been used in the literature to study the behavior of fires in closed environments, such as CFX [27–29], FLUENT [30,31] and ISIS CODE [32–34].

Qin et al. [20] investigated the influence of different exhaust systems for a fire in a gymnasium with a capacity for 18,000 people, using the FDS software. The authors observed that an increase in the speed of the mechanical exhaust fan positioned on the ceiling, in the range between 2.0 and 3.0 m/s, does not necessarily promote a more efficient smoke exhaustion, thus, there is a critical speed that depends on the heat release rate (HRR) from the fire. Further, the downward vertical displacement of smoke layer for the mechanical exhaust fan at a speed of 3.0 m/s occurred more quickly as compared to the natural exhaust system located in the same position. For the cases in which the mechanical exhaust fans were installed on the walls, the smoke exhaust occurred much more efficiently as compared to the use of natural exhaust fans in the same position, and a critical speed was not obtained, that is, the higher the fan speed, the lower the downward vertical displacement of the smoke. In this last analysis, an increase in speed from 1.5 to 3.0 m/s of the mechanical exhaust fan provided a more efficient smoke exhaust.

Qin et al. [12] validated the FDS software by comparing numerical results with experimental data for a fire in an atrium with internal dimensions of 22.40 m × 11.90 m and 27.00 m height, and cases with low (560 kW) and high (4 MW) HRR. For both cases, it was observed that the natural smoke exhaust vents are more efficient when located on the roof of the atrium. On the other hand, when the exhaust vents are located on the walls of the atrium, higher positions are preferred. Subsequently, the authors evaluated the influence of the positioning of the burners, noting that the smoke layer descends more rapidly when the burner is located in the center of the atrium.

Xiao [24] compared numerical results using FDS with experimental data of temperature and mass flow rate in the doorway of a room with dimensions of 9.75 m × 4.88 m × 2.44 m (length × width × height). This compartment has only one opening and a 0.46 m^2 propane burner with different heat release rates. From the obtained results, the author observed a reasonable agreement between the results of temperature and mass flow rate in the opening for the cases with and without sprinkler. A greater discrepancy was observed between the numerical and experimental results of temperature and mass flow rate in the opening, for the case with sprinkler, due to the stronger turbulence, uncertainties in the fire spread rate and in the water behavior (spray angles, number of drops per second, initial speed and average drop diameter).

Ayala et al. [26] showed good agreement between the numerical results obtained by the FDS software and the experimental data of 1.36 MW and 2.34 MW pool fires burning inside a 20 m cubic atrium with a natural ventilation system. In addition, the authors showed that the area-to-height-squared ratio of the atrium, in the range of 0.3 to 3.8, does not present significant effects on the temperature and smoke layer growth.

Abotaleb [16] performed a numerical analysis using the FDS software to evaluate the influence of smoke management techniques in a building with dimensions of 10.00 m × 10.00 m and 12.00 m in height. The author observed that using six mechanical make-up air on the walls and a mechanical exhaust fan on the roof with total volumetric flow rates equivalent to 36 and 40 air changes per hour (ACH), respectively, decreasing the downward vertical displacement of the smoke layer interface height and the average temperature by 71.18% and 31.6%, respectively.

Shih et al. [35] performed a numerical simulation using the FDS software, proving that make-up air has a significantly influences on the effectiveness of a natural smoke exhaust system in a tall space, with dimensions of 8.00 m × 1.00 m and 10.00 m height, under fire scenario. In the research, the authors used the Schlieren photography technique, that allows visualization of the post-combustion hot gas distribution in the model space, to validate the simulation results.

Yuen et al. [17] used the FDS software to numerically evaluate the efficiency of natural exhaust fans and a smoke curtain in an atrium. In this research, the values of temperature and smoke layer interface height were compared with experimental measures reported by Hägglund et al. [36], obtaining good agreement for both cases (with and without a smoke curtain). The authors observed that the smoke curtain is efficient to compartmentalize the smoke, as long as its height is sufficient to completely block the spread of smoke to the other side of the environment.

Huang et al. [37] performed a numerical analysis using the FDS to evaluate the relationship between the obscuration ratio, the main parameter of smoke detectors, and soot yield, which is defined as the mass of soot produced per mass of fuel reacted. The simulated compartment has dimensions 10.00 m × 7.00 m × 4.00 m (length × width × height) without openings to the outside and a with fire source in the center. After analyzing several fire scenarios, the authors observed that the smoke speed, in the vertical, was 0.54 m/s and, the higher the soot yield, the higher the obscuration rate. In addition, the results of the simulation indicate that, at a height of 3.00 m from the floor, the diameter of the smoke plume varied between 0.30 and 0.60 m during the first 300 s of firing.

Tan et al. [38] numerically investigated the influence of HRR and ambient pressure on the efficiency of the smoke extraction system in road tunnel fires using FDS software. In addition, the authors showed that there is a critical exhaust rate at which there is an excessive fresh air discharge from the exhaust vent, decreasing the efficiency of the system.

More recently, Wang et al. [39] performed several numerical simulations to investigate the influence of different smoke control systems on smoke flow, temperature and visibility in a subway station, assisting passenger evacuation and firefighting. As results, the authors presented the best scheme for air control and smoke exhaustion for different fire locations.

Despite the importance, no studies were found to evaluate, jointly, the influence of the type and dimensions of the exhaust system, HRR, natural ventilation (openings in the lower region of the compartment for air intake) and smoke curtain in the temperature distribution and smoke dispersion during a fire in an enclosed space.

Thus, complementing the cited works, the main purpose of this work is to evaluate the thermo-fluid dynamic behavior of smoke originated from a fire in an enclosed space using the FDS software. The studied cases were elaborated in order to verify the influence of the HRR, natural exhaust fans, mechanical exhaust fans, smoke curtain and ventilation (opening windows) in the lower compartment at the smoke layer interface height, in the temperature distribution in the simulated compartment and in the exhaust volumetric flow rate. In addition, the influence of the smoke curtain and opening windows on the pressure and smoke velocity vector fields inside the compartment under analysis are also evaluated.

2. Methodology

2.1. The Physical Problem and the Computational Domain

The physical problem under study consists in evaluating the fluid dynamic behavior, spread and exhaust of the smoke generated from a burner located in a closed compartment. The compartment has dimensions 30.00 m × 15.00 m × 6.00 m (length × width × height), containing a door and four windows (of the same dimension), four exhaust fans, a smoke curtain and a burner centered in the right quadrant of the compartment, as shown in Figure 1.

The FDS software developers recommend that the computational domain should be extended beyond the physical domain when there are openings (doors, windows and exhaust vents), in order to guarantee a pressure boundary condition in the openings that is closer to reality [40]. In view of this recommendation, Wang et al. [41] carried out a numerical study and proved that the values predicted by the FDS software for the mass flow through a door were closer to the experimental data for greater distances between the limit of the computational domain and the opening of the physical domain.

Thus, in this research, the computational domain was extended 2 m beyond the dimensions of the compartment on the three faces where there are openings to the external environment. After extension, the computational domain started to have the following dimensions: 32.00 m × 17.00 m × 8.00 m, as shown in Figure 1.

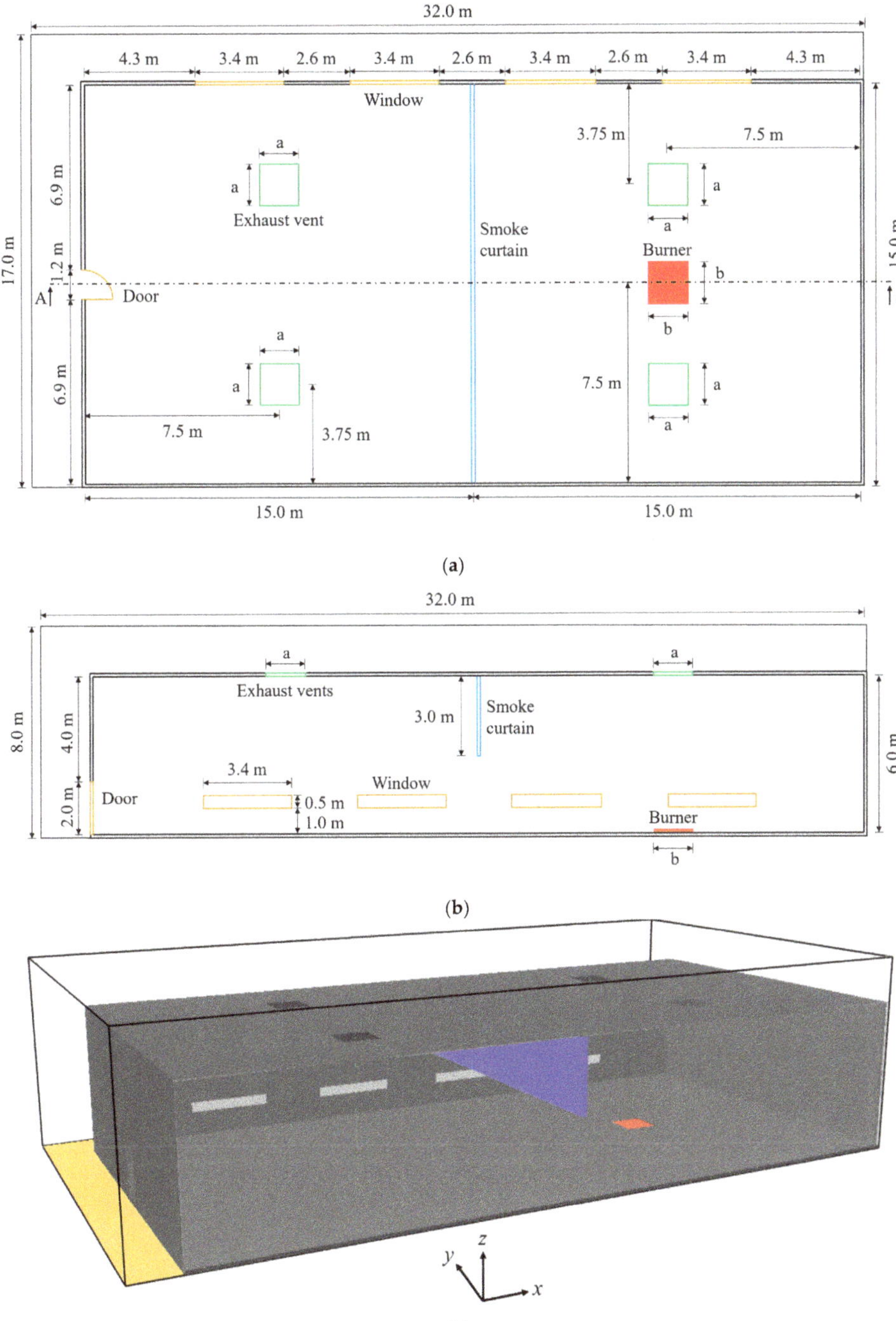

Figure 1. Physical domain under study (**a**) top view, (**b**) section A-A and (**c**) isometric view with transparent front wall.

The representative numerical mesh of the physical domain used in the simulations is composed entirely of structured, hexahedral, evenly spaced elements with aspect ratio equal to 1 (one), that is, all sides of each element have the same dimension, as illustrated in Figure 2.

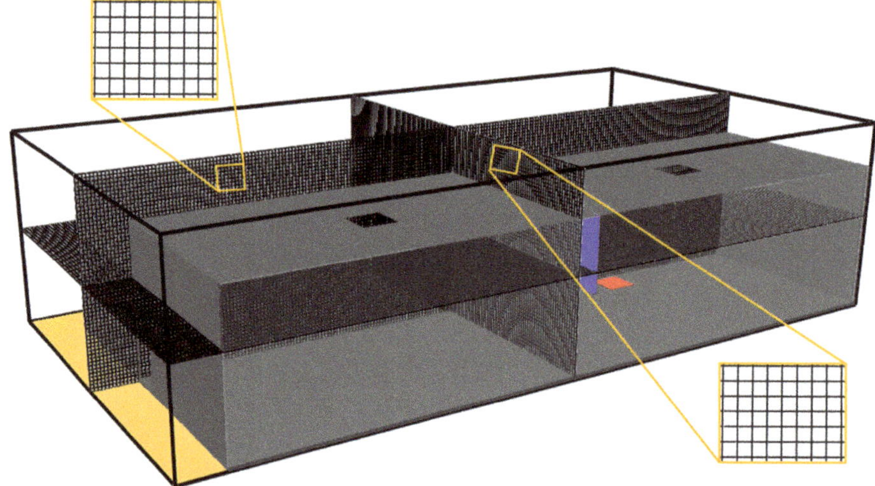

Figure 2. Details of the numerical mesh used in the simulations.

2.2. The Mathematical Model

For the numerical analysis, we used the FDS software, version 6.7.4, which solves the conservation equations of mass, species, linear momentum and energy, and of turbulence, with emphasis on the transport of smoke and heat. The FDS uses an approximation for low Mach numbers, developed by Rehm and Baum [42], large eddy simulation (LES) model to treat turbulence, and the Deardorff model [43] to calculate the turbulent viscosity. The FDS software also includes combustion, evaporation, pyrolysis and radiation heat transfer models.

2.2.1. The Governing Equations

The governing equations (equation of state, conservation of mass, species, linear momentum and energy) that describe the physical problem under study are presented in Equations (1)–(5), as follows:

(a)   Mass conservation:

$$\frac{\partial \rho}{\partial t} + \nabla \cdot (\rho \boldsymbol{u}) = \dot{m}_b''', \tag{1}$$

where ρ is the density, $\boldsymbol{u}$ is the velocity vector and $\dot{m}_b'''$ is the source term associated with the addition of mass from evaporating droplets or other subgrid-scale particles that represent, for example, sprinkler and fuel sprays, vegetation, and any other type of small, unresolvable object.

(b)   Species conservation:

$$\frac{\partial (\rho Y_\alpha)}{\partial t} + \nabla \cdot (\rho Y_\alpha \boldsymbol{u}) = \nabla \cdot (\rho D_\alpha \nabla Y_\alpha) + \dot{m}_\alpha''' + \dot{m}_{b,\alpha}''', \tag{2}$$

where Y_α and D_α are the mass fraction and diffusivity of species α. $\dot{m}_\alpha'''$ and $\dot{m}_{b,\alpha}'''$ are the mass production rate per unit volume of species α by chemical reactions and evaporating droplets/particles, respectively.

(c)   Linear momentum conservation:

$$\frac{\partial(\rho \boldsymbol{u})}{\partial t} + \nabla\cdot\rho\boldsymbol{u}\boldsymbol{u} = -\nabla p + \rho\boldsymbol{g} + \boldsymbol{f_b} + \nabla\tau_{ij},\tag{3}$$

where p is the pressure, $\boldsymbol{g}$ is the gravitational acceleration vector, $\boldsymbol{f_b}$ is the external force vector (excluding gravity) and τ_{ij} is the viscous stress tensor.

(d)   Energy conservation:

$$\frac{\partial(\rho h_s)}{\partial t} + \nabla\cdot(\rho h_s\boldsymbol{u}) = \frac{Dp}{Dt} + \dot{q}''' - \dot{q}_b''' - \nabla\cdot\dot{q}'',\tag{4}$$

where h_s is the enthalpy, $\dot{q}'''$ is the heat release rate per unit volume from a chemical reaction and $\dot{q}_b'''$ is the energy transferred to subgrid-scale droplets and particles (for example, sprinkler), and $\dot{q}''$ represents the conductive, diffusive and radiative heat fluxes.

(e)   Equation of state (ideal gas law):

$$p = \frac{\rho \overline{R} T}{M},\tag{5}$$

where $\overline{R}$ is the universal constant, T is the absolute temperature and M is the molecular weight of the gas mixture.

More information on the mathematical model and submodels used in this research can be found in the FDS Technical Reference Guide [44].

2.2.2. Initial and Boundary Conditions

As an initial condition, atmospheric pressure P_0, temperature T_0, air relative humidity, RH_0, velocity $\boldsymbol{u}_0$, and mass fractions for air $Y_{air,0}$, and soot $Y_{soot,0}$ were considered. The values of these parameters are shown in Table 1.

Table 1. Initial condition for the problem under study.

P_0 (Pa)	T_0 (°C)	RH_0 (%)	u_0 (m/s)	$Y_{air,0}$ (-)	$Y_{soot,0}$ (-)
101,325	28	70	0	1	0

Open boundary conditions were used at the maximum and minimum extremes of the computational domain, as shown in Figure 3. This means that the fluid is allowed to flow into or out of the computational domain depending on the local pressure gradient (upwind boundary condition). Typically, in this kind of boundary condition, the gradients of the tangential velocity components are set to zero [44].

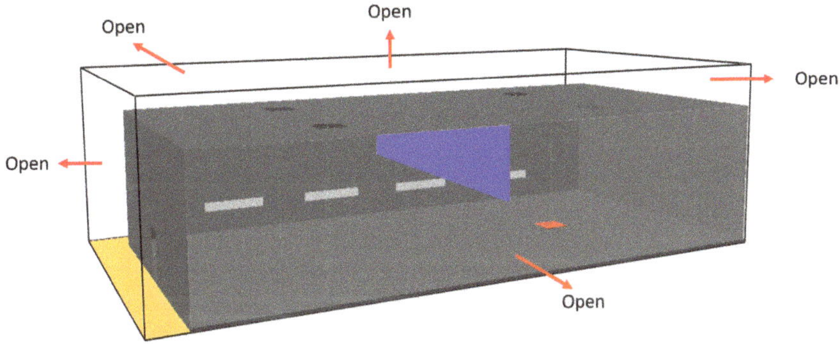

Figure 3. Boundary conditions used in the numerical simulation.

2.2.3. Heat Release Rate

The heat release rate (HRR) is the amount of energy per unit time that a material releases into the environment when it undergoes combustion. Once started, the fire goes through three stages: growth, fully developed (in which the HRR remains constant) and decay [45,46]. Normally the growth phase of the fire is modeled in such a way that the HRR is directly proportional to the time squared (*t*-squared fires) [47,48], that is:

$$HRR_{growth} = \alpha \times t^2,$$ (6)

where α is the fire growth coefficient in kW/s^2 and t is the time in s.

2.3. Numerical Solution Method

The algorithm for solving the governing equations uses an explicit predictor-corrector finite difference scheme, with second-order precision in space and time. At each time step, between the predictor and corrector procedures, the algorithm checks whether the Courant-Friedrichs-Lewy (CFL) stability criterion is satisfied, that is:

$$CFL = \delta t \times max\left(\frac{|u|}{\delta x}, \frac{|v|}{\delta y}, \frac{|w|}{\delta z}\right) < CFL_{max},$$ (7)

where CFL_{max} varies between 0.8 and 1.0; δt is the time step, u, v and w are the components of the velocity vector in the x, y and z directions, respectively. If the criterion is not satisfied, the time step is adjusted (reduced), returning to the beginning of the predictor procedure. If the stability criterion is satisfied, the procedure continues to the corrective procedure. In this way, the time step in the numerical simulation is not constant.

In the FDS software, the spatial variables are discretized using a staggered grid [49], that is, scalar quantities (e.g., pressure, temperature, density), velocity components and vorticity components are assigned to the centers, faces and edges of each cell, respectively. The radiation heat transfer is quantified using the finite volume method and assuming gray gas radiation model [44].

2.4. Thermo-Physical Properties of Materials

The material used to model the floor, walls and ceiling of the compartment was concrete, 10 cm thick. The density ρ, specific heat c_p, thermal conductivity k, and emissivity ε of the concrete are shown in Table 2.

Table 2. Thermo-physical properties of materials used in the simulations.

Parameter	Material		Source
	Concrete	**Wood (Fuel)**	
ρ (kg/m^3)	2100	-	[50]
c_p (kJ/(kg·K))	0.88	-	[50]
k (W/(m·K))	1.37	-	[50]
ε (-)	0.92	-	[51]
y_{soot} (kg/kg)	-	0.0015	[52]
y_{CO} (kg/kg)	-	0.004	[52]
Δh	-	16,400	[52]
HRRPUA (kW/m^2)	-	100	[53]

The source term of heat and smoke release was obtained considering the burning of wood, with chemical formulation $CH_{1.7}O_{0.74}N_{0.002}$. The yields of soot ($y_{soot}$) and carbon monoxide ($y_{CO}$), heat of combustion ($\Delta h$) and Heat Release Rate Per Unit Area (HRRPUA) are shown in Table 2.

The thermo-physical properties of the air are temperature-dependent (ideal gas law) and calculated by the software at each control volume and for each instant of time.

2.5. Cases Studied

Table 3 shows the operational conditions of the door, windows (open or closed), and smoke curtain (with or without), type of exhaust system, HRR from the burner and dimensions of the exhaust vents (a) and of the burner (b) for each of the twelve simulated cases.

Table 3. Cases studied numerically in this research.

Case	Door (2.20 m^2)	Windows (6.80 m^2 Total)	Exhaust System	Smoke Curtain	HRR (kW)	a * (m)	b * (m)
1	Open	Closed	Without	Without	900	0.00	3.0
2	Open	Open	Without	Without	900	0.00	3.0
3	Open	Closed	Natural (2.25 m^2 total)	Without	900	0.75	3.0
4	Open	Closed	Natural (9.00 m^2 total)	Without	900	1.50	3.0
5	Open	Open	Natural (2.25 m^2 total)	Without	900	0.75	3.0
6	Open	Open	Natural (9.00 m^2 total)	Without	900	1.50	3.0
7	Open	Closed	Natural (9.00 m^2 total)	Without	225	1.50	1.5
8	Open	Open	Natural (9.00 m^2 total)	Without	225	1.50	1.5
9	Open	Closed	Natural (9.00 m^2 total)	With	225	1.50	1.5
10	Open	Open	Natural (9.00 m^2 total)	With	225	1.50	1.5
11	Open	Open	Mechanical (18.00 m^3/h total)	Without	900	1.50	3.0
12	Open	Open	Mechanical (36.00 m^3/h total)	Without	900	1.50	3.0

* Geometrical parameters $\underline{a}$ and $\underline{b}$ are specified in the Figure 1.

As HRRPUA = 100 kW/m^2 (Table 2), burners with negligible height and surface areas of 9.00 m^2 (b = 3.00 m) and 2.25 m^2 (b = 1.50 m) have HRR of 900 kW and 225 kW, respectively. For all analyzed cases, a fast growth rate (α = 0.0469 kW/s^2) was considered, according to Alpert [54], with a fully developed HRR as indicated in Table 3, that is equivalent to the maximum value of the growth phase. The simulations were carried out in the initial 600 s of the fire process. Thus, the decay phase was not analyzed. Furthermore, it was considered that the smoke layer interface height is the distance from the floor to the point where the mass fraction of soot is approximately two orders of magnitude lower than in the fire zone, as suggested by Sinclair [55] and validated by Qin et al. [20] and Qin et al. [12]. Based on the mass fraction results obtained in the fire zone, from the simulations, a value of 10^{-6} for the mass fraction of soot in the other locations of the environment was adopted, to determine the interface between the two layers; one with low temperatures and smoke concentrations, located in the lower region of the compartment, and the other with high temperatures and smoke concentrations, located in the upper region of the compartment. To monitor the temperature and smoke layer interface height transient history, 29 temperature and soot mass fraction measurement points were defined below each of the four exhaust fans, with a vertical spacing of 0.20 m, as shown in Figure 4.

At a given instant of time, the smoke layer interface height below a given exhaust fan was established when the soot mass fraction at a given measurement point reached the value of 10^{-6}. Thus, for cases in which there is no smoke curtain (cases 1–8, 11 and 12), the smoke layer interface height was obtained from the arithmetic mean of the values found for each of the four measurement columns. For cases with a smoke curtain (cases 9 and 10), there are two different smoke layer interface height, each obtained from the arithmetic mean of the values found for each of the two measurement columns located on each side of the compartment's smoke curtain.

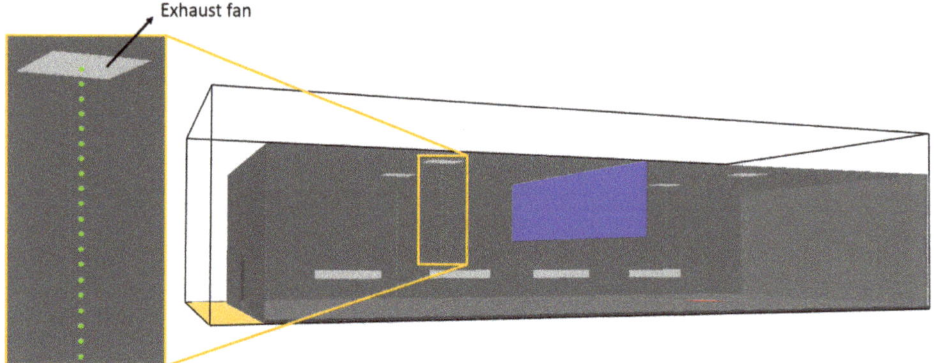

Figure 4. Location details of the temperature and soot mass fraction measurement points inside the system.

3. Results and Discussion

3.1. Mesh Convergence Analysis

To ensure that the results obtained in the simulations were independent of the number of control volumes, a mesh convergence study was made, for case 6 (Table 3) considering 04 (four) distinct meshes, named M_1, M_2, M_3 and M_4, as reported in Table 4. Figure 5 presents the results of this mesh convergence analysis for four investigated parameters: smoke layer interface height, average temperature (t = 600 s), exhaust volumetric flow rate and HRR. From the analysis of Figure 5, it can be seen that the transient results obtained for the smoke layer interface height as a function of time and the temperature profile at t = 600 s for the coarsest mesh (M_4) show significant variations as compared to the obtained results for the more refined mesh (M_1). It is also observed that the results obtained for the M_3 mesh shows good concordance as compared with that to the M_1 mesh, except for the temperature profile at t = 600 s, indicating only an acceptable agreement. Comparing the meshes M_1 and M_2, there is a good agreement for all analyzed parameters.

In order to reduce the computational effort and maintain a good accuracy of the obtained results, a mesh with 1,492,736 elements (M_2) was chosen to be used in the simulations of the other cases under study. Thus, the present work uses a uniform mesh with an element size of approximately 14.26 cm, more refined than the values used by Qin et al. [12], Abotaleb [16] and Yuen et al. [17] in their researches.

Table 4. Characteristics of the analyzed meshes.

Mesh	Number of Elements in Each Direction	Total Number of Elements	Element Size (cm)
M_1	$320 \times 170 \times 80$	4,352,000	10.00
M_2	$224 \times 119 \times 56$	1,492,736	14.26
M_3	$160 \times 85 \times 40$	544,000	20.00
M_4	$128 \times 68 \times 32$	378,528	25.00

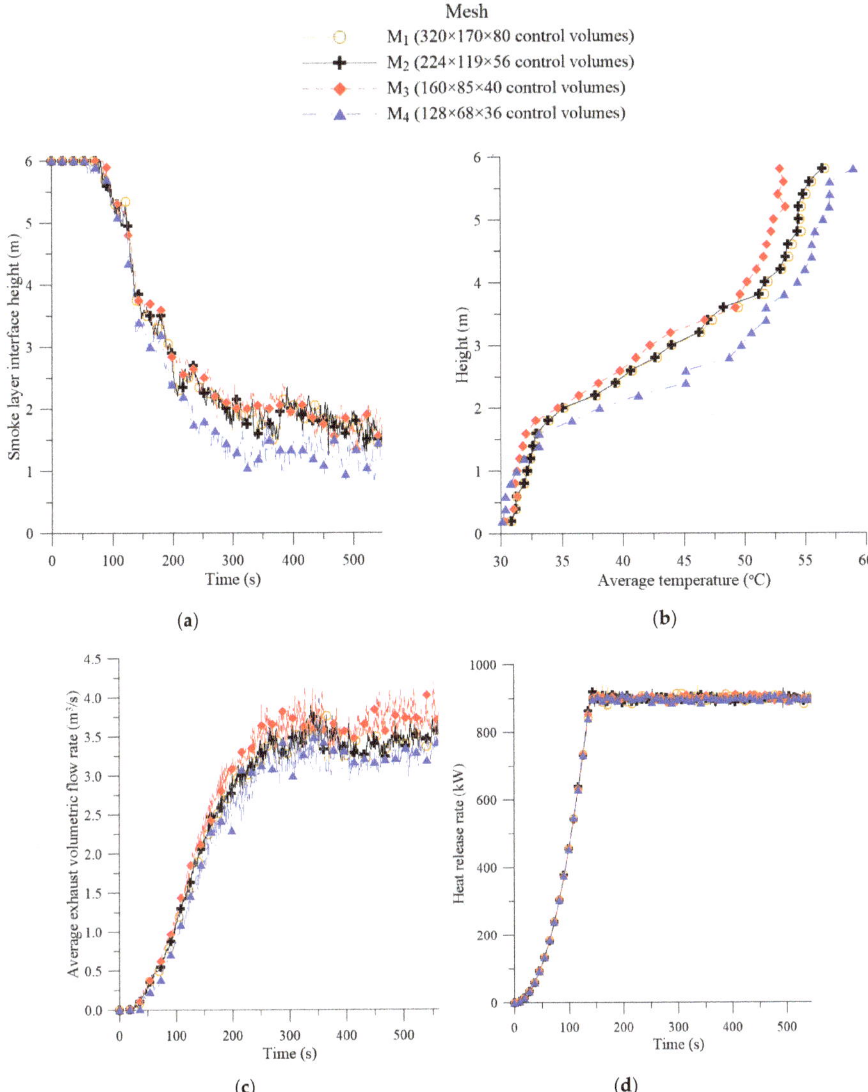

Figure 5. Mesh convergence analysis for case 6: (**a**) smoke layer displacement; (**b**) average temperature profile at t = 600 s; (**c**) average exhaust volumetric flow rate and (**d**) HRR.

3.2. Thermo-Fluid Dynamic Analysis of Processes

3.2.1. Natural Exhaust with Closed Windows

Figure 6 shows the influence of the natural exhaust area, for the situation of the compartment with closed windows and HRR = 900 kW, in the smoke layer interface height as a function of time, in the average temperature profile for the time t = 600 s, in the average exhaust volumetric flow rate as a function of time and in the HRR as a function of time.

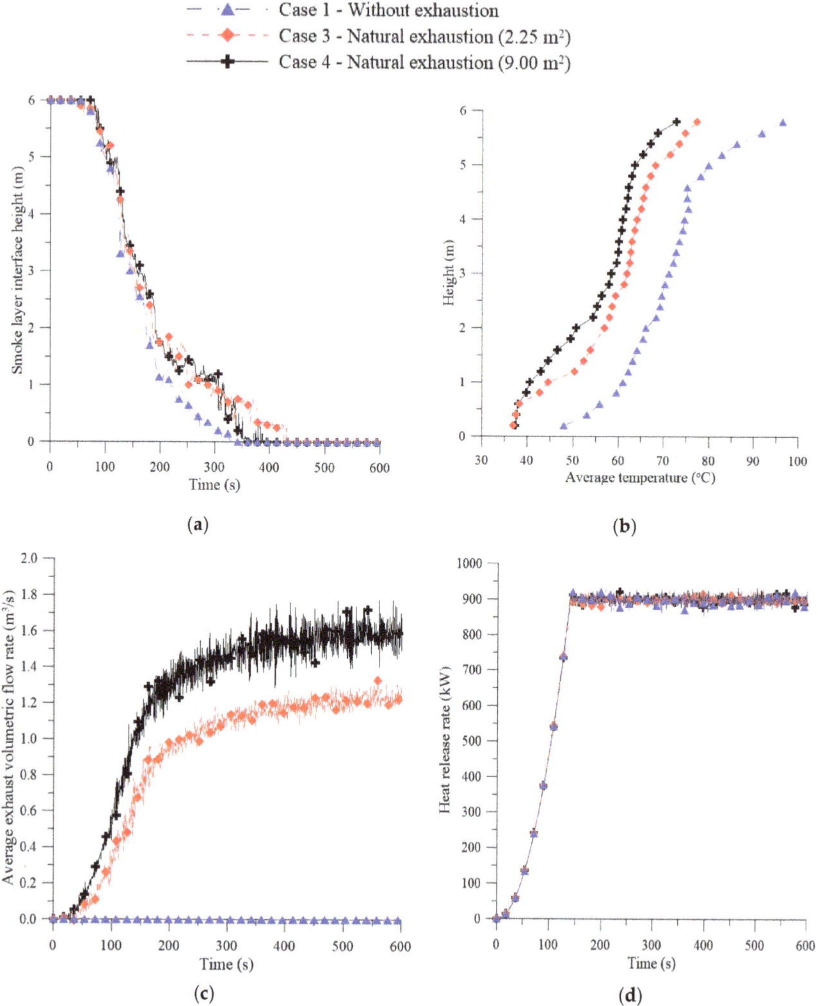

Figure 6. Influence of the natural exhaust area in the (**a**) smoke layer vertical displacement; (**b**) average temperature profile at t = 600 s; (**c**) average exhaust volumetric flow rate and (**d**) HRR (compartment with closed windows and HRR = 900 kW).

Upon examining Figure 6a–c, it may be observed that although the volume of gases exhausted in case 4 is approximately 35% greater than the value presented in case 3, during the initial 600 s of fire, the increase in the exhaust area for the poorly ventilated compartment (closed windows in the lower region of the compartment, cases 1, 3 and 4) does not provide great variations in the smoke layer interface height along most of the time and presents a slightly lower result between 320 and 400 s. A similar result was obtained by Qin et al. [12]. This behavior can be explained by the exhaust of gases with a lower soot concentration through the exhaust system with a larger area.

The smoke layer takes 180 s to reach a height of 1.80 m from the floor, to the compartment without an exhaust system (case 1), and, for other cases in that a natural exhaust system is used, this time is approximately 195 s, that is, a delay of only 15 s.

A higher volumetric exhaust flow rate provokes suction of the environment air outside of the compartment (T = 28 °C) through the door, justifying the achievement of lower temperatures for case 4

(Figure 6b). At time t = 600 s, the temperature, at a height of 1.80 m from the floor, is reduced from 65.50 °C (without exhaust) to 55.96 °C and 49.46 °C, when using exhaust systems with total areas of 2.25 and 9.00 m^2, respectively.

From Figure 6d, it can be seen that the heat release rates in the growth phase, for the three cases analyzed, are in accordance with the formulation presented in Equation (6) and that in the fully developed phase (after 138.53 s) there were small oscillations around the value of 900 kW.

3.2.2. Natural Exhaust with Open Windows

For physical situation where the compartment has open windows and HRR = 900 kW, Figure 7 illustrates the transient effect of the natural exhaust area in the smoke layer interface height, average exhaust volumetric flow rate and in the HRR, and in the average temperature profile for the time t = 600 s.

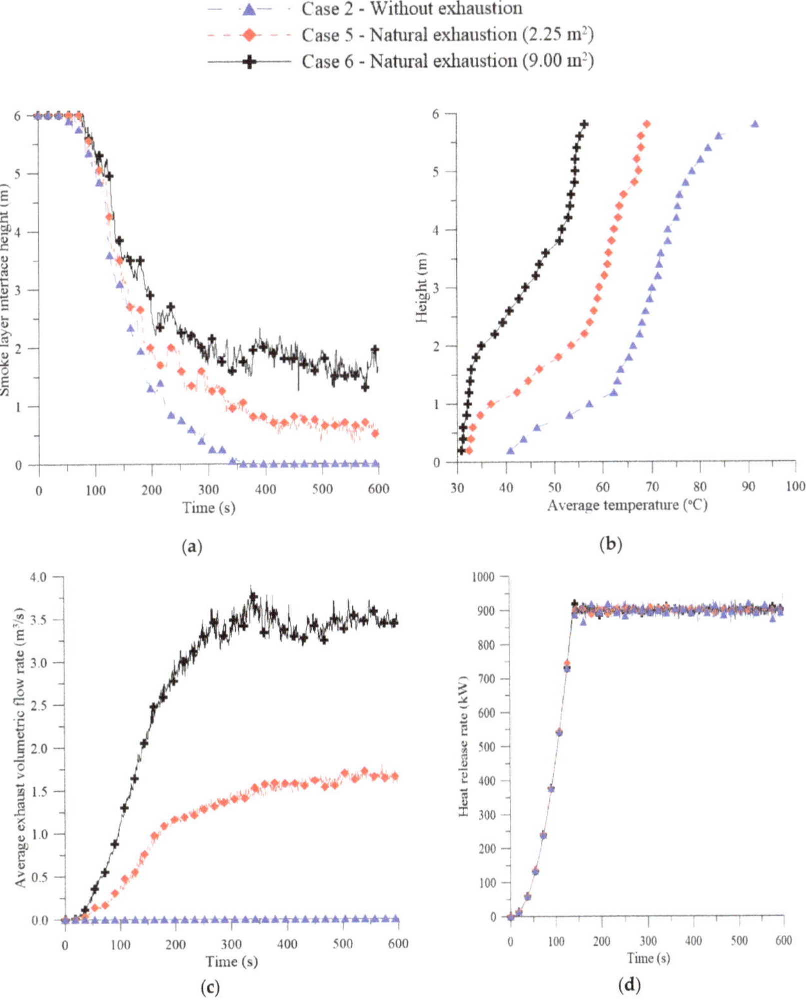

Figure 7. Influence of the natural exhaust area in the (**a**) smoke layer vertical displacement; (**b**) average temperature profile at t = 600 s; (**c**) average exhaust volumetric flow rate and (**d**) HRR (compartment with open windows and HRR = 900 kW).

After analyzing Figure 7a, one can observe that, for the well-ventilated compartment (with open windows), the area of the exhaust fans significantly influences the smoke layer interface height, the temperature profile and the average exhaust volumetric flow rate.

The exhaust volume for case 6, during the first 600 s of fire, is approximately 130% higher comparing to case 5, delaying the downward displacement of the smoke layer, especially after 160 s, and reducing the temperature inside the entire compartment.

The smoke layer takes 188 s to reach a height of 1.80 m from the floor for the compartment without exhaust system (case 2), while this time is 206 s and 290 s for cases 5 and 6, respectively. It is observed that, even after 600 s, the smoke layer does not reach the floor, for both compartments with exhaust systems (cases 5 and 6), differently from what was previously presented for the cases with closed windows (cases 3 and 4). For case 2, the smoke reaches the floor at time t = 352 s and the minimum values obtained for the smoke layer interface height for cases 5 and 6 were 0.35 and 1.25 m, respectively.

For time t = 600 s, the temperature at a height of 1.80 m from the floor is reduced from 65.24 °C (without exhaust) to 50.72 °C and 33.81 °C when using exhaust systems with total areas of 2.25 and 9.00 m^2, respectively.

Figure 7d indicates that the results of the heat release rates for the cases 2, 5 and 6 are in accordance with Equation (6) (growth phase) and Table 3 (HRR for fully developed phase).

Figure 8 illustrates the temperature distribution for the compartment with natural exhaust (9.00 m^2 total), HRR = 900 kW, door and windows open and without smoke curtain (case 6) at different moments of the process.

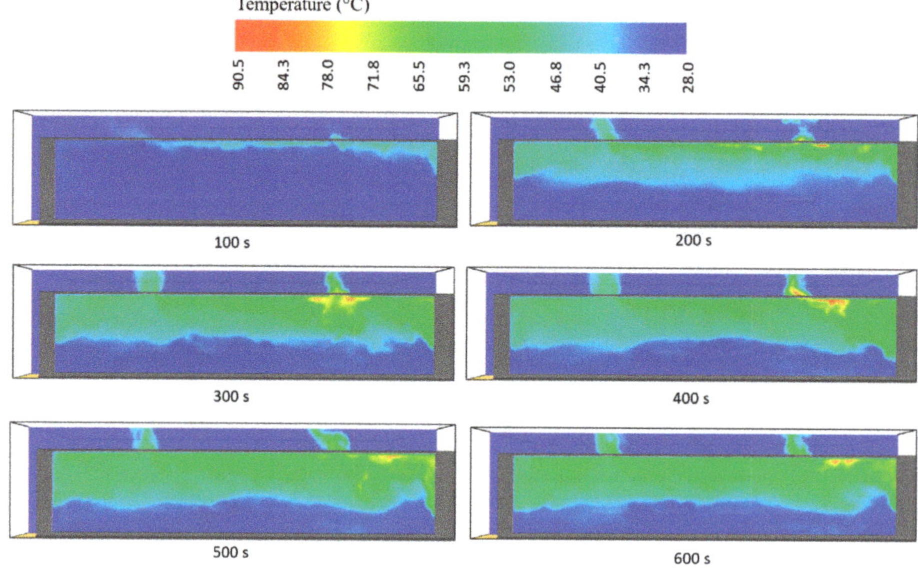

Figure 8. Temperature distribution in an xz plane that crosses two exhaust vents (y = 3.75 m) for different time periods (case 6).

Upon examining Figure 8, it is observed that 100 s after the start of the fire, the temperature in the plane y = 3.75 m did not vary significantly. This is because the fire is still growing. It can be seen, between 100 and 200 s, that there is a very significant variation in temperature in the analyzed plane, due to the fact that the fire reached the fully developed stage at t = 138.53 s and, consequently, the HRR is maximum from that instant. It is also observed that the temperature doesn't vary significantly in the analyzed plane, after 300 s. This fact indicates that the heat released by the burner is balanced by the enthalpy of the mixture that leaves the compartment through the exhaust fans.

Also, from the analysis of Figure 8 and t ≥ 200 s, it is evident the formation of an interface between two distinct layers: one with high temperature and concentration of soot, in the upper region of the compartment, and the other with low temperature and concentration of soot in the lower compartment. Further, there is a greater variation in temperature in the vertical direction (z axis) of the plane, when compared to that variation in the horizontal direction (x axis) of the plane under analysis. In this plane, the highest temperatures are located, predominantly, on the right side of the compartment, region where is placed the burner (source term of heat).

3.2.3. Natural Exhaust with Different Heat Release Rates

For compartments with natural exhaust area of 9.00 m^2, Figure 9 shows the transient behavior of the smoke layer interface height, average exhaust volumetric flow rate and HRR, and the average temperature profile in different height at t = 600 s, for two heat release rates HRR = 225 and 900 W.

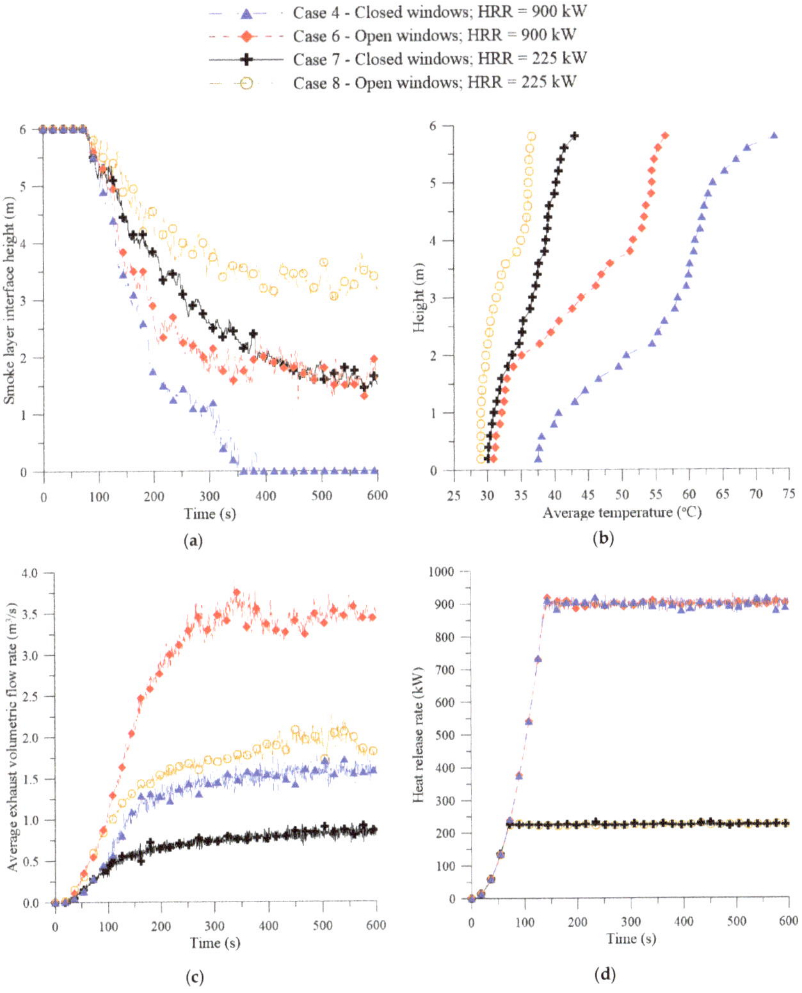

Figure 9. Influence of the HRR in the (**a**) smoke layer vertical displacement; (**b**) average temperature profile at t = 600 s; (**c**) average exhaust volumetric flow rate and (**d**) HRR (compartments with natural exhaust areas of 9.00 m^2).

Upon analyzing Figure 9, it is evident that the HRR from the burner affects the smoke layer interface height, temperature profile and the average exhaust volumetric flow rate, for both compartments (with open and closed windows).

For the compartment with closed windows (cases 4 and 7), reducing the HRR from 900 kW to 225 kW reduces the downward vertical displacement of the smoke layer. Whereas for case 4 the smoke reaches the floor at time t = 360 s, the minimum smoke layer interface height for case 7 is 1.45 m, occurred in the instant of time t = 574 s. For the compartment with open windows (cases 6 and 8), the reduction of the HRR makes the minimum smoke layer interface height to increase from 1.25 m to 2.95 m, in cases 6 and 8, respectively.

While the total energy released during the initial 600 s of fire (integral of the HRR over time) for cases with HRR = 900 kW is 3.66 times greater than that for cases with HRR = 225 kW, the volume of gases removed from the compartment is 1.88 times higher than that for compartments with closed windows (comparing cases 4 and 7), and 1.79 times higher than that for environments with open windows (comparing cases 6 and 8), respectively, at the same time interval.

From the analysis of Figure 9a,b, it is evident the importance of open windows to promote smoke extraction, increase the smoke layer interface height, and reduce the temperature of the compartment, regardless of the HRR. Besides, it is also observed that even with a HRR 4 times higher during most part of the fire process (fully developed phase), the minimum smoke layer interface height (Figure 9a) and the temperature below 2.00 m at t = 600 s (Figure 9b), both for case 6, are very similar to those obtained for case 7. Thus, it is clearly shown the importance of a well-ventilated environment, in order to promote the control of the room temperature and a more efficient smoke outlet through natural exhaust fans.

Analyzing Figure 9d, it can be seen that, up to 69.26 s, the HRR is the same for all cases, according to the growth rate established for the fire (α = 0.0469 kW/s^2). For the cases 7 and 8, at this instant of time, the fully developed phase is reached, with small oscillations around the value of 225 kW, as defined in Table 3.

3.2.4. Natural Exhaust with Smoke Curtain and Closed Windows

Figure 10 illustrates the influence of the smoke curtain, for compartments with natural exhaust system (total area of 9.00 m^2), closed windows and HRR = 225 kW, in the smoke layer interface height as a function of time, in the average temperature profile for the time t = 600 s, in the average exhaust volumetric flow rate as a function of time and in the HRR as a function of time.

Upon examining this figure, it can be seen that the 3.00 m smoke curtain was efficient to restrict the smoke on the right side of the compartment with four natural exhaust vents of 2.25 m^2 each, closed windows and maximum HRR = 225 kW during the initial 600 s of fire. For a higher HRR, a larger total exhaust area and/or a higher smoke curtain would probably be needed to ensure smoke confinement in the upper right region of the compartment. Now, from the analysis of the Figure 10b, it is possible to observe that the use of the smoke curtain promotes a significant reduction in temperature on the left side of the compartment and a moderate reduction in temperature on the right side, up to a height of 3.40 m.

Figures 11 and 12 illustrate, respectively, the pressure field and the velocity vector field for the compartment with smoke curtain, open windows and HRR = 225 kW. After analyzing Figure 11, it was observed that the greatest pressures occur, precisely, in the regions closest to the exhaust vents located on the right side of the compartment.

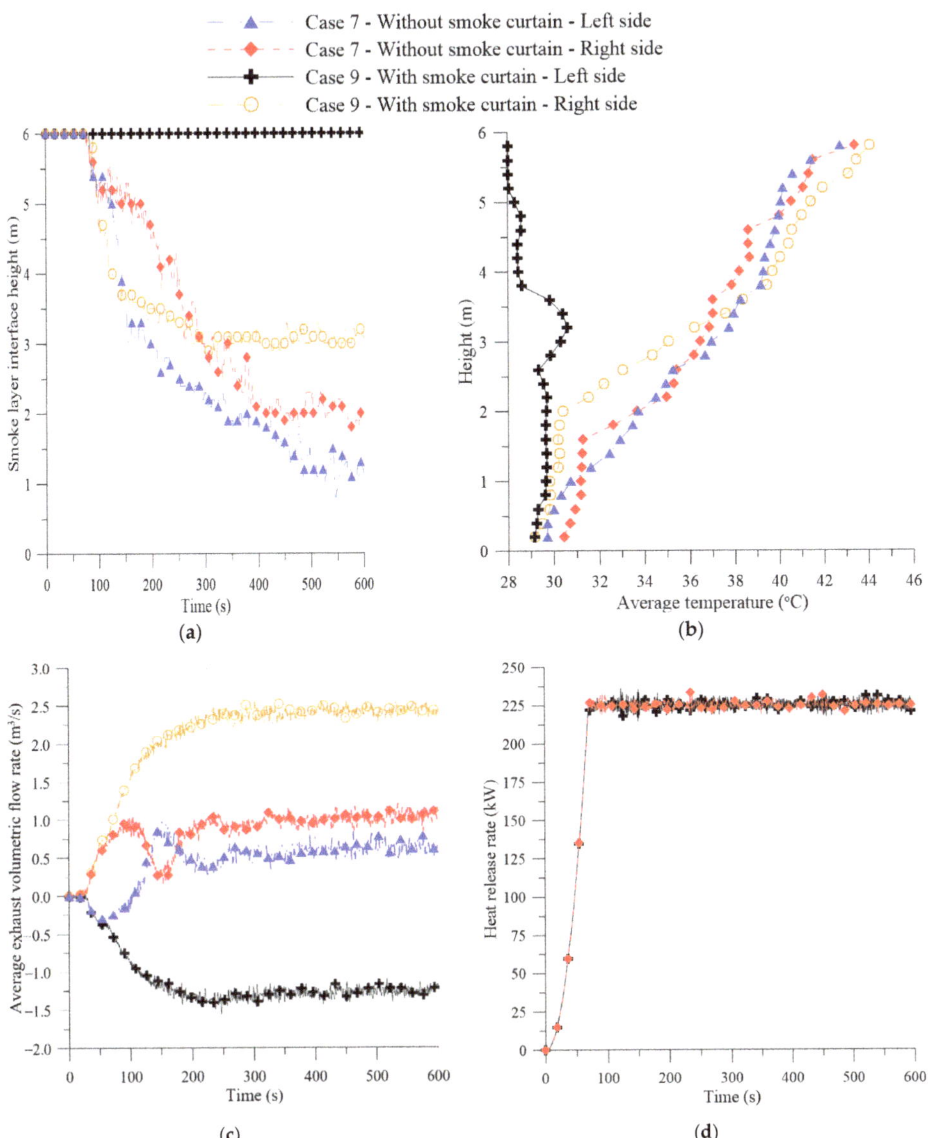

Figure 10. Influence of the smoke curtain in the (**a**) smoke layer vertical displacement; (**b**) average temperature profile at t = 600 s; (**c**) average exhaust volumetric flow rate and (**d**) HRR (compartments with natural exhaust areas of 9.00 m², closed windows and HRR = 225 kW).

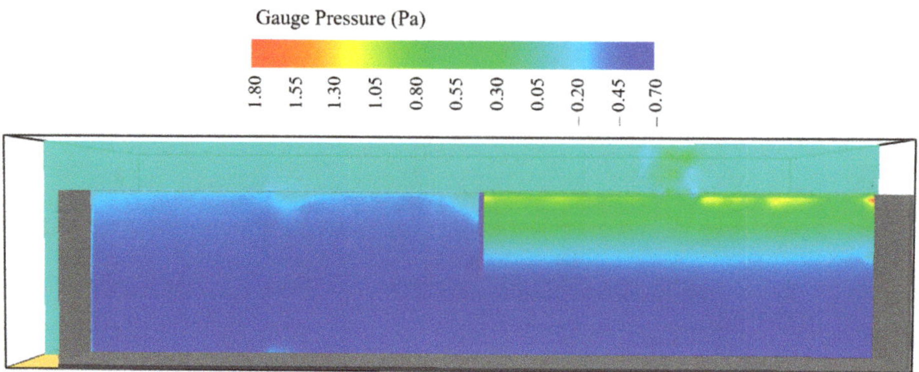

Figure 11. Pressure field in a plane that crosses two exhaust vents (y = 3.75 m) at time t = 600 s (case 9).

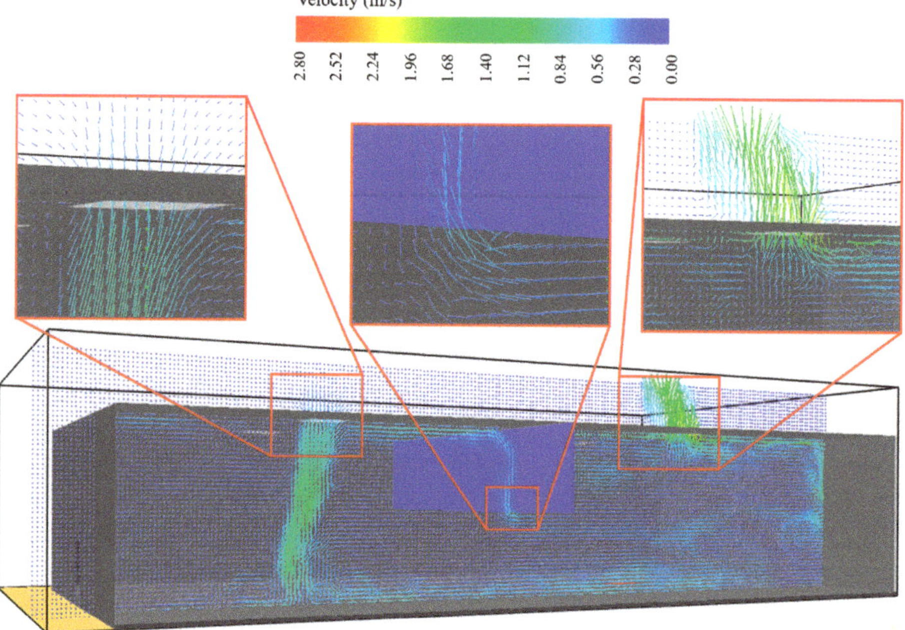

Figure 12. Velocity vector field in a plane that crosses two exhaust vents (y = 3.75 m) at time t = 600 s (case 9).

Furthermore, it can be seen that the smoke curtain, in addition to functioning as a physical barrier, is also responsible for maintaining positive pressure in the upper right region of the compartment (Figure 11), forcing hot gases out of the internal environment through the natural exhaust fans located at the right side of the smoke curtain, clearly seen in Figure 12.

The pressure gradient between the upper right region of the compartment and the external environment (outside of the compartment) justifies the higher exhaust volumetric flow rate obtained for case 9 when compared to case 7 (Figure 10c). For a better understanding, the volume of gases removed by the two exhaust fans located on the right side of the compartment with smoke curtain (case 9), during the first 600 s of the fire, is 60% higher than the total volume that came out of the four exhaust fans in the compartment without a smoke curtain (case 7).

As the amount of gases removed from the right side of the compartment (case 9) is greater than the amount of air entering the compartment, much of the environment is depressurized (below atmospheric pressure), as seen in Figure 11, provoking the entry of more external air, at 28 °C, through the door and exhaust fans on the left side (Figure 12). All these physical phenomena justify the negative values for the average exhaust volumetric flow rate on the left side (Figure 10c) and low temperatures obtained between 3.80 and 6.00 m for the left side of the compartment (case 9), as shown in Figure 10b. Besides, in Figure 12, at approximately 600 s after the start of the fire, it can be seen that a small amount of gases begins to flow around the smoke curtain, moving from the right side to the left side of the compartment. As the volumetric flow rate and soot concentration of the gases surrounding the smoke curtain are lower, the smoke layer interface height on the left side of the compartment wasn´t affected (Figure 10a).

3.2.5. Natural Exhaust with Smoke Curtain and Open Windows

As considering the compartments with natural exhaust (9.00 m^2), open windows and HRR = 225 kW, Figure 13 shows the influence of the smoke curtain in the smoke layer interface height as a function of time, in the average temperature profile for the time t = 600 s, in the average exhaust volumetric flow rate as a function of time and in the HRR as a function of time. Figures 14 and 15 illustrate, respectively, the pressure field and the velocity vector field for the compartment with smoke curtain, open windows and HRR = 225 kW.

From the analysis of Figure 13a, it can be noticed that the 3.00 m smoke curtain is sufficient to restrict the smoke on the right side of the compartment, given that the smoke layer interface height on the left side (case 10) remains in 6.00 m, in 600 s elapsed time.

Since the minimum smoke layer interface height for case 8 is 2.80 m (left side) and considering that the purpose of smoke control systems is to avoid contact between smoke and people, the smoke curtain is not so useful for the well-ventilated compartment (open windows), for the physical situation with four natural exhaust fans of 2.25 m^2 each and maximum HRR of 225 kW. In contrast, it was observed for the poorly ventilated compartment (closed windows) that the smoke layer interface height went from 0.80 m (case 7-left side) to 3.00 m (case 9-right side) when using smoke curtain.

Upon analyzing Figure 13b, it is observed that the use of the smoke curtain does not promote significant improvements in the temperature distribution in the compartment up to a height of 2.00 m. For heights above 3.00 m, the temperature on the right side of the compartment with a smoke curtain is, on average, 4.2 °C higher than the values obtained for case 8, due to the compartmentalization of hot gases. On the other hand, the average temperature on the left side of the compartment, between 3.00 and 6.00 m in height, is reduced by 5.3 °C when using the smoke curtain.

Analyzing Figure 13c, one can observe that air enters through the exhaust fans located on the left side of the compartment, due to a slightly negative pressure in the region, up to approximately 370 s. During this time interval, the volume of air entering the exhaust fans, for case 10 (with smoke curtain and open windows), corresponds to approximately 25% of the value obtained for case 9 (with smoke curtain and closed windows). After this time interval, the upper left region of the compartment has a slightly positive pressure, due to the air flow, with low concentration of soot, that surrounds the smoke curtain (Figure 15) and the effect of the natural convection that promotes upward vertical displacement of the hotter air, causing the escape of gases through the exhaust fans. Even so, in the interval between 370 and 600 s the volume of gases that was exhausted from the left side of the compartment corresponds to only 6% of the total exhaust volume in the same period of time.

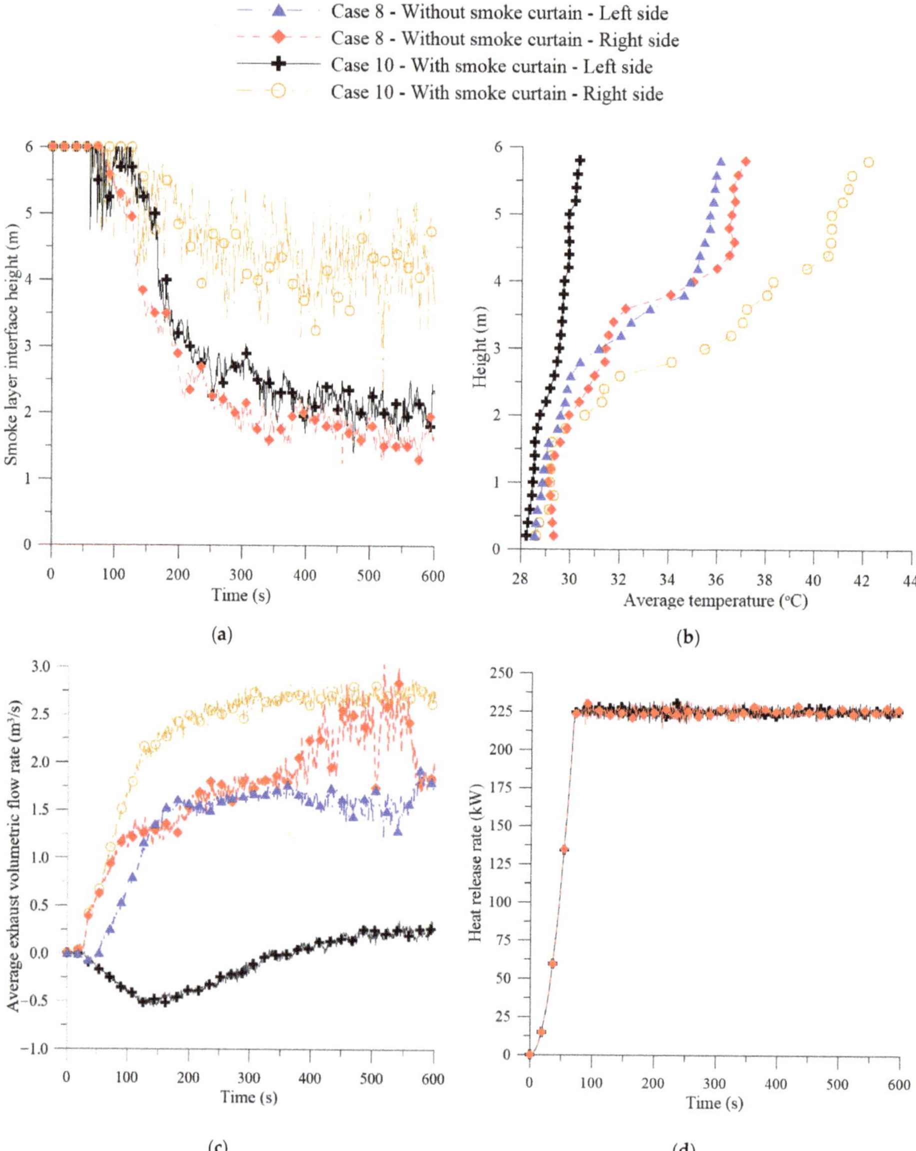

Figure 13. Influence of the smoke curtain in the (**a**) smoke layer vertical displacement; (**b**) average temperature profile at t = 600 s; (**c**) average exhaust volumetric flow rate and (**d**) HRR (compartments with natural exhaust areas of 9.00 m^2, open windows and HRR = 225 kW).

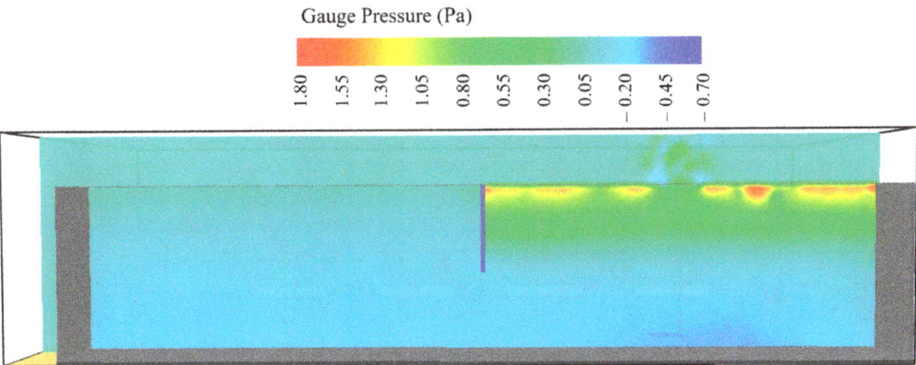

Figure 14. Pressure field in a plane that crosses two exhaust vents (y = 3.75 m) at time t = 600 s (case 10).

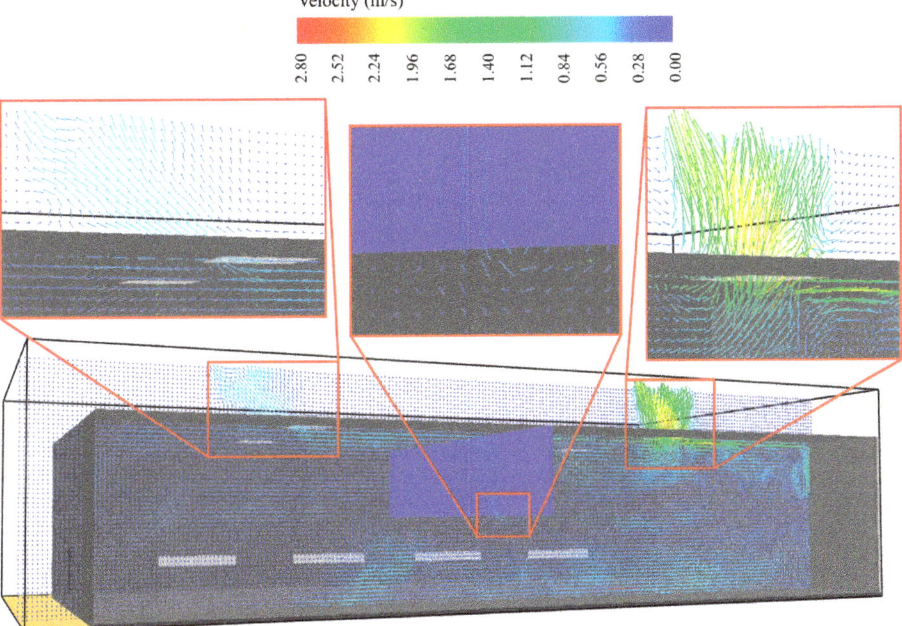

Figure 15. Velocity vector field in a plane that crosses two exhaust vents (y = 3.75 m) at time t = 600 s (case 10).

Regarding the total volume of gases removed from the compartment by the exhaust fans during the 600 s of fire, the value obtained for case 10 (with smoke curtain) represents only 77% of the value obtained for case 8 (without smoke curtain), proving the low efficiency of the smoke curtain in the smoke protection for the well-ventilated compartment (open windows).

Now analyzing Figure 14, a slightly higher pressure gradient is observed between the upper right region of the compartment and the external environment, for case 10, when compared to case 9 (Figure 11), promoting a exhaust volumetric flow rate of gases 12% higher at the analyzed time, evidenced by the velocity vectors in Figure 15, and a 10% higher volume of gases exhausted during the first 600 s of fire.

There is also a lower pressure gradient between the other regions of the compartment and the external environment (outside of the compartment) when compared to case 9 (Figure 11) due to a larger ventilation area (open windows), which facilitates air renewal. For the instant of time t = 600 s, the lower region has a slightly negative pressure, promoting the entry of air through the windows and door, while the upper left region of the compartment has a slightly positive pressure, promoting the exit of air through the exhaust fans, as seen in Figure 15.

By the inclined direction of the velocity vectors at the exhaust fan on the left side of the compartment (Figure 15), it can be verified that the air, with a low concentration of soot, surrounding the smoke curtain, is the main responsible for the exhaust phenomenon that occurs on the left side.

3.2.6. Mechanical Exhaust without Smoke Curtain and Open Windows

In this section, we compared two mechanical exhaust systems, with total volumetric flow rates of 18.00 m^3/h (case 11) and 36 m^3/h (case 12), with the natural exhaust system (case 6). The results of this analysis for the compartment with open windows and a HRR of 900 kW are shown in Figure 16.

Although it presents results with a lot of oscillation, the mechanical exhaust system with a total volumetric flow rate of 36.00 m^3/h has the best results for the smoke layer interface height (Figure 16a). The downward vertical displacement of the smoke layer for cases 6 and 11 are very close, with the mechanical exhaust system showing a slight advantage compared to the natural exhaust system.

With respect to Figure 16b, it is observed that the greater the distance from the floor, the greater the temperature difference between the three systems. The mechanical exhaust system with a total volumetric flow rate of 36.00 m^3/h presented the lowest temperatures and, in contrast, the natural exhaust system presented the highest temperatures in all height.

From Figure 16c, it can be seen that the average exhaust volumetric flow rates of the natural and mechanical exhaust systems increase gradually and abruptly, respectively, over time. The volumetric flow rates adopted for mechanical exhaust systems are greater than the maximum value reached by the natural exhaust system. The total volumes of gases exhausted during the initial 600 s of fire for cases 11 and 12 are, respectively, 1.68 and 3.36 times greater than those obtained for case 6.

Despite presenting advantages in delaying the downward vertical displacement of the smoke layer and in reducing the internal temperature, the mechanical exhaust system reported in case 12, promotes very high speeds (3.00 m/s) in the occupation area of people inside the compartment (below a height of 2.00 m), more than double the values obtained for cases 6 and 11. Another disadvantage is that the unpredictability of the HRR of a real fire makes it difficult the effective design of suitable mechanical exhaust systems with fixed volumetric flow rate, making the use of a natural exhaust system recommended [20].

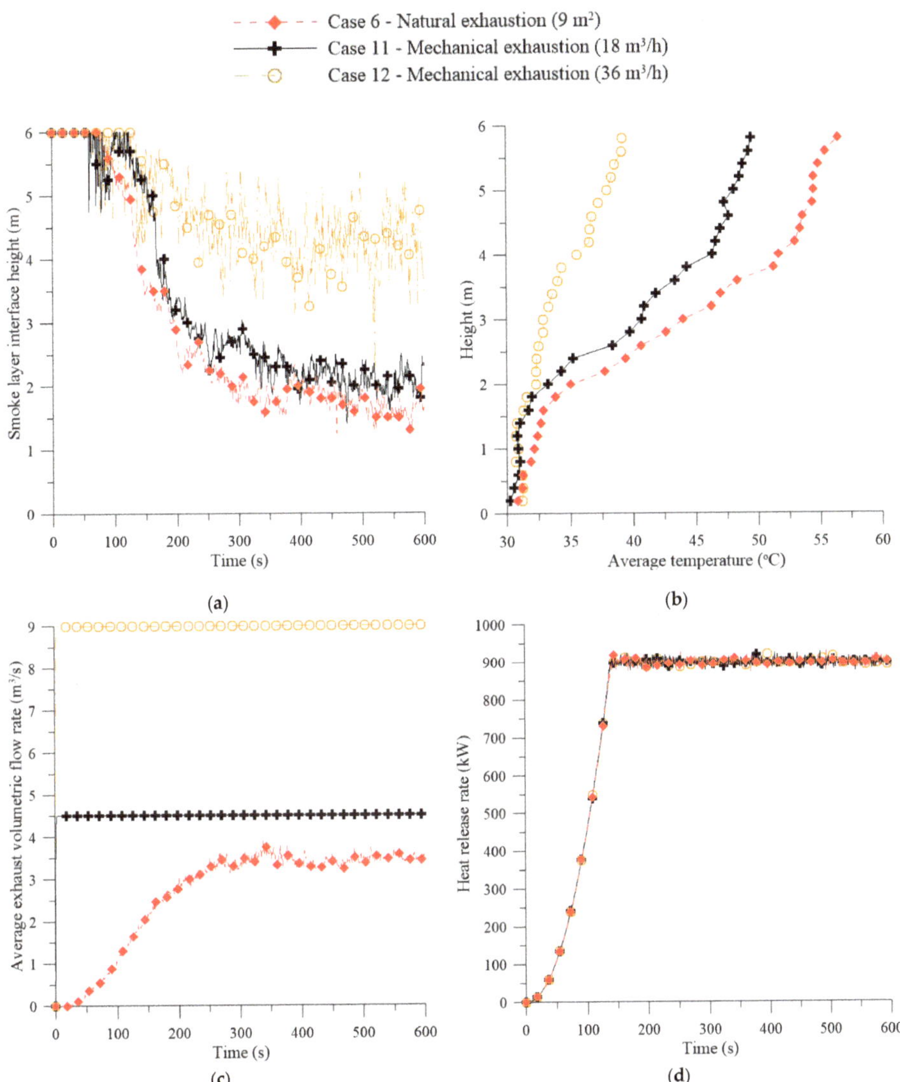

Figure 16. Influence of the mechanical exhaust system in the (**a**) smoke layer vertical displacement; (**b**) average temperature profile at t = 600 s; (**c**) average exhaust volumetric flow rate and (**d**) HRR (compartment with open windows and HRR = 900 kW).

4. Conclusions

This work aimed to perform a numerical analysis, using the FDS software, to evaluate the thermo-fluid dynamic behavior of the smoke generated by a fire in an enclosed space. From the obtained results, it can be concluded that:

(a) For a poorly ventilated compartment (closed windows) and a HRR of 900 kW, the increase in the natural exhaust area did not improve the smoke layer interface height. In all cases, the smoke reached the floor in less than 450 s and when using any of the natural exhaust systems, there was a delay of only 15 s in the time necessary for the smoke to reach a height of 1.80 m from the floor.

(b) For the well-ventilated compartment (open windows) and HRR of 900 kW, the increase in the natural exhaust area significantly increased the volume of gases exhausted, delaying the vertical displacement of the smoke layer and reducing the temperature in the entire compartment. For the case without an exhaust system, smoke reaches the floor and, for the cases with natural exhaust areas of 2.25 and 9.00 m^2, the minimum values obtained for the smoke layer interface height were 0.35 and 1.25 m, respectively.

(c) The increase in the HRR from the burner accelerates the downward vertical displacement of the smoke layer and increases the volume of gases exhausted and the compartment temperature, regardless of the ventilation (open or closed windows).

(d) Although the use of the 3.00 m smoke curtain was sufficient to restrict the smoke on the right side of the compartment, with HRR = 225 kW, for both cases analyzed. The use of a smoke curtain was efficient in smoke exhaustion, only for the case of poorly ventilated environment (closed windows), in which the exhaust fans on the left side acted as air intakes.

(e) Although the mechanical exhaust system with a total volumetric flow rate of 36.00 m3/h has shown good results in the downward displacement of the smoke layer and in the reduction of the internal temperature; it promotes very high speeds in the occupation area of people inside the compartment.

Finally, it is concluded that the FDS is a very useful tool for analyzing real fire situations, making it possible to carry out several comparative analyzes without inherent risks of experimental tests, with reduction of costs and time, assisting designers to develop more efficient smoke control systems for each type of building.

Author Contributions: Conceptualization, R.S.G., H.L.F.M., A.C.Q.S., V.H.V.V., K.C.G. and A.G.B.L.; Formal analysis, R.S.G., T.R.N.P., H.L.F.M., A.C.Q.S., V.H.V.V., K.C.G. and A.G.B.L.; Investigation, R.S.G. and T.R.N.P.; Methodology, R.S.G., T.R.N.P., H.L.F.M., K.C.G. and A.G.B.L.; Software, R.S.G., T.R.N.P. and H.L.F.M.; Supervision, A.C.Q.S., V.H.V.V., K.C.G. and A.G.B.L.; Writing—original draft, R.S.G., K.C.G. and A.G.B.L.; Writing—review & editing, R.S.G., A.C.Q.S., V.H.V.V., K.C.G. and A.G.B.L. All authors have read and agreed to the published version of the manuscript.

Funding: This research was funded by CNPq (grant number: 313531/2019-6), CAPES, and FINEP (Brazilian research agencies).

Acknowledgments: The authors would like to thank the Computational Laboratory of Thermal and Fluids, Mechanical Engineering Department, Federal University of Campina Grande (Brazil), for the research infrastructure and the references cited in the manuscript.

Abbreviations

FDS	Fire Dynamic Simulator
CFD	Computational Fluid Dynamic
CFL	Courant-Friedrichs-Lewy constraint
CFL_{max}	Maximum allowed value for CFL
c_p	Specific heat
f_b	External force vector (excluding gravity)
g	Gravitational acceleration vector
HRR	Heat Release Rate
HRRPUA	Heat Release Rate Per Unit Area
HRR_{growth}	Heat Release Rate in the growth phase
h_s	Enthalpy
K	Thermal conductivity
$\dot{m}_b'''$	Mass production rate per unit volume by evaporating droplets/particles
$\dot{m}_\alpha'''$	Mass production rate per unit volume of species α by chemical reactions

$\dot{m}_{b,\alpha}'''$	Mass production rate per unit volume of species α by evaporating droplets/particles
NIST	National Institute for Standards and Technology
p	Pressure
P_0	Initial pressure
$\dot{q}'''$	Heat release rate per unit volume from a chemical reaction
$\dot{q}_b'''$	Energy transferred to subgrid-scale droplets/particles
$\dot{q}''$	Heat flux vector
RH_0	Initial relative humidity
T	Temperature
T_0	Initial temperature
T	Time
u	Velocity vector
$\mathbf{u}_0$	Initial velocity vector
u	Velocity component in the x direction
v	Velocity component in the y direction
w	Velocity component in the z direction
Y_α	Mass fraction of species α
$Y_{air,0}$	Initial mass fraction of air
$Y_{soot,0}$	Initial mass fraction of soot
α	Fire growth coefficient
Δh	Heat of combustion
ε	Emissivity
ρ	Density
τ_{ij}	Viscous stress tensor

References

1. Khan, E.A.; Ahmed, M.A.; Khan, E.H.; Majumder, S.C. Fire Emergency Evacuation Simulation of a shopping mall using Fire Dynamic Simulator (FDS). *J. Chem. Eng.* **2017**, *30*, 32–36. [CrossRef]
2. Stefanidou, M.; Athanaselis, S.; Spiliopoulou, C. Health impacts of fire smoke inhalation. *Inhal. Toxicol.* **2008**, *20*, 761–766. [CrossRef] [PubMed]
3. Thompson, O.F.; Galea, E.R.; Hulse, L.M. A review of the literature on human behaviour in dwelling fires. *Saf. Sci.* **2018**, *109*, 303–312. [CrossRef]
4. Xiong, L.; Bruck, D.; Ball, M. Preventing accidental residential fires: The role of human involvement in non-injury house fires. *Fire Mater.* **2017**, *41*, 3–16. [CrossRef]
5. Mihailidou, E.K.; Antoniadis, K.D.; Assael, M.J. The 319 major industrial accidents since 1917. *Int. Rev. Chem. Eng.* **2012**, *4*, 529–540.
6. Miller, I. *Human Behaviour Contributing to Unintentional Residential Fire Deaths, 1997–2003*; New Zealand Fire Service Commission, 2005; Available online: https://fireandemergency.nz/assets/Documents/Research-and-reports/Report-47-Human-Behaviour-Contributing-to-Unintentional-Residential-Fire-Deaths-1997-2003.pdf (accessed on 30 July 2020).
7. Shen, T.-S.; Huang, Y.-H.; Chien, S.-W. Using fire dynamic simulation (FDS) to reconstruct an arson fire scene. *Build. Environ.* **2008**, *43*, 1036–1045. [CrossRef]
8. Hu, L.H.; Huo, R.; Li, Y.Z.; Wang, H.B.; Chow, W.K. Full-scale burning tests on studying smoke temperature and velocity along a corridor. *Tunn. Undergr. Space Technol.* **2005**, *20*, 223–229. [CrossRef]
9. Barillo, D.J.; Goode, R. Fire fatality study: Demographics of fire victims. *Burns* **1996**, *22*, 85–88. [CrossRef]
10. Antonio, A.C.P.; Castro, P.S.; Freire, L.O. Smoke inhalation injury during enclosed-space fires: An update. *J. Bras. Pneumol.* **2013**, *39*, 373–381. [CrossRef]
11. Black, W.Z. Smoke movement in elevator shafts during a high-rise structural fire. *Fire Saf. J.* **2009**, *44*, 168–182. [CrossRef]
12. Qin, T.X.; Guo, Y.C.; Chan, C.K.; Lin, W.Y. Numerical simulation of the spread of smoke in an atrium under fire scenario. *Build. Environ.* **2009**, *44*, 56–65. [CrossRef]

13. Klote, J.H.; Ferreira, M.J.; Kashef, A.; Turnbull, P.G.; Milke, J.A. *Handbook of Smoke Control Engineering*; American Society of Heating Refrigerating and Air-Conditioning Engineers: Atlanta, GA, USA, 2012; ISBN 1936504243.
14. Anseeuw, K.; Delvau, N.; Burillo-Putze, G.; De Iaco, F.; Geldner, G.; Holmström, P.; Lambert, Y.; Sabbe, M. Cyanide poisoning by fire smoke inhalation: A European expert consensus. *Eur. J. Emerg. Med.* **2013**, *20*, 2–9. [CrossRef] [PubMed]
15. Król, M.; Król, A. Multi-criteria numerical analysis of factors influencing the efficiency of natural smoke venting of atria. *J. Wind Eng. Ind. Aerodyn.* **2017**, *170*, 149–161. [CrossRef]
16. Abotaleb, H.A. Numerical study on smoke exhaust system in a mall with mechanical make-up techniques. *Alexandria Eng. J.* **2018**, *57*, 2961–2974. [CrossRef]
17. Yuen, A.C.Y.; Chen, T.B.Y.; Yang, W.; Wang, C.; Li, A.; Yeoh, G.H.; Chan, Q.N.; Chan, M.C. Natural ventilated smoke control simulation case study using different settings of smoke vents and curtains in a large Atrium. *Fire* **2019**, *2*, 7. [CrossRef]
18. Gupta, K.; Mehrotra, M.; Kumar, P.; Gogia, A.R.; Prasad, A.; Fisher, J.A. Smoke inhalation injury: Etiopathogenesis, diagnosis, and management. *Indian J. Crit. Care Med. Peer-Rev. Off. Publ. Indian Soc. Crit. Care Med.* **2018**, *22*, 180. [CrossRef]
19. Chen, T.B.Y.; Yuen, A.C.Y.; Yeoh, G.H.; Timchenko, V.; Cheung, S.C.P.; Chan, Q.N.; Yang, W.; Lu, H. Numerical study of fire spread using the level-set method with large eddy simulation incorporating detailed chemical kinetics gas-phase combustion model. *J. Comput. Sci.* **2018**, *24*, 8–23. [CrossRef]
20. Qin, T.X.; Guo, Y.C.; Chan, C.K.; Lin, W.Y. Numerical investigation of smoke exhaust mechanism in a gymnasium under fire scenarios. *Build. Environ.* **2006**, *41*, 1203–1213. [CrossRef]
21. McGrattan, K.; Hostikka, S.; McDermott, R.; Floyd, J.; Vanella, M.; Weinschenk, C.; Overholt, K. *Fire Dynamics Simulator Technical Reference Guide Volume 2: Verification*, 6th ed.; National Institute of Standards and Technology (NIST): Gaithersburg, MD, USA, 2017.
22. Zalok, E.; Hadjisophocleous, G.V. Assessment of the use of fire dynamics simulator in performance-based design. *Fire Technol.* **2011**, *47*, 1081–1100. [CrossRef]
23. Gutiérrez-Montes, C.; Sanmiguel-Rojas, E.; Viedma, A.; Rein, G. Experimental data and numerical modelling of 1.3 and 2.3 MW fires in a 20 m cubic atrium. *Build. Environ.* **2009**, *44*, 1827–1839. [CrossRef]
24. Xiao, B. Comparison of numerical and experimental results of fire induced doorway flows. *Fire Technol.* **2012**, *48*, 595–614. [CrossRef]
25. McGrattan, K.; Hostikka, S.; McDermott, R.; Floyd, J.; Weinschenk, C.; Overholt, K. *Fire Dynamics Simulator Technical Reference Guide Volume 3: Validation*, 6th ed.; National Institute of Standards and Technology (NIST): Gaithersburg, MD, USA, 2017.
26. Ayala, P.; Cantizano, A.; Gutiérrez-Montes, C.; Rein, G. Influence of atrium roof geometries on the numerical predictions of fire tests under natural ventilation conditions. *Energy Build.* **2013**, *65*, 382–390. [CrossRef]
27. Liu, Y.; Moser, A.; Sinai, Y. Comparison of a CFD fire model against a ventilated fire experiment in an enclosure. *Int. J. Vent.* **2004**, *3*, 169–181. [CrossRef]
28. Hasib, R.; Kumar, R.; Kumar, S. Simulation of an experimental compartment fire by CFD. *Build. Environ.* **2007**, *42*, 3149–3160. [CrossRef]
29. Barsim, M.M.; Bassily, M.A.; El-Batsh, H.M.; Rihan, Y.A.; Sherif, M.M. Numerical simulation of an experimental atrium fires in combined natural and forced ventilation by CFD. *Int. J. Vent.* **2020**, *19*, 1–24. [CrossRef]
30. Meroney, R.N. Wind effects on atria fires. *J. Wind Eng. Ind. Aerodyn.* **2011**, *99*, 443–447. [CrossRef]
31. Gutiérrez-Montes, C.; Sanmiguel-Rojas, E.; Kaiser, A.S.; Viedma, A. Numerical model and validation experiments of atrium enclosure fire in a new fire test facility. *Build. Environ.* **2008**, *43*, 1912–1928. [CrossRef]
32. Lapuerta, C.; Suard, S.; Babik, F.; Rigollet, L. Validation process of ISIS CFD software for fire simulation. *Nucl. Eng. Des.* **2012**, *253*, 367–373. [CrossRef]
33. Suard, S.; Koched, A.; Pretrel, H.; Audouin, L. Numerical simulations of fire-induced doorway flows in a small scale enclosure. *Int. J. Heat Mass Transf.* **2015**, *81*, 578–590. [CrossRef]
34. Pretrel, H.; Suard, S.; Audouin, L. Experimental and numerical study of low frequency oscillatory behaviour of a large-scale hydrocarbon pool fire in a mechanically ventilated compartment. *Fire Saf. J.* **2016**, *83*, 38–53. [CrossRef]

35. Shih, C.; Chen, Y.; Su, C.; Wang, S.; Yang, Y. Analysis of makeup air in a natural smoke vent system in a tall space using numerical simulation and Schlieren technique. *Int. J. Numer. Methods Heat Fluid Flow* **2019**. [CrossRef]

36. Hagglund, B. *Effects of Inlets on Natural Fire Vents: An Experimental Study*; DIANE Publishing: Stockholm, Sweden, 1996; ISBN 0788137298.

37. Huang, Y.; Chen, X.; Zhang, C. Numerical simulation of the variation of obscuration ratio at the fire early phase with various soot yield rate. *Case Stud. Therm. Eng.* **2020**, *18*, 100572. [CrossRef]

38. Tan, T.; Yu, L.; Ding, L.; Gao, Z.; Ji, J. Numerical investigation on the effect of ambient pressure on mechanical smoke extraction efficiency in tunnel fires. *Fire Saf. J.* **2020**, 103136. [CrossRef]

39. Wang, K.; Cai, W.; Zhang, Y.; Hao, H.; Wang, Z. Numerical simulation of fire smoke control methods in subway stations and collaborative control system for emergency rescue. *Process Saf. Environ. Prot.* **2021**, *147*, 146–161. [CrossRef]

40. McGrattan, K.; Hostikka, S.; McDermott, R.; Floyd, J.; Vanella, M. *Fire Dynamics Simulator: User's Guide*, 6th ed.; National Institute of Standards and Technology (NIST): Gaithersburg, MD, USA, 2017.

41. Wang, L.; Lim, J.; Quintiere, J.G. On the prediction of fire-induced vent flows using FDS. *J. Fire Sci.* **2011**, *30*, 110–121. [CrossRef]

42. Rehm, R.G.; Baum, H.R. The equations of motion for thermally driven, buoyant flows. *J. Res. NBS* **1978**, *83*, 297–308. [CrossRef]

43. Deardorff, J.W. Stratocumulus-capped mixed layers derived from a three-dimensional model. *Boundary-Layer Meteorol.* **1980**, *18*, 495–527. [CrossRef]

44. McGrattan, K.; Hostikka, S.; McDermott, R.; Floyd, J.; McDermott, R.; Vanella, M. *Fire Dynamics Simulator Technical Reference Guide Volumes 1: Mathematical Model*, 6th ed.; National Institute of Standards and Technology, Building and Fire Research: Gaithersburg, MD, USA, 2017.

45. Chisum, W.J.; Turvey, B.E. *Crime Reconstruction*; Academic Press: Waltham, MA, USA, 2011; ISBN 0123864615.

46. Maevski, I.Y. *Design Fires in Road Tunnels*; National Academy of Sciences: Washington, DC, USA, 2011; ISBN 0309143306.

47. Wu, G.-Y.; Chen, R.-C. The analysis of the natural smoke filling times in an atrium. *J. Combust.* **2010**, *2010*. [CrossRef]

48. Hurley, M.J.; Rosenbaum, E.R. *Performance-Based Fire Safety Design*; CRC Press: Boca Raton, FL, USA, 2015; ISBN 1482246562.

49. Harlow, F.H.; Welch, J.E. Numerical calculation of time-dependent viscous incompressible flow of fluid with free surface. *Phys. Fluids* **1965**, *8*, 2182–2189. [CrossRef]

50. Hurley, M.J.; Gottuk, D.T.; Hall, J.R., Jr.; Harada, K.; Kuligowski, E.D.; Puchovsky, M.; Watts, J.M., Jr.; WIECZOREK, C.J. *SFPE Handbook of Fire Protection Engineering*; Springer: New York, NY, USA, 2015; ISBN 1493925652.

51. Yang, J.; Wong, M.S.; Menenti, M.; Nichol, J. Modeling the effective emissivity of the urban canopy using sky view factor. *ISPRS J. Photogramm. Remote Sens.* **2015**, *105*, 211–219. [CrossRef]

52. Tewarson, A. Generation of heat and gaseous, liquid, and solid products in fires. In *The SFPE Handbook of Fire Protection Engineering*; Springer: New York, NY, USA, 2008; Volume 4, pp. 3–109.

53. de Freitas Rocha, M.A.; Landesmann, A. Combustion properties of Brazilian natural wood species. *Fire Mater.* **2016**, *40*, 219–228. [CrossRef]

54. Alpert, R.L. Ceiling jet flows. In *SFPE handbook of Fire Protection Engineering*; Springer: New York, NY, USA, 2016; pp. 429–454.

55. Sinclair, R. CFD simulation in atrium smoke management system design. *ASHRAE Trans.* **2001**, *107*, 711.

Publisher's Note: MDPI stays neutral with regard to jurisdictional claims in published maps and institutional affiliations.

Article

In Situ Monitoring of Drying Process of Masonry Walls

Łukasz Cieślikiewicz , **Piotr Łapka *** and **Radosław Mirowski**

Institute of Heat Engineering, Faculty of Power and Aeronautical Engineering, Warsaw University of Technology, 21/25 Nowowiejska St., 00-665 Warsaw, Poland; lukasz.cieslikiewicz@pw.edu.pl (Ł.C.); radoslaw.mirowski@gmail.com (R.M.)
* Correspondence: piotr.lapka@pw.edu.pl

Received: 4 November 2020; Accepted: 18 November 2020; Published: 25 November 2020

Abstract: The in situ hygro-thermal behavior of a wet masonry wall during its drying process is presented in this paper. The considered wall is a part of a basement of a historic building that was subjected to renovation works. The building is located in the City of Łowicz (Poland). The drying process was implemented by applying the thermo-injection method and a novel prototype of the drying device used for this method. The dedicated acquisition system was developed to in situ monitor parameters of the drying process. The air temperature and relative humidity in various locations in the basement, temperatures and moisture contents at several points of the wet wall as well as the electrical parameters of the drying device were registered. Based on variations of the monitored parameters, the hygro-thermal behavior of the wall during drying was studied. After 6 days of drying, the wall temperature in the drying zone was increased to approximately 40–55 °C, while the moisture content was reduced to the mean level of 3.76% vol. (2.35% wt.). These wall parameters allowed for effective impregnation of the wall with the hydrophobic silicone micro-emulsion, which created horizontal and vertical waterproofing. Moreover, the specific energy consumption during the drying process defined as energy consumption divided by the mean volumetric moisture content drop (MC) between the initial and final state in the wall and by the length of the dried wall section was estimated to be 11.08 kWh/MC%/m.

Keywords: drying; heat and moisture transfer; hygro-thermal behavior; masonry walls; wet wall; in situ monitoring

1. Introduction

In old and historical buildings, the problem of rising moisture (groundwater) in masonry walls is very common. The lack or incorrect implementation or worn out horizontal and vertical damp-proof insulation in the underground parts of buildings are the sources of the dampness problem. The excessive moisture in the masonry walls in the basement may results in severe damage, e.g., material degradation, dissolution of compounds, salt migration and crystallization, cracking and spalling. Moreover, the dampness may penetrate higher parts of buildings, which causes damage in plasters, facings or precious ornaments covering the walls. These effects are accompanied by serious deterioration of mechanical and hygro-thermal properties of the material of walls as well as sanitary conditions inside the building. Therefore, when excessive moisture is detected in the building due to capillary rising from the ground, retrofitting must be conducted. One of the methods of historic building renovation is drying damp masonry walls in the basement and ensuring new horizontal and vertical waterproofing. Other renovation activities should also be planned [1].

In order to correctly dry masonry walls, knowledge about the hygro-thermal phenomena in building elements and in the surrounding humid air is crucial. However, the transport of moisture in porous building structures is a very complex process which is influenced by many factors. The investigation of the moisture and temperature behavior during the drying of masonry walls may be helpful in understanding and properly designing this process. This is also very important for the efficiency optimization of the drying process, which is highly energy consuming.

Most of the research presented in the literature has been devoted to the problem of monitoring and analyzing the moisture content and moisture transfer in various elements of buildings, e.g., in masonry walls or floors, during different periods of their normal/standard use. In situ moisture monitoring is very important for both the diagnosis of building problems and the acquisition of knowledge on the moisture transport to predict building performance.

Several researches related to development and/or application of in situ monitoring techniques and systems for the investigation of moisture behavior in the different elements of buildings were identified. These systems were based on different techniques [2]. For example, Válek et al. [3] analyzed the drying behavior of laboratory models of masonry walls made of fired clay brick, sandstone and spongilite that were flooded. The following techniques were investigated: infrared thermography, complex resistivity, ground penetrating radar and ultrasonics. The measured moisture contents were compared with those obtained with the gravimetric method. Good qualitative matching of the results obtained, applying different non-destructive testing methods, was observed. Fidríková et al. [4] developed a new sensor principle for moisture monitoring in the masonry walls. This system was based on changes in the thermal conductivity of porous structures when they were filled with moist air, water or ice depending on the existing thermodynamic conditions in the building. The system was placed in the northern, southern and western masonry wall of St. Martin's Cathedral tower in Bratislava and tested in situ. Zegowitz et al. [5] investigated drying process of different components and layers of the floor construction and walls. They built a large-scale laboratory tests site with four rooms, each with three types of masonry walls made of internally insulated clay bricks and a typical floor for intermediate storeys. Rooms were located within a climate simulator, and water damage was artificially simulated. A measuring system with more than 300 sensors for moisture content, relative humidity and temperature accompanied by thermography was applied to monitor the drying behavior. They demonstrated the advantages and disadvantages of the tested drying systems, i.e., underfloor drying equipment, infrared heating panels and lance wall drying system. Walker and Pavía [6] studied the in situ moisture behavior of a solid brick wall with different internal insulations of varying thermal and moisture characteristics. The measurements were carried out over a two-and-a-half-year period. They monitored the internal temperature and relative humidity near the internal interface between the insulation and solid brick wall. They found that the vapor permeability of the insulation had a great impact on the moisture behavior of walls. Rymarczyk et al. [7] proposed a system for the analysis of moisture content in walls, including historical buildings, which was based on a hybrid tomography, original measurement sensors and a solution of the inverse problems. This system was able to estimate moisture not only on the surface but also on the inside of the wall. Cieślikiewicz et al. [8] developed an experimental stand for the investigation of the drying process in a model of a brick wall (small scale). The stand was equipped with a precise platform balance to measure the total moisture content in the specimen as well as several time-domain reflectometry (TDR) probes to register the local variations of the volumetric moisture content and resistance temperature detectors (RTD) to measure the temperature changes in a test wall. They presented preliminary results of the monitoring of the drying process of a small wall. Hoła and Sadowski [9] developed the method of neural identification of the moisture content in saline brick walls in historic buildings. The method utilized artificial neural networks, which were trained on a set of data obtained using non-destructive methods and collected on a selected representative group of masonry historic buildings from different historical periods. The data set contained two parameters that described the moisture content assessed by the dielectric and microwave methods and three parameters that described the concentration of basic salts in the damp walls assessed by the semi-quantitative method and the mass moisture content estimated by the gravimetric method. The test that was carried out showed that it is possible to reliably identify the moisture content of saline brick walls using the proposed approach without the necessity of interference in the wall structure. Next, the method proposed in [9] was successfully validated by investigating the moisture content in two historic buildings other than those used for the learning and testing of the artificial neural networks [10]. Orr et al. [11] evaluated the performance of microwave sensors and radar for monitoring the moisture movement in the masonry walls subjected to wind-driven rain. Tests were carried out for two objects, i.e., the laboratory model of a granite wall and a tower constructed from sandstone, which were typical for Oxford in England and Edinburgh in Scotland, respectively. It was demonstrated that both measurement techniques are useful for monitoring the moisture in stone masonry systems. Recently, Hoła [12] proposed an original methodology of testing the moisture content of brick walls in buildings. The methodology was developed based on the investigation of many excessively wet buildings erected in various historical periods. It used various non-destructive methods for the measurements of moisture content in walls. The methodology was tested on a facility from the 14th century. It was concluded that it is important to not only determine the moisture content, but also to find its distribution along the length and height of individual walls. The heat and moisture behavior of the whole buildings, elements of the buildings and building materials was also investigated by applying numerical modeling. For example, Cabrera et al. [13] simulated two conventional tests used for the characterization of the hygro-thermal behavior of heritage elements to check the capacity and quality of the moisture transfer simulation with the new functions, i.e., with the hydraulic conductivity and diffusion resistance factor. These tests were carried out using a moisture transport module implemented in COMSOL Multiphysics. Belleghem et al. [14] developed a three-dimensional (3D) heat and mass transfer model which allowed for simultaneously accounting for both the convective conditions surrounding a porous material and the heat and moisture transport in the porous material. The model was implemented in the ANSYS Fluent software and was validated for the case of convective drying of a saturated ceramic brick. In [15], two models of heat and moisture transfer in building materials, i.e., the equilibrium and non-equilibrium one, have been proposed, and their accuracies were verified. As indicated, these models may be applied for analysis of hygro-thermal behavior during drying of masonry walls including historical buildings. However, to carry out reliable numerical simulations, detailed knowledge about hygro-thermal properties of building materials is crucial. Luckily, such data may be found in the literature [16,17].

The quoted investigations dealt mainly with moisture monitoring in buildings or with the simulation of hygro-thermal behavior of building elements under their normal/standard use. There is a lack of studies related to the analysis of the hygro-thermal behavior of the masonry wall during its drying in the real scale. Moreover, there is no in situ analysis of power consumption during the drying process of real masonry walls. These gaps in the knowledge are filled in this work.

The paper is organized as follows. First, descriptions of the renovated building, drying and waterproofing techniques as well as measurement system are given. Next, the results of monitoring of drying process are shown and discussed. Finally, the study is summarized.

2. Materials and Methods

2.1. Renovated Building

The building under consideration is shown in Figure 1. It is a former monastic custody located in the city of Łowicz in Poland, which was built at the end of the 18th century. Today, it is a rectory. The building belongs to a group of historic buildings located at the northern frontage of the Old Market Square in Łowicz. The facility is entered in the provincial register of monuments.

Figure 1. The renovated building.

The building is made of red brick, has two floors with a high basement and is covered with a gable roof. The plan of the basement is shown in Figure 2. The masonry walls and the floor in the basement were also made of fired red bricks manufactured locally from local constituents. A lime-based joint was used. The thickness of the masonry walls in the basement varies from approximately 60 up to 80 cm. Moreover, the walls in the basement on the internal side are covered with lime-based plaster of the thickness approximately 2–3 cm (see Figure 3b with the wall structure). On the external side the walls are in the direct contact with the soil. Visual inspection of the building revealed the plaster degradation and salt crystallization sites, which were found both inside and outside the building. This means that the groundwater arose from the basement by capillary force action and evaporated at the internal and external side of the wall located above the ground. Nevertheless, before the renovation works, the moisture content in the masonry walls in the basement was measured by using the calcium carbide method (C-M Hydrometer by Gann). The measurements were carried out in six locations in the basement, which are presented in Figure 2. In each location, the holes were drilled approximately 10 cm above the basement floor, and the drilling residue from a depth of approximately 30–35 cm was analyzed. The measurements showed that the moisture content in the masonry walls was in the range of 5–11% wt. According to the building code, the level of moisture in the masonry walls was moderate to high. The recommended level is below 3% wt. The measurements show the necessity of drying masonry walls in the basement and ensuring horizontal and vertical waterproofing.

During the renovation, the building was subjected to several types of works. First of all, the masonry walls were dried and new waterproofing was ensured by using the thermo-injection method, which is described in the next subsection. Moreover, the old plaster was removed and the walls in the basement were covered with a special renovation plaster, which consists of three layers (i.e., Silten Renobase, Silten Renotop and Silten Renofine by Silten Terbud) of different properties. This lime-based plaster has a special composition and is characterized by hydrophobic properties, high porosity and high vapor permeability. Moreover, there are plans to remove the old floor and make a new one with thermal insulation from the ground.

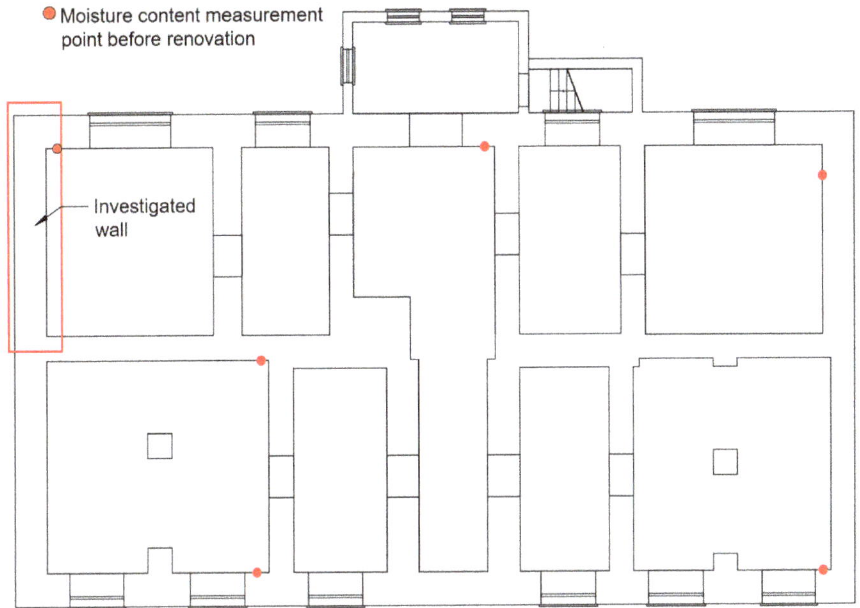

Figure 2. Plan of the renovated basement with locations of the initial moisture content measurement points.

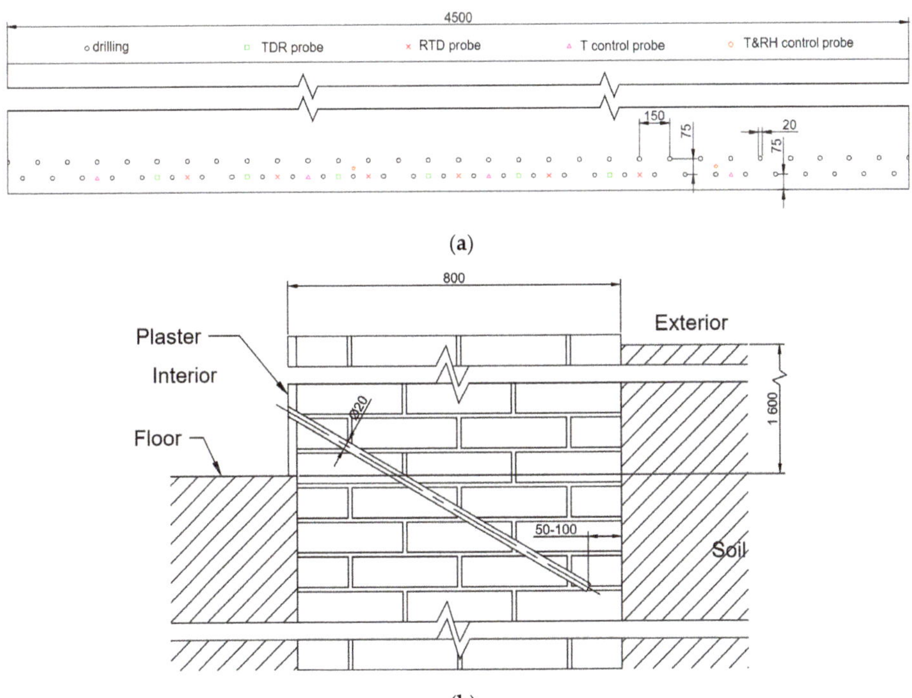

Figure 3. Schemas with locations of drilling holes and monitoring sensors: (**a**) Positions of the drilling holes and sensors in the wet masonry wall and (**b**) the cross-section of the masonry wall along the upper drilling hole with the wall structure (dimensions in mm).

2.2. Thermo-Injection Method for the Drying and Sealing of Masonry Walls

2.2.1. Method Description

The thermo-injection method [8,18–20] was used to carry out the renovation of the considered building. This method is classified as a convective method [8]. In this method, before the onset of drying, 20 mm holes are drilled in the wet wall in two staggered rows every 15 cm at a 30° angle from the horizon. The distribution of the drillings is presented in Figure 3a, while the cross-section of the upper drillings is presented in Figure 3b. The holes decrease the strength of the wall by approximately 5–10%. However, after drying and sealing, they are filled with special mortar (Silten Renoflow by Silten Terbud), injected into the holes using a membrane pump. This mortar has a special composition which prevents it from foaming and creating bubbles inside the holes. After this process, the strength of the wall is restored. Drying probes are placed in the upper drillings. These probes act as heaters and also supply fresh air close to the bottom of the hole. Air then flows out through the gap between the probe and hole to the surroundings. Next, through the probes and also by air, the wall is heated up and air that leaves the hole removes moisture from the wall. When the moisture content in the wall decreases bellow 3–4% wt. heating is stopped and a special hydrophobic silicone micro-emulsion SMK type (Silten Me by Silten Terbud) is immediately injected under elevated pressure to all holes. This process leads to the creation of the waterproof membrane, which prevents rewetting of the wall [18–20]. The silicone micro-emulsion in the aqueous solution has hydrophilic properties; therefore, it mixes well with water present in the building materials. After drying, it becomes hydrophobic. Moreover, the micro-emulsion does not react with the wall material. Based on the observation, it was found that the penetration depth is approximately 8–10 cm. Therefore, holes for injection are drilled in a staggered configuration with horizontal and vertical pitches as shown in Figure 3a and with the bottom approximately 5–10 cm from the external side of the wall as shown in Figure 3b. It should be noticed that the wall heating also improves the impregnation efficiency. The thermal energy accumulated in the hot wall increases the temperature of the micro-emulsion and reduces its viscosity during injection, which results in the deeper penetration into wall structure.

The thermo-injection method, even though it is an invasive method, is approved for use by historic preservation officers. It was successfully applied during the renovation of many historic and very precious buildings in Poland. The highland gate in Gdańsk, the Sejm building in Warsaw and Chopin's birthplace in Żelazowa Wola are, among others, examples of the application of this method. This method has been used by the company Silten Terbud since the 1990s, and the waterproofing recreated in the first applications almost 20 year ago are still working well, effectively preventing damping in the walls. This proves that this method of waterproofing renovation is very durable.

The most important drawbacks of the thermo-injection method are the interference into the wall structure and the large consumption of energy for drying. However, the thermo-injection method is usually applied in the basement, where valuable plasters and ornaments are not present. Alternatively, the method may be applied on the external side; however, to do so, the foundations must be excavated, which increases the costs of works. The advantage of the method, compared to non-invasive methods, which are usually based on surface heating (e.g., convective, thermal radiative or microwave heating), is that it effectively dries and impregnates the wall through its whole thickness, and therefore, efficiently stops the capillary water rise. Moreover, in the case of microwave heating, there is the risk of a too high wall temperature increase or the occurrence of intensive water evaporation inside the wall. Chemically bonded water in building materials may also be removed. These effects might seriously deteriorate the wall structure. Microwaves are also dangerous for people. Such problems are not present in the thermo-injection method.

2.2.2. Drying Device Description

The prototype of the device used for drying of the wet wall was developed within the project DryWall (no. POIR.04.01.02-00-0099/16) founded by the National Centre for Research and Development

(Poland). Simplified schematics of the device and a picture of the device and probes mounted to the wet wall as well as a picture of the data acquisition and control system are presented in Figures 4 and 5, respectively. This device is designed to carry out the drying of 5 m of masonry wall at one time. In order to achieve this goal, the device is equipped with two DC 24 V fans providing air stream to 35 heating probes through a distribution collector and metal spiral hose with outer electric insulation. Fans are controlled using a Bluetooth switch that allows to set three fan speeds corresponding to the air velocities and flow rate at the exit from the probe, namely, 2.7, 3.1 and 3.8 m/s and 1.95, 2.25 and 2.75 m^3/h, respectively. Heating probes are powered by 24 V AC and controlled by a Bluetooth proportional–integral–derivative (PID) regulator. Probes are made of brass tubes with a diameter and length of 16 mm and 50 cm, respectively. Brass tubes act as 40 watts electric heaters that generate heat, and as air ducts that lead air to the bottom of the drilled holes. For the automatic control and monitoring of the drying process, five resistive PT1000 temperature detectors (one per meter of wet wall) and two temperature and relative humidity sensors were installed in the wall (see Figure 3a). Heating probes were grouped in five sections containing seven probes. Each section is supplied by an individual transformer and may have different temperature control values. Each probe has a state signalization which allows for the detection of malfunction. Moreover, a standalone dehumidifier was used to reduce the humidity of air sucked up by the device.

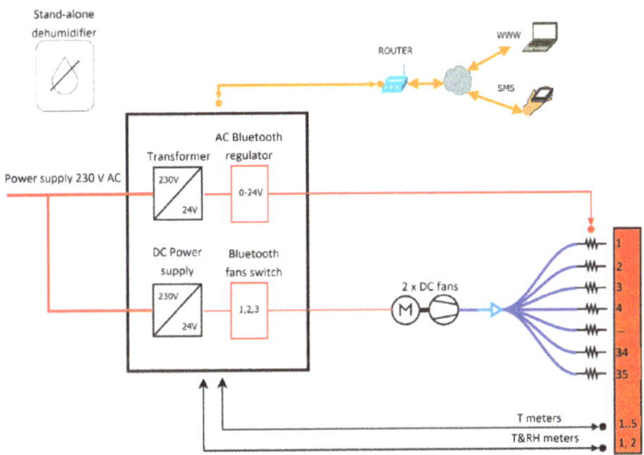

Figure 4. Schematics of the drying device.

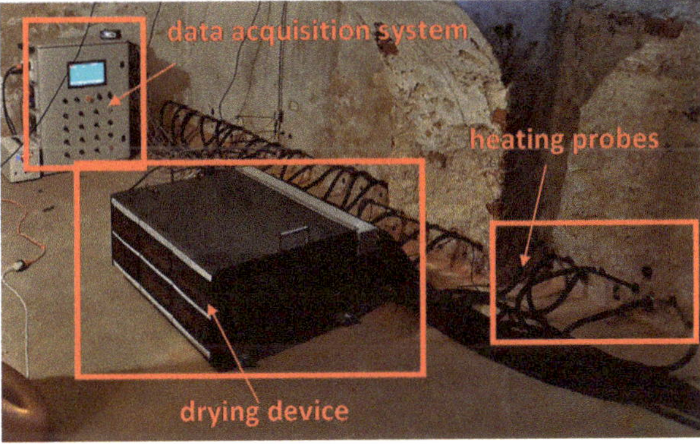

Figure 5. The drying device with probes mounted to a wet wall, as well as the data acquisition and control system, in the basement of the considered building in Łowicz.

2.3. Measurements Method

For investigating and monitoring the drying process, a dedicated data acquisition system was developed. The schematics of the system are presented in Figure 6. The main element of the system was the programmable logic controller (PLC) controller (XV 303 by Eaton Industries GmbH) equipped with a 10″ screen. The PLC controller collected data from all sensors, saved them on the internal memory and allowed for remote contact via LTE GSM modem. The following sensors were used by the data acquisition system:

- For air temperature and relative humidity measurements: three transducers (AR 252 by APAR) with an accuracy of ±0.3 °C for temperature and ±2% for relative humidity. The communication with the PLC controller was via the Modbus RS 485 protocol. Measurements were carried out in three locations, i.e., in the middle of dried room, approximately 0.5 m from the ceiling, close to the drying device inlet and close to the wet wall, approximately 10 cm from the floor.

- For wall volumetric moisture content measurements: the time-domain reflectometry device (TDR/MUX/MPTS by E-TEST) with six field probes with a maximum error of ±2% for moisture content. The locations of the probes are presented in Figure 3a. The numbering of the probes starts from the right to left. Probes were made from two sharpened acid-resistant steel rods and PCV tube and were placed in slots of 20 mm diameter and 30 cm depth filled with drilling residue for easier installation. The communication with the PLC controller was via the Modbus RS 485 protocol.

- For wall temperature measurements: six three-wire resistance temperature detectors were used with the A tolerance class, with the maximum acceptable error defined in standards lower than ±0.3 °C for the whole temperature range measured during investigation, connected through two analog input modules (XN-322-4AI-PTNI by Eaton Industries GmbH) and linked to the communication module (XN-312-GW-CAN by Eaton Industries GmbH). Stainless steel sheathed probes with a diameter of 4.5 mm were placed in 30 cm-deep slots in locations presented in Figure 3a. The numbering of the probes starts from the right to left. The communication with the PLC controller was via the CANopen protocol.

- For electrical parameters measurements: the 1-phase power network meter (N27P by LUMEL) with 32/63 A measurement range was used. The meter allowed for the measurements of: current (basic error ±0.2%), voltage (basic error ±0.2%), frequency (basic error ±0.2%), active power (basic error ±0.5%), reactive power (basic error ±0.5%), apparent power (basic error ±0.5%), active energy (basic error ±0.5%) and reactive energy (basic error ±0.5%). Communication with the PLC controller was via the Modbus RS 485 protocol.

The whole system and sensors were checked for possible interaction. Test measurements were carried out for each sensor working separately and for all sensors working together. During these tests, the drying device was turned off and on. The only detected possible interference problem was between TDR probes. However, measurements applying these sensors were carried out in sequence, so the interferences were eliminated.

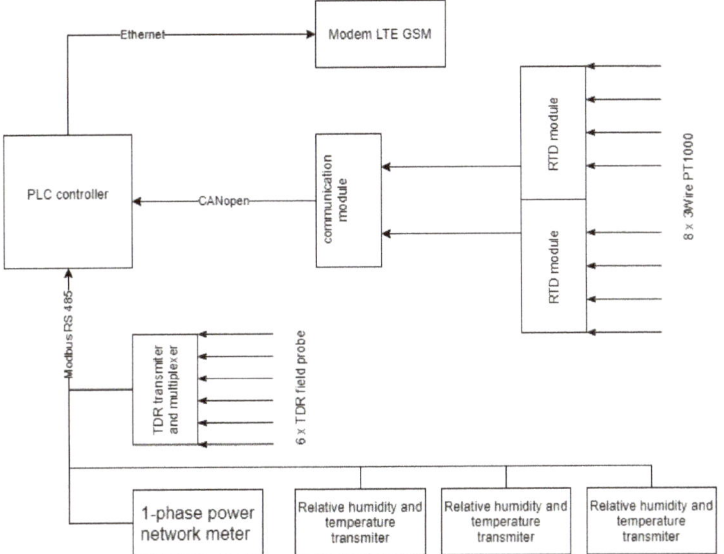

Figure 6. Schematics of the data acquisition system.

3. Results and Discussion

The drying process using the drying device and data acquisition and control system described above was carried out in the basement of the considered building. The renovation works were conducted in August 2020. In this paper, only the results of in situ monitoring of the process for one 5-m-long part of the external masonry wall are presented. The wall had a thickness of 80 cm and a structure shown in Figure 3b. For the other masonry walls, similar behavior was observed. The process begun without heating, with fans switched on and set at the second speed. This stage lasted 24 h and was characterized by quite stable air temperatures, presented in Figure 7a. The slight decrease of the absolute air humidity shown in Figure 7b was a combined effect of moisture gained from the wet wall and moisture captured by the dehumidifier. In the first stage of drying, moisture was removed mainly from the surface of the wall as well as surfaces of drillings and their surroundings. The temperature drop of the wall seen in Figure 8a in the first hours was connected to achieving the local saturation temperature of the wall.

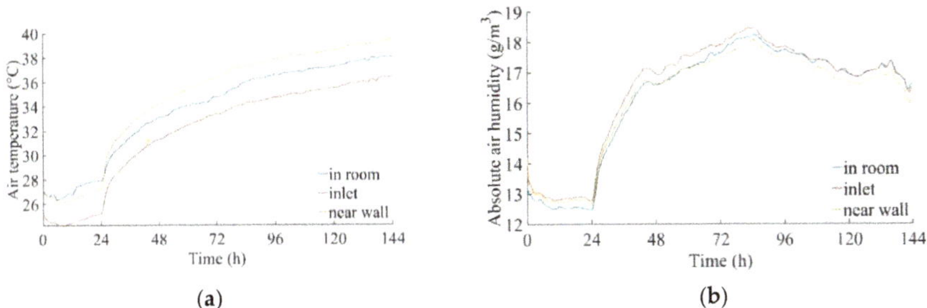

Figure 7. Air parameters during the drying process. (**a**) Temperature, (**b**) absolute humidity.

The second period of drying started after 24 h when heating was turned on and the fan was operated without changes. This affected both the air and wall temperature, which started to increase—see Figures 7a and 8a. Differences between temperature variations for different points shown in Figure 8a resulted from different distances from sensors to heating holes and differences in the wall structure—it was very difficult to precisely drill the holes for probes and sensors. The highest temperature observed for point T2 in Figure 8a was a result of an accidental connection during the drilling of slots for the drying probe and temperature sensor. This sensor showed a temperature that was very close to the temperature of the heating probe. As the process progressed, the wall temperature stabilized. Differences in the behavior of moisture content in the wall, presented in Figure 8b, were connected to the variation in the local content of the water in the wall as well as with the material which surrounded the slots. The TDRs 3–5 were in locations surrounded by a much higher amount of mortar than TDR 1, 2 and 6, which were in locations surrounded by ceramic brick. In the first stage of drying, the reduction of moisture content was rapid even without heating; after 24 h, additional heating was necessary to maintain the drying rate. Finally, after 144 h of drying, TDR probes 1–6 indicated a moisture content inside the wall of 2%, 3.6%, 4.8%, 5.9%, 3.4% and 3% vol., respectively.

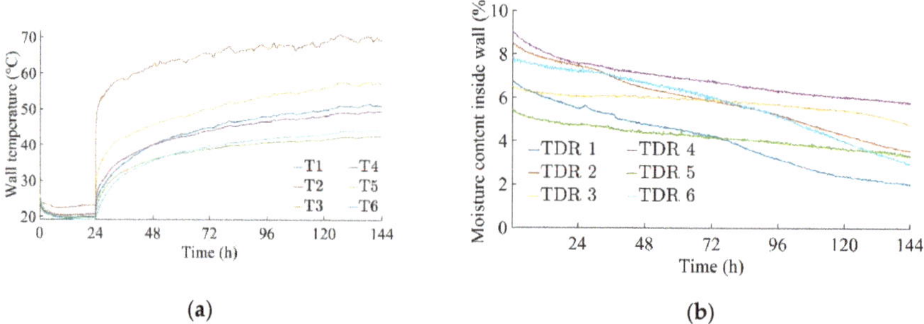

(a)   (b)

Figure 8. Wet wall parameters. (**a**) Temperature and (**b**) volumetric moisture content.

Figure 9 presents the results of the infrared image of the surrounding of two heating slots for the end of the drying process. The surface temperature of the wall between the heating probes oscillated around approximately 43 °C, while the surface temperature close to the heating probes reached approximately 53 °C. The thermogram clearly shows the drying and heating area, which was a strip of approximately 12 cm width.

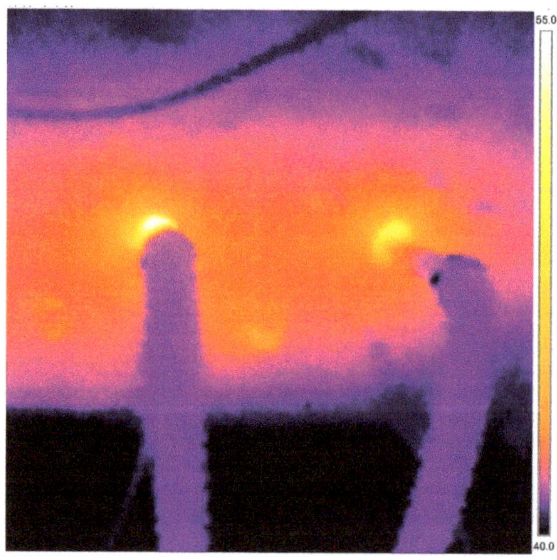

Figure 9. Thermogram of two drying probes and their surrounding at end of drying process.

During the 6 days of the drying process, the drying device consumed 208.6 kWh of energy with a mean active power of 241 W and 1688 W for the drying stage without and with heaters, respectively. At the same time, a 3.77% vol. moisture content reduction was observed on average in the wall. The new parameter, i.e., the specific energy consumption, was introduced to express the relation between the energy consumption and the amount of moisture removed from the wall. The specific energy consumption during the drying process was defined as the energy consumption divided by the mean volumetric moisture content (MC) difference between the initial and final state in the wall and by the length of the dried wall section. For the considered wall, it was equal to 11.08 kWh/MC%/m. This value is considered as satisfactory. The reactive power consume by the device was capacitive, which was due to the installed DC power supplies. When heaters were turned on, the power factor was high, so the device did not need reactive power compensation. Detailed energy parameters measured during the drying process are shown in Table 1.

Table 1. Energy parameters during the drying process.

Parameter	Unit	Value
Specific energy consumption	kWh/MC%/m	11.08
Mean moisture content reduction	% vol. (% wt.)	3.77 (2.35)
Energy consumption	kWh	208.6
Mean active power without heaters	W	241
Mean apparent power without heaters	VA	405
Mean reactive power without heaters	VAr	−325
Mean power factor without heaters	-	0.595
Mean active power with heaters	W	1 688
Mean apparent power with heaters	VA	1 725
Mean reactive power with heaters	VAr	−356
Mean power factor with heaters	-	0.978

During the six days of the drying process, the moisture content of the wall decreased to a value allowing for effective infiltration of the wall by the hydrophobic silicone micro-emulsion, i.e., to a mean level of 3.76% vol. (2.35% wt.). The increased temperature of the wall not only facilitated moisture removal but also had a positive effect on the impregnation process, i.e., high wall temperature decreased the viscosity of the micro-emulsion, which encouraged its depth infiltration and generation of waterproof membrane.

4. Conclusions

In this paper, hygro-thermal behavior of the wet wall during its drying process was monitored in situ. The considered wall was part of the basement of a historical building which was renovated. The drying process was conducted by applying the thermo-injection method and the novel protype of the device to its implementation developed with the project DryWall. By applying the dedicated data acquisition system developed to in situ monitor parameters of the drying process, the following quantities were registered: the air temperature and relative humidity in various locations in the basement, temperatures and moisture contents in several points in the wet wall and electrical parameters of the drying device. The obtained results may be concluded as follows:

- The drying process resulted in the increase of the drying zone temperature up to 40–55 °C depending on the distance from the heating probes. After six days of drying, the wall temperature attained stabilization.
- The moisture was effectively removed from the wall. After six days of drying, the moisture content dropped to a mean level of 3.76% vol. (2.35% wt.), which is a value that allows effective impregnation of the wall with the hydrophobic silicone micro-emulsion. This fluid is used to create horizontal and vertical waterproofing in the wall. Moreover, the high temperature of the

wall not only facilitated moisture removal, but also helped in the infiltration of the wall by the hydrophobic fluid.

- The energy consumption during the process was equal to 208.6 kWh. This value included heating and working of fans. The new parameter, i.e., the specific energy consumption, was also defined to refer to the energy consumption to the amount of moisture removed from the wall. For the considered wall, it was equal to 11.08 kWh/MC%/m.

Author Contributions: Conceptualization, Ł.C. and P.Ł.; methodology, Ł.C., P.Ł. and R.M.; software, Ł.C. and R.M.; validation, Ł.C. and P.Ł.; formal analysis, Ł.C. and P.Ł.; data curation, Ł.C. and R.M.; investigation, Ł.C. and R.M.; writing—original draft preparation, Ł.C. and P.Ł.; writing—review and editing, P.Ł.; visualization, Ł.C. and R.M.; supervision, P.Ł.; project administration, P.Ł.; funding acquisition, P.Ł. All authors have read and agreed to the published version of the manuscript.

Funding: This work was supported by the European Union within the European Regional Development Fund under project no. POIR.04.01.02-00-0099/16 "Development of innovative technology of drying and moisture sealing of masonry walls, DryWall" granted by the National Centre for Research and Development (Poland).

Conflicts of Interest: The authors declare no conflict of interest.

References

1. De Santoli, L.; Mancini, F.; Rossetti, S.; Nastasi, B. Energy and system renovation plan for Galleria Borghese, Rome. *Energy Build.* **2016**, *129*, 549–562. [CrossRef]
2. Phillipson, M.C.; Baker, P.H.; Davies, M.; Ye, Z.; McNaughtan, A.; Galbraith, G.H.; McLean, R.C. Moisture measurement in building materials: An overview of current methods and new approaches. *Build. Serv. Eng. Res. Technol.* **2007**, *28*, 303–316. [CrossRef]
3. Válek, J.; Kruschwitz, S.; Wöstmann, J.; Kind, T.; Valach, J.; Köpp, C.; Lesák, J. Nondestructive Investigation of Wet Building Material: Multimethodical Approach. *J. Perform. Constr. Facil.* **2010**, *24*, 462–472. [CrossRef]
4. Fidríková, D.; Greif, V.; Dieška, P.; Štofanik, V.; Kubičár, L.; Vlčko, J. Monitoring of the temperature-moisture regime in St. Martin's Cathedral tower in Bratislava. *Environ. Earth Sci.* **2013**, *69*, 1481–1489. [CrossRef]
5. Zegowitz, A.; Renzl, A.; Hofbauer, W.; Meyer, J.; Kuenzel, H. Drying behaviour and microbial load after water damage. *Struct. Surv.* **2016**, *34*, 24–42. [CrossRef]
6. Walker, R.; Pavía, S. Thermal and moisture monitoring of an internally insulated historic brick wall. *Build. Environ.* **2018**, *133*, 178–186. [CrossRef]
7. Rymarczyk, T.; Sikora, J.; Tchórzewski, P. Implementation of electrical impedance tomography for analysis of building moisture conditions. *COMPEL Int. J. Comput. Math. Electr. Electron. Eng.* **2018**, *37*, 1837–1861. [CrossRef]
8. Cieślikiewicz, Ł.; Łapka, P.; Mirowski, R.; Wasik, M.; Kubiś, M.; Pietrak, K.; Furmański, P.; Seredyński, M.; Wiśniewski, T. Development of the experimental stand for investigation of the drying process in moist walls. *IOP Conf. Ser. Mater. Sci. Eng.* **2019**, *660*, 012021. [CrossRef]
9. Hoła, A.; Sadowski, Ł. A method of the neural identification of the moisture content in brick walls of historic buildings on the basis of non-destructive tests. *Autom. Constr.* **2019**, *106*, 102850. [CrossRef]
10. Hoła, A.; Sadowski, Ł. Verification of a nondestructive method for assessing the humidity of saline brick walls in historical buildings. *Appl. Sci.* **2020**, *10*, 6926. [CrossRef]
11. Orr, S.A.; Fusade, L.; Young, M.; Stelfox, D.; Leslie, A.; Curran, J.; Viles, H. Moisture monitoring of stone masonry: A comparison of microwave and radar on a granite wall and a sandstone tower. *J. Cult. Herit.* **2020**, *41*, 61–73. [CrossRef]
12. Hoła, A. Methodology for the in situ testing of the moisture content of brick walls: An example of application. *Arch. Civ. Mech. Eng.* **2020**, *20*, 1–13. [CrossRef]
13. Cabrera, V.; López-Vizcaíno, R.; Yustres, Á.; Ruiz, M.Á.; Torrero, E.; Navarro, V. A functional structure for state functions of moisture transfer in heritage building elements. *J. Build. Eng.* **2020**, *29*. [CrossRef]
14. Van Belleghem, M.; Steeman, M.; Janssen, H.; Janssens, A.; De Paepe, M. Validation of a coupled heat, vapour and liquid moisture transport model for porous materials implemented in CFD. *Build. Environ.* **2014**, *81*, 340–353. [CrossRef]

15. Seredyński, M.; Wasik, M.; Łapka, P.; Furmański, P.; Cieślikiewicz, Ł.; Pietrak, K.; Kubiś, M.; Wiśniewski, T.S.; Jaworski, M. Analysis of Non-Equilibrium and Equilibrium Models of Heat and Moisture Transfer in a Wet Porous Building Material. *Energies* **2020**, *13*, 214. [CrossRef]
16. Berardi, U.; Tronchin, L.; Manfren, M.; Nastasi, B. On the effects of variation of thermal conductivity in buildings in the Italian construction sector. *Energies* **2018**, *11*, 872. [CrossRef]
17. Kubiś, M.; Pietrak, K.; Cieślikiewicz, Ł.; Furmański, P.; Wasik, M.; Seredyński, M.; Wiśniewski, T.S.; Łapka, P. On the anisotropy of thermal conductivity in ceramic bricks. *J. Build. Eng.* **2020**, *31*. [CrossRef]
18. Matuszewska, M.; Matuszewski, T. Method of Uniformly Heating Holes in a Masonry Wall during Drying Same and Protecting It against Absorption of Moisture as well as Heating Element Therefor. Polish Patent 184012, 3 October 1997.
19. Matuszewski, T. Method for Drying and Protecting the Walls against Repeated Rising Damp. Polish Patent 219284, 30 September 2009.
20. Matuszewski, T.; Olędzki, M. Method for Drying Walls and Protect Them from Moisture Afresh. Polish Patent 223771, 31 October 2012.

Article

A New Generation of Thermal Energy Benchmarks for University Buildings

Salah Vaisi [1,*], **Saleh Mohammadi** [1,2], **Benedetto Nastasi** [3] **and Kavan Javanroodi** [4]

[1] Department of Architecture, Faculty of Art and Architecture, University of Kurdistan (UOK), Sanandaj 0871, Iran; saleh.mohammadi@tudelft.nl

[2] Department of Architectural Engineering + Technology, Faculty of Architecture and the Built Environment, Delft University of Technology (TU Delft), 2628BX Delft, The Netherlands

[3] Department of Planning, Design & Technology of Architecture, Sapienza University of Rome, Via Flaminia 72, 00196 Rome, Italy; benedetto.nastasi@outlook.com

[4] Solar Energy and Building Physics Laboratory (LESO-PB), Ecole Polytechnique Fédérale de Lausanne (EPFL), 1015 Lausanne, Switzerland; kavan.javanroodi@epfl.ch

* Correspondence: svaisi@uok.ac.ir or vaisis@tcd.ie; Tel.: +98-918-871-5086

Received: 10 November 2020; Accepted: 11 December 2020; Published: 14 December 2020

Abstract: In 2008, the Chartered Institution of Building Services Engineers (CIBSE TM46 UC) presented an annual-fixed thermal energy benchmark of 240 kWh/m^2/yr for university campus (UC) buildings as an attempt to reduce energy consumption in public buildings. However, the CIBSE TM46 UC benchmark fails to consider the difference between energy demand in warm and cold months, as the thermal performance of buildings largely depends on the ambient temperature. This paper presents a new generation of monthly thermal energy benchmarks (MTEBs) using two computational methods including mixed-use model and converter model, which consider the variations of thermal demand throughout a year. MTEBs were generated using five basic variables, including mixed activities in the typical college buildings, university campus revised benchmark (UCrb), typical operation of heating systems, activities impact, and heating degree days. The results showed that MTEBs vary from 24 kWh/m^2/yr in January to one and nearly zero kWh/m^2/yr in June and July, respectively. Based on the detailed assessments, a typical college building was defined in terms of the percentage of its component activities. Compared with the 100% estimation error of the TM46 UC benchmark, the maximum 21% error of the developed methodologies is a significant achievement. The R-squared value of 99% confirms the reliability of the new generation of benchmarks.

Keywords: energy benchmarking; university campus; energy performance certificate; CIBSE TM46; thermal energy efficiency

1. Introduction

There has been a global trend in the recent years to reduce energy demand and greenhouse gas (GHG) emissions in the higher educational institution buildings [1]. The trend is even more accelerated by the new policies and regulations such as the European Green Deal with ambitious goals to achieve neutral GHG cities and areas by 2050 [2]. In this regard, energy benchmarking is a useful tool to evaluate the energy performance of buildings [3]. The higher educational buildings (university buildings) are important in terms of high energy demand (kWh/m^2) and the variety of activities in the buildings.

Chartered Institution of Building Services Engineers (CIBSE) TM46:2008 [4] is one of the fundamental references for energy performance certification, and benchmarking in buildings. Despite the improvement of the energy performance of university buildings in recent years, the CIBSE TM46

UC (university campus) benchmark has remained unchanged [5]. The CIBSE TM46 UC benchmark significantly overestimates the thermal demand compared with the actual measurements [6]. Most of the benchmarking methodologies such as "Energy Star" and CIBSE TM46 have focused on the annual scale [7], while failed to consider the differences in thermal energy consumption in the cold and warm months. This leads to a notable gap in the energy demand estimations where the annual benchmark is incapable to provide detailed information based on outdoor temperature [8]. This can be even more critical considering the convoluted urban microclimate conditions around buildings [9] and complex interactions between outdoor temperature and other climate variables. Although the benchmarking methodology is not feasible to take into account detailed climate variations, it is vital to investigate for finer temporal resolution (e.g., seasonal or monthly) models to assess energy consumption profiles of university buildings. This paper addressed this research gap by introducing a novel method, namely, monthly thermal energy benchmarks (MTEBs). MTEBs aim to represent the monthly variations of mixed-use campus buildings as an accurate tool to move towards sustainable transition pathways in educational buildings.

This paper is structured as follows. First, the background of energy benchmarking systems is assessed (Section 1.1) to highlight the major research gaps in the field. The study of the related works and the discussion of the TM46 benchmarking method are presented in Sections 1.2 and 1.3, respectively. In Section 1.4, the contributions of this study are discussed. The methods and material adopted and developed in the paper are explained thoroughly in Section 2. The application of major benchmarking methods, including mixed-use and converter models are assessed in Sections 3 and 4, respectively. The novel benchmarking model (MTEBs) is presented in Section 5, followed by the conclusion to highlight the major findings of the study.

1.1. Background of Energy Benchmarking Systems

The "energy benchmarking" term was used in the 1990s to refer to the knowledge of comparing energy consumption in similar building types (peer buildings) [10]. The top-down benchmarking method uses real consumption data to calculate the energy benchmark of peer buildings. This is a comprehensive method applying officially in the EU, US, Australia, Japan, Canada, and other countries to manage the end-use energy consumption in buildings [11]. Benchmarking is a cornerstone of the European Council Directive 93/76/CEE [12] to improve energy efficiency and reduce CO_2 emissions in buildings. Energy benchmarking compares the annual total primary energy required (TPER) per unit area (m^2) in a building with the median consumption of peers [13].

Based on Chapter 20 of the original CIBSE Guide F: "Energy efficiency in buildings" and Energy Consumption Guide ECG 19: "Energy efficiency in offices", the CIBSE TM46 energy benchmark was updated by the Chartered Institution of Building Services Engineers (CIBSE) in 2008. CIBSE TM46 [4] and TM47 [14] explain the statutory energy benchmarks in buildings, which are used as predominant references in the EU and UK to calculate the building energy ratio (BER). BER is the main part of a display energy certificate (DEC).

According to the CIBSE TM46, 237 building types were classified into 29 benchmark categories based on the building's dominant function (single function). TM46 presumes the buildings as a single function and neglects other functions (activities) in the buildings, while many of them are multifunctional (mixed-use) particularly in city centers. According the CIBSE TM46, a university campus building (a typical educational building on/off campus) needs 240 kWh/m^2/yr of thermal energy per year [4].

There are fundamental modifications in thermal demand during a year; however, TM46 and Energy Star methodology cannot explain such variations. The majority of heat demand (80%) in winters is used for space heating purposes, whereas in summers the energy is consumed to prepare domestic hot water [15]. The accuracy of TM46 UC benchmarks has been studied recently by several researchers and a series of problems, such as a significant discrepancy between the benchmark and actual measurements have been reported frequently [16,17]. For example, Vaisi et al. discovered a

30% gap between the actual consumption and TM46 UC benchmark [8]. Based on the actual data of four university buildings in Dublin, the authors revised the CIBSE TM46 thermal benchmark of 240 kWh/m^2/yr and introduced a university campus revised benchmark (UCrb) of 130 kWh/m^2/yr as a validated annual index. In addition, the reviewed studies not only highlighted the requirement for revising the TM46 benchmarks [18], but also suggested the necessity for renaming the UC category [19]. The majority of current energy models present the annual-fixed benchmark, which take into account buildings as single-use (single function, single activity) because the data on mixed activities usually are unavailable or hard to collect.

1.2. Display Energy Certificate (DEC)

Display energy certificate (DEC) is an authentic certificate that shows the annual energy performance of buildings (Figure 1). The DEC dataset is used frequently for energy management in buildings. In summer 2008, for the first time, DECs were introduced in the EU under the Energy Performance of Building Directive (EPBD) regulation [20]. DEC presents the building energy efficiency, which is calculated using the total primary energy requirement (TPER). TPER is the overall quantity of all energies (electricity, oil, coal, gas, renewables, etc.) delivered to a building, including the energy that is used or lost beyond the boundary of the building during energy transformation, transmission, and distribution processes. The other index displayed on DECs is total primary fossil energy required (TPEFR), which shows the annual fossil thermal energy delivered to the boundary of buildings (Figure 1).

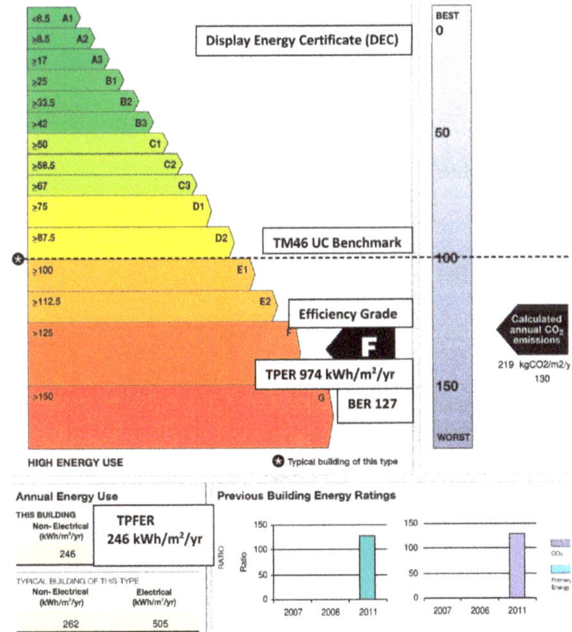

Figure 1. The main data presented on a display energy certificate (DEC).

Total final consumption (TFC) or actual consumption (recorded consumption) is the amount of energy consumed in a building. TFC is measured by meters and it is typically the quantity shown on bills [21]. If other types of bulk energy such as oil and coal are used, for calculation of TPFER they must be converted into kgCO$_2$ or kWh of energy. Generally, TPFER is approximately 20% greater than TFC [22].

On DEC, the quantity of TPER (kWh/m^2/yr) is divided by the annual benchmark, the consumption of 50% of samples, and the percentage of the result is called BER, which is graded. The alphabetical grades range from "A$_1$" to "G" and show the best to worst efficiency, respectively. The TPER, TPFER, and BER displayed on a DEC are presented in Figure 1.

1.3. Related Works

The literature in the field of benchmarking can be divided into four categories including (1) benchmarking methods and data assessment, (2) underlining the discrepancy between the energy benchmarks and actual consumption, (3) energy performance over time, and (4) reviewing the policy and presenting new recommendations. This study falls into the first and second categories.

Pasichnyi et al. [23] recommended the display energy certificate system as a new opportunity for data-enabled urban energy policy instruments. However, the certificate systems are mostly limited to annual scale rather than monthly. Burman et al. [24] compared the annual fossil–thermal performance of five new educational buildings in the UK against the operational benchmarks at the annual scale and discovered a significant discrepancy between the heating energy use and the design expectations. Papadopoulos et al. [25] assessed the energy use intensity between 2011 and 2016 and used approximately 15,000 energy consumption data of New York City properties based on an annual period.

To address the role of mixed activities on energy consumption, a study was conducted based on quantile regression model. The authors analyzed the electricity consumption of nearly 1000 buildings and found that cooling degree days and the presence of gyms, spas, and elevators were significant factors affecting the energy use. Moreover, the number of employees per unit area had a great effect on the total electricity consumption in poorly performing buildings [26].

Liu et al. [27] developed a systematic methodology as well as an energy consumption rating (ECR) system to create dynamic energy benchmarks for an individual office building with very limited information. Based on outdoor temperature, relative humidity, and daily energy consumption, the authors, at an hourly scale analysis, presented four typical energy benchmarks, including 272, 427, 497, and 592 kWh, which represent the momentary operation of the studied building. Another study identified three fundamental energy consumption periods, i.e., morning, noon, and evening peak energy consumption patterns using K-means clustering and load shape profile [28]. The authors discovered how energy consumption is changed during the daytime and consequently, they plotted the typical consumption patterns of four groups of buildings. Those patterns are the basis for modeling higher resolution profiles from monthly bills [29] or to evaluate flexibility potential of the built environment [30].

Papadopoulos and Kontokosta [31] developed a building GREEN energy grading methodology by adopting machine learning and city-specific energy use and building data to enable more precise, reasonable, and contextualized individual building energy profiles [31]. They indicated how different factors such property value (cost/square ft), unit density, bedroom density, built year, etc. affected the energy use intensity. Finally, they proposed a graded (alphabetical) annual benchmark instead of the 0–100 rating system of Energy Star. A large number of studies have frequently adopted statistical benchmarking models using machine-learning algorithms that can illustrate multifaceted relationships between energy uses and building characteristics, such as floor area and functions [32–34].

Khoshbakht et al. [35] adopted stochastic frontier analysis (SFA) to determine benchmark values for various activities and disciplines in higher educational buildings. They classified the educational buildings into different activities (e.g., research, academic offices, administration, library, teaching spaces) but did not look into the monthly or seasonal consumption patterns. In another work conducted in 81 residential buildings in Singapore [36], the authors proposed a framework to categorize the buildings by their operational similarities using data mining obtained from smart meters. They highlighted the impact of the mixed-use operation on energy demand and discovered that the activity plays a key role in energy consumption. For instance, the residential buildings had fewer facilities and lower energy load density compared to the buildings with research centers. Therefore, the EUI (Energy Use Intensity) was much smaller than the mixed-use buildings due to the galleries and laboratories that require energy in 24 h. However, the impact of each activity on energy consumption and their weight were not addressed.

Arjunan et al. [37] developed a method based on both linear and nonlinear models to increase the accuracy of energy benchmarking of office buildings in the US. They applied several building attributes such as gross floor area, cooling gross floor area, number of employees, computers, and cooling degree days, and determined the features affecting energy consumption.

1.4. The Novelty of the Proposed Method

Based on the reviewed literature, there are still unexplored particular areas, even not addressed by the renowned benchmarking systems such as CIBSE (worldwide approved benchmarking system) and Energy Star (US benchmarking system). Most of the research reviewed focused on analyzing static snapshots of buildings, i.e., annual fixed energy benchmark rather than dynamic performance trends over time, and considered buildings as a single activity [38]. Applying an annual-fixed benchmark and considering the buildings as single-use are the major research gaps in the field. This paper moves beyond the current state-of-art by proposing a new generation of thermal energy benchmarks, monthly thermal energy benchmarks (MTEBs), instead of a fixed-annual benchmark. The MTEBs benchmarking method improves the CIBSE TM46 UC benchmark of 240 kWh/m^2/yr by incorporating monthly variables, which are sensitive to ambient temperature and environmental conditions. Moreover, this study considers the impacts of various activities such as computer rooms, offices, library, laboratory, seminar and research rooms, workshop, stores, and restaurant and coffee shops on the energy consumption in typical college buildings using a revised benchmark (UCrb) model. Readers are referred to an earlier study by the authors [8] for more information about the UCrb benchmark.

Moreover, five fundamental parameters were applied in the mixed-use and converter models, including conditioned area of buildings, heating degree days (HDD), mixed-use, a recently revised benchmark (UCrb), and typical operation hours of heating systems. Finally, this study aims to fill the discrepancy between the TM46 UC benchmark and actual heat consumption highlighted in the literature, which is a step beyond the model introduced by Vaisi et al. [8] in 2018. For the first time, a definition of typical college buildings based on their mixed activities is presented.

Figure 2 is a schematic ideogram that shows the gap between CIBSE TM46 benchmark and the actual consumptions during a year, and it illustrates how a curved line benchmark can be better adapted to reality. The CIBSE TM46 UC benchmark is a horizontal line, an index for a whole year, while the methodology of MTEBs has focused on transforming the horizontal TM46 UC into a monthly dynamic benchmark (a curved line) that delivers valuable information.

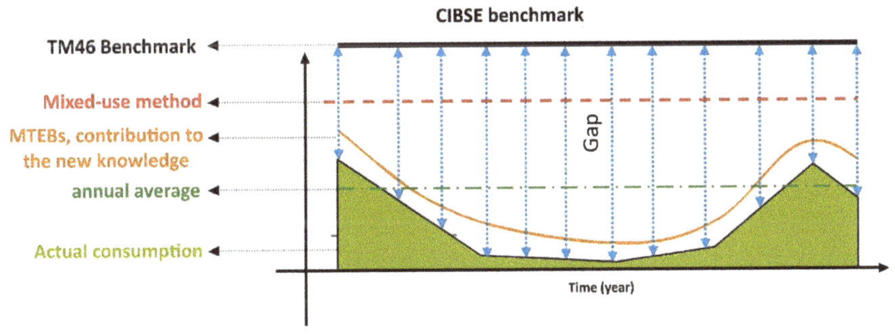

Figure 2. Monthly thermal energy benchmarks (MTEBs) ideogram.

2. Methodology

To create the monthly thermal energy benchmarks (MTEBs), the actual thermal consumption data and the operational hours of the heating systems of 52 buildings in four university campuses (Trinity College Dublin, University College Dublin, Dublin City University, Dublin Institute of Technology) were analyzed. The actual energy consumption data were obtained from the Cylon Active Energy Management online dataset [39]. The heating degree day data were collected from Degree Days.net [40]. To discover the mixed activities in the case study buildings, a survey was conducted at the floor scale. According to the assessment of energy consumption of 52 UC buildings, five key parameters that affect the thermal energy demand were found to be:

1. Area (m^2)—building useful area and activities area;
2. Mixed-use activities—this factor considers all activities in a building and calculates the value of each activity based on its area—the composite benchmark is one of the results of the mixed-use method;
3. UCrb (university campus revised benchmark)—the revised benchmark of 130 kWh/m^2/yr [8] was used instead of 240 kWh/m^2/yr as suggested by CIBSE TM46;
4. Heating degree days (HDD);
5. Typical operation hours of heating systems—usually influenced by the college's energy policy, not occupants' behavior.

The area of all activities in the surveyed buildings was calculated based on the architectural plans of the buildings. The impact of various activities on thermal energy consumption in the college buildings was determined based on the percentage area of activities. Based on the actual thermal consumption data recorded at the quarter-hour scale [39], the typical operating hours of the heating systems were calculated and the results presented in Table 1.

Table 1. Typical operation hours of heating systems.

Months	Jan	Feb	Mar	Apr	May	Jun	Jul	Aug	Sep	Oct	Nov	Dec	Total Year
Mean operation of 10 buildings	300	280	260	250	240	85	45	35	80	223	249	229	2276

Two models were developed to generate the MTEBs: (1) mixed-use model and (2) converter model. The mixed-use model relies upon the impact of all activities in a building on thermal consumption. Accordingly, a composite benchmark that considers the role of mixed activities in terms of thermal energy demand was progressed. The converter model, developed based on the annual thermal consumption, presenting on DECs. The accuracy of both models was validated against the actual thermal consumption.

To assess the impact of various activities on thermal demand, the area of all the activities of the case study buildings was surveyed, and then the area of each activity calculated in AutoCAD precisely. Ten activities were identified in 52 analyzed college buildings, while among them, 7 activities were common in all cases. Based on the analysis, a typical college building in terms of mixed activities is defined for the first time: a typical college building is a type of educational building, comprising seven typical mixed activities, including computer rooms and laboratories (31%), offices (29%), seminar and research rooms (18%), library (14%), workshop (4%), stores (3%), and restaurant or coffee shop (1%).

The energy demand estimation based on TM46 UC benchmark against the actual consumption data of "Aras An Phiarsaigh" building at the Trinity College Dublin (TCD) campus was analyzed as a sample and the results, as well as the estimation of the mixed-use model, are presented in Figure 3. Both estimations were assessed against the actual data. Lines (a) and (M) show the mean annual estimations of TM46 UC benchmark (240 kWh/m^2/yr) and the mixed-use model, respectively, while line (b) presents the mean of annual actual data.

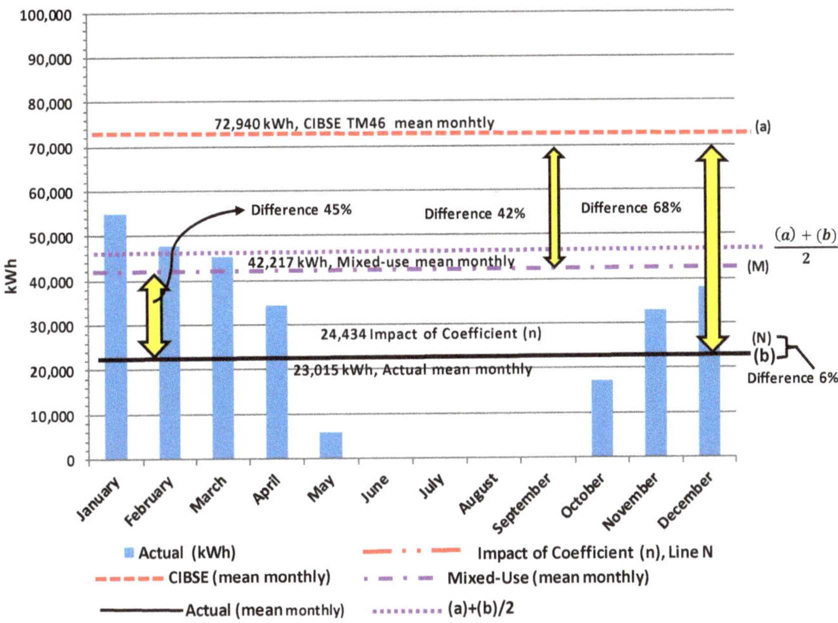

Figure 3. CIBSE TM46 UC and mixed-use model for thermal estimation against the actual data, Aras An Phiarsaigh building, Trinity College Dublin (TCD) campus 2014.

Considering the Aras An Phiarsaigh building as an example, the differences between thermal demand estimations of TM46 (mean annual) and the mixed-use model with the actual consumption were 68% and 45%, respectively (Figure 3). The result shows the mixed-use model improved the thermal demand estimation, approximately 42% compared with TM46. Coefficient (n) was defined to improve the accuracy of the mixed-use model as the ratio of the composite benchmark to the TM46 UC benchmark (240 kWh/m^2/yr). Coefficient (n) reduced the errors of the mixed-use model to 6%. At this stage, the mixed-use model presents an annual-fixed estimation (line M); however, the aim is to convert this horizontal line into monthly figures. To generate the monthly thermal benchmarks, two models were improved using further drivers. Additional information about the generation of the models is presented in Sections 2.1 and 2.2.

2.1. Mixed-Use Model

The mixed-use methodology is applicable to existing buildings and buildings at the construction stage. The method relies on CIBSE TM46 benchmarks, including 29 building categories, especially those categories found mostly in a typical college building such as "general office", "restaurant", "cultural activities", "classrooms", and "general retail". Based on the analysis, most of the college buildings comprise seven typical activities, i.e., mixed-use functions. In fact, activity plays a key role in thermal demand; for example, a general office needs 120 kWh/m^2/yr of thermal energy while a restaurant needs 370 kWh/m^2/yr [4,14].

Using Equation (1) and the architectural maps, the quantity of thermal demand of a mixed-use college building can be calculated. By dividing the annual thermal demand by 12 (Equation (2)) the mean monthly thermal demand can also be calculated. To calculate the composite benchmark, Equation (1) is divided by the total useful floors area (*TUFA*) of the buildings; therefore, Equation (3) indicates how to calculate a composite benchmark. The mixed-use method to estimate the annual thermal demand follows:

$$[\,f_1 \times A_1 + f_2 \times A_2 + f_3 \times A_3 + \ldots f_n \times A_n\,] = \sum_{i=1}^{n} (Ai \times fi) \tag{1}$$

$$\text{Mixed} - \text{Use (mean monthly heat demand)} = \frac{\sum_{i=1}^{n}(Ai \times fi)}{12} = \frac{\text{Equation (1)}}{12} \tag{2}$$

$$\text{Composite benchmark} = \frac{\sum_{i=1}^{n}(Ai \times fi)}{A(TUFA)} = \frac{\text{Equation (1)}}{A\,(TUFA)} \tag{3}$$

$$\text{Coefficient (n)} = \frac{\text{Equation (3)}}{TM46\ UC\ benchmark} \tag{4}$$

where (f_i) is the *CIBSE TM*46 benchmark of activity (*i*), (A_i) is the relevant area of activity (*i*), and *A* (m^2) is the total useful floor area of the building.

To indicate how the mixed-use method was developed, further discussion is presented in the following sections. As a sample, the model was applied in the Aras An Phiarsaigh building. The energy benchmarks of various activities are presented in Table 2. For example, the energy benchmark of a library is 200 kWh/m^2/yr while the benchmark of a laboratory is 160 kWh/m^2/yr. The weight of each benchmark is normalized based on its area in the building. The other necessary data to run the model are presented in Table 2.

Table 2. Mixed activities value in the Aras An Phiarsaigh building.

Activity	Area (m^2)	% of Total Useful Floor Area	Category Name	Category No	TM46 Benchmarks
Seminar and research room	817	22	UC	18	UCrb:130
Office	1651	45	General office	1	120
Computer rooms and Laboratory	1014	29	Laboratory	24	160
workshops	48	1	Workshop	27	180
Coffee shop	47	1	Restaurant	7	370
Library	70	2	Cultural activities	10	200
Total	3647	100	—	—	—

The annual thermal demand estimation using the mixed-use model equals:

$$[160 \times 1014 + 130 \times 817 + 120 \times 1651 + 370 \times 47 + 180 \times 48 + 200 \times 70] = 506600\ \text{kWh/yr}$$

$$\text{Mixed} - \text{Use estimation (mean monthly)} = 506600 \div 12 = 42217\ \text{kWh/yr}$$

$$\text{Composite benchmark} = 506600 \div 3647 = 139\ \text{kWh/m}^2/\text{yr}$$

$$\text{Coefficient (n)} = \frac{139}{240}$$

The assessments demonstrated that by considering the role of mixed activities (Equation (4)) in a building, the accuracy of thermal demand estimation can be improved. Comparing the results of estimations with the actual records proved this progress.

To develop the annual model into a monthly model, a series of other drivers were taken into account. One of the important factors is the heating degree days (HDD). The HDD is sensitive to the outdoor conditions. The weather data of Dublin Airport, IE (6.30° W, 53.42° N) was applied in the calculations and the base temperature of 15.5 °C chosen to determine the HDDs. In Table 3, the HDD data of 2014 are reported.

Table 3. Heating degree days (HDD) for 2014.

Months	Jan	Feb	Mar	Apr	May	Jun	Jul	Aug	Sep	Oct	Nov	Dec
HDD	303	274	267	182	133	63	32	70	72	132	225	316
Annual						2069						

Through multiplying Equations (1) and (2) by the result obtained from the division of the monthly *HDD* by annual *HDD* ($\frac{HDD\ month}{HDD\ annual}$), Equation (5) was created. Then, using Equation (5), the primary version of the monthly thermal models was generated. The primary model was applied in 10 buildings and its accuracy was calibrated using the actual thermal measurements; nevertheless, the Aras An Phiarsaigh building is discussed in detail.

$$\text{Equation (5)} \ = \ \frac{\left[\sum_{i=1}^{n}(Ai \times fi)\right]^2 \times HDD\ month}{240 \times A \times HDD(annual)} \tag{5}$$

where (f_i) is the *CIBSE TM*46 benchmark of activity (*i*), (*Ai*) is the relevant area of activity (*i*), *A* (m^2) is the total useful floor area of a building, and the *HDD* is the heating degree days at both annual and monthly scale.

The analysis showed there were significant differences between the estimations of the primary version (Equation (5)) of the model and the actual monthly consumption data. The differences, especially in the summer season, were notable. The reason for the lower accuracy of the primary version of the model refers to the local energy efficiency policies in universities. For example, it was found that despite heating degree days, which shows the thermal demand even during summer in Dublin (Table 3), the Estates and Facilities Office at TCD turns off the heating systems during summer. This policy drastically reduced the actual thermal consumption during the summer at TCD. Therefore, another factor, i.e., typical operation hours of heating systems, was taken into account and multiplied by Equation (5) to create Equation (6). In public buildings such as colleges, the operation hours of heating systems are not affected by occupant behavior, but controlled by energy managers at universities.

$$\text{Equation (6)} \ = \ \left[\frac{\left[\sum_{i=1}^{n}(Ai \times fi)\right]^2 \times HDD\ month}{240 \times A \times HDD(annual)} \right] \times \frac{Monthly\ typical\ operation\ (hours)}{Standard\ monthly\ opration\ (CIBSE,\ hours)} \tag{6}$$

where (f_i) is the *CIBSE TM*46 benchmark of activity (*i*), (*Ai*) is the relevant area of activity (*i*), *A* (m^2) is the total useful floor area of a building, and *HDD* is heating degree day at both annual and monthly scale.

The mean absolute percentage error (MAPE) on a monthly scale evaluated the accuracy of the final mixed-use model (Equation (6)). Besides, the accuracy of the model was calibrated by R-squared value, which indicates the error between the modeled values and the recorded values. The model applied to the other case study buildings. In all of the analyzed buildings, the maximum MAPE at the monthly level was under 21%, whereas it was 18% at the annual level. Compared with the best result (22%) of other annual estimation models [14], the result is acceptable.

2.2. Converter Model

Display energy certificates (DECs) present annual thermal consumption. If DEC documents are available, the converter model is more user-friendly compared to the mixed-use method to convert the annual heat demand into the monthly profiles. Normally the TPFER (Figure 1) is presented on DECs in kWh/m^{2}yr. To create a monthly thermal energy model using TPFER, *HDD* and the operation hours of heating systems play a key role. Equation (7) shows the final version of the converter model:

$$\text{Equation (7)} \ = \ \left[\text{TPFER} \times m \times A \times \frac{HDD\ month}{Total\ HDD(annual)}\right] \times \frac{Monthly\ typical\ operation\ (hours)}{Standard\ monthly\ opration\ (CIBSE,\ hours)} \tag{7}$$

where A (m^2) is the total useful floor area of the building and HDD is heating degree day at both annual and monthly scale.

The maximum unit interval of 20%, presented by the coefficient (m) in which m $\in$ [0.80, 1] was considered in the model and refers to the difference between TPFER and TFC. This difference was also shown by other scholars [22]. To increase the accuracy of simulations this difference was considered. Using the converter model, the annual thermal demand of a typical college building can be converted into the monthly figures. To understand how both mixed-use and converter models can be applied in practice, a flowchart is presented in Appendix A.

3. Application of the Mixed-Use Model

The Museum Building on the TCD campus is located on the south of the New Square, just beside the Berkeley Library. The building is a mixed-use, typical college building where the Geology and Engineering Departments are housed. TM46 predicts that the building needs 240 kWh/m^2 of thermal energy per year. The actual consumption, HDD, and the mean of monthly thermal demand based on TM46 and the mixed-use model are presented in Figure 4. Compared with TM46, the mixed-use model improved the accuracy of estimation by 42%. The data were used to run the mixed-use model for the Museum Building, as presented in Table 4.

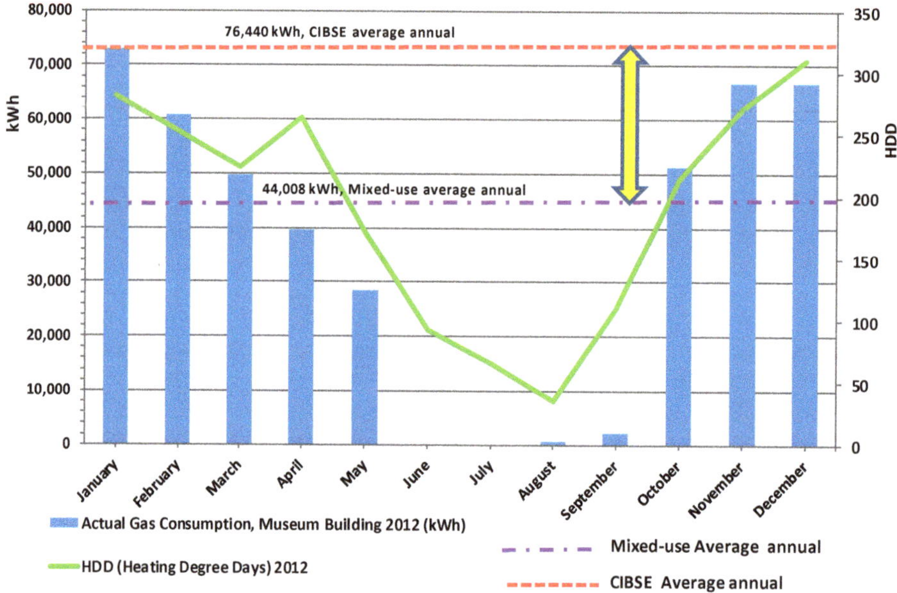

Figure 4. Comparison of actual heat consumption with CIBSE and mixed-use model.

Table 4. Museum Building data.

Activities	Area (m^2)	% Area of Activities (m^2)
Computer rooms and Laboratory	683	19
Office	1553	43
Seminar, class, and Research room	965	26
Library	324	9
Stores	120	3
Total	3645	100

Based on the data presented in Table 4 and using Equation (6), the monthly thermal demand of the Museum Building was generated (Table 5). The MAPE (mean absolute percentage error) of the mixed-use model and TM46 (mean monthly) compared with the actual consumption and the results are presented in Table 5.

Table 5. Monthly heat demand and the percent of errors.

Months	Actual Gas Consumption, Museum Building 2012 (kWh/yr)	HDD 2012	Typical Operation of Heating Systems (Hours)	Mixed-Use Model (kWh/yr)	TM46 Mean Monthly (kWh/yr)	MAPE of the Mixed-Use Model	MAPE of TM46 (Mean Monthly)
January	64,200	281	300	57,414	72,900	11	14
February	51,374	253	280	48,247	72,900	6	42
March	47,607	224	260	39,666	72,900	17	53
April	39,534	264	250	44,951	72,900	14	84
May	28,433	171	240	27,951	72,900	2	156
June	0	93	85	5383	72,900	*	*
July	0	66	45	2023	72,900	*	*
August	751	36	35	858	72,900	14	9607
September	5276	110	80	5993	72,900	14	1282
October	40,697	214	223	32,502	72,900	20	79
November	53,484	272	249	46,128	72,900	14	36
December	56,758	310	229	48,349	72,900	15	28
Total	388,114	2294	2276	359,466	874,800	7	125

The overall difference in thermal demand using the mixed-use model with actual annual consumption was 7%, while the error of TM46 was 125% (Table 5). The greatest error of the mixed-use model was 20% in October, while the lowest error of 2% was observed in May. In April, August, and September, the model shows 14% overestimation. However, the greatest monthly MAPE of TM46 was 9607%. The high estimation errors of TM46 in summer months means that this benchmarking system cannot reliably predict the thermal demand at smaller temporal resolutions.

Adopting linear regression model [41], the energy demand prediction results of the model were assessed versus the actual energy demand (Figure 5). R-squared (R^2) is a statistical measure that represents the proportion of the variance for a dependent variable that is explained by the independent variables in a regression model. It is the percentage of the response variable variation that is explained by a linear model. In our models, the R-squared of 0.971 shows a strong relationship between the actual data and the predicted figures. Therefore, it proves the high level of accuracy of the mixed-use model.

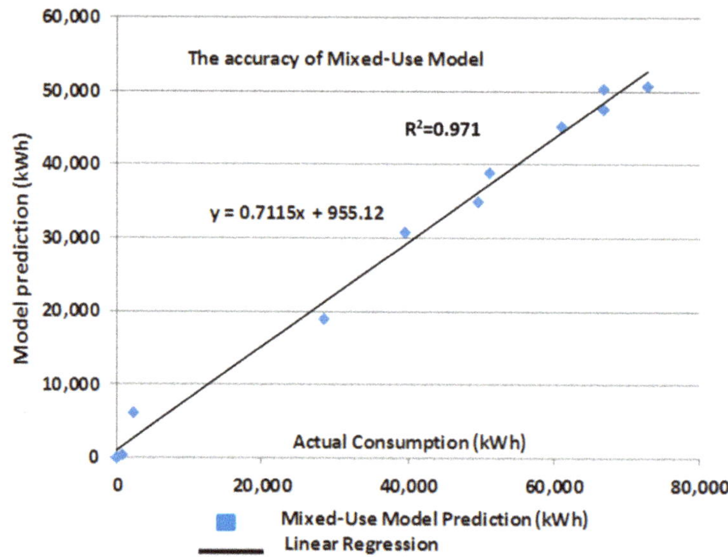

Figure 5. R-squared assessment to control the accuracy of the model.

4. Application of the Converter Model

The converter model is applicable when DECs are available. In fact, this approach relies upon the total primary fossil (nonelectrical) energy required (TPFER) displaying on DECs. In the converter model, the TPFER number, an annual index, was converted into monthly thermal figures, which are more informative for the energy efficiency planning and management. Using Equation (7), the TPFER number on DECs can be converted into the monthly thermal demand values.

As an example, using five key parameters, a monthly thermal demand profile was generated for the Nova Building at the UCD (University College Dublin) campus (Figure 6). According to the Nova's DEC certificate, the building requires 122 kWh/m^2/yr of total primary fossil energy and the building's total useful area is 4066 m^2. Both approaches, mixed-use model and converter model, were applied to the Nova Building and the results compared with the actual records (Figure 6). It can be seen that the actual consumption is located between the estimated values generated by the both models.

Table 6 shows the results of monthly thermal demand prediction generated by both models in the Nova Building. Furthermore, the MAPE of the two models was compared with TM46 estimations. The accuracy of TM46 and the monthly models was assessed against the actual figures. The differences of errors between TM46 and the predictions of the two models were significant. The maximum monthly MAPE of the mixed-use model and converter model was under 22%, while the maximum MAPE of TM46 in August was 7187% (Table 6). This huge error of TM46 in August means that the CIBSE benchmarking system overestimates the energy demand 71 times more than the actual energy consumption, which indicates the weakness and inability of the CIBSE TM46 benchmarking system. The minimum error of the mixed-use model was 5% and that of the converter model was only 1%, while the minimum error of TM46 was 13%. The annual errors of the monthly models were 11% and 14%, respectively. In contrast, the annual error of TM46 was 116%. The comparison methodology indicates a substantial development of the accuracy for both the mixed-use and converter models.

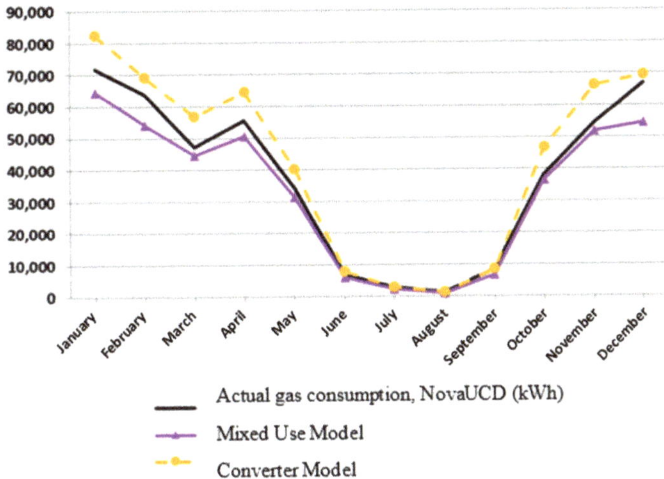

Figure 6. Monthly thermal demand profiles, mixed-use model and converter model, Nova Building, University College Dublin (UCD).

Table 6. Recorded data and monthly profiles and percent of errors compared with mean annual of CIBSE for the Nova Building, UCD.

Months	Actual Gas Consumption (kWh)	Mixed Use Model (kWh)	Converter Model (kWh)	TM46 Estimation(Mean Annual) (kWh)	MAPE of Mixed Use Model	MAPE of Converter Model	MAPE of TM46
January	71,907	64,550	82,407	81,320	10	15	13
February	63,696	54,244	69,249	81,320	15	9	28
March	47,268	44,538	56,859	81,320	6	20	72
April	55,451	50,538	64,518	81,320	9	16	47
May	34,113	31,425	40,118	81,320	8	18	138
June	6739	6053	7727	81,320	10	15	1107
July	2784	2274	2903	81,320	18	4	2821
August	1116	965	1232	81,320	14	10	7187
September	8544	6738	8602	81,320	21	1	852
October	39,015	36,569	46,685	81,320	6	20	108
November	54,489	51,895	66,252	81,320	5	22	49
December	66,876	54,438	69,497	81,320	19	4	22
Total	451,998	404,227	516,051	975,840	11	14	116

5. Monthly Thermal Energy Benchmarks (MTEBs)

Using the mixed-use and converter models, the monthly thermal energy benchmarks (MTEBs) for typical college buildings were generated. This new generation of thermal energy benchmarks varies during a year, following the outdoor conditions. The MTEBs methodology can extrapolate into other weather conditions as well as building types. If in Equations (6) and (7) the total useful area of buildings is assumed to be 1 m^2 (the definition of benchmark), then the monthly benchmarks per unit area can be determined accordingly. The annual-fixed benchmark was proposed by TM46 in 2008; i.e., 240 kWh/m^2/yr was developed through the models into 12 monthly thermal energy benchmarks.

The MTEBs (Figure 7) show various thermal demand in each month. For example, in January, a typical college building needs 24 kWh/m^2/month, and the demand was reduced regularly when the outdoor temperature was decreased; therefore in June, the benchmark is 1 kWh/m^2/month. Likewise, the benchmark from nearly 0 kWh/m^2/month in July increased to 19 kWh/m^2/month in December.

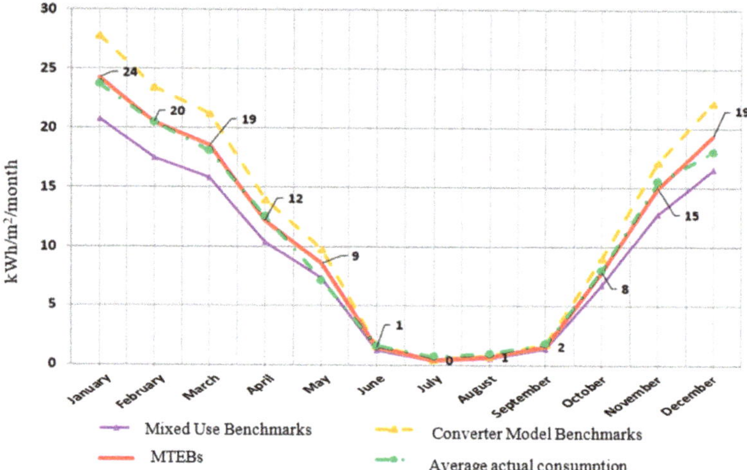

Figure 7. Monthly thermal energy benchmarks (MTEBs) for typical college buildings.

Table 7 shows the MTEBs indexes which were validated against the mean of monthly actual consumption (kWh/m²/month) of 10 college buildings obtained from the AEM (Active Energy Management dataset) [39]. Using the mean of actual thermal consumption of the buildings belonging to the four case study universities, the accuracy of MTEBs was assessed and the results are presented in Figure 7.

Table 7. MTEBs against TM46 UC benchmark and actual thermal consumptions.

Months	MTEBs Based on Mixed-Use Model (kWh/m²/month)	MTEBs based on Converter Model (kWh/m²/month)	MTEBs Mean of Both Models (kWh/m²/month)	Mean of Actual Thermal Consumption of 10 Buildings (kWh/m²/month)	TM46 Benchmark (kWh/m²/yr)
January	21	28	24	24	-
February	17	23	20	20	-
March	16	21	19	18	-
April	10	14	12	13	-
May	7	10	9	7	-
June	1	2	1	2	-
July	0	0	0	1	-
August	1	1	1	1	-
September	1	2	2	2	-
October	7	9	8	8	-
November	13	17	15	15	-
December	17	22	19	18	-
Total	111	149	130	128	240

In addition, the values of MTEBs were compared with the TM46 annual benchmark. According to the analysis, the predictions of MTEBs were very close to the actual measurements. The mean annual actual thermal consumption was 128 kWh/m²/yr and the developed MTEBs predicted 130 kWh/m²/yr, while the TM46 method predicted 240 kWh/m²/yr. The overall MTEB was 130 kWh/m²/yr. The R-squared of 0.995 shows the high level of accuracy for MTEBs, as presented in Figure 8.

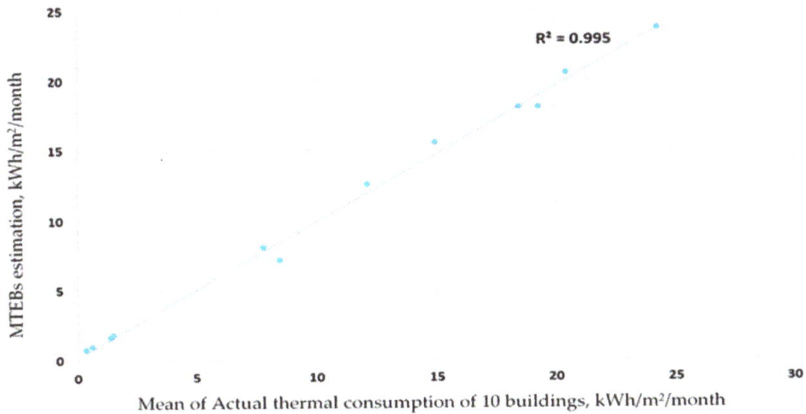

Figure 8. Accuracy assessment of the MTEBs.

6. Conclusions

Due to the excessive dependence of heat consumption on the ambient temperature, the annual-fixed thermal benchmark (240 kWh/m^2/yr) suggested by CIBSE TM6 for the category of UC is not very effective. Instead, the concept of monthly thermal energy benchmarks (MTEBs) for typical college buildings was developed, which are more informative, especially for managing the thermal consumption/efficiency at the community scale. Unlike other benchmarking methodologies that consider buildings as having a single function, in this study the mixed activities in buildings were taken into account. Two methods, including mixed-use model and converter model, were adopted to generate the MTEBs. MTEBs present information that is more detailed and therefore more applicable compared to the annual benchmarks such as TM46. This detailed information from the viewpoint of heat efficiency and planning, as well as the energy supplying and financial policy, is vital.

The accuracy of the developed models at a monthly scale was validated against the actual thermal consumption using the mean absolute percentage error (MAPE). In addition, the truthfulness of the new generation of the developed benchmarks was examined by linear regressions.

While the discrepancy of the CIBSE TM46 benchmark with the actual consumption was radically significant (e.g., 7187%), the maximum monthly error of the progressed models was lower than 22%. The MTEBs show that a typical college building needs 24 kWh/m^2/month in January and the demand reduces regularly in summer months. In June, only 1 kWh/m^2/month of heat is needed while in July it is nearly zero. The monthly benchmarks from July increased gradually to 19 kWh/m^2/month in December. The overall annual MTEBs is 130 kWh/m^2/yr, which shows a significant improvement compared with 240 kWh/m^2/yr suggested by TM46. The benchmarking methodology developed presents a curved line instead of an annual-fixed horizontal line as proposed by TM46. In this paper, 12 thermal energy benchmarks at the monthly level were presented instead of a TM46 annual benchmark. Finally, the R-squared of 0.995 indicated the high level of reliability of MTEBs. Planners, energy suppliers, and professionals for detailed heat planning at the community scale can use MTEBs. Since the benchmarks play a key role in energy action plans at the national scale, the new generation of proposed benchmarks can improve the accuracy of national action plans by sharing more information at the monthly level.

Author Contributions: Conceptualization, S.V.; formal analysis S.V., S.M., and K.J.; investigation, B.N. and S.V.; methodology, S.V., B.N., and S.M.; writing, S.M., K.J., and B.N.; supervision, S.V. and B.N. All authors have read and agreed to the published version of the manuscript.

Funding: This research received no external funding.

Acknowledgments: The authors acknowledge the State Office of four universities in Dublin for sharing DECs and data provided, which facilitated this research. The authors acknowledge in particular Kieron McGovern,

Mark Argue, Stephen Folan, and Kieran Brassil for their kind help and providing data. We also acknowledge the Degree Days.net website for sharing the data for free.

Conflicts of Interest: The authors declare no conflict of interest.

Nomenclature

BER: building energy ratio; CIBSE: Chartered Institution of Building Services Engineers; DEC: display energy certificate; HDD: heating degree days; MAPE: mean absolute percentage error; MTEBs: monthly thermal energy benchmarks; TFC: total final consumption or actual consumption is the amount of energy consumed in the buildings measured by meters and displayed on energy bills; TPER: total primary energy required in a building including thermal and electricity; TPFER: total primary fossil energy required in a building; UC: university campus, refers to the category number 18 of CIBSE TM46:2008 benchmark

Appendix A The Flowchart of Developed Models

The following flowchart shows how both mixed-use model and converter model can be applied in practice step-by-step, given available energy data.

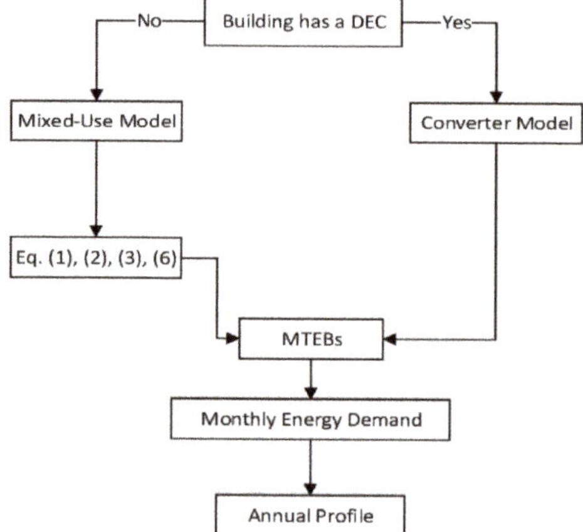

Figure A1. The flowchart of model application.

References

1. Evangelinos, K.I.; Jones, N.; Panoriou, E.M. Challenges and opportunities for sustainability in regional universities: A case study in Mytilene, Greece. *J. Clean. Prod.* **2009**, *17*, 1154–1161. [CrossRef]
2. Locmelis, K.; Blumberga, D.; Blumberga, A.; Kubule, A. Benchmarking of industrial energy efficiency. outcomes of an energy audit policy program. *Energies* **2020**, *13*, 2210. [CrossRef]
3. Kim, H.G.; Kim, S.S. Development of energy benchmarks for office buildings using the national energy consumption database. *Energies* **2020**, *13*, 950. [CrossRef]
4. *Chartered Institution of Building Services Engineers*; CIBSE TM46; Energy Institue, University College, CIBSE: London, UK, 2008.
5. Hong, S.-M.; Steadman, P. *An Analysis of Display Energy Certificates for Public Buildings, 2008 to 2012, a Report to Chartered Institution of Building Services Engineering*; Energy Institue, University College, CIBSE: London, UK, 2013.
6. Hawkins, D.; Hong, S.M.; Raslan, R.; Mumovic, D.; Hanna, S. Determinants of energy use in UK higher education buildings using statistical and artificial neural network methods. *Int. J. Sustain. Built Environ.* **2012**, *1*, 50–63. [CrossRef]
7. Bruhns, H.; Jones, P.; Cohen, R. CIBSE review of energy benchmarks for display energy certificates. In Proceedings of the CIBSE Technical Symposium, Leicester, UK, 6–7 September 2011.
8. Vaisi, S.; Pilla, F.; McCormack, S.J. Recommending a thermal energy benchmark based on CIBSE TM46 for typical college buildings and creating monthly energy models. *Energy Build.* **2018**, *176*, 296–309. [CrossRef]

9. Javanroodi, K.; Mahdavinejad, M.; Nik, V.M. Impacts of urban morphology on reducing cooling load and increasing ventilation potential in hot-arid climate. *Appl. Energy* **2018**, *231*, 714–746. [CrossRef]

10. Pérez-Lombard, L.; Ortiz, J.; González, R.; Maestre, I.R. A review of benchmarking, rating and labelling concepts within the framework of building energy certification schemes. *Energy Build.* **2009**, *41*, 272–278. [CrossRef]

11. International Energy Agency. *Energy Performance Certification of Building, a Policy Tool to Improve Energy Efficiency*; International Energy Agency: Paris, France, 2010.

12. The Council of the European Commuties. *European Council Directive 93/76/EEC, 13 September 1993*; European Commuties, Ed.; The Council of the European Commuties: Brussels, Belgium, 1993.

13. Sustainable Energy Authority Ireland (Seai). What is the Total Primary Energy Requirement (TPER)? Available online: http://www.seai.ie/Your_Business/Public_Sector/FAQ/Calculating_Savings_Tracking_Progress/What_is_the_Total_Primary_Energy_Requirement.html (accessed on 28 December 2015).

14. *Chartered Institution of Building Services Engineers*; CIBSE TM47; CIBSE: London, UK, 2009.

15. Carbon Trust. *The Carbon Trust Annual Report 2012/2013, Accelerating the Move to a Sustainable, Low Carbon Economy*; Carbon Trust: London, UK, 2013.

16. Hong, S.-M.; Paterson, G.; Burman, E.; Steadman, P.; Mumovic, D. A comparative study of benchmarking approaches for non-domestic buildings: Part 1—Top-down approach. *Int. J. Sustain. Built Environ.* **2013**, *2*, 119–130. [CrossRef]

17. Burman, E.; Hong, S.-M.; Paterson, G.; Kimpian, J.; Mumovic, D. A comparative study of benchmarking approaches for non-domestic buildings: Part 2—Bottom-up approach. *Int. J. Sustain. Built Environ.* **2014**, *3*, 247–261. [CrossRef]

18. Burman, E.; Mumovic, D.; Kimpian, J. Towards measurement and verification of energy performance under the framework of the European directive for energy performance of buildings. *Energy* **2014**, *77*, 153–163. [CrossRef]

19. Salah, V.; Mark, D.; Francesco, P. Energy requirement mapping for university campus using CIBSE benchmarks and comparing CIBSE to display energy certificate (DEC) to extract a new criterion. In Proceedings of the Energy Systems Conference, London, UK, 24–25 June 2014.

20. EPBD Energy Performance of Building Directive. Directive 2002/91/CE of the European Parliament and Council from 16 of December 2002. *Off. J. Eur. Communities* **2003**. Available online: https://eur-lex.europa.eu/legal-content/EN/TXT/PDF/?uri=CELEX:32002L0091&from=IT (accessed on 12 September 2020).

21. Borgstein, E.H.; Lamberts, R.; Hensen, J.L.M. Evaluating energy performance in non-domestic buildings: A review. *Energy Build.* **2016**, *128*, 734–755. [CrossRef]

22. Lee, W.-S.; Lee, K.-P. Benchmarking the performance of building energy management using data envelopment analysis. *Appl. Therm. Eng.* **2009**, *29*, 3269–3273. [CrossRef]

23. Pasichnyi, O.; Wallin, J.; Levihn, F.; Shahrokni, H.; Kordas, O. Energy performance certificates—New opportunities for data-enabled urban energy policy instruments? *Energy Policy* **2019**, *127*, 486–499. [CrossRef]

24. Burman, E.; Kimpian, J.; Mumovic, D. Building schools for the future: Lessons learned from performance evaluations of five secondary schools and academies in England. *Front. Built Environ.* **2018**, *4*. [CrossRef]

25. Papadopoulos, S.; Bonczak, B.; Kontokosta, C.E. Pattern recognition in building energy performance over time using energy benchmarking data. *Appl. Energy* **2018**, *221*, 576–586. [CrossRef]

26. Roth, J.; Rajagopal, R. Benchmarking building energy efficiency using quantile regression. *Energy* **2018**, *152*, 866–876. [CrossRef]

27. Liu, J.; Chen, H.; Liu, J.; Li, Z.; Huang, R.; Xing, L.; Wang, J.; Li, G. An energy performance evaluation methodology for individual office building with dynamic energy benchmarks using limited information. *Appl. Energy* **2017**, *206*, 193–205. [CrossRef]

28. Park, J.Y.; Yang, X.; Miller, C.; Arjunan, P.; Nagy, Z. Apples or oranges? Identification of fundamental load shape profiles for benchmarking buildings using a large and diverse dataset. *Appl. Energy* **2019**, *236*, 1280–1295. [CrossRef]

29. Lamagna, M.; Nastasi, B.; Groppi, D.; Nezhad, M.M.; Garcia, D.A. Hourly energy profile determination technique from monthly energy bills. *Build. Simul.* **2020**, *13*, 1235–1248. [CrossRef]

30. Mancini, F.; Nastasi, B. Energy retrofitting effects on the energy flexibility of dwellings. *Energies* **2019**, *12*, 2788. [CrossRef]

31. Papadopoulos, S.; Kontokosta, C.E. Grading buildings on energy performance using city benchmarking data. *Appl. Energy* **2019**, *233–234*, 244–253. [CrossRef]
32. Seyedzadeh, S.; Rahimian, F.P.; Glesk, I.; Roper, M. Machine learning for estimation of building energy consumption and performance: A review. *Vis. Eng.* **2018**, *6*, 5. [CrossRef]
33. Koo, C.; Hong, T. Development of a dynamic operational rating system in energy performance certificates for existing buildings: Geostatistical approach and data-mining technique. *Appl. Energy* **2015**, *154*, 254–270. [CrossRef]
34. Yan, C.; Wang, S.; Xiao, F.; Gao, D.-C. A multi-level energy performance diagnosis method for energy information poor buildings. *Energy* **2015**, *83*, 189–203. [CrossRef]
35. Khoshbakht, M.; Gou, Z.; Dupre, K. Energy use characteristics and benchmarking for higher education buildings. *Energy Build.* **2018**, *164*, 61–76. [CrossRef]
36. Zhan, S.; Liu, Z.; Chong, A.; Yan, D. Building categorization revisited: A clustering-based approach to using smart meter data for building energy benchmarking. *Appl. Energy* **2020**, *269*, 114920. [CrossRef]
37. Arjunan, P.; Poolla, K.; Miller, C. EnergyStar++: Towards more accurate and explanatory building energy benchmarking. *Appl. Energy* **2020**, *276*, 115413. [CrossRef]
38. Wei, Z.; Xu, W.; Wang, D.; Li, L.; Niu, L.; Wang, W.; Wang, B.; Song, Y. A study of city-level building energy efficiency benchmarking system for China. *Energy Build.* **2018**, *179*, 1–14. [CrossRef]
39. Cylon. Cylon Active Energy Management. Available online: https://cylonaem.com/energy/#v=1&t=9&c=0 (accessed on 10 March 2015).
40. Heating Degree Days. BizEE Software, Degree Days.Net—Custom Degree Day Data. Available online: http://www.degreedays.net/ (accessed on 2 January 2016).
41. Manfren, M.; Nastasi, B.; Tronchin, L. Linking design and operation phase energy performance analysis through regression-based approaches. *Front. Energy Res.* **2020**, *8*, 557649. [CrossRef]

Publisher's Note: MDPI stays neutral with regard to jurisdictional claims in published maps and institutional affiliations.

Article

Dynamic Approach to Evaluate the Effect of Reducing District Heating Temperature on Indoor Thermal Comfort

Benedetta Grassi , Edoardo Alessio Piana * , Gian Paolo Beretta and Mariagrazia Pilotelli

Department of Mechanical and Industrial Engineering, University of Brescia, 25123 Brescia, Italy; benedetta.grassi@unibs.it (B.G.); gianpaolo.beretta@unibs.it (G.P.B.); mariagrazia.pilotelli@unibs.it (M.P.)
* Correspondence: edoardo.piana@unibs.it

Abstract: To reduce energy consumption for space heating, a coordinated action on energy supply, building fabric and occupant behavior is required to realize sustainable improvements. A reduction in district heating supply temperature is an interesting option to allow the incorporation of renewable energy sources and reduce distribution losses, but its impact on the final users must be considered. This aspect is especially critical as most European countries feature an old building stock, with poor insulation and heating systems designed for high-temperature operation. In this study, a complete methodology is devised to evaluate the effect of district heating temperature reduction on the end users by modeling all the stages of the system, from the primary heat exchanger to the indoor environment. A dynamic energy performance engine, based on EN ISO 52016-1:2017 standard and completed with a heat exchanger model, is implemented, and its outputs are used to calculate thermal comfort indicators throughout the heating season. As a practical application, the method is used to evaluate different scenarios resulting from the reduction of primary supply temperature of a second-generation district heating network in Northern Italy. Several building typologies dating back to different periods are considered, in the conservative assumption of radiator heating. The results of the simulations show that the most severe discomfort situations are experienced in buildings built before 1990, but in recent buildings the amount of discomfort occurrences can be high because of the poor output of radiators when working at very low temperatures. Among the possible measures that could help the transition, actions on the primary side, on the installed power and on the building fabric are considered. The investigation method requires a limited amount of input data and is applicable to different scales, from the individual building to entire urban areas lined up for renovation.

Keywords: dynamic model; energy performance of buildings; low temperature district heating; indoor comfort; renovation; urban scale

Citation: Grassi, B.; Piana, E.A.; Beretta, G.P.; Pilotelli, M. Dynamic Approach to Evaluate the Effect of Reducing District Heating Temperature on Indoor Thermal Comfort. *Energies* **2021**, *14*, 25. https://dx.doi.org/10.3390/en14010025

Received: 27 November 2020
Accepted: 17 December 2020
Published: 23 December 2020

Publisher's Note: MDPI stays neutral with regard to jurisdictional claims in published maps and institutional affiliations.

1. Introduction

Buildings are a key element in energy and environmental policies worldwide. According to the International Energy Agency [1], despite the increased awareness of governments and population, direct and indirect energy-related CO_2 emissions from buildings have been rising again in the last three years after the encouraging flattening in 2013–2016. With many European countries still lacking mandatory building codes, the benefit in terms of energy savings deriving from renovations of existing buildings should grow from the current 15% to at least 30%–50% to be in line with the sustainable development scenario by 2030 [1]. These figures show that the development of strategies, methodologies and tools are essential to rapidly promote effective upgrades of the current stock.

The need to devise strategies at the urban scale is widely documented in the recent literature. In a 2007 article, Lowe [2] focuses on measures to reduce CO_2 emissions in the UK residential sector, pointing out that the performance of housing is the result of the synergies between the different subsystems, and for a given level of investment the combination of several even modest improvements is likely to produce more benefit than a single major retrofit. The three pillars on which any upgrading approach is based are input data collection, choice of the appropriate calculation method, and identification of relevant indicators. The input information required for the analysis is related, among others, to building details and technical systems. Ballarini et al. [3] present the Italian results of the European project TABULA (Typology Approach for Building Stock Energy Assessment), in which the participants defined the characteristic building typologies of their countries and possible sets of retrofitting measures. A frequently used criterion for the identification of reference buildings is the age of construction as obtained from municipal databases or censuses, as in Di Turi and Stefanizzi [4] and in Delmastro et al. [5]. However, these sources generally do not include information about refurbished buildings, which poses additional challenges to the accurate analysis of the existing building stock [6]. Another possible source of data are energy performance certificates, although some uncertainty in the compilation of the reports must be accepted [7]. The trend is to look for automated data collection methods as in Nageler et al. [8], who propose a methodology to automatically create building models of entire urban districts, based on publicly available datasets and tools, ready to be fed as input to the energy calculation engine. Mutani and Todeschi [9] add urban context parameters by means of a geographic information system (GIS) tool to the calculation model, whereas Torabi Moghadam et al. [10] describe the construction of a GIS database from different sources, including district heating meter data.

Building modeling for energy calculation purpose is the subject of several literature works. Conversely, not many papers focus on the building "circulatory system": indeed, heating, ventilation, and air conditioning (HVAC) systems are often modeled by assuming the technical equipment typical of the building construction period; however, such models cannot incorporate the variability of the actual installations and the comfort preferences of the occupants [11]. Noussan and Nastasi [12] analyze data from the regional registry of heating plants in Lombardy, Italy, while Mancini and Nastasi [13] adopt a survey approach to collect information about building features, technical services and appliances. Westermann et al. [14] develop a method to automatically infer building type and heating system type from the energy signature, expressed as the electricity consumption for heating and/or cooling detected by smart meters as function of outside temperature recordings.

The evolution of water-based systems points in the direction of reducing supply and return temperatures. It is well known that condensing boilers are fully exploited with low return temperatures, whereas electrically-powered heat pumps work best with low supply temperature systems. Moreover, low-temperature heating systems are associated to lower energy consumption [15], which expands the range of usable sources. In the case of district heating (DH), several benefits can be obtained by lowering the temperature regime. A summary of enabling technologies, best practices and pilot installations for 4th generation DH can be found in the "Guidelines for Low-Temperature District Heating" report [16], from which the amount of variables involved at different scales and the complexity of possible conversion processes emerge. In their work on the optimization of a distributed DH network, Sameti et al. [17] act on piping layout, power flow among the buildings, and type, location, and rated heat and electricity production of generators to minimize the annual cost function, including the cost of CO_2 emissions. The main tool for decarbonization is the integration of renewable energy sources (RES), but this is a challenging task, mainly because of their variability in space and time. Solar energy, for instance, is an abundant source at most European latitudes, but it requires reliable tools to evaluate both the actual energy potential [18] and the correct sizing and operation of the system [19]. The mismatch between demand and supply is usually overcome through flexible-oriented solutions, like the installation of thermal energy storages (TES) [20]). As an example, the long-standing DH system in Brescia, located in Northern Italy, has recently started to incorporate TES and aims to quickly deploy an overall stored energy of 550 MWh, which is estimated to save 1000 toe per year of energy produced by fossil fuels. Another RES-related issue is the correct allocation of the produced energy between fossil and RES share [21], which is important for example with regard to incentivizing policies.

The incorporation of RES is one of the driving forces towards the progressive decrease in the heat transfer medium temperatures. Other benefits associated to this action are the reduction of grid losses [22,23], and the improvement of the powerplant efficiency when the heat is produced in cogeneration facilities (such as in the Brescia municipal system). However, whether this action is feasible (and to which extent) depends on several factors. The installed power and the original design conditions are key elements for the transition to lower temperature systems. Østergaard et al. [24] provide an insight on the radiators installed in Denmark by means of a survey, showing that, as per common belief, most of them are oversized. Tunzi et al. [25] correlate the logarithmic mean temperature difference (LMTD) between the radiator and the surrounding space with the part-load duration, showing that even the radiators sized exactly to fulfill the design heat load are actually oversized for most of the time. Both papers suggest that there is a large potential for low temperature DH with minor or no modification of the building or its internal HVAC systems. Such potential depends strongly on the original design temperatures, which are typical of a specific time period and of a given country. For example, in the 1970s, common design supply/return temperatures for the heating systems were 80/40 °C in Denmark [26] and 75/50 °C in Sweden [27], while 85/75 °C was still common in Italy [28]. Millar et al. [29] investigate the limitations of connecting a traditional 20th century tenement to a low temperature district heating (LTDH). The authors calculate that the minimum supply temperature at the radiator inlet required to meet 80% of the demand is around 85 °C for dwellings with no insulation nor double glazing. They assume common practice radiators' sizes and design temperatures (82/71/20 °C) similar to those typical in Italy until the 1990s. Their conclusion is that internal improvements are likely to be necessary for the connection to a third generation DH to become a viable option. Likewise, Ashfaq and Ianakiev [30] review the challenges associated to the transition from boiler-base heating to LTDH and, while focusing on the low temperature difference deriving from hydraulic imbalance, they take the need of building retrofitting for granted.

The temperature lowering potential also depends on the type of substation connecting the DH network to the building internal system. When DH is directly connected to the space heating user pipes, the only difference between DH supply temperature and actual temperature at the emitter inlets is caused by thermal losses along the pipe. However, when the user system is separated from DH network by a heat exchanger (indirect connection), the actual radiator supply temperature strongly depends on the type of control and on the characteristics of the heat exchanger. Neirotti et al. [31] explore the effect of reducing DH temperature on a typical North-Italian building with radiator-based heating, comparing two different refurbishing strategies based on continuous operation and on building insulation improvement. They assume a coil heat exchanger modeled by means of a Modelica library component, although DH substations more often use plate heat exchangers.

With regard to the choice of the mathematical modeling framework, given the need to examine transient periods and to incorporate controls and schedules, it is reasonable to opt for a dynamic energy calculation engine. In the past, the level of detail of the models were somehow limited by the available computational capabilities. Crabb et al. [32] develop a single-zone resistor-capacitor (RC) model based on two capacity nodes (indoor air and thermal inertia of the building), which despite its simplicity provides a fair representation of the energy performance of a school. The first international standard including a dynamic hourly model is EN ISO 13790:2008, in which the building is described by five resistors and only one capacitor (5R1C). A validation of the model can be found in the works by Michalak [33]. The EN ISO 13790:2008 was withdrawn in 2017, when it was replaced by the current reference standard for dynamic hourly energy calculations, EN ISO 52016-1:2017 [34], in which the number of resistances and capacitors depends on the geometry and thermal characteristics of the building features and fixtures. The standard is open to the inclusion of other validated methods on a national basis. It is the case of Annex A in the Italian implementation, in which a procedure is provided to calculate the number of nodes of each opaque element instead of the fixed number of layers prescribed by the general framework [35]. The EN ISO 52016-1:2017 has multiple usages: it is the common platform for energy calculation softwares and for national building regulations; it can be employed as a tool to validate other methods (see for instance Campana and Morini [36]); it provides a complete calculation framework that can serve as a basis to build affordable codes. In a recent paper, Lundström et al. [37] introduced some simplifications to the EN ISO 52016-1:2017 building model, with the aim of reducing the complexity of input data and the required computational effort. Their focus is the supervision of thousands of dwellings by few managers, who must be able to quickly construct the building model and run swift simulations to identify present and potential energy performances. An alternative to physical modeling is represented by gray box [38] or black box [39] approaches, which nowadays can rely on increasing amounts of available information, on a wide range of efficient algorithms and on improved performances of calculation machines. Open models and data are expected to play a key role in the development of new energy-related methodologies [40].

The final stage of the analysis is the definition of performance indicators to quantitatively evaluate and compare different scenarios. As pointed out by Tronchin et al. [41], an effective energy modeling strategy should feature modularity and possibility of representation across multiple scales, including the user perspective. The classical approach to indoor comfort was developed by Fanger in the 1970s and relies upon the calculation of parameters such as predicted mean vote (PMV) and predicted percentage of dissatisfied (PPD), whose formulations have been included in standards worldwide, such as ANSI/ASHRAE 55 and EN ISO 7730. Derived long-term indices are the number of discomfort hours, used for example in the work by Braulio-Gonzalo et al. [42] to analyze the indoor thermal comfort of residential building stocks, and percentage outside range (POR), among the several indices compared by Carlucci et al. [43] in the cooling assessment of office building variants. Stasi et al. [44] use PMV, PPD and POR for the evaluation of indoor comfort of a nearly zero-energy building in Southern Italy. References [42–44] all derive the comfort indices from dynamic energy simulations.

To summarize, the scientific and technical literature features dynamic energy calculation models, studies on the collection and processing of the required input data, and indoor thermal comfort assessment methodologies and case studies. The theoretical applicability of LTDH to existing buildings is also discussed in several papers, but with some limitations in primary/secondary interface modeling and with results mainly presented in terms of temperature variations. To the best of the authors' knowledge, no large-scale dynamic analysis of a building-thermal systems complex has been carried out that includes an accurate model of the primary heat exchanger and a quantitative estimation of the occupants' comfort response. The aim of the present study is to devise a complete methodology to assess the impact of DH supply temperature reduction on the indoor thermal comfort in existing buildings, including a detailed model of the DH heat exchanger and expressing the results in terms of thermal comfort indices. A practical application is presented to demonstrate the potential of the method on buildings connected a typical Northern Italian DH network. The evaluation is extended to the buildings stock of the selected area, to represent which the Italian reference building matrix elaborated in the TABULA project is used. The main assumptions in this application are the indirect connection of the user heating system to the DH through a heat exchanger and the presence of radiator-based user heating systems, being this type of emitter more critical than low-temperature radiant panel heating from the viewpoint of supply temperature reduction.

The paper is organized as follows. Section 2 summarizes the modules composing the proposed simulation framework. In particular, the dynamic model recently proposed by Lundström et al. [37] has been chosen as a central energy calculation engine. Primary heat exchanger and radiator heating system models are also described in this Section. The application of the method is presented in Section 3, where the climate-related input data, the heating system, the building typology matrix, and the selected comfort performance indices are described. The results and discussion are reported in Section 4, followed by conclusions and future developments in Section 5.

2. Methodology

2.1. Calculation Framework

The proposed method relies upon the modular structure represented in Figure 1, whose center is the energy calculation model. The model takes as input climate information, building details and heating system preferences, and provides time-dependent output information that can be used for energy consumption, performance and comfort evaluations. Several calculation options are available to perform different tasks—for example, the user can simulate free-floating operation, operation with ideal heating/cooling (thermostat) or with actual heating/cooling systems.

Input data are provided in two ways:

- A text general input file, in which the main settings and input parameters are listed, and
- Database spreadsheet files for buildings and primary heat exchangers

The code is developed in MATLAB.

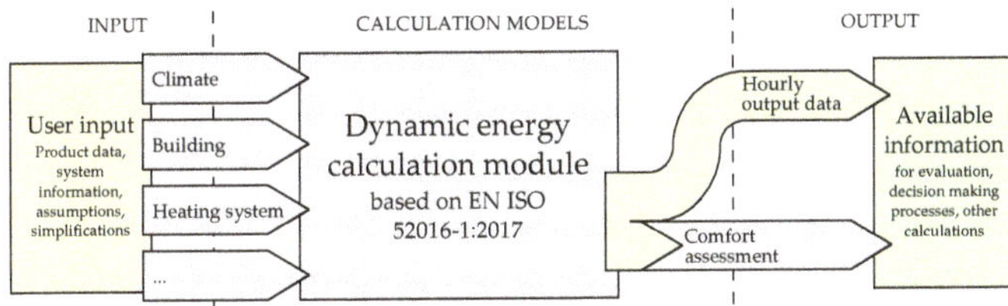

Figure 1. Modular calculation framework.

2.2. Energy Calculation Model

To calculate the energy balance of the building, the approach proposed by Lundström et al. [37] is chosen. The model is largely based on the dynamic hourly method of EN ISO 52016-1:2017, the main difference being that all the elements of the same type (for example external walls) are lumped into a single element. Moreover, each opaque element is modeled with only three layers, with respect to the five prescribed by the standard. These simplifications lead to a building model with a fixed number of nodes (14, hence the model name "ISO14N" adopted by the authors), which greatly reduces the complexity of the required input data. ISO14N is developed for a single thermal zone. The calculation matrix, tying the previous time-step values to the present time-step outputs, is compiled following the EN ISO 52016-1:2017 approach, that is, by writing the heat balances for the thermal zone and its different elements. In the balances for the external surface nodes, where radiant heat gains depend upon surface orientation, the actual expositions of the elements are considered by calculating average irradiances weighted for the fractions of the overall surface exposed at different angles. In Reference [37], building simulation software IDA ICE is used to validate the method. In the present work, an additional comparison between the results obtained from ISO14N code, EN ISO 52016-1:2017 supporting calculation sheets, and EnergyPlus software for the reference BESTest cases provided by EN ISO 52016-1:2017 standard is reported in the Supplementary Materials, where the statistical indices described in Reference [45] are used to evaluate the prediction capabilities of the simplified approach.

In the study presented in the original paper in Reference [37], a model was included for a radiator-based heating system, which is illustrated and expanded in the next Subsection.

2.3. Heating System Model

The indirect connection between DH and user system is modeled in this paper according to the most common configuration in Italy, sketched in Figure 2, that is, by means of a plate heat exchanger. Hence, the base configuration assumes the following equipment:

- Plate heat exchanger (HE) to transfer heat from the primary (DH) side to the secondary (user, U) side; it is modeled based on the technical information provided by the manufacturer, as explained below.
- Primary control valve (V1), to modulate the primary flow rate in order to obtain the desired supply temperature on the secondary side; it is assumed as an ideal component that instantaneously performs the required action.
- Thermostatic radiator valve (TRV), a proportional controller that modulates the secondary flow rate according to the difference between room temperature and average radiator temperature.
- Radiator (RAD) with a given nominal power referred to the design temperatures typical of the period of installation; due to the single-zone assumption, all the heating emitters are lumped into a single component, whose installed power is calculated in a preliminary simulation run based on the peak load of the building.

The primary control valve can be associated with either weather-compensated or fixed-point control, according to the user preferences. In the former case, the heating curve relating the secondary supply temperature to the outdoor temperature must be set, whereas only a fixed value is required in the latter case. It is worth noting that no hydraulic calculation is performed on the primary or the secondary circuit. However, a maximum flow rate can be set for both sides, to simulate the maximum capacity of a pump.

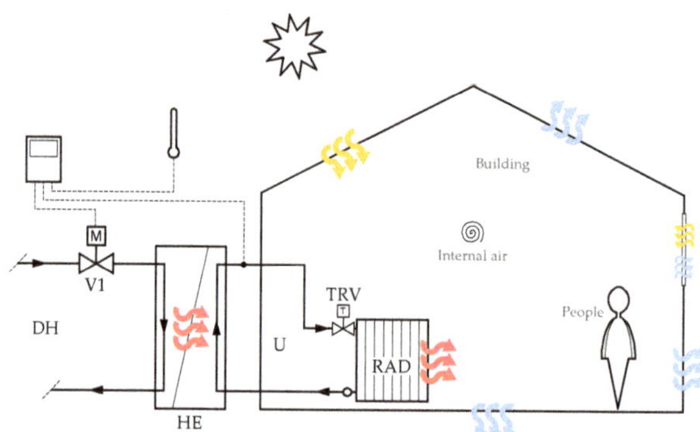

Figure 2. Schematic representation the heating system within the general model.

The following equations describe the whole heating system model. They are listed starting from the room and backwards to the DH network.

Power released to the room. The equation describing the heat release Φ from a radiator with thermostatic control to the room is

$$\Phi = \Phi_{\text{nom}} \times \left(\frac{\Delta\theta}{\Delta\theta_{\text{nom}}} \right)^{n_{\text{rad}}} \times u_{\text{TRV}}, \tag{1}$$

where Φ_{nom} is the nominal thermal power referred to the conditions expressed by the nominal logarithmic mean temperature difference (LMTD) $\Delta\theta_{\text{nom}}$, and n_{rad} is the characteristic exponent of the radiator. u_{TRV} is the modulation factor introduced by the action of a thermostatic valve with proportional band PB:

$$u_{\text{TRV}} = \max\left\{ 0, \min\left\{ 1, (\theta_{\text{ref,set}} - \theta_{\text{ref},t-1})/PB \right\} \right\}. \tag{2}$$

$\Delta\theta$ is the actual LMTD between the radiator mean temperature and the room, calculated at each time step and given by

$$\Delta\theta = \frac{\theta_{\text{u,s}} - \theta_{\text{u,r}}}{\ln(\theta_{\text{u,s}} - \theta_{\text{ref},t-1}) - \ln(\theta_{\text{u,r}} - \theta_{\text{ref},t-1})}, \tag{3}$$

where $\theta_{u,s}$ and $\theta_{u,r}$ are the supply and return temperatures of the radiator at the current time step, and $\theta_{ref,t-1}$ is the reference internal temperature at the previous time step. Either the air or the operating temperature, defined as the arithmetic average of air and mean radiant temperatures, can be taken as reference internal temperature. The latter option has been chosen in this work.

Return temperature. In their paper, Lundström et al. [37] include an empirical correlation for the return temperature from the radiator which has also been used in the present work:

$$\theta_{u,r} = \theta_{u,s} - b \cdot (\theta_{u,s} - \theta_{ref,t-1,})^{a} \tag{4}$$

where a and b are functions of the nominal supply, return and reference internal temperatures and of the radiator exponent. Other equations are available in the literature, in which the dependencies on other quantities like instantaneous flow rate and radiator inertia are also taken into account (see for example Teskeredzic and Blazevic [46]).

Secondary supply temperature control. The secondary supply temperature is controlled through the modulation of the primary flow rate according to a certain logic. In case of fixed-point option, the control system acts to keep the secondary supply temperature at the set value, $\overline{\theta}_{u,s}$:

$$\theta_{u,s} = \overline{\theta}_{u,s}. \tag{5}$$

In the case of outdoor temperature reset, the secondary supply temperature set-point is a function of the current external temperature, θ_e, via a heating curve:

$$\theta_{u,s} = f(\theta_e). \tag{6}$$

Radiator energy balance. Neglecting the thermal inertia of the radiator, the heating power transferred from the water flowing through the radiator is given by

$$\Phi = \dot{m}_u \cdot c_{w,u} \cdot (\theta_{u,s} - \theta_{u,r}), \tag{7}$$

where $\dot{m}_u$ is the water flow rate through the radiator and $c_{w,u}$ the specific heat of the water circulating on the user side.

Power supplied by DH. Neglecting the thermal inertia of the water circulating on the user side, the heating power transferred from the DH to the secondary side of the heat exchanger is

$$\Phi = \dot{m}_{DH} \cdot c_{w,DH} \cdot (\theta_{DH,s} - \theta_{DH,r}), \tag{8}$$

where $\dot{m}_{DH}$ is DH flow rate and $c_{w,DH}$ is the specific heat of water on the primary side. In this simplified model, the heat losses through the secondary distribution network have been neglected, therefore, the heat output from the DH interface is equal to the heat output from the radiator. For a general analysis and for comparative studies like the one proposed in this paper, this set of assumptions is considered acceptable. However, if a real building is to be simulated, one might deem appropriate to include heat losses and inertia of the user distribution piping network, and inertia of the room radiators.

DH heat exchanger model. To close the system, a set of equations must be derived that represents the heat transfer behavior of the interface heat exchanger. This model is fully described in the next Subsection.

According to the type of calculation required, these equations can be rearranged to obtain the desired outputs in the light of the given inputs. For example, if no limit is set to the primary flow rate (or if the calculated primary flow rate is below the set limit), the secondary supply temperature is an input and the primary flow rate is an output of the simulation. However, if the calculated primary flow rate exceeds the upper limit, the calculation is repeated taking the maximum primary flow rate as an input and the secondary supply temperature as an output.

2.4. Heat Exchanger Model

The most common heat exchangers that can be found in DH-supplied dwellings in Italy are of plate type. Gasketed heat exchangers feature a frame in which a pack of thin metal sheet plates are sandwiched between gaskets and pressed together by cover plates and bolts. This type of plate heat exchanger is particularly common in apartment blocks, where the possibility to disassemble the device to replace only the damaged parts is an economic advantage. For smaller installations, such as single family houses, brazed-plate heat exchangers are frequently encountered because they are more compact and relatively cheap. In both cases, each plate features a corrugation obtained by stamping a given surface pattern on the metal sheet. The corrugations are essential for many reasons. Besides making the plates more rigid, the corrugations of consecutive plates result in the formation of the flow channels in patterns studied so to promote high heat transfer coefficients while reducing the risk of clogging (fouling). Different types of corrugations are constantly being developed by manufacturers, like the very popular "chevron-type" pattern.

In the present analysis, a rating (or performance) calculation is required. Therefore, an effectiveness-number of transfer units model (ε-NTU) is chosen, where the exchanged power is expressed as a function of the inlet temperatures [47]:

$$\Phi = \varepsilon \cdot C_{min} \cdot (\theta_{DH,s} - \theta_{u,r}), \tag{9}$$

where $C_{min/max} = \min / \max\{\dot{m}_{DH}\, c_{w,DH}\,;\, \dot{m}_u\, c_{w,u}\}$. The effectiveness parameter ε represents the ratio between the actual exchanged power and the maximum power that can be theoretically exchanged given the two temperature endpoints (primary hot inlet and secondary cold inlet). The effectiveness can be written as a function of two parameters: the number of heat transfer units, NTU, and the ratio between heat capacity rates of the two streams, R:

$$NTU = \frac{U \cdot A}{C_{min}} \qquad R = \frac{C_{min}}{C_{max}}, \tag{10}$$

where A is the overall heat transfer surface and U is the overall heat transfer coefficient:

$$U = \left(\frac{1}{h_h} + \frac{1}{h_c} + \frac{\delta_p}{k_p} + R_{foul,h} + R_{foul,c} \right)^{-1}. \tag{11}$$

In this expression, h_h and h_c are the heat transfer coefficients of the hot and cold flows, δ_p is the plate thickness and k_p its thermal conductivity, while $R_{foul,h}$ and $R_{foul,c}$ are the additional resistances introduced by the possible fouling of the plates on the hot (subscript "h") and cold (subscript "c") side, respectively. In case of new or well-maintained exchangers, the fouling resistances are set to zero.

Several studies in the literature provide empirical correlations for the estimates of these parameters, based on experimental data available for specific commercial products (see for instance the review in [47], chapter 7). Recently, methods based on artificial neural networks have been proposed which could be particularly interesting when very large amount of data are available (see for example Longo et al. [48]). In the framework of the present analysis, the system of equations is closed by a Dittus-Boelter type correlation [49] as a function of the channel-based Reynolds number Re and the Prandtl number Pr of the flow:

$$Nu = c1 \cdot Re^{c2} \cdot Pr^{c3}, \tag{12}$$

where

$$Nu = \frac{h}{k/d} \tag{13}$$

is the Nusselt number, that is, the ratio between convective heat transfer, embodied by heat transfer coefficient h, and conductive heat transfer, represented by the fluid conductivity k

divided by a characteristic length d. Combining Equations (12) and (13), one obtains for the heat transfer coefficient in the i-th channel:

$$h_i = c1 \cdot \left(\frac{h_i}{d}\right) \cdot \left(\frac{\rho_i u_i d}{\mu_i}\right)^{c2} \cdot \left(\frac{\mu_i c_{w,i}}{k_i}\right)^{c3}. \tag{14}$$

The key parameters to be estimated in each channel are u_i and d. The flow velocity u_i can be expressed as a function of the flow rate once the channel area is obtained from the manufacturer. The characteristic length d is the equivalent diameter defined as $4 \cdot W\,H/[2 \cdot (W + H)] \approx 2 \cdot H$, where W is the useful plate width and H the inter-plate height, which can be estimated from the channel area and the number of channels of the considered side. The latter can be found by the number of plates, n_p, which is usually even: therefore, one side has $n_p/2$ channels (cold fluid side) and the other side $n_p/2 - 1$ channels (hot fluid side). This is the preferred configuration because the channels flown through by the hot fluid are separated from the surrounding environment by the two external cold fluid channels, thus the heat losses are limited.

The tuning parameters of the model are $c1$, $c2$ and $c3$ in Equation (14), which should be obtained for every real heat exchanger by fitting experimental data and/or simulation results from advanced proprietary software that manufacturers often make available for their customers.

2.5. Evaluation of Indoor Comfort

EN 16798-1:2019 recently replaced EN 15251:2007 as the reference standard for the parameters required to design and assess the indoor air quality of a building. The standard prescribes that the design criteria for any mechanically conditioned environment be established through comfort indices PMV (predicted mean vote) and PPD (predicted percentage of dissatisfied), described in EN ISO 7730:2005. This comfort assessment model is known as "PMV/PPD", "static", or "heat balance" model.

The PMV is an index predicting the mean value of the votes given by a large number of people in a thermal sensation scale ranging from -3 (cold) to $+3$ (hot), neutrality being at 0. The formula for PMV is obtained from the thermal balance for the human body, thus it is a function of several input parameters:

- metabolic rate M, expressing a typical level of activity of the occupants;
- thermal insulation for clothing, I_{cl};
- internal air temperature, θ_{air};
- mean radiant temperature of the internal surfaces, θ_{mr};
- air velocity, v_{air};
- relative humidity, RH.

$$PMV(t) = f(M, I_{cl}, \theta_{air}, \theta_{mr}, v_{air}, RH). \tag{15}$$

The PPD index is a function of the sole PMV index representing the percentage of thermally dissatisfied people that can be expected. Once the PMV and/or the PPD for a given condition have been calculated, they must be compared to acceptability limits. These are given by the so-called indoor environmental quality (IEQ) classes (Table 1). If used for design purposes, the limit comfort indices for the selected class allow to determine a range of design operative temperatures, θ_{op}.

Table 1. Indoor environmental quality classes and relative index limits ($v_{air} < 0.1$ m/s; RH $\approx 50\%$; $I_{cl} \approx 1$ clo).

Category	Level of Expectation	Accepted PMV	Accepted PPD	Minimum θ_{op} (Heating)
I	High	<6	$-0.2 < \text{PMV} < +0.2$	21.0
II	Medium	<10	$-0.5 < \text{PMV} < +0.5$	20.0
III	Moderate	<15	$-0.7 < \text{PMV} < +0.7$	18.0
IV	Low	<25	$-1.0 < \text{PMV} < +1.0$	16.0

For long-term evaluation of the comfort conditions, individual index values can be conveniently aggregated [43]. The simplest aggregation can be made by calculating the percentage of hours during which the PMV is outside the comfort range.

It is worth mentioning that a possible alternative (or complementary) approach to evaluate indoor thermal comfort is the application of adaptive methodologies linking the indoor comfort temperature to the outdoor running mean temperature. This category of methods account for the ability of people to adapt to the environment over time, thus coming to accept lower indoor temperatures in winter and higher indoor temperatures in summer with respect to the static approach. They were first introduced in the 1970s for "free running" (or "naturally ventilated") buildings, in which the occupants have direct control over their environment. An adaptive model is also included in EN 16798-1:2019 standard.

3. Application

3.1. Presentation of the Case

The methodology outlined in the previous Section has been applied to a practical assessment instance. First of all, the location is identified to provide the energy calculation core described in Section 2.2 with the required climate-related input data. Relevant characteristics of the DH network and of the user heating systems, such as working temperature, installed heating power, and heat exchanger features are also collected to fill the input file and the component database. The buildings to simulate are fully described starting from the TABULA typology matrix with the addition of mass distribution, heat capacity, and heating system design information.

Once the input files are completed, the dynamic simulation is run over a year and all the outputs described in EN ISO 52016-1:2017 are obtained. Among these, the hourly indoor air and mean radiant temperatures are extracted and fed as input to the comfort assessment module, and other quantities provided by the dynamic hourly calculation, such as heating system flow rates, supply temperatures and return temperatures, are stored for further analysis.

Details on input information, main assumptions, and calculation choices are given below.

3.2. Heating System Information

According to the 2019 annual report on Italian district heating networks [50], the extension of urban heating distribution network has reached 4446 km, over 95% of which are in Northern Italy. Heat-only boiler stations and cogeneration facilities based on burning fossil fuels, biomass, and urban wastes are the most common types of DH power stations, occasionally in combination with geothermal/groundwater, heat pumps, industrial heat recovery and, more rarely, solar generation systems. Among these, only a few systems operate with wintertime supply temperatures around 70 °C, almost half of the networks operate around 90 °C, and the other half above 110 °C. This study focuses on the city of Brescia, characterized by the second largest DH network in Italy after Turin. Based on declared data, the DH supply temperature is assumed to be 115 °C.

Multiple buildings have been analyzed (see next Subsection) and a heat exchanger had to be assigned to each of them to perform the calculations. The heat exchanger database has been built according to the following steps:

1. Full definition of the building.

2. Calculation of heating load according to EN ISO 52016-1:2017, clause 6.5.5.2.
3. Definition of user supply/return temperatures based on period of construction, which are set as the outlet and inlet flow temperatures on the secondary side of the heat exchanger.
4. Calculation of design user flow rate from the calculated load and temperature difference: the secondary side of the heat exchanger is fully characterized.
5. Sizing of the heat exchanger with the complete information on the secondary side and the DH supply temperature on the primary side: the DH return temperature and the primary flow rate are outputs of the calculation, together with the heat exchanger model. Here, this step is entirely performed by means of the proprietary software made available by the heat exchanger manufacturer.
6. Collection of the heat exchanger parameters necessary for the model described in Section 2.4, including number of plates, heat transfer area, channel areas and so on. Parameters $c1$, $c2$ and $c3$ in Equation (12) are derived by fitting the equation on the results of simulations performed with the manufacturer software.

The heat exchanger database comes in the form of a plain spreadsheet. New plate heat exchangers can seamlessly be added, which are recalled by the building entries in the building database through a unique identifier.

Although the code allows to set a heating curve to simulate the outdoor temperature reset control of the user supply temperature, the simulations in this application case are performed assuming a fixed-point control. This choice is motivated by the need to clearly identify the impact of the DH supply temperature variation without introducing further overlapping effects. Similarly, continuous operation is assumed for the heating system with no night time set-back. As concerns the thermostatic radiator valve, a proportional band $PB = 1.0\,\text{K}$ has been assumed in the calculations. The radiator characteristic exponent n_{rad} has been set to 1.3.

3.3. Building Typology Matrix and Additional Data

To show the potential of the calculation framework even on a large scale, the whole Italian building stock as represented by the Italian building typology matrix developed in the framework of the IEE-TABULA project [3] and of its follow-up IEE-EPISCOPE [51] has been considered. The matrix, originally built based on the analysis of Piedmont region data, well represents the building stock of Italian middle climatic zone, including most of Lombardy region and the city of Brescia in particular. The matrix defines four types of buildings:

1. single family house (SFH)
2. terraced house (TH)
3. multifamily house (MFH)
4. apartment block (AB)

and describes them as a function of eight periods of construction:

1. ≤ 1900
2. 1901–1920
3. 1921–1945
4. 1946–1960
5. 1961–1975
6. 1976–1990
7. 1991–2005
8. ≥ 2006

Therefore, each building typology can be identified by a type/period pair. For example, SFH.05 will indicate a single family house built between 1961 and 1975. No renovations are considered in the baseline matrix.

In the TABULA/EPISCOPE project, the energy performances of the buildings were assessed through the quasi-steady monthly model of the Italian reference standards, the

UNI/TS 11300 set of standards. Therefore, most of the necessary information is given in the national deliverables of the project [52], such as heated volumes, floor areas, and U-values. However, some additional data are required to perform dynamic hourly calculations with a EN ISO 52016-1-based tool:

- mass distribution of opaque and ground floor elements (Table 1). The reference is Table 1 in Lundström et al. [37], where five possible arrangements of the three layers are outlined: mass on interior side (I), mass on exterior side (E), mass divided on interior and exterior side (IE), equally distributed mass (D) and inside/centered mass (M);
- specific heat capacities of opaque and ground floor elements (Table 1). The reference is Table A.14 of EN ISO 52016-1:2017 standard, where five classes are identified: very light (VL), light (L), medium (M), heavy (H), very heavy (VH). The assigment has been made based on the detailed layer structure description given in the TABULA/EPISCOPE project;
- exposition fractions (Table 2);
- nominal supply and return temperature of the user heating system (Table 1); assuming a design internal set-point, the calculation of the nominal LMTD is straightforward;
- nominal (installed) power, referred to the nominal conditions assumed at the previous point (Table 1).

Finally a heat exchanger has been assigned to each building according to the procedure described in Section 3.2. The original design flow rate on the primary side has also been set as an upper limit to the actual DH flow rate—therefore, it is assumed that the primary control valve can modulate the DH flow rate up to this value to obtain the desired user supply temperature; when a higher DH flow rate would be required, it is capped at the maximum value and the actual user supply temperature is re-calculated.

3.4. Climate-Related Input Data

The city of Brescia, Italy, has been considered in the simulations. Brescia belongs to Italian climate zone E (2101–3000 degree-days with base temperature 20 °C) [53]. Climate-related data for the selected location have been collected from two sources: the Photovoltaic Geographical Information System (PVGIS) in the form of typical meteorological year 2007–2016, and the climate reanalysis dataset ERA5.

ISO14N model requires several climate-related input data:

- outdoor air temperature;
- wind speed at 10 m height;
- beam normal and diffuse horizontal solar irradiance;
- ground temperature;
- surface infrared thermal irradiance on horizontal plane.

All data have been obtained from PVGIS except the ground temperature, for which the variable "Soil temperature level 4", that is, the temperature of the soil in the layer between 1 m and 2.89 m underground, has been retrieved from ERA5 dataset.

Table 1. Additional information for the Italian building typology matrix (MDc: mass distribution class, according to Reference [37], Table 1; SHCc: specific heat capacity class, according to Reference [34], Table A.14). Subscripts "rf", "ew" and "gf" indicate roof, external walls and ground floor elements, respectively.

Quantity	Unit	SFH.01	SFH.02	SFH.03	SFH.04	SFH.05	SFH.06	SFH.07	SFH.08
$SHCc_{rf}$	-	VL	VL	VL	H	H	H	VH	VH
MDc_{rf}	-	E	E	E	M	M	M	D	IE
$SHCc_{ew}$	-	VH	VH	VH	VH	VH	VH	VH	VH
MDc_{ew}	-	D	D	D	D	D	IE	M	I
$SHCc_{gf}$	-	VH	VH	VH	VH	VH	VH	VH	VH
MDc_{gf}	-	D	D	D	D	D	D	D	E
$\theta_{u,s,nom}$	°C	85	85	85	85	85	75	70	60
$\theta_{u,r,nom}$	°C	75	75	75	75	75	65	55	40
Φ_{nom}	W	20,280	18,120	16,630	20,710	22,280	11,870	7000	4480

Quantity	Unit	TH.01	TH.02	TH.03	TH.04	TH.05	TH.06	TH.07	TH.08
$SHCc_{rf}$	-	VL	VL	VH	H	VH	VL	VH	VH
MDc_{rf}	-	E	E	E	M	D	E	IE	IE
$SHCc_{ew}$	-	VH	VH	VH	H	VH	H	VH	VH
MDc_{ew}	-	D	D	D	IE	D	IE	M	I
$SHCc_{gf}$	-	H	VH	VH	VH	VH	VH	VH	VH
MDc_{gf}	-	D	D	D	D	D	D	IE	E
$\theta_{u,s,nom}$	°C	85	85	85	85	85	75	70	60
$\theta_{u,r,nom}$	°C	75	75	75	75	75	65	55	40
Φ_{nom}	W	12,310	11,400	9400	9960	7830	6690	4510	3500

Quantity	Unit	MFH.01	MFH.02	MFH.03	MFH.04	MFH.05	MFH.06	MFH.07	MFH.08
$SHCc_{rf}$	-	H	VL	VH	VH	VH	VH	VH	VH
MDc_{rf}	-	D	D	D	D	D	D	IE	IE
$SHCc_{ew}$	-	VH	VH	VH	VH	H	H	H	VH
MDc_{ew}	-	D	D	D	D	IE	M	IE	M
$SHCc_{gf}$	-	H	VH	VH	VH	VH	VH	VH	VH
MDc_{gf}	-	D	D	D	D	D	D	IE	IE
$\theta_{u,s,nom}$	°C	85	85	85	85	85	75	70	60
$\theta_{u,r,nom}$	°C	75	75	75	75	75	65	55	40
Φ_{nom}	W	51,000	70,090	94,330	63,200	61,620	46,150	34,660	18,750

Quantity	Unit	AB.01	AB.02	AB.03	AB.04	AB.05	AB.06	AB.07	AB.08
$SHCc_{rf}$	-	VL	VH	VH	VH	VH	VH	VH	VH
MDc_{rf}	-	D	D	E	D	D	D	IE	IE
$SHCc_{ew}$	-	VH	VH	VH	H	H	H	VH	VH
MDc_{ew}	-	D	D	D	IE	IE	IE	M	M
$SHCc_{gf}$	-	VH	VH	VH	VH	VH	VH	VH	VH
MDc_{gf}	-	D	D	I	D	D	D	IE	IE
$\theta_{u,s,nom}$	°C	85	85	85	85	85	75	70	60
$\theta_{u,r,nom}$	°C	75	75	75	75	75	65	55	40
Φ_{nom}	W	52,550	224,740	139,630	116,810	164,000	123,030	93,170	46,160

Table 2: Fraction of vertical elements exposed to South, West, East and North.

Element	South	West	East	North
External walls	0.3	0.2	0.2	0.3
Glazing (SFH, MFH, AB)	0.4	0.3	0.3	0
Glazing (TH)	0.6	0	0	0.4

3.5. Comfort Indicators

The framework presented in Section 2.5 can be adopted to evaluate indoor comfort in buildings in the different retrofitting scenarios. In the choice of the evaluation method, it has been considered that adaptive concepts can be applied when the occupants have direct control on the surrounding environment and are aware of it, which might be a borderline assumption in large multi-family buildings. Some studies report that the use of adaptive methodologies for design purposes led to too-cold buildings, as pointed out by Halawa et al. [54]. Therefore, PMV/PPD method has been chosen over the adaptive approach in a conservative perspective, and the following indicators have been used:

- Operative temperature, defined according to EN ISO 52016-1:2017 as:

$$\theta_{op} = \frac{\theta_{air} + \theta_{mr}}{2} \tag{16}$$

for a qualitative evaluation of the transient periods.

- Minimum PMV calculated in the reference period T:

$$PMV_{min} = \min_{\forall t \in T}\{PMV(t)\}. \tag{17}$$

- Percentage of time steps in the reference period in which the PMV is below the minimum threshold for the selected IEQ class, $PMV_{IEQ,l}$, which is a modified version of the "percentage of outside the range" parameter:

$$POR = \frac{\sum_{\forall t \in T}\left(PMV(t) \leq PMV_{IEQ,l}\right)}{N_T} \cdot 100, \tag{18}$$

where N_T is the total number of time steps in the reference period.

The indoor thermal comfort calculation module has also been written in MATLAB. The reference period is the heating season for Italian climate zone E [53] (15 October–15 April). IEQ class II has been selected. Input parameters have been fixed to $M = 1.2$ met, corresponding to a standing/relaxed activity, $v_{air} = 0.1$ m/s, a reasonably low value expected with radiator-based heating, $I_{cl} = 1.0$ clo, corresponding to typical winter indoor clothing, and RH 50%.

4. Results and Discussion

4.1. Baseline and Pre-Retrofitting

Hourly indoor air and mean radiant temperatures estimated by the energy calculation core are given as input to the thermal comfort module. The PMV for each time step is calculated with Equation (15), assuming the parameter values reported at the end of Section 3.5. Hourly PMVs are then used to obtain the desired aggregated indicators described in the same Section 3.5. This process is repeated for all the identified scenarios, namely:

- original building with design DH supply temperature;
- original building with reduced DH supply temperature;
- renovated building with reduced DH supply temperature,

where different DH temperature reduction levels and building renovation measures are explored. This process is performed for all the considered building types.

The modified POR parameter can be used to quantify the possible increase of discomfort occurrences when the DH temperature is decreased. It is expected that the lower the DH temperature with respect to the original (design) value, the higher the percentage of discomfort occurrences over the heating season. Figure 3 shows the variation of POR parameter with the reduction of DH supply temperature. Two decrease levels have been considered: $\theta_{DH,s} = 115\ ^\circ C$ to $\theta_{DH,s} = 95\ ^\circ C$ (representative of the shift from second to third generation DH) and $\theta_{DH,s} = 115\ ^\circ C$ to $\theta_{DH,s} = 88\ ^\circ C$ (which is slightly higher than the maximum user design supply temperature, set to 85 $^\circ C$ for older buildings). It is worth noting that, since the POR associated to the original design conditions is zero, the ΔPOR represented in the bar charts numerically coincide with the POR of the reduced-temperature scenarios. As expected, the POR increases with decreasing DH supply temperature for all the building types. Although it rarely exceeds 5% in case of the smaller reduction ($\theta_{DH,s} = 95\ ^\circ C$), the POR is non-negligible when the temperature is decreased to 88 $^\circ C$, especially for multifamily houses and recent buildings.

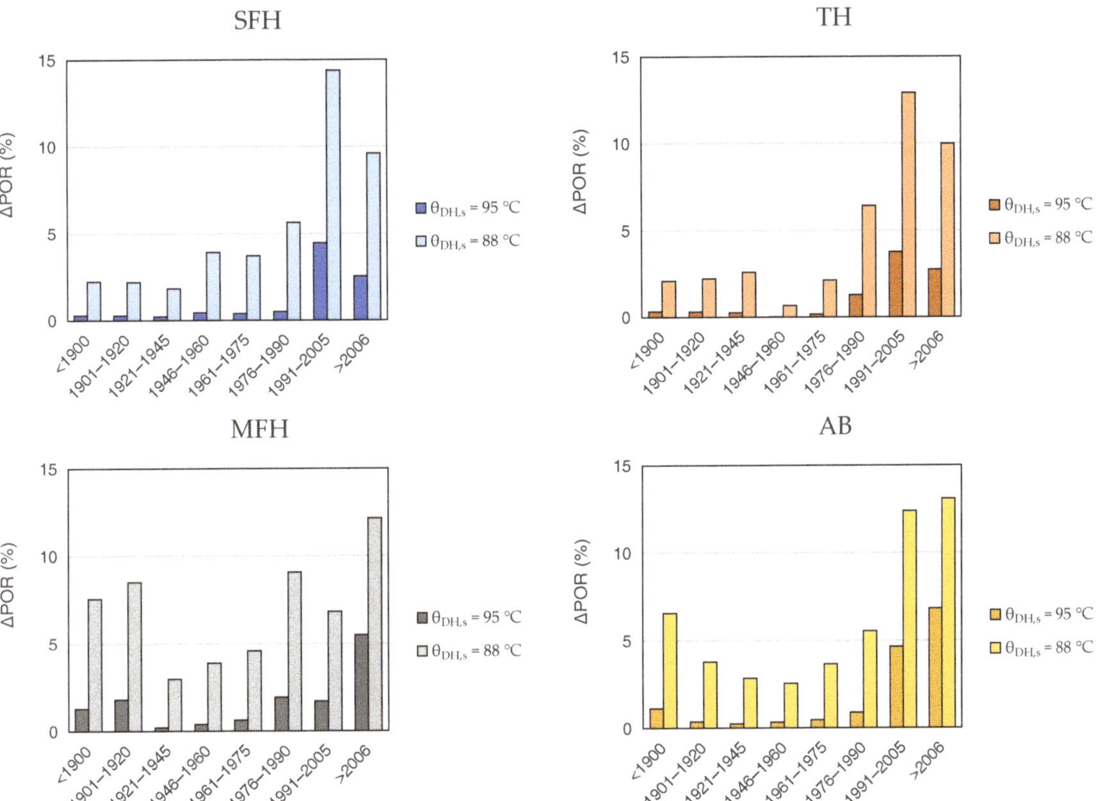

Figure 3. Variation of percentage outside range (POR) parameter with decreased district heating (DH) temperature (115 °C to 95 °C and 115 °C to 88 °C).

The latter result might look counter-intuitive. However, it can be explained in the light of the driving parameter for radiator heat transfer, that is, the LMTD between radiator and surrounding environment. Figure 4 shows an example of user supply and return temperatures obtained with $\theta_{DH,s} = 88$ °C for a single-family house (SFH) built before 1900 and the same type of building built after 2006. It is worth reminding that the design user supply/return temperatures have been set as a function of the period of installation (see Table 1). In particular, they are 85/75 °C and 60/40 °C for the selected older building and newer building, respectively. It can be observed that the user flow/return temperatures following the reduction of DH supply temperature are still compatible with a decent radiator output in the older plant. On the other hand, in the recent building the decrease from design temperatures that are already low might quickly lead to insufficient heat outputs.

The above observation of the different role of the LMDT reduction in the two cases confirms that attention must be paid to low temperature systems (Figure 5). The nominal heat emission of a radiator scales with the ratio between actual and nominal LMTD, $\Delta\theta / \Delta\theta_{nom}$, raised to the power of the radiator exponent, about 1.3. It can be seen how this ratio tends to be low for longer periods in case of low-temperature regime systems, as they are sized near the limit and, therefore, have little margin for further temperature reductions.

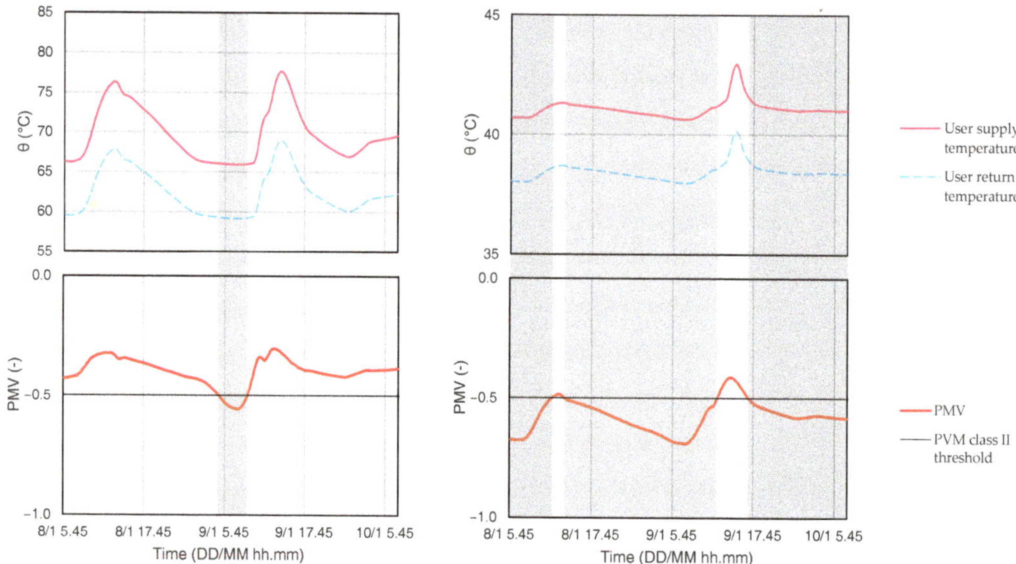

Figure 4. Top: user supply/return temperatures in a winter period in a single family house built before 1900 (**left**) and a single family house built after 2006 (**right**) after decreasing the DH temperature from 115 °C to 88 °C. Bottom: corresponding PMVs. Discomfort periods are indicated by the gray areas.

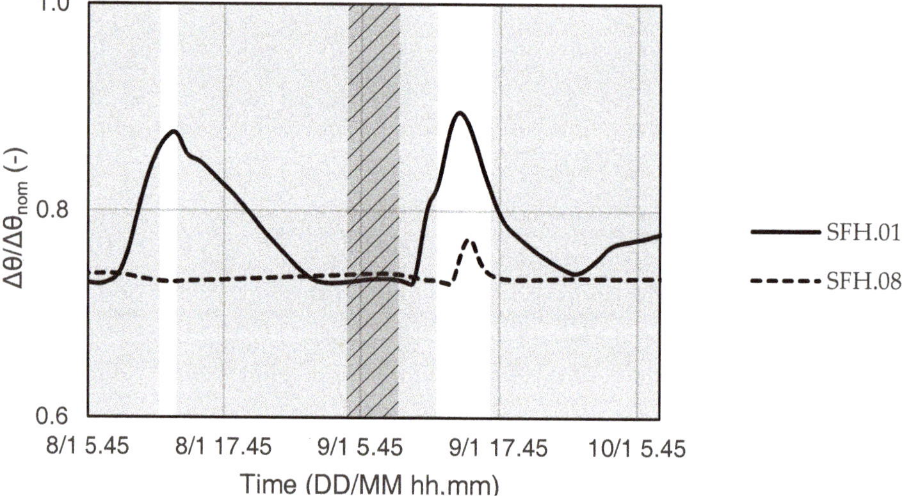

Figure 5. Ratio between actual and design LMTD for an old single family house built before 1900 (solid line) and a new single family house built after 2006 (dashed line), for the same period of the year as in Figure 4. Discomfort periods are indicated by the hatched area (old building) and the gray area (new building).

Since the percentage of time steps with too low PMV does not give any indication on the severity of the discomfort, the minimum value of PMV, PMV_{min}, recorded during the heating season has also been taken into account. Figure 6, which refers to the case of $\theta_{DH,s} = 88$ °C, shows indeed that lower PMV values are obtained on average for older buildings.

Further qualitative information about the severity of discomfort can be extracted from the analysis of the PPD distribution during the heating season. Figure 7 shows two examples of the evolution of PPD probability density distribution with decreasing DH supply temperatures for two single-family houses built in different periods. The mean PPD value does not change very much with the period of construction or with DH supply temperature, but the probability distribution does: the more recent building shows higher frequencies of low PPD (5–6%) for all DH supply temperatures, indicating a general higher degree of comfort. However, it also displays higher frequencies of high PPD (9–12%) when the temperature decreases from 115 °C to 95 °C and 88 °C. On the other hand, very few occurrences of PPD below 6% can be observed in the older building irrespective of the DH supply temperature, but even after decreasing the latter to 95 °C or 88 °C the occurrence of PPD values above 10% remains relatively low.

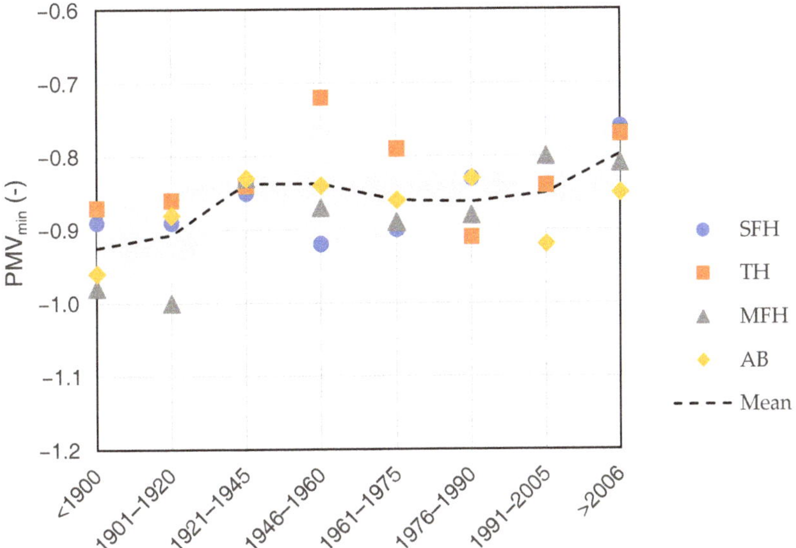

Figure 6. Minimum predicted mean vote (PMV) recorded in the heating season for the different types of buildings as a function of construction period. Gray area: standard deviation of the PMVs for the different building types.

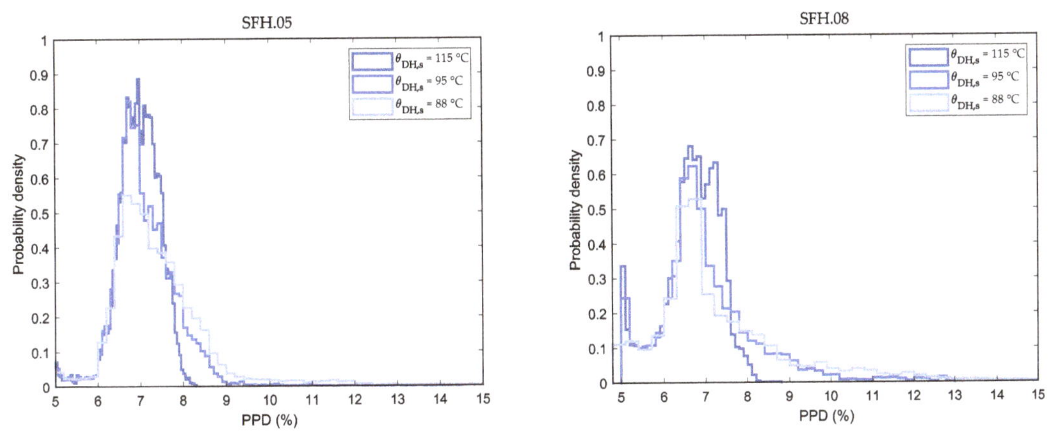

Figure 7. Distribution of predicted percentage of dissatisfied (PPD) over the heating season for single family house built between 1961–1975 (**left**) and built after 2006 (**right**), and decreasing DH supply temperatures.

Finally, dynamic calculations allow the direct observation of indoor transients, such as early morning warm-up. In Figure 8 a representative period is isolated in which the operative temperature in the design and in the reduced DH supply temperature conditions are compared for a single family house built in the period 1961–1975. It is expected that this effect be even magnified in case of intermittent control of the heating system (for example, on/off operation or set-back periods).

The results of the simulations show that, with the assumed design temperatures and interface heat exchangers, further actions are likely to be required not to affect the comfort of the occupants. A comparison between three possible scenarios is described in the next Subsection.

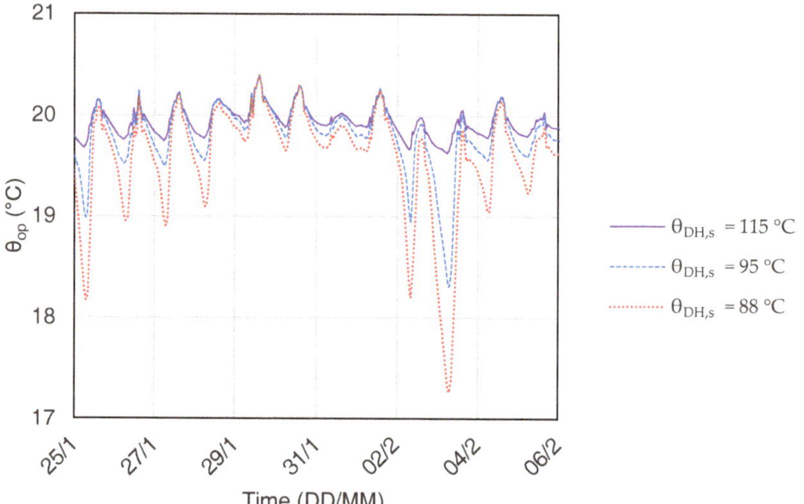

Figure 8. Indoor operative temperature in a winter period for a single family house built between 1961 and 1975 at different DH supply temperatures. Original design temperature: $\theta_{DH} = 115$ °C. Operative temperature set-point: 20.5 °C.

4.2. Three Retrofitting Scenarios

In the following, three possible sets of measures are compared. The first set consists of two actions on the primary (DH) side: heat exchanger replacement and flow rate increase. The new heat exchanger is sized so to provide twice the original thermal power, which is achieved with a larger number of plates and/or a higher heat exchange length (longer plates). For this set of simulations, the DH flow rate has not been capped to any value. Figure 9 shows the ratio between $\dot{m}_{DH}$ and $\dot{m}_{DH,nom}$, where $\dot{m}_{DH}$ is the average DH flow rate that would be required to obtain the user supply temperature set-point with $\theta_{DH} = 88$ °C and the oversized heat exchanger, and $\dot{m}_{DH,nom}$ is the design flow rate in the initial state ($\theta_{DH} = 115$ °C and original heat exchanger). Once again, buildings built after the 1980s display a different behaviour than older buildings—in this case the effect of a lower design heating temperature regime is positive, in that the mere installation of a generously sized heat exchanger covers most of the new requirements. As expected, in older plants where the design conditions are close to the new DH supply temperature (85/75 °C versus 88 °C) the heat transfer on the heat exchanger is limited, and a larger increment of DH flow rate is required.

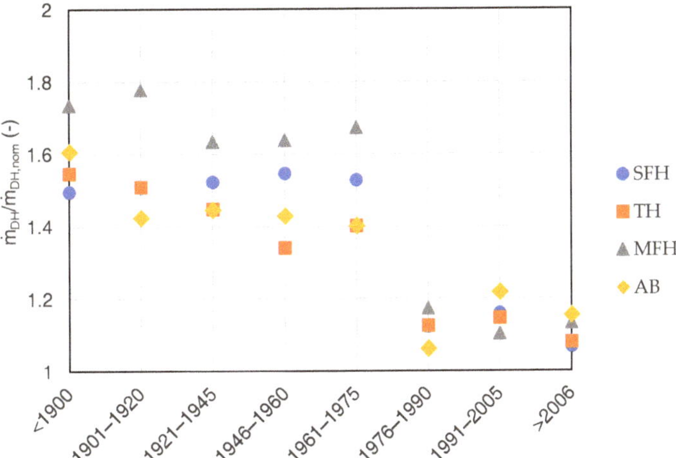

Figure 9. Ratio between the secondary flow rate need with decreased DH temperature ($\theta_{DH} = 88\,°\text{C}$) and the original design flow rate ($\theta_{DH} = 115\,°\text{C}$) for the different types of buildings as a function of construction period.

On the user side, two basic options are available—acting on the heating system or on the building fabric. Figure 10 shows the oversizing factor of the radiator, calculated as the ratio between the minimum LMTD recorded during the heating season with $\theta_{DH} = 88°\text{C}$ and the original design LMTD, raised to the power of the radiator exponent, $k_{rad} = (\min_{\forall t \in T}\{\Delta\theta_{88°\text{C}}(t)\}/\Delta\theta_{nom})^{n_{rad}}$. The value looks quite stable for all the building types, irrespective of the construction period. According to this simulation, around 55% extra exchange area should be installed in private dwellings to fulfill the requirements, either by replacing existing radiators or by adding new ones. This measure is invasive and expensive, thus it does not appear to be the most indicated solution except maybe in public houses within the framework of a renovation campaign charged to the public owner.

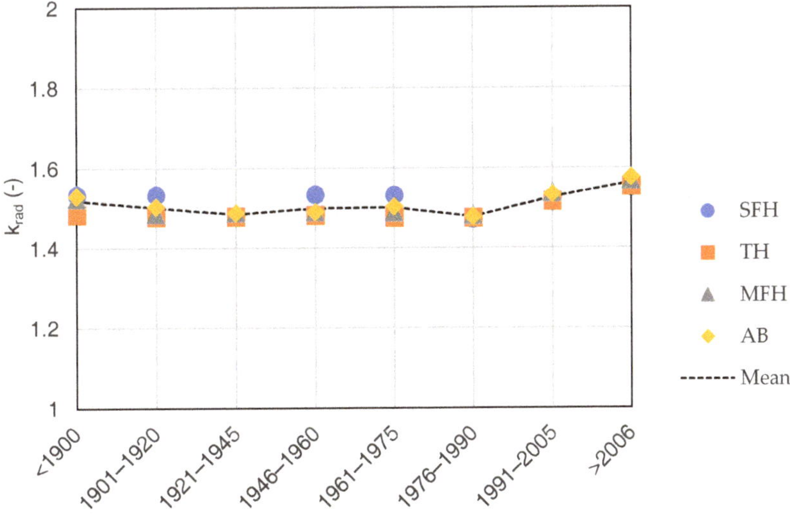

Figure 10. Radiator oversizing factor, k_{rad}, for the different types of buildings as a function of construction period.

The other obvious option is the partial or full renovation of the building fabric. Figure 11 shows the comparison of the POR parameters calculated with reduced DH supply temperature ($\theta_{\mathrm{DH,s}} = 88\,°\mathrm{C}$) for all the building typologies in the following fabric conditions:

- Building in the original state
- Building with replaced windows
- Building with replaced windows and state-of-the-art insulation applied to the opaque elements

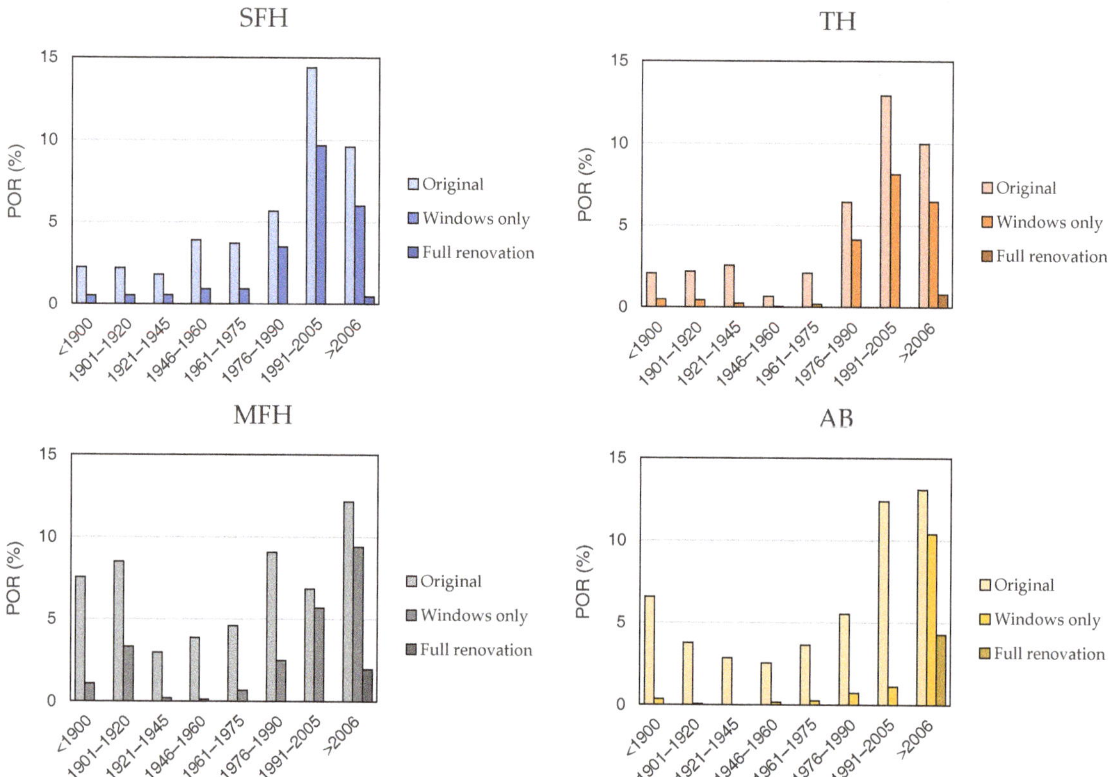

Figure 11. Variation of POR parameter with decreased DH temperature (115 °C to 88 °C) for different retrofit measures on the buildings: original building with no renovation (lightest shade); building with original opaque elements and state-of-the-art windows (intermediate shade); fully renovated building (darkest shade).

The reference for the retrofit measures is the "Usual" (or "Standard") refurbishment set of measures described in the Italian TABULA/EPISCOPE deliverables [52] and defined on the basis of national and local regulatory framework [3]. It can be observed that the replacement of single-glass windows with high performance double-glazed windows greatly improves the comfort conditions in old buildings, especially in single family dwellings. For buildings built in the 1980s–1990s, improving the insulation of the opaque elements would be an effective solution. Conversely, none of these measures would be advisable for more recent buildings where the improvement margin is thin and the prospect of expensive works would face more resistance from the dwelling owners.

The renovation of poor energy performance buildings is clearly visible in the temperature transients: the improved insulation helps limiting the temperature decrease in the night-time and speeds up the reprise (Figure 12).

On an ending note, it is worth remarking that quantitative results depend on the heat exchanger models and the original design heating temperature regimes assumed for each building typology. If different hypotheses are made, for example, due to the availability of further information, the calculation may lead to different numerical values, although the key points of attention emerged from the present simulations are expected to be confirmed.

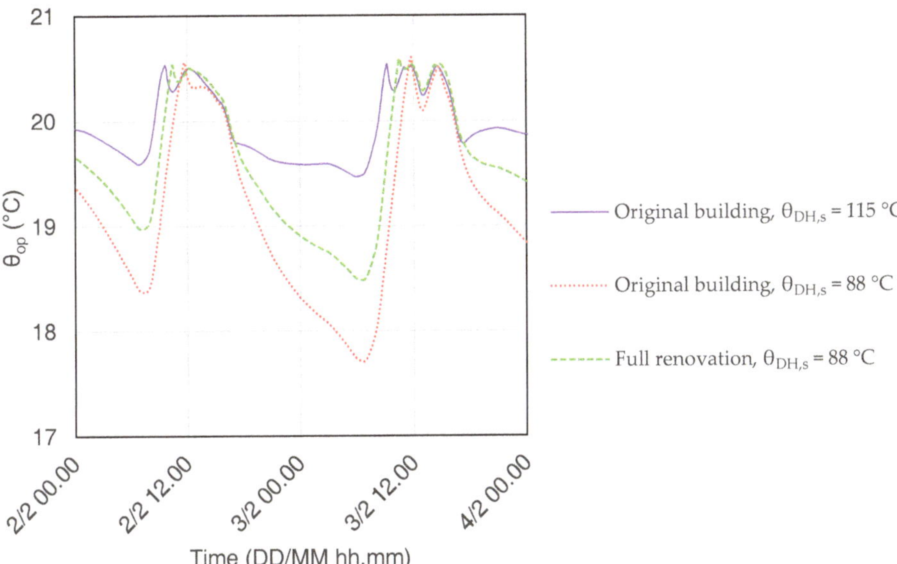

Figure 12. Indoor operative temperature in a winter period for a single family house built between 1961 and 1975. Operative temperature set-point: 20.5 °C.

5. Conclusions

To evaluate the impact of district heating supply temperature reduction on the users, a comprehensive methodology has been presented that exploits the energy simulation results from a literature dynamic calculation method to assess indoor thermal comfort. The main innovative elements are the incorporation of an accurate model for the primary plate heat exchanger, and the use of comfort indicators to express the acceptability of district heating temperature reduction. Benefits of the proposed methodology are the limited input details to the energy calculation engine and a modular structure that allows for further expansions and tailoring.

The method has been applied to the case of existing buildings connected to a typical Northern Italian district heating network. The building stock has been represented by the TABULA/EPISCOPE Italian building typology matrix, in the conservative hypothesis of radiator-based heating systems. In the first set of simulations, the impact of decreasing district heating supply temperature has been analyzed. The results showed that key aspects to consider are the original sizing of the heating components and the original design conditions. As a consequence, the reduction of district heating supply temperature may result in longer periods of moderate discomfort in more recent buildings, and shorter periods of severe discomfort in older heating systems.

The second set of simulations different improvement scenarios have been evaluated. In particular, actions on the primary side appear to be suitable to more recent buildings, in which the fabric thermal performances are already fair to good, and the lower design heating temperature regime has a positive effect. For older buildings, improving opaque element insulation and/or replacing obsolete windows seems a wiser solution than installing additional heating power. Both strategies are indeed quantified as potentially invasive and expensive, but the former aims at reducing the energy needs and is therefore to be preferred.

The heating system model does not take into account distribution heat losses and original system oversizing, whose influence could be the subject of further studies. In this respect, the proposed methodology could also be applied to an actual urban area for which more detailed information is available, as well as measurements before and after the changing event. In the analysis of large buildings, the assumption of a single thermal zone for the whole building could be simplistic for the purpose of comfort assessment. A possible solution would be to simulate the behavior of the individual dwelling with the actual boundary conditions. This option is already included in the simulation framework. Future developments include the use of this method for studies related to district heating return temperature, heating control settings, and cost-benefit assessment on both the primary and the secondary heating side.

Supplementary Materials: The following are available online at https://www.mdpi.com/1996-1073/14/1/25/s1, File S1: Validation of ISO14N model.

Author Contributions: Conceptualization, methodology, software, all authors; writing–original draft preparation, B.G.; revision and editing, E.A.P. and G.P.B.; supervision, M.P. All authors have read and agreed to the published version of the manuscript.

Funding: This study was sponsored by the Regione Lombardia project "Smart Grid Pilot: Banco EnergETICO" (CUP E89I17000410009, Asse I, POR FESR 2014-2020).

Acknowledgments: The authors would like to thank Luca Rigoni, Alessandro Gnatta, Leonardo Zanoni, Ettore Filippini and Massimo Girelli of di A2A Calore & Servizi, Brescia for the support and the cooperation.

Conflicts of Interest: The authors declare no conflict of interest.

Abbreviations

The following abbreviations are used in this manuscript:

AB	Apartment Block
DH	District Heating
GIS	Geographical Information System
HE	Heat Exchanger
HVAC	Heating, Ventilation, and Air Conditioning
LMTD	Logarithmic Mean Temperature Difference
LTDH	Low Temperature District Heating
MFH	Multi-Family House
PB	Proportional Band
PMV	Predicted Mean Vote
POR	Percentage Outside Range
PPD	Predicted Percentage of Dissatisfied
RC	Resistor-Capacitor
RES	Renewable Energy Source
RH	Relative Humidity
SFH	Single Family House
TES	Thermal Energy Storage
TH	Terraced House
TRV	Thermostatic Radiator Valve
air	internal air
c	cold fluid
cl	clothing
DH	district heating
e	external, outdoors
ew	external walls

foul	fouling
gf	ground floor
gl	glazing
h	hot fluid
int	internal
max	maximum
min	minimum
nom	nominal
op	operative
p	plate
r	return
rad	radiator
rf	roof
rm	mean radiant
s	supply
set	set-point
u	user
w	water

References

1. Abergel, T.; Delmastro, C. Tracking Buildings 2020. Available online: https://www.iea.org/reports/tracking-buildings-2020 (accessed on 18 October 2020).
2. Lowe, R. Technical options and strategies for decarbonizing UK housing. *Build. Res. Inf.* **2007**, *35*, 412–425. [CrossRef]
3. Ballarini, I.; Corgnati, S.P.; Corrado, V. Use of reference buildings to assess the energy saving potentials of the residential building stock: The experience of TABULA project. *Energy Policy* **2014**, *68*, 273–284. [CrossRef]
4. Di Turi, S.; Stefanizzi, P. Energy analysis and refurbishment proposals for public housing in the city of Bari, Italy. *Energy Policy* **2015**, *79*, 58–71. [CrossRef]
5. Delmastro, C.; Mutani, G.; Schranz, L. The evaluation of buildings energy consumption and the optimization of district heating networks: a GIS-based model. *Int. J. Energy Environ. Eng.* **2016**, *7*, 343–351. [CrossRef]
6. Zirak, M.; Weiler, V.; Hein, M.; Eicker, U. Urban models enrichment for energy applications: Challenges in energy simulation using different data sources for building age information. *Energy* **2020**, *190*, 116292. [CrossRef]
7. Tronchin, L.; Fabbri, K. Energy Performance Certificate of building and confidence interval in assessment: An Italian case study. *Energy Policy* **2012**, *48*, 176–184. [CrossRef]
8. Nageler, P.; Zahrer, G.; Heimrath, R.; Mach, T.; Mauthner, F.; Leusbrock, I.; Schranzhofer, H.; Hochenauer, C. Novel validated method for GIS based automated dynamic urban building energy simulations. *Energy* **2017**, *139*, 142–154. [CrossRef]
9. Mutani, G.; Todeschi, V. Building energy modeling at neighborhood scale. *Energy Effic.* **2020**, *13*, 1353–1386. [CrossRef]
10. Torabi Moghadam, S.; Toniolo, J.; Mutani, G.; Lombardi, P. A GIS-statistical approach for assessing built environment energy use at urban scale. *Sustain. Cities Soc.* **2018**, *37*, 70–84. [CrossRef]
11. Tronchin, L.; Manfren, M.; James, P. Linking design and operation performance analysis through model calibration: Parametric assessment on a Passive House building. *Energy* **2018**, *165*, 26–40. [CrossRef]
12. Noussan, M.; Nastasi, B. Data Analysis of Heating Systems for Buildings—A Tool for Energy Planning, Policies and Systems Simulation. *Energies* **2018**, *11*, 233. [CrossRef]
13. Mancini, F.; Nastasi, B. Energy retrofitting effects on the energy flexibility of dwellings. *Energies* **2019**, *12*. [CrossRef]
14. Westermann, P.; Deb, C.; Schlueter, A.; Evins, R. Unsupervised learning of energy signatures to identify the heating system and building type using smart meter data. *Appl. Energy* **2020**, *264*, 114715. [CrossRef]
15. Tronchin, L.; Fabbri, K. Analysis of buildings' energy consumption by means of exergy method. *Int. J. Exergy* **2008**, *5*, 605–625. [CrossRef]
16. Kaarup Olsen, P.; Christiansen, C.H.; Hofmeister, M.; Svendsen, S.; Thorsen, J.E. *Guidelines for Low-Temperature District Heating*; Technical report; Danish Energy Agency and EUDP 2010-II; 2014. Available online: https://www.danskfjernvarme.dk/-/media/danskfjernvarme/gronenergi/projekter/eudp-lavtemperatur-fjv/guidelines-for-ltdh-final_rev1.pdf (accessed on 10 November 2020).
17. Sameti, M.; Haghighat, F. Optimization of 4th generation distributed district heating system: Design and planning of combined heat and power. *Renew. Energy* **2019**, *130*, 371–387. [CrossRef]
18. Neri, M.; Luscietti, D.; Pilotelli, M. Computing the Exergy of Solar Radiation From Real Radiation Data. *J. Energy Resour.-ASME* **2017**, *139*, 061201. [CrossRef]
19. Piana, E.A.; Grassi, B.; Socal, L. A Standard-Based Method to Simulate the Behavior of Thermal Solar Systems with a Stratified Storage Tank. *Energies* **2020**, *13*, 266. [CrossRef]
20. Nuytten, T.; Claessens, B.; Paredis, K.; Van Bael, J.; Six, D. Flexibility of a combined heat and power system with thermal energy storage for district heating. *Appl. Energy* **2013**, *104*, 583–591. [CrossRef]

21. Iora, P.; Beretta, G.P.; Ghoniem, A.F. Exergy loss based allocation method for hybrid renewable-fossil power plants applied to an integrated solar combined cycle. *Energy* **2019**, *173*, 893–901. [CrossRef]
22. Lund, H.; Østergaard, P.A.; Chang, M.; Werner, S.; Svendsen, S.; Sorknæs, P.; Thorsen, J.E.; Hvelplund, F.; Mortensen, B.O.G.; Mathiesen, B.V.; et al. The status of 4th generation district heating: Research and results. *Energy* **2018**, *164*, 147–159. [CrossRef]
23. Buffa, S.; Cozzini, M.; D'Antoni, M.; Baratieri, M.; Fedrizzi, R. 5th generation district heating and cooling systems: A review of existing cases in Europe. *Renew. Sustain. Energy Rev.* **2019**, *104*, 504–522. [CrossRef]
24. Østergaard, D.; Svendsen, S. Are typical radiators over-dimensioned? An analysis of radiator dimensions in 1645 Danish houses. *Energy Build.* **2018**, *178*, 206–215. [CrossRef]
25. Tunzi, M.; Østergaard, D.S.; Svendsen, S.; Boukhanouf, R.; Cooper, E. Method to investigate and plan the application of low temperature district heating to existing hydraulic radiator systems in existing buildings. *Energy* **2016**, *113*, 413–421. [CrossRef]
26. Østergaard, D.; Svendsen, S. Theoretical overview of heating power and necessary heating supply temperatures in typical Danish single-family houses from the 1900s. *Energy Build.* **2016**, *126*, 375–383. [CrossRef]
27. Wang, Q.; Ploskić, A.; Holmberg, S. Retrofitting with low-temperature heating to achieve energy-demand savings and thermal comfort. *Energy Build.* **2015**, *109*, 217–229. [CrossRef]
28. Bo, M. Riqualificazione dei vecchi impianti di riscaldamento a radiatori. *Aicarr J.* **2012**, *12*, 20–28.
29. Millar, M.A.; Burnside, N.; Yu, Z. An investigation into the limitations of low temperature district heating on traditional tenement buildings in Scotland. *Energies* **2019**, *12*. [CrossRef]
30. Ashfaq, A.; Ianakiev, A. Investigation of hydraulic imbalance for converting existing boiler based buildings to low temperature district heating. *Energy* **2018**, *160*, 200–212. [CrossRef]
31. Neirotti, F.; Noussan, M.; Riverso, S.; Manganini, G. Analysis of Different Strategies for Lowering the Operation Temperature in Existing District Heating Networks. *Energies* **2019**, *12*, 321. [CrossRef]
32. Crabb, J.A.; Murdoch, N.; Penman, J.M. A simplified thermal response model. *Build. Serv. Eng. Res. Technol.* **1987**, *8*, 13–19. [CrossRef]
33. Michalak, P. The simple hourly method of EN ISO 13790 standard in Matlab/Simulink: A comparative study for the climatic conditions of Poland. *Energy* **2014**, *75*, 568–578. [CrossRef]
34. CEN/TC 89. *EN ISO 52016-1:2017 Energy Performance of Buildings— Energy Needs for Heating and Cooling, Internal Temperatures and Sensible and Latent Heat Loads—Part 1: Calculation Procedures*; European Committee for Standardization, CEN: Brussels, Belgium, 2017.
35. Mazzarella, L.; Scoccia, R.; Colombo, P.; Motta, M. Improvement to EN ISO 52016-1:2017 hourly heat transfer through a wall assessment: the Italian National Annex. *Energy Build.* **2020**, *210*, 109758. [CrossRef]
36. Campana, J.P.; Morini, G.L. BESTEST and EN ISO 52016 Benchmarking of ALMABuild, a New Open-Source Simulink Tool for Dynamic Energy Modelling of Buildings. *Energies* **2019**, *12*, 2938. [CrossRef]
37. Lundström, L.; Akander, J.; Zambrano, J. Development of a space heating model suitable for the automated model generation of existing multifamily buildings—A case study in Nordic climate. *Energies* **2019**, *12*, 485. [CrossRef]
38. Déqué, F.; Ollivier, F.; Poblador, A. Grey boxes used to represent buildings with a minimum number of geometric and thermal parameters. *Energy Build.* **2000**, *31*, 29–35. [CrossRef]
39. Yu, W.; Li, B.; Jia, H.; Zhang, M.; Wang, D. Application of multi-objective genetic algorithm to optimize energy efficiency and thermal comfort in building design. *Energy Build.* **2015**, *88*, 135–143. [CrossRef]
40. Manfren, M.; Nastasi, B.; Groppi, D.; Astiaso Garcia, D. Open data and energy analytics-An analysis of essential information for energy system planning, design and operation. *Energy* **2020**, *213*, 118803. [CrossRef]
41. Tronchin, L.; Manfren, M.; Nastasi, B. Energy efficiency, demand side management and energy storage technologies—A critical analysis of possible paths of integration in the built environment. *Renew. Sustain. Energy Rev.* **2018**, *95*, 341–353. [CrossRef]
42. Braulio-Gonzalo, M.; Bovea, M.D.; Ruá, M.J.; Juan, P. A methodology for predicting the energy performance and indoor thermal comfort of residential stocks on the neighbourhood and city scales. A case study in Spain. *J. Clean. Prod.* **2016**, *139*, 646–665. [CrossRef]
43. Carlucci, S.; Pagliano, L.; Sangalli, A. Statistical analysis of the ranking capability of long-term thermal discomfort indices and their adoption in optimization processes to support building design. *Build. Environ.* **2014**, *75*, 114–131. [CrossRef]
44. Stasi, R.; Liuzzi, S.; Paterno, S.; Ruggiero, F.; Stefanizzi, P.; Stragapede, A. Combining bioclimatic strategies with efficient HVAC plants to reach nearly-zero energy building goals in Mediterranean climate. *Sustain. Cities Soc.* **2020**, *63*, 102479. [CrossRef]
45. Ballarini, I.; Costantino, A.; Fabrizio, E.; Corrado, V. The Dynamic Model of EN ISO 52016-1 for the Energy Assessment of Buildings Compared to Simplified and Detailed Simulation Methods. *Build. Simul.* **2019**, *2019*, 3847–3854. [CrossRef]
46. Teskeredzic, A.; Blazevic, R. Transient Radiator Room Heating—Mathematical Model and Solution Algorithm. *Buildings* **2018**, *8*, 163. [CrossRef]
47. Wang, L.; Sunden, B.; Manglik, R.M. *Plate Heat Exchangers: Design, Applications and Performance*, 2nd ed.; Wit Pr/Computational Mechanics: Southampton, UK; Boston, MA, USA, 2007.
48. Longo, G.A.; Righetti, G.; Zilio, C.; Ortombina, L.; Zigliotto, M.; Brown, J.S. Application of an Artificial Neural Network (ANN) for predicting low-GWP refrigerant condensation heat transfer inside herringbone-type Brazed Plate Heat Exchangers (BPHE). *Int. J. Heat Mass Transf.* **2020**, *156*, 119824. [CrossRef]
49. Wright, A.; Heggs, P. Rating Calculation for Plate Heat Exchanger Effectiveness and Pressure Drop Using Existing Performance Data. *Chem. Eng. Res. Des.* **2002**, *80*, 309–312. [CrossRef]
50. Associazione Italiana Riscaldamento Urbano, AIRU. *Il Riscaldamento Urbano-Annuario 2019*; Editrice Alkes: Milan, Italy, 2019.

51. Ballarini, I.; Corrado, V. A New Methodology for Assessing the Energy Consumption of Building Stocks. *Energies* **2017**, *10*, 22. [CrossRef]
52. Politecnico di Torino, Energy Department. Joint EPISCOPE and TABULA Website—Italy Country Page. Available online: https://episcope.eu/building-typology/country/it/ (accessed on 14 May 2020).
53. D.P.R. 26/08/1993, n. 412—Regolamento Recante Norme per la Progettazione, L'installazione, L'esercizio e la Manutenzione Degli Impianti Termici Degli Edifici ai Fini del Contenimento dei Consumi di Energia, in Attuazione Dell'art. 4, Comma 4, Della Legge 9 Gennaio 1991, n. 10, 1993. https://www.gazzettaufficiale.it/eli/id/1993/10/14/093G0451/sg (accessed on 30 July 2020). (In Italian)
54. Halawa, E.; van Hoof, J. The adaptive approach to thermal comfort: A critical overview. *Energy Build.* **2012**, *51*, 101–110. [CrossRef]

energies

MDPI

Article

Support Decision Tool for Sustainable Energy Requalification the Existing Residential Building Stock. The Case Study of Trevignano Romano

Fabrizio Cumo [1], Federica Giustini [1], Elisa Pennacchia [2,*] and Carlo Romeo [3]

[1]   Interdepartmental Center for Territory, Building, Conservation and Environment, Sapienza University of Rome, Via A. Gramsci, 53-00197 Rome, Italy; fabrizio.cumo@uniroma1.it (F.C.); federica.giustini@uniroma1.it (F.G.)
[2]   Department of Planning Design and Technology of Architecture, Sapienza University of Rome, Via Flaminia, 72-00197 Rome, Italy
[3]   ENEA, Italian National Agency for New Technologies, Energy and Sustainable Economic Development, Via Anguillarese, 301-00123 Rome, Italy; carlo.romeo@enea.it
*   Correspondence: elisa.pennacchia@uniroma1.it

Abstract: The control and improvement of energy-environmental quality in buildings are responsible for almost 40% of the emissions related to energy and processes, and are essential to achieve the commitment of the Paris Agreement and the Sustainable Development Goals (SDGs) United Nations (UN). This paper provides a support tool to planners and administrators of the territory for the identification of interventions aimed at the energy requalification of the existing Italian building heritage, mainly for residential use. The purpose of this tool is to reduce energy consumption by intervening on the building envelope with specific solutions that are identified through a matrix resulting from the study. In the first part of the study, an analysis was carried out on various factors such as the existing residential building, the building and construction types and the materials of the envelope typical of each construction period, which are critical for energy efficiency issues. In the second part of the study, the analysis of the state of the art of the insulating materials existing on the international and national market was carried out, in order to standardize the efficiency interventions of the building envelope. By exploiting the potential of the proposed matrix, and integrating it with Geographic Information System (GIS) technology, it would be possible to create a database containing information regarding the characteristics of the building envelope of the residential building stock and to identify a set of insulation interventions more suited to each specific case near Rome, Italy.

Keywords: energy requalification; building envelope; sustainable development and planning; standardized interventions of requalification; Geographic Information System

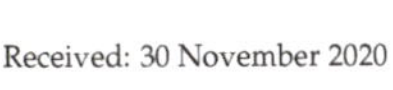
check for updates

Citation: Cumo, F.; Giustini, F.; Pennacchia, E.; Romeo, C. Support Decision Tool for Sustainable Energy Requalification the Existing Residential Building Stock. The Case Study of Trevignano Romano. *Energies* **2021**, *14*, 74. https://dx.doi.org/10.3390/en14010074

Received: 30 November 2020
Accepted: 20 December 2020
Published: 25 December 2020

Publisher's Note: MDPI stays neutral with regard to jurisdictional claims in published maps and institutional affiliations.

1. Introduction

Climate change is a major global phenomenon today. In recent decades, attention to the environment has grown more and more, in particular to global warming, which has caused enormous quantities of greenhouse gas emissions released into the atmosphere, deriving from anthropogenic activities that question natural balances. The environmental issue is closely linked to the energy issue, the way in which energy is produced, distributed and consumed. According to the IPCC (Intergovernmental Panel on Climate Change), the energy supply sector is the largest contributor to greenhouse gas emissions [1], which is one of the issues that must be addressed and solved in the short term to limit the damage caused to ecosystem and human health.

Global residential energy demand has steadily increased over the past decades [2].

The Global Status Report for Buildings and Construction 2019 highlights the importance of decarbonizing the building and construction sector to achieve the Paris Agreement commitment and the United Nations (UN) Sustainable Developments Goals (SDGs);

as these are responsible for nearly 40% of energy and related emissions process, taking climate action in buildings and constructions is among the most cost-effective method [3]. These data highlight the need to identify new strategies to make the construction sector more sustainable. Two-thirds (about 65%) of the European building stock was built before 1980; around 97% of the EU buildings need to be refurbished to reach the 2050 decarbonization target, but only 0.4–1.2% are updated every year [4].

In the Green Deal, Europe defines the improvement of the energy efficiency of buildings as the key to achieving the ambitious goal of eliminating carbon emissions by 2050. [5]. The EU Directives implementing the energy and climate package therefore promote an energy policy aimed at decarbonizing the economy in order to achieve the goal of climate neutrality by 2050.

Energy efficiency is an effective tool for implementing this policy and addressing the recent challenges of climate change due to the scarcity of resources, emissions of climate-altering gases and dependence on energy imports, mainly from non-renewable sources.

The term energy efficiency generically refers to the ability of a physical system to obtain a certain result using the least amount of energy compared to other less efficient systems, increasing its performance and consequently obtaining energy savings, lower costs and significant environmental benefits. The energy requalification of buildings and the rational use of energy in all phases of the construction process are key interventions in the latest international policy documents aimed at decarbonizing real estate assets by 2050, with intermediate stages in 2030 and 2040. For example, the European Directive 2018/844, which replaces the previous Directive 2010/31/EU on the energy performance of buildings and the Directive 2012/27/EU on energy efficiency, obliges Member States to develop long-term national strategies to promote the energy renovation of residential, non-residential, private and public buildings. The aim is to reduce emissions in the EU by 80–85% compared to 1990 levels, by promoting the transformation of existing buildings into nearly zero energy buildings (NZEB) [6]. The thermal performance of buildings determines the amount of energy used for heating and cooling buildings, which profoundly affects energy efficiency. Far and Far [7] state that the use of sustainable design principles and the effective use of building materials can play a crucial role in improving the thermal performance of new and existing buildings.

Several research activities have focused on the evaluation of the energy consumptions of existing buildings stock. Paraschiv et al. [8] highlighted that most of the energy saving potential, and therefore a possible reduction of greenhouse gas emissions is the thermal renovation of existing buildings, which require thermal efficiency improvements in order to reduce the heating needs of the building. According to the data for 2018 reported in [9], the residential sector accounted for 26.1% of final energy consumption, or 16.6% of gross inland energy consumption in the European Union. In particular, the main energy consumption in EU households is for heating their homes (63.6%) and space cooling (0.4%).

The energy performance of the entire building depends mainly on the efficiency of the envelope, which establishes the boundary between indoor and outdoor environments [10]. In particular, 50% of a building's total energy consumption for general use is dissipated through its envelope [11]. As clarified by Poel et al., [12] given the very low renovation rate of the building stock, the refurbishment of existing buildings can be a good solution to improve the environmental performance of the building sector. Numerous researches conducted in the field of energy efficiency have started from the study of the characteristics and conditions of the existing heterogeneous building stock, considering only stationary thermal transmittance values, in order to identify methodologies for evaluating the energy performance of the building stock, such as EPIQR [13], IFORE [14], SUSREF [15], TABULA [16] and others [17,18]. These renovation research projects do not assess the environmental sustainability of the proposed energy efficiency interventions. In recent years, social housing societies in Europe have presented and implemented many retrofitting projects. For example, various initiatives are currently underway in the Netherlands to explore the financial and energy bill consequences of insulating terraced houses up to Passivhaus Standards, and renewable energy technologies for domestic hot water and space heating. Garufi et al. [19] developed a decision support tool for sustainable renovation projects in the Dutch housing corporations, applicable to all building complexes with same archetype (terraced houses 1945–1965). This tool is intended to provide an estimate of the total investment and operating cost applicable in large scale projects involving terraced and non-residential houses.

Understanding the energy performance of existing buildings in an urban cell, an entire neighborhood or an entire municipality is very important for sustainable energy planning strategies aimed at accelerating the energy renewal process.

O' et al. [20] and Moghadam et al. [21] developed a methodology based on the use of tools in a Geographic Information System (GIS) platform, which allows a complete and low-cost picture of the energy performance of buildings. These methodologies are mainly based on the information that is already available on the building stock from data collection and literature (e.g., energy auditors, municipality technical department, web and others), which is subsequently transferred to the GIS.

That said, the recent literature on the subject has strongly focused on the performance of the components of the building envelope that degrades due to environmental conditions over time, and on the enormous potential for energy savings even with basic energy retrofit actions. The use of a GIS platform is limited to monitoring the energy consumption of buildings, but does not offer any indication on improvement measures to be implemented for each building [22].

In this paper, the authors propose and describe their tool to immediately identify the most suitable standardized insulated solutions for the building envelope, according to the age of construction, the building type and the construction type of the existing building in the Italian territory.

The solutions were chosen considering the entire production chain (design, production of components, assembly, installation, evaluation of energy, environmental, seismic and economic performance), in order to reduce the environmental impacts during the execution, management/maintenance and future disposal as much as possible.

The proposed methodology also provides for the georeferencing of existing building envelope types and related interventions to improve efficiency through a GIS platform. This system is proposed to facilitate compliance with the obligation to improve the energy performance of buildings imposed by current legislation. The methodology was developed by the research center of Sapienza University of Rome—CITERA, as part of the "Research of the Electricity System" program with ENEA and the Ministry of Economic Development on "Improving the energy efficiency of production processes and management of the built environment". In the end, from the literature review, it emerged how that most methodology concerning the energy efficiency of buildings did not include a study of insulation materials to make the production chain more efficient.

For that reason, the authors believe that their contribution to the knowledge in this topic consists of: the development of an abacus of standardized envelope modules as the basis of an industrial production process for Deep Renovation interventions in the national residential building stock; a tool for identifying the most effective insulation solution for each existing building; an integration of the solutions identified through the tool in the GIS platform to the geolocation of the improvement interventions to be carried out on the existing building heritage; and to record the interventions already carried out.

2. Materials and Methods

Improving the energy performance of production and management processes in the built environment by developing a catalogue of standard construction layouts for insulation systems to be applied to the vertical opaque envelope of existing buildings is the aim of the research.

An outline of the methodology developed to obtain a tool for the identification of the efficiency solution for the existing building envelope, is shown in Figure 1.

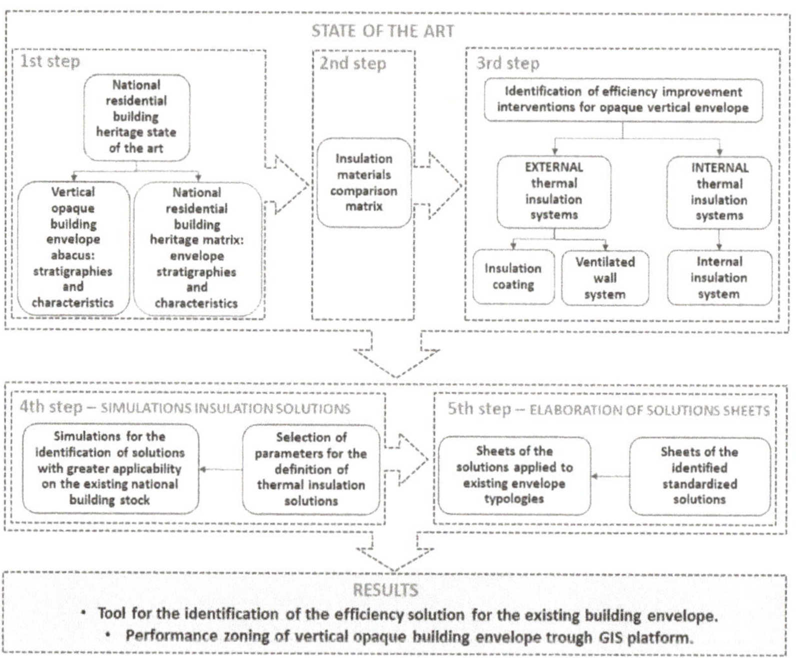

Figure 1. Research methodology flow chart.

The methodological approach is based on the search for the information needed to identify a range of main representative features to recognize such a rich national residential housing stock. In this context, the research aims to identify a set of buildings according to parameters considered relevant for the classification and to allow a subsequent identification of standardized interventions of deep sustainable redevelopment of the envelope system:

- Climatic area
- Seismic area
- Class of construction age
- Building type
- Construction type
- Layers and characteristics of the building envelope

2.1. State of the Art of the National Residential Building Stock

The investigation of the state of the art has made it possible to identify the most widespread construction types, and recognize the construction types that need to be upgraded in terms of energy efficiency, as made before any energy consumption regulations. The Italian territory is divided into six climatic areas, from A to F, according to D.P.R. no.412/1993, based on the number of degrees day (HDD—EN ISO 15927-6:2007). A comparison between the day degrees of each municipality present in each Italian region clearly shows a prevalence of climate zone E. The seismic areas of each Italian municipality have been analyzed using the classification updated in January 2019 by the Civil Protection Department [23]. This classification has the purpose of establishing the best interventions for both the reduction of seismic risk and the energy requalification of the national residential building envelope. 44% of the national territory is characterized by a high seismic risk, according to the seismic classification of the Italian municipalities of the Civil Protection (Seismic Zone 1 or Seismic Zone 2), mainly in the Center and South of the Italian peninsula.

The national residential building stock has been classified according to the age of construction into eight classes; these have been identified in connection with significant historical events and with the emanation of energy regulations that affected the use of building types and construction techniques:

- Class 1: until 1900
- Class 2: from 1901 to 1920
- Class 3: from 1921 to 1945
- Class 4: from 1946 to 1960
- Class 5: from 1961 to 1975
- Class 6: from 1976 to 1990
- Class 7: from 1991 to 2005
- Class 8: after 2005

According to ISTAT data, the national residential building stock is 14,515,795 units, of which 12,187,698 are buildings or residential building complexes (about 84%). More than half of the residential buildings, about 60%, were built after World War II until the 1990s, 15% before 1919, 14% after 1990 and 11% between 1919 and 1945. Most of the residential building stock was built before the first energy saving law No. 373 of 1976. More than 25% of the buildings constructed before this law record annual consumption from a minimum of 160 kWh/m^2 per year to more than 220 kWh/m^2 per year. [24]. As far as building types are concerned, there are two main types: isolated and aggregated. The first type includes the building organizations with prevalent residential use located within an urban context, usually situated in a marginal position with respect to the historical center, provided with a surrounding green area of different dimensions. The single-family housing development has been transformed over time into the minimum associable unit of the multi-family residential building, complex and significant in size. The residential building heritage can be classified into seven main types of buildings: isolated building single-family or duplex; multi-family linear; tower; terraced; with balcony access; Palazzina; or block building [25]. The prevalence of single-family buildings characterized by one or two floors above ground is clearly revealed by the analysis of the latest data published by the National Institute of Statistics. Construction techniques remained almost completely the same for several centuries, until the beginning of the 20th century, when there was a transformation in terms of the use of materials and construction systems thanks to the industrialization process.

In order to define the main construction systems, the classification, provided for by Law no. 64 of 2 February 1974, "Measures for constructions with special requirements for seismic zones" in Article 54, was used, which lists them as follows:

- masonry structure
- framed structure made of normal or prestressed reinforced concrete, steel or combined systems of the above-mentioned materials
- load-bearing panel structure
- timber structure

According to the 15th population and housing Italian census of 2011, load-bearing masonry is the most widespread type of construction, with an average of 57%, compared to 29% for reinforced concrete and 13% for other materials. The regions with the highest percentage of buildings made of load-bearing masonry are Molise, Sardinia and Tuscany, with percentages of around 70%, followed by Friuli-Venezia Giulia, Lombardy and Sicily with 50%. There are no substantial differences between the northern and southern Italian regions. The largest number of residential buildings made of reinforced concrete between 1946 and 1960 was in Lombardy (61,608), followed by Sicily (33,855) and Piedmont (32,054). The largest number of residential buildings built with a type of construction other than load-bearing masonry and reinforced concrete before 1918 was in Piedmont (19,362) and Lombardy (16,504). Knowledge of the different building types and masonry textures plays a fundamental role in defining the interventions for energy requalification of residential buildings, especially for historical buildings.

2.2. Opaque Building Envelope Type Abacus

A bibliographical investigation has been carried out into architectural technology, which revealed a very wide and heterogeneous stratigraphy used in residential buildings all over Italy.

A selection of the most representative ones has been carried out, which can be referred to a series of other similar stratigraphies, for the most significant parameters considered within this research for the energetic performance of the envelope package, thickness and transmittance. An abacus of the types of opaque building envelope packages emerged from the selection, reflecting the national residential building stock, according to the period of greatest diffusion, in order to facilitate the following identification of standardized energy requalification interventions. For example, with regard to most of the older masonry made with the use of natural materials, there is a wide variation related to the type of manufacturing, to the aggregation of the elements, to the historical period and to the geographical location [26]. The stratigraphies of these historical masonry are almost all related to stone masonry, since with regard to this research, the evaluation and comparison parameter used for the classification of the building envelope is the energy performance, the transmittance which, in this case, depends mainly on the thickness of the wall itself. The selection of the most representative stratigraphies and construction age classes was also made basing it on an existing internationally recognized research, the TABULA project, funded by the European Programme Intelligent Energy Europe (2009–2012) [16].

The building envelope types have been divided according to their position and function within the building system into:

- Roof (R)
- External Wall (EW)
- Floor (F)

A progressive numeric code has been assigned to each of them according to a chronological and typological order. In the abacus for each stratigraphy, it has been reported: the period in which it was most widespread; the total thickness; the description and thickness of each component numbered; and the energetic performance parameter in winter and summer regimes (Table 2).

The stratigraphies of the upper horizontal envelope selected are 12, while those of the vertical envelope are 28 and those of the lower horizontal envelope are 14.

The periodic and stationary transmittance values of each type of building envelope closure have been assessed.

The study carried out makes it possible to highlight the main cases on which *Deep Renovation*'s interventions should be focused on and to identify the best solutions to be adopted, both on the national and international market, linked to the concept of off-site construction according to the different characteristics of existing degraded buildings.

An example of Deep Renovation intervention is the replacement of the existing façade with a pre-fabricated sunspace in framed buildings. A comparison was made between the limit value of transmittance required by the regulations (Ecobonus Requirements Decree GU 05/10/2020) and that of walls (Table 1); a deviation from the limit value of up to 0.5 was deemed acceptable.

Table 1. The transmittance limit values required by the Ecobonus Requirements Decree GU 05/10/2020 (calculation according to UNI EN ISO 6946).

UNI EN ISO 6946	
Climate zone A e B	≤0.38
Climate zone C	≤0.30
Climate zone D	≤0.26
Climate zone E	≤0.23
Climate zone F	≤0.22

The following table shows the performance level of the wall according to climate zones, indicated by the letters A to F according to D.P.R. no.412/1993, red when the deviation from the standard transmittance value is greater than or equal to 0.5 and orange if it is smaller (Table 2). The other tables of the abacus are shown in Supplement 1.

Table 2. Extract from the abacus of building envelope types.

Construction Year Class		EW.01—Plastered Stone Masonry		
Until 1920				
Stratigraphy		Thermal Performance		Level of Performance
1. Exterior plaster: 2 cm 2. Stone: 41 cm 3. Interior plaster: 2 cm Total thickness: 45 cm	Winter	Thermal transmittance (U) = 2.40 W/m^{2}K		A–B
	Summer	Periodic thermal transmittance (Yie) = 0.362 W/m^{2}K		
		Internal areal heat capacity (Cip) = 77.6 kJ/m^{2}K		C
		Thermal phase shift (Φ) (h) = 11.23		D
		Attenuation factor (fa) = 0.142		E
		Surface mass (Ms) = 1089 kg/m^2		F

The Thermal Transmittance is expressed in W/m^{2}K. The measure unit indicates watts of energy that are lost across one square meter of surface for a temperature difference of one degree Kelvin.

2.3. National Residential Building Heritage Matrix

A matrix of buildings, mainly for residential use, was drawn up, summarizing the results of the preliminary study of the state of the art (the main source of data used is National Statistical Institute—ISTAT), starting from the selection proposed in the abacus and in the data sheets where the possible building configurations that can be found within the national heritage have been outlined. The configurations shown are defined by the features that can commonly be found in buildings of the respective class of construction age, building type and construction type. The lines that make up the matrix represent the configurations of the envelope, characterized by a certain type of upper horizontal closure, vertical closure and lower horizontal closure (which can then be against the ground or towards an external space, such as on pilotis), relating to the class of construction age, building type and construction type most widespread. The order of the different configurations follows the evolution of construction techniques in chronological order. The configurations present in the matrix are 346 (only an extract is given due to the length of the text in Figure 2). At the end of the analysis of the state of the art of the national residential building patrimony, once the relative matrix has been set up, some data sheets representative of the designer's way of using the matrix were created. The National Residential Building Stock Matrix was elaborated by the authors, in the program for "Research of the Electricity System" in cooperation with ENEA on the project "Energy efficiency of industrial products and processes", 2019–2020.

	NATIONAL RESIDENTIAL BUILDING HERITAGE MATRIX																									
	Class of construction age								Building type								Construction type							Building envelope		
									Isolated buildings	Multi-family buildings							Load Bearing Masonry				Frame structure					
Ref. Matrix Data Sheet	1. Until the 1900	2. From 1901 to 1920	3. From 1921 to 1945	4. From 1946 to 1960	5. From 1961 to 1975	6. From 1976 to 1990	7. From 1991 to 2005	8. After 2005	Single-family	Duplex	Linear	Tower	Terraced	Balcony access	Palazzina	Block	Stone	Stone and brick	Clay-brick/block	Reinforced concrete	Wood	Reinforced concrete	Steel	Roof	External Wall	Floor
SM.01	x								x		x		x	x		x	x							R.01 - Pitched roof with wooden structure and planking 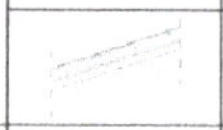	EW.01 - Plastered stone masonry (45 cm) 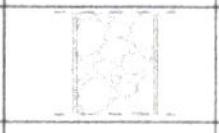	F.02 - Clay vaulted slab 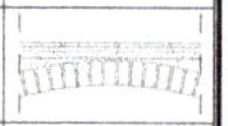
SM.01	x								x		x		x	x		x	x							R.01 - Pitched roof with wooden structure and planking 	EW.01 - Plastered stone masonry (45 cm) 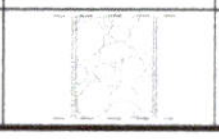	F.01 - Wooden beam and hollow tile slab 

Figure 2. Extract from the National residential building stock matrix.

2.4. Building Envelope Configuration Data Sheets

A series of data sheets have been elaborated basing on the abacus to provide a summary of the different configurations (and combinations) of the characteristics of the building envelope components belonging to the main residential building categories. A detailed study on the individual case histories is offered in the overview of building and construction types according to the class of construction age, also allowing a quick evaluation of the energy performance (the thermal performance in stationary and dynamic regimes, calculation according to UNI EN ISO 6946) of the individual components by both experienced technicians and simple users. After selecting the building that could be involved in the renovation, the operator can identify the case history that best matches the case by using the data sheets. In this way, in order to carry out a qualitative evaluation of the possible interventions for the redevelopment of the building envelope, it will be possible to set out from a reference baseline, knowing its components and performance qualities. Especially in the case of existing buildings for which there are only a few indications, it will still be possible to base on some data references in order to quickly provide an early qualitative assessment of the energy performance of the building. There are 125 total sheets (The 125 sheets are in the report elaborated by the authors, in the program for "Research of the Electricity System" in cooperation with ENEA on project "Energy efficiency of industrial products and processes", 2019–2020).

2.5. Insulation Material Comparison Matrix

The regulations concerning the buildings' energy efficiency focus on the performance characteristics of thermal insulating materials, also in compliance with the principles introduced by the circular economy. This has led to the introduction on the market of different types of high-performance materials, which can be differentiated into four macro categories regarding their origin: natural organic insulation; natural inorganic insulation; synthetic organic insulation; or synthetic inorganic insulation [27,28]. The choice of the type of insulation depends on many considerations, which in addition to the satisfaction thermal comfort respecting the limits imposed by the reference standards, also concern the geographical location and orientation of the building under intervention and the environmental impact indicators. Another fundamental aspect for the selection of the insulation material is related to the prefabrication and therefore to the way the product is installed, which should allow a quick and dry application, in order to reduce the work site time, the environmental impacts associated with it and the possibility of future reuse. The building industry is the world's largest consumer of raw materials [29]. Therefore, the entire methodological approach is based on the principles of circular economics right from the design phase. The use of industrial processes of technologies and products is preferred, thinking from the very beginning about their use at the end of their life, having characteristics that will allow their disassembly or restructuring [30].

Priority is given to the modularity, versatility and adaptability of the product, and to the replacement of virgin raw materials with secondary raw materials coming from being recycled/reused, which preserves their qualities.

Through the key of circular economy, it is consequently possible to reduce the environmental impact of building interventions by preferring dry or mixed construction techniques, prefabricated components and materials from recycling, renewable and recyclable or reusable processes [31].

A matrix has been developed to compare the various types of insulation materials present on the national and international market for the following identification of the most suitable insulation system based on the features of the building to be energetically improved.

In order to allow a comparison between the numerous varieties of insulation materials, parameters relating to insulation capacities in winter and summer regime, environmental impact and direction for use were selected [32] (the table of the comparison of insulation materials is in Supplement 2).

After comparing the energy and environmental performance values, the selected insulation materials are: wood fiber among the natural organic insulators, rock wool and aerogel among the synthetic inorganic insulators, synthetic organic insulators include expanded polyurethane and expanded polystyrene.

The different types of insulation materials identified have been associated with the main insulation solutions: insulation coating; ventilated wall system; and internal insulation system.

The development of a methodology in which the performance of insulation scenarios is precalculated using an energy performance certificate software, can direct the production chain towards more efficient and environmentally sustainable standardized solutions.

Simulations with different insulation thicknesses were carried out on each type of wall of the abacus, for six standardized insulation solutions (simulations can be found at the end of the text as a Supplement 2 of the paper). Simulations have been carried out to identify the thicknesses of the selected insulating materials that allow the greatest application on the existing walls in all climate zones (Figure 3).

Through this study, it was possible to assess the percentage of applicability of insulation solutions on the national building stock according to climatic zones. The standardized insulation solutions, their characteristics and their applicability are shown in Table 3.

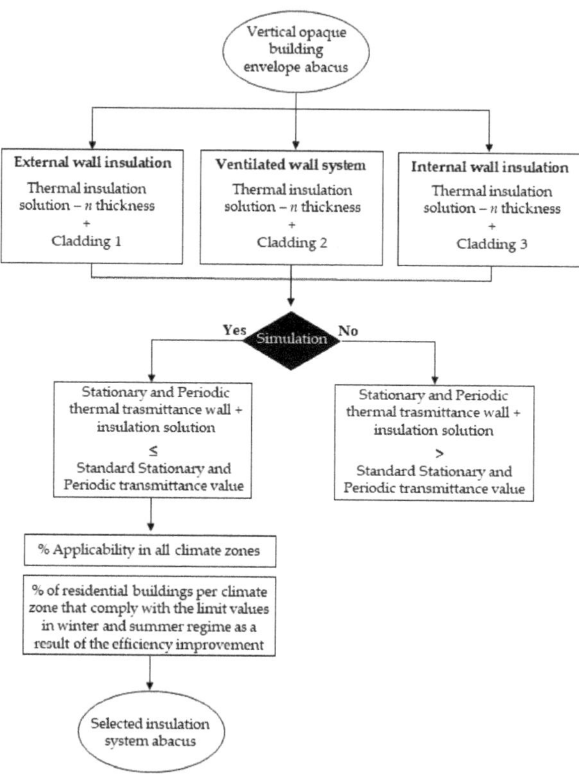

Figure 3. Flow chart of the methodology to identify standardized insulation solutions.

Table 3. Selected insulation systems abacus.

External thermal insulation systems	Insulation coating	**EIS.01** Synthetic inorganic **Rock wool**	• Semi-rigid rock wool insulation panel 0.12 m (λ = 0.032); • Calcium silicate slab 0.012 m (λ = 0.039); • Render/cladding 0.005 m (λ = 0.7)	% applicability in all climate zones: Winter regime: 91% Summer regime: 100%
		EIS.02 Natural organic **Wood fiber**	• Semi-rigid wood fiber insulation panel 0.14 m (λ = 0.04); • Cement mortar 0.005 m (λ = 0.9); • Render/cladding 0.015 m (λ = 0.9)	% applicability in all climate zones Winter regime: 81% Summer regime: 100%
	Ventilated wall system	**VWS.01** Synthetic organic **Rigid expanded polyurethane**	• Semi-rigid wood fiber insulation panel 0.08 m (λ = 0.022); • Air gap 0.04 m (λ = 0.25); • Facing panel 0.02 m (λ = 1.3)	% applicability in all climate zones Winter regime: 91% Summer regime: 100%
		VWS.02 Synthetic inorganic **Rock wool**	• Semi-rigid rock wool insulation panel 0.12 m (λ = 0.032); • Calcium silicate slab 0.012 m (λ = 0.039); • Air gap 0.04 m (λ = 0.25); • Facing panel 0.02 m (λ = 1.3)	% applicability in all climate zones Winter regime: 93% Summer regime: 100%

Table 3. *Cont.*

Internal thermal insulation systems	Interior insulation system			
		IIS.01 Synthetic inorganic **Aerogel**	• Semi-rigid aerogel insulation panel 0.05 m (λ = 0.015); • Gypsum plasterboard 0.012 m (λ = 0.7); • Render/cladding 0.005 m (λ = 0.7)	% applicability in all climate zones Winter regime: 75% Summer regime: 97%
		IIS.02 Synthetic organic **Extruded polystyrene** **(XPS)**	• Semi-rigid extruded polystyrene (XPS) insulation panel 0.10 m (λ = 0.031); • Gypsum plasterboard 0.013 m (λ = 0.21) • Render/cladding 0.005 m (λ = 0.7)	% applicability in all climate zones Winter regime: 73% Summer regime: 97%

3. Results

The research has led to the development of a tool to identify the insulation solution best suited to the characteristics of the wall, building and climate zone.

Below the design per tool's menu is described in Figure 4, with the required inputs per step and outputs. In the Menu, the user is invited to go through three categories of selection: context data; general building data; and additional details.

INPUTS

1. SELECT *Context data* (Region -> Province -> Municipality) ⟶ Climatic zone

2. SELECT *General building data* (Period of construction -> Building type -> Construction typology -> Existing external Wall)

3. OPTIONAL SELECTION *Additional details* (Insulation position -> Cladding typology -> Price range)

OUTPUTS

• INSULATION SOLUTION SYSTEM

• PERFORMANCE ACHIEVED AFTER THE INTERVENTION

Figure 4. The steps involved in the tool.

The first category regards geographical localization and allows to select Region, Province and Municipality. The tool automatically determines the climate zone. General building data allows to select, through a drop-down menu, the *period of construction* (until 1900; from 1901 to 1920; from 1921 to 1945; from 1946 to 1960; from 1961 to 1975; from 1976 to 1990; from 1991 to 2005; or after 2005), the *building type* (isolated buildings: single family or duplex; multifamily buildings: linear; tower; terraced; balcony access; palazzina; or block), the *construction typology* (load-bearing masonry: stone, stone and brick, clay-brick/block or reinforced concrete; frame structure: wood, reinforced concrete or steel). There is also the possibility to choose further details: *insulation position* (External or internal thermal insulation system); *cladding typology* (plaster, facing panel); or *price range* (<150.00 €/m^2; >150.00 €/m^2). After the selection of the various parameters, the tool indicates the most suitable solution for the building to be treated, providing the technical data sheet containing the characteristics of the insulation system applied to the existing wall (wall and insulation system stratigraphy, performance in winter and summer regime and interstitial condensation verification).

The acquired information has been used for the realization of a GIS [33]; it is organized according to different layers, in order to be able to individually process the different categories of information collected. The layers developed regards *general building data* and the *level of performance*. The program can be questioned, and it also allows to simulate interventions based on the results obtained through the data entered by the technician. Once the data have been simulated, the program shows the various solutions of suitable type of insulation to have a better energy efficiency for each building.

With the use of a GIS software, it is possible to create an archive of general information such as the age of construction, type of construction and level of performance of each building in the studied area. This archive leads to the creation of a map that shows the status of the real estate assets of the area. The collection of this information allows the user to interrogate each building, and the program shows all the results that it has in its archive.

Therefore, it has been created a tool with all the information of each building that allows the user to interact with the characteristics of the building, and also have the technical sheet of the best simulation solution.

The platform development is based on open software, which aims to be an interoperable and flexible tool for administrator of the territory and for planners. Performance zoning of a vertical opaque building envelope through the GIS platform allows the administrators of the territory to view the actual state of performance of the building heritage, plan improvements and record the interventions carried out.

4. Case Study: Trevignano Romano A Small Town Near Rome, Italy

The methodology described has been applied and verified in the small town of Trevignano Romano, of about 6000 inhabitants, in the Lazio region, about 30 kilometers from Rome. Trevignano is part of the Regional Natural Park of the Bracciano-Martignano lake complex. There are 878 buildings in this city, 844 of which are residential buildings. The study was carried out in the old town center of the municipality and on a part of the lake promenade, since it is a particularly significant area for the purpose of this research.

The classification of the current state of the existing buildings was carried out by cross-referencing data from high-resolution satellite orthophotos, the municipal administration, ISTAT data and on-site surveys. The visual inspection, supported by the matrix of the national residential building heritage, has made it possible to identify the time of construction, the types of buildings and the characteristics of the building envelope. Almost 70% of the residential building heritage of the area investigated in Trevignano Romano was built between 1976 and 1990, and consists mainly of masonry buildings in the historical center and reinforced concrete with solid brick infill, and in a smaller number, of perforated bricks in the rest of the area examined (Figure 5).

The GIS has therefore made it possible to locate the different buildings in space, linking them to specific alphanumeric attributes saved in the relational database (Archive), and to manage them as "information layers" (Layers) that identify their spatial relations. Each building has been associated with a data sheet containing the information described above to guide the designers' choices towards the most suitable and sustainable energy efficiency systems (Figure 6) [34].

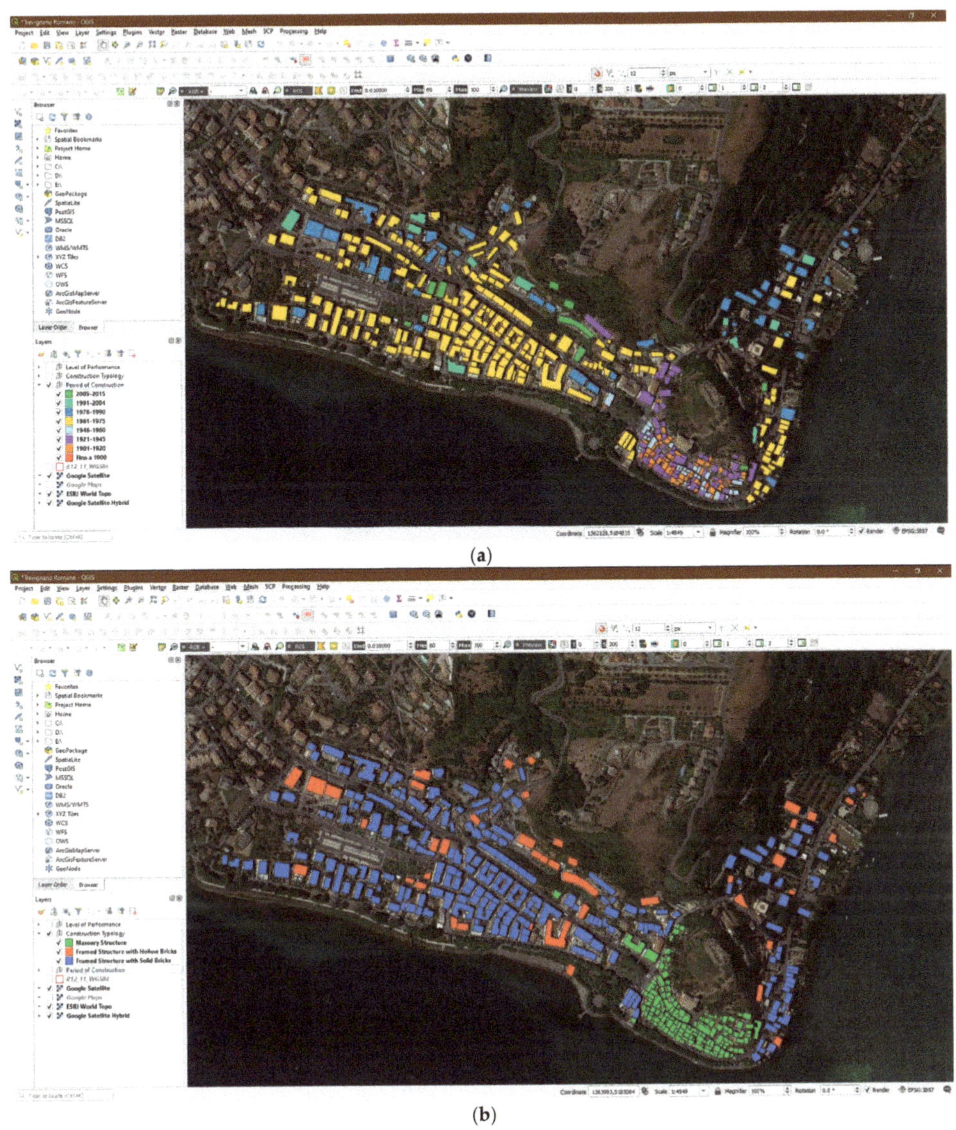

Figure 5. Extract from the Geographic Information System (GIS). (**a**) Class of construction age; (**b**) Construction type.

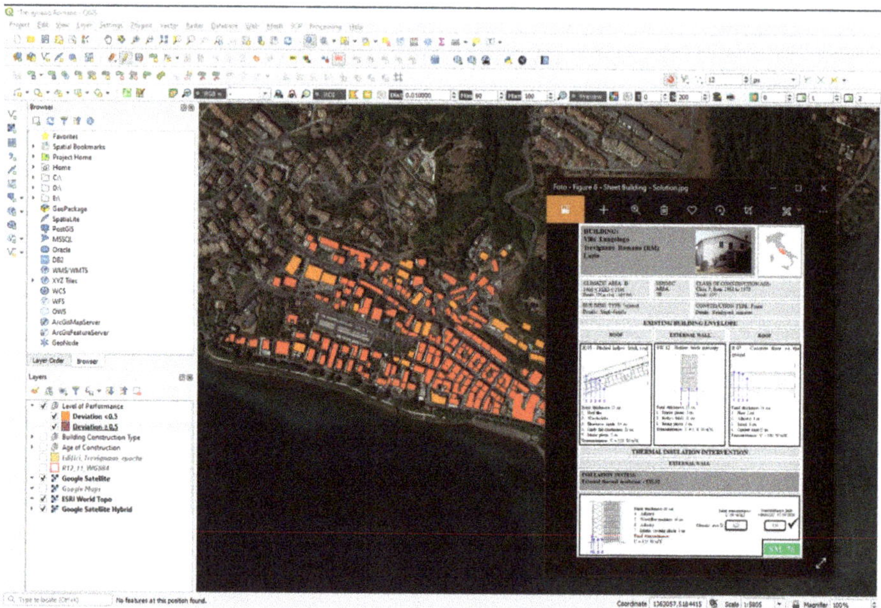

Figure 6. Example of a building data sheet with the insulation solution suggested and final performance.

5. Conclusions and Future Developments

With a view to reducing environmental impacts related to the construction industry, which strongly influence global warming due to greenhouse gas emissions caused by the excessive use of air conditioning systems, the residential construction sector will face major challenges in the coming years and decades addressing these issues.

The proposed methodology aims to support sustainable planning of interventions to improve the energy efficiency of production processes and management of the built environment. The methodology currently applied to external walls can also be applied to roofs and floors and to plant systems from the perspective of Deep Renovation.

With a view to applying the digitization to building and urban regeneration, as a future development, a Building Information Modeling (BIM) tool could be created to automatically identify the improvement interventions for the specific case study to be energy-efficient according to the climate. For building retrofit projects, the integration of BIM and GIS supports decision-making. Putting as-is geometric BIM data and other necessary data into GIS, it becomes possible to create a pre-retrofit simulation model to perform building data and corresponding insulation solutions for existing building [35].

This development leads to an important design and process innovation for territorial information systems playing a fundamental role in addressing the development of built environment, providing both the congruity information management in GIS environment and the information produced through BIM processes [36].

Supplementary Materials: The following are available online at https://www.mdpi.com/1996-1073/14/1/74/s1, Supplement 1, Abacus of building envelope types; Supplement 2, Comparison of insulation materials; Supplement 3, Simulations with different insulation thicknesses.

Author Contributions: Conceptualization, F.C. and C.R.; methodology, E.P.; tool, E.P.; validation, F.G., E.P. and C.R.; formal analysis, F.G.; investigation, E.P.; resources, E.P.; data curation, F.G.; writing—original draft preparation, E.P.; writing—review and editing, E.P. and F.G.; visualization, F.C.; supervision, F.C. and C.R.; and funding acquisition, F.C. All authors have read and agreed to the published version of the manuscript.

Funding: This research was funded by MINISTRY OF ECONOMIC DEVELOPMENT, in the program for "Research of the Electricity System" in cooperation with ENEA on project "Energy efficiency of industrial products and processes", 2019–2020.

Institutional Review Board Statement: Not applicable.

Informed Consent Statement: Not applicable.

Data Availability Statement: Data is contained within the article supplementary material.

Conflicts of Interest: The authors declare no conflict of interest.

References

1. Bruckner, T.; Bashmakov, I.A.; Mulugetta, Y.; Chum, H.; de la Vega Navarro, A.; Edmonds, J.; Faaij, A.; Fungtammasan, B.; Garg, A.; Hertwich, E.; et al. Energy Systems. In *Climate Change 2014: Mitigation of Climate Change. Contribution of Working Group III to the Fifth Assessment Report of the Intergovernmental Panel on Climate Change*; Edenhofer, O.R., Pichs-Madruga, Y., Sokona, E., Farahani, S., Kadner, K., Seyboth, A., Adler, I., Baum, S., Brunner, P., Eickemeier, B., et al., Eds.; Cambridge University Press: Cambridge, UK; New York, NY, USA, 2014.
2. Tsemekidi-Tzeiranaki, S.; Bertoldi, P.; Labanca, N.; Castellezzi, L.; Serrenho, T.; Economidou, M.; Zangheri, P. *Energy Consumption and Energy Efficiency Trends in the EU-28 for the Period 2000–2016*; EUR 29473 EN; Publications Office of the European Union: Luxembourg, 2018.
3. Global Alliance for Buildings and Construction; International Energy Agency; United Nations Environment Programme. 2019 Global Status Report for Buildings and Construction: Towards a Zero-Emission, Efficient and Resilient Buildings and Construction Sector. Available online: https://www.unenvironment.org/resources/publication/2019-global-status-report-buildings-and-construction-sector (accessed on 23 November 2020).
4. Bean, F.; Volt, J.; Dorizas, V.; Bourdakis, E.; Staniaszek, D.; Roscetti, A.; Pagliano, L. Future-Proof Buildings for All Europeans. A Guide to Implement the Energy Performance of Buildings Directive (2018/844). Buildings Performance Institute Europe (BPIE), 2019. Available online: https://www.bpie.eu/wp-content/uploads/2019/04/Implementing-the-EPBD_BPIE_2019.pdf (accessed on 23 November 2020).
5. Communication from The Commission to The European Parliament, The European Council, The Council, The European Economic and Social Committee and The Committee Of The Regions. The European Green Deal, COM/2019/640 Final. Available online: https://eur-lex.europa.eu/legal-content/EN/TXT/?uri=COM%3A2019%3A640%3AFIN (accessed on 23 November 2020).
6. Magrini, A.; Lentini, G.; Cuman, S.; Bodrato, A.; Marenco, L. From nearly zero energy buildings (NZEB) to positive energy buildings (PEB): The next challenge—The most recent European trends with some notes on the energy analysis of a forerunner PEB example. *Dev. Built Environ.* **2020**, *3*, 100019. [CrossRef]
7. Far, C.; Far, H. Improving energy efficiency of existing residential buildings using effective thermal retrofit of building envelope. *Indoor Built Environ.* **2018**, *28*, 1–17. [CrossRef]
8. Paraschiv Lizica, S.; Paraschiv, I.S.; Ion, V.I. Increasing the energy efficiency of buildings by thermal insulation. *Energy Procedia* **2017**, *128*, 393–399.
9. Eurostat. Energy Consumption in Households. Available online: https://ec.europa.eu/eurostat/statistics-explained/index.php/Energy_consumption_in_households (accessed on 23 November 2020).
10. Mavromatidis, L.E.; Bykalyuk, A.; Lequaya, H. Development of polynomial regression models for composite dynamic envelopes thermal performance forecasting. *Appl. Energy* **2013**, *104*, 379–391. [CrossRef]
11. Sala Lizarraga, J.M.P.; Picallo-Perez, A. 12—Design and optimization of the envelope and thermal installations of buildings. In *Exergy Analysis and Thermoeconomics of Buildings. Design and Analysis for Sustainable Energy Systems*; Butterworth-Heinemann: Oxford, UK, 2020; pp. 911–1005.
12. Poel, B.; Cruchten, G.; Balaras, C.A. Energy performance assessment of existing dwellings. *Energy Build* **2007**, *39*, 393–403. [CrossRef]
13. Flourentzou, F.; Genre, J.-L.; Roulet, C.-A. Epiqr-tobus: A new generation of refurbishment decision aid methods. In *Towards Sustainable Building*; Maiellaro, N., Ed.; Springer Science+Business Media: Dordrecht, The Netherlands, 2001; Volume 61, pp. 161–169.
14. Häkkinen, T. Systematic method for the sustainability analysis of refurbishment concepts of exterior walls. *Constr. Build. Mater.* **2012**, *37*, 783–790. [CrossRef]
15. Sdei, A.; Tittelein, P.; Lassue, S.; McEvoy, M.E. Dynamic Thermal Modeling of Retrofitted Social Housing in England and France. In *Proceedings of the CLIMA 2013: 11th REHVA World Congress and the 8th International Conference on Indoor Air Quality, Ventilation and Energy Conservation in Buildings, Prague, Czech Republic, 16–19 June 2013*; Karel Kabele, M.U., Suchý, K., Lain, M., Eds.; Society of Environmental Engineering (STP): Prague, Czech Republic, 2013; p. 6882.
16. Ballarini, I.; Corgnati, S.P.; Corrado, V. Use of reference buildings to assess the energy saving potentials of the residential building stock: The experience of TABULA project. *Energy Policy.* **2014**, *68*, 273–284. [CrossRef]
17. Nemry, F.; Uihlein, A.; Colodel, C.M.; Wetzel, C.; Braune, A.; Wittstock, B.; Hasan, I.; Kreiflig, J.; Gallon, N.; Niemeier, S.; et al. Options to reduce the environmental impacts of residential buildings in the european union—Potential and costs. *Energy Build.* **2010**, *42*, 976–984. [CrossRef]

18. Konstantinou, T. A Methodology to Support Decision-Making towards an Energy-Efficiency Conscious Design of Residential Building Envelope Retrofitting. *Buildings* **2015**, *5*, 1221–1241. [CrossRef]

19. Garufi, D.; de Vries, B. Decision support tool for sustainable renovation projects in the dutch housing corporations. In *Proceedings of the CISBAT 2015, International Conference, Lausanne, Switzerland, 9–11 September 2015*; Future Buildings & Dis-tricts —Sustainability from Nano ot Urban Scale (blz. 987–992); Centre de Recherches en Physique des Plasmas: Lausanne, Switzerland, 2015; pp. 987–992.

20. Dall'O', G.; Galante, A.; Torri, M. A methodology for the energy performance classification of residential building stock on an urban scale. *Energy Build.* **2012**, *48*, 211–219. [CrossRef]

21. Moghadam, S.T.; Mutani, G.; Lombardi, P. GIS-Based Energy Consumption Model at the Urban Scale for the Building Stock. In *Proceedings of the 9th International Conference Improving Energy Efficiency in Commercial Buildings and Smart Communities (IEECB&SC'16), Frankfurt, Germany, 16–18 March 2016*; Paolo Bertoldi; Publications Office of the European Union: Luxembourg, 2016.

22. Gobakis, K.; Mavrigiannaki, A.; Kalaitzakis, K.; Kolokotsa, D.-D. Design and development of a Web based GIS platform for zero energy settlements monitoring. *Energy Procedia* **2017**, *134*, 48–60. [CrossRef]

23. Seismic Classification, Protezione CIvile. Available online: www.protezionecivile.gov.it/attivita-rischi/rischio-sismico/attivita/classificazione-sismica. (accessed on 28 February 2020).

24. Predisposizione del PAEE2014, ALLEGATO 1—Riqualificazione Energetica del Parco Edilizio Nazionale. Available online: www.ec.europa.eu/energy/sites/ener/files/documents/it_building_renov_2017_annex_1_neeap_it.pdf (accessed on 28 February 2020).

25. Mattogno, C. *Ventuno Parole per l'urbanistica*; Aracne: Rome, Italy, 2015; pp. 79–100.

26. Astiaso Garcia, D.; Amori, M.; Giovanardi, F.; Piras, G.; Groppi, D.; Cumo, F.; De Santoli, L. An identification and a prioritisation of geographic and temporal data gaps of Mediterranean marine databases. *Sci. Total Environ.* **2019**, *668*, 531–546. [CrossRef] [PubMed]

27. Asdrubali, F.; D'Alessandro, F.; Schiavoni, S. A review of unconventional sustainable building insulation materials. *Sustain. Mater. Technol.* **2015**, *4*, 1–17. [CrossRef]

28. Adityaa, L.; Mahliaa, T.M.I.; Rismanchic, B.; Nge, H.M.; Hasane, M.H.; Metselaare, H.S.C.; Murazaf, O.; Aditiyab, H.B. A review on insulation materials for energy conservation in buildings. *Renew. Sustain. Energy Rev.* **2017**, *73*, 1352–1365. [CrossRef]

29. Finch, G.; Marriage, G.; Pelosi, A.; Gjerde, M. Building Envelope Systems for the Circular Economy; evaluation parameters, current performance and key challenges. *Sustain. Cities Soc.* **2020**, *64*, 102561. [CrossRef]

30. Tokbolat, S.; Nazipov, F.; Kim, J.R.; Karaca, F. Evaluation of the Environmental Performance of Residential Building Envelope Components. *Energies* **2020**, *13*, 174. [CrossRef]

31. Finch, G.; Marriage, G.; Pelosi, A.; Gjerde, M. Circular economy optimised energy efficient building skins for residential construction in New Zealand. In Proceedings of the Advanced Building Skins, Bern, Switzerland, 28–29 October 2019.

32. Piras, G.; Pennacchia, E. *Materiali e Component per l'Efficienza Energetica Degli Edifici*; Legislazione Tecnica: Rome, Italy, 2018; pp. 71–72.

33. Majidi Nezhad, M.; Groppi, D.; Marzialetti, P.; Piras, G.; Laneve, G. Mapping sea water surface in Persian Gulf, oil spill detection using Sentinal-1 Images. In Proceedings of the 4th World Congress on New Technologies (NewTech'18), Madrid, Spain, 18–20 August 2018.

34. Cumo, F.; Curreli, F.R.; Pennacchia, E.; Piras, G.; Roversi, R. Enhancing the urban quality of life: A case study of a coastal city in the metropolitan area of Rome. *WIT Trans. Built Environ.* **2017**, *170*, 127–137.

35. Wang, H.; XiaochunLuo, Y. Integration of BIM and GIS in sustainable built environment: A review and bibliometric analysis. *Autom. Constr.* **2019**, *103*, 41–52. [CrossRef]

36. Agostinelli, S.; Cumo, F.; Guidi, G.; Tomazzoli, C. The Potential of Digital Twin Model Integrated with Artificial Intelligence Systems. In *Proceedings of the 2020 IEEE International Conference on Environment and Electrical Engineering and 2020 IEEE Industrial and Com-mercial Power Systems Europe, EEEIC/I and CPS Europe 2020, Madrid, Spain, 9–12 June 2020*; Institute of Electrical and Electronics Engineers Inc.: Piscataway Township, NJ, USA, 2020.

Review

Biometric Data as Real-Time Measure of Physiological Reactions to Environmental Stimuli in the Built Environment

Sandra G. L. Persiani *, Bilge Kobas, Sebastian Clark Koth and Thomas Auer

Chair of Building Technology and Climate Responsive Design, Department of Architecture, Technical University of Munich, 80333 München, Germany; bilge.kobas@tum.de (B.K.); sebastian.koth@tum.de (S.C.K.); thomas.auer@tum.de (T.A.)
* Correspondence: sandra.persiani@tum.de

Abstract: The physiological and cognitive effects of environmental stimuli from the built environment on humans have been studied for more than a century, over short time frames in terms of comfort, and over long-time frames in terms of health and wellbeing. The strong interdependence of objective and subjective factors in these fields of study has traditionally involved the necessity to rely on a number of qualitative sources of information, as self-report variables, which however, raise criticisms concerning their reliability and precision. Recent advancements in sensing technology and data processing methodologies have strongly contributed towards a renewed interest in biometric data as a potential high-precision tool to study the physiological effects of selected stimuli on humans using more objective and real-time measures. Within this context, this review reports on a broader spectrum of available and advanced biosensing techniques used in the fields of building engineering, human physiology, neurology, and psychology. The interaction and interdependence between (i) indoor environmental parameters and (ii) biosignals identifying human physiological response to the environmental stressors are systematically explored. Online databases ScienceDirect, Scopus, MDPI and ResearchGate were scanned to gather all relevant publications in the last 20 years, identifying and listing tools and methods of biometric data collection, assessing the potentials and drawbacks of the most relevant techniques. The review aims to support the introduction of biomedical signals as a tool for understanding the physiological aspects of indoor comfort in the view of achieving an improved balance between human resilience and building resilience, addressing human indoor health as well as energetic and environmental building performance.

Keywords: biometric data; biosignals; non-intrusive sensing; physiological metrics; environmental stimuli; stress detection; health; comfort

Citation: Persiani, S.G.L.; Kobas, B.; Koth, S.C.; Auer, T. Biometric Data as Real-Time Measure of Physiological Reactions to Environmental Stimuli in the Built Environment. *Energies* **2021**, *14*, 232. https://doi.org/10.3390/en14010232

Received: 1 December 2020
Accepted: 28 December 2020
Published: 4 January 2021

Publisher's Note: MDPI stays neutral with regard to jurisdictional claims in published maps and institutional affiliations.

1. Introduction

Urban living is on the rise. The majority of the population worldwide lives in urban areas and expectations are that the number of global urban inhabitants will grow up to two thirds in the next 30 years [1]. At the same time, research shows that people also spend a great majority of their time in closed environments made by humans; over the past few decades people on average spent between 90% and 98% of their time indoors depending on season and holiday time [2–5]. These numbers can also be expected to rise further during crisis events forcing people indoors—be it heat waves, hurricanes, or global pandemics.

The progressive detachment of humans from the natural world has proven to have a series of negative impacts on users ranging from low productivity [6], disruption of the circadian rhythm [7,8] and the hormonal system [9], resulting in diseases such as sleep disorders, immune system disorders, macular degeneration, cardiovascular diseases, diabetes and osteoporosis to cancer [10]. Furthermore, as we increasingly become an "indoor species", the indoor built environment is known to pose serious health hazard risks, commonly known as sick building syndrome (SBS) or "building-related illnesses" (BRI) [11,12]. They derive from a lack of good design and proper maintenance and the use of new hazardous chemicals in building materials as well as in furnishings and consumer products [9]. Being largely undocumented, exposure levels to the chemicals in everyday life should seemingly increase with people spending more time indoors and air exchange rates decreasing as buildings are made more airtight to improve energy efficiency [9,12].

Achieving healthier indoor conditions also becomes relevant from the perspective of energy and environmental building efficiency, if the bigger picture of overall sustainability is considered. Not only have researchers claimed that improved indoor conditions affecting user productivity can end up with massive savings on a company's operational costs [13,14], but the overall energy use has been clearly described as a consequence of trying to attain comfort (in terms of homeostasis) [15].

Therefore, we urgently need to improve our understanding of short- and long-term impacts of the indoor environments on us, in order to not only design energy and resource efficient spaces, but also ones that improve the health and wellbeing of people [16].

1.1. Measuring Human Wellbeing and Health in Indoor Environments

Wellbeing and health are naturally related concepts. While these can be considered to consistently impact each other, they are neither synonyms nor consequent [17]. The World Health Organization (WHO) defines health as a "state of complete physical, mental and social wellbeing", in other words, a state that goes beyond the mere absence of disease [18]. While "wellbeing" is an easy concept to grasp when contextualized, it remains hard to define and predict [19]. It is widely used across a great number of disciplines, setting focus on different aspects such as physical health (absence of disease) and comfort, psychological aspects (individual internal generation of meaning and sense of self), ethical aspects, social aspects (comparison with others as a means of assessing one's own abilities and functioning), economical aspects (standard of living, the real opportunities available to individuals), etc. [19,20]. In many cases, wellbeing, wellness [21] or also subjective wellbeing (SWB) [20] are interchangeably used and measured in terms of "happiness" [19,20,22] or "life satisfaction" [20]. These concepts are not only very broad but are also to a big part based on subjective judgments of satisfaction, as well as individual aims and values, which can change over time and between contexts. This creates critical challenges in terms of measurement and comparison due to differences in type and magnitudes of the chosen parameters and experimental conditions along with notorious problems of information bias even in studies with good reliability and validity [22].

Needing more objective parameters to evaluate the quality and efficacy of wellbeing conditions and interventions, SWB has been further broken down into concepts that include more measurable and objective factors [20,22].

In general, the human reactions that can be observed in response to environmental agents can be classified into three kinds: (i) behavioral, (ii) physiological and (iii) psychological [23]. Behavioral aspects are coordinated responses of individuals or groups to internal and/or external stimuli [24] including objectively observable activities and nonconscious processes (habits) [25]. These actions, reactions, or mannerisms (habitual gestures or way of speaking or behaving [26]) mostly refer—in the context of building physics—to physical (body) actions. Physiological aspects refer to the autonomic responses of the bodily parts of an organism to a stimulus, whether as a form of homeostasis, withstanding changes in environmental conditions that are outside their optimal range [27], or to trigger a behavioral reaction in response to an immediate threat [28]. The rising availability of sensing technologies and data processing methodologies has turned biometric sensors into interesting high-precision tools to study the physiological effects of environmental stimuli on humans [29]. Psychological aspects are functional processes, operations, and changes that relate to the mental and emotional state of an individual (or a group of individuals in the case of group dynamics). Depending on what is measured and, on the methods used, different combinations of these three aspects are taken into account.

Which parameters and methods are appropriate to reliably measure the effect of environmental conditions on users?

The answer to the question is inevitably complex as it depends, among many factors, on the timeframe taken into account and the overall aim of the study. While health and psychologically oriented studies tend to focus on measuring outcomes as individual stress or satisfaction [30–34], economically oriented research tends to use more performance-based criteria focusing on productivity [35,36].

For what concerns the built environment, the significance of the environmental context and conditions in determining human health has mainly been the object of environmental psychology studies [37,38], healthcare environment design [15,38–41], and studies on sick building syndrome [15]. Research in the field of building physics on the other hand has largely focused on indoor comfort [15,16,30,42–51] and user satisfaction [30,32] as a measure of the indoor environmental conditions, and in some cases also on productivity and performance [2,14,35,52–54]. These methods follow, at heart, radically different visions of human wellbeing and/or health, and can be grouped into three major categories:

- Self-selection metrics include preference, acceptance, satisfaction and comfort;
- Performance-based metrics include attention, distractibility, productivity and mental workload;
- Physiological metrics include discomfort and stress.

1.2. Self-Selection Metrics

Self-selection refers to the act of individually opting "in or out of something (such as a group, activity, or category) in accordance with one's personality, interests, etc." [14]. The term is used in economic and social statistics referring to a type of selection bias that can arise when following rules of non-probability sampling, as decisions deriving from self-selection are the object of study [55]. By distorting the selection rules, self-selection makes determination of causation more difficult, resulting in influencing the data.

The term "self-selection" is generally not used in the field of building physics, although three main methods used to measure the quality of indoor environments (preference, satisfaction, and comfort) are overtly known to be highly subjective [22,50,56–58]. Experiments in the field of architecture largely rely on soft data assessed by questioning test subjects about their preferences, which is the case in a great majority of thermal comfort studies [6,16,30,42,50,59–62].

1.2.1. Preference, Acceptance and Satisfaction

Preference involves a choice between two or more alternatives. In preference testing, users or consumers are typically given a choice and asked to indicate their most liked option [63]. Acceptance (or liking) involves rating a specific option or aspect on a scale and can also be achieved without the need of comparing the solution with any other one [63]. What fundamentally connects preference and acceptance as measures is that none of these concepts reflect an ideal solution (the best possible), but at most an optimized solution. Both terms are largely used in business-oriented fields for consumer testing. Research in the context of the built environment also employs these terms [30] to estimate the quality of indoor environments.

Satisfaction is a more complex type of testing as it involves, from the side of the test subject, the rating of a specific option against one's individual scale of values and expectations. As compared to preference and acceptance, satisfaction does in fact reflect a positive assessment. Despite the innate complexity of the valuation method, satisfaction is a well-established metric for assessing quality in the indoor environments [32,57,64]. In building engineering, an indoor environment is generally considered acceptable when 80% of the occupants are satisfied with the conditions [58]. Occupants' environmental satisfaction is directly related to the amount of perceived comfort [32,64], and both of these aspects sequentially impact user performance [14].

All three concepts are often, and sometimes interchangeably, used in indoor environmental research to describe user feedback and assess environmental conditions. While part of the assessment truly depends on the physical conditions of the context, a large part of the estimation is determined by individual psychological conditions.

1.2.2. Comfort

Comfort is defined as the "absence of unpleasant sensations" or as described within the most widespread definition in the field of building physics "the condition of mind that expresses satisfaction with the (thermal) environment and is assessed by *subjective* evaluation" [58]. It is therefore, by definition, a lack of discomfort: a neutral state with no stress. From a biological perspective, comfort can be effectively compared to the maintenance of homeostasis, indicating the absence of environmental stressors [15]. Comfort is broadly recognized as a multidimensional and subjective construct that varies across contexts [15] and depends on functional and environmental factors, on personal health and mood and is recognized as being highly subjective [14,15,47,65].

A large majority of comfort studies are based on the unvoiced assumption that physical health is a direct consequence of comfort. The gap existing between the definition of indoor comfort conditions and long-term health conditions however becomes apparent when looking at studies showing correlations between time spent in thermoneutral environments and the likeness of developing obesity [66–69]. Similarly, exposure to conditions outside the thermoneutral zone have shown to reduce the susceptibility to developing type 2 diabetes [70].

Measuring comfort is a difficult task. Comfort studies rely largely on self-reports and other types of user-feedback methods [6,16,30,42,50,59–62]. Though subjective measurement remains a major research method in many disciplines, the reliability of the results is contested by many authors [56]. If on one hand self-reports allow taking into account the mental state and mood of users, the data also involve a large risk of information bias [22,71]. The feedback is not only strongly influenced by the individual character of the perceived conditions, which can be minimized if analyzed over larger numbers of individuals, but psychological components are also known to consistently alter users' perceptions and thermal expectations of an environment [50,56]. Moreover, the data typically lack precision in timing, are generally not scalable and are usually difficult to reproduce [56]. Moreover, measuring comfort involves the collection of processed and combined data, such as physiological factors (thermoregulation), potentially random or non-repeating variables such as climatic behavior (subjects turn on the desk fan, wear another piece of clothing, etc.), and psychological or other highly individual factors (emotional stress levels, hormonal levels, etc.). Although current practice statistically normalizes individual differences in the datasets by collecting data on a large scale, it also involves significant physiological aspects not being easily parsed from the datasets.

1.2.3. Considerations on Self-Selection Metrics

Investigating the *perceived* indoor environmental conditions is on one hand an important task, as it takes the objective physical measures of environmental data into account, as well as the psychological mindset of the occupants [72]. However, from a health perspective, the importance of users' perception is relative for two main reasons:

1. Indoor comfort and indoor health are related concepts but refer to two essentially different timeframes. While comfort is perceived in an immediate and narrowly defined point in time, potentially changing within hours, minutes or seconds, health is the result of a multitude of actions happening over a much longer lapse of time, and can as such not be assessed by users in the present time but only retrospectively;
2. The use of self-selection measures, such as preference, satisfaction and to some degree also comfort, to determine the quality of an environment (often implying also health), is essentially driven by the idea that users are able to distinguish the conditions that are positive from those that are detrimental to their health. This is however not always the case as even in the cases where the long-term health effects of a specific action are

known, the choice between an immediate gratification (e.g., smoking) and a delayed gratification (health) is in psychological terms not always obvious [73].

In all self-selection metrics described, issues concerning potential systematic data bias have been raised. Surveys, questionnaires, and other self-reporting systems can also be argued as to discard neutrality as the subjects are made temporarily consciously aware of their surroundings when they are questioned about them. Additionally, these highly individual pieces of information pose problems with replicability, scalability and comparability of the said data.

1.3. Performance-Based Metrics

Performance-based metrics are most often aimed at measuring the efficiency of a system in economic terms, whether the outcome is expressed as environmental, financial or process excellence. Such metrics are effective to express usability and to inform key decisions [74] as they reduce complex measurements and result in a single value that can be tracked, managed, and improved. While complexity is reduced, looking only at the outcome of a network of interacting factors, it is important to keep in mind that these shortcuts can also become misleading when used for process improvement.

In the context of the built environment, performance-based metrics are used to express functionality in domains such as energy performance, indoor environmental quality, environmental impact, capital, and operating costs [75–77]. Few studies have focused on performance metrics in the occupant domain, and even those mostly focus on the built environment's qualities rather than on the effect on users [75].

Typical user-based performance metrics measure task success (binary success or level-based success), time on task (completion of a task within a time limit), error-based measure (single or multiple error opportunities), efficiency (number of actions required to complete a task, ratio of task success rate to the average time per task) and learnability (how any efficiency metric changes over time) [2,14,33,35,36,38,74,78,79].

1.3.1. Productivity

A number of studies have attempted to measure the effect of indoor environmental conditions by evaluating user performance in terms of *productivity*, mostly focusing on the thermal aspect [2,14,51,78] and on indoor air quality [14,53,78]. The underlying concept is that as environmental conditions exceed the range of comfort, the human body adapts to sustain the level of task performance. This is achieved through homeostatic adaptation on one hand and through psychological adaptive behavior (as attentional focus) on the other [51]. If the environmental conditions exceed the body's maximum range of adaptability, the attentional resources are depleted, and human performance ultimately deteriorates [51]. Assessing productivity can be achieved either using (i) subjective approaches, with self-assessing methods to rate perceived performance over a specific period of time [2]; (ii) quantifying productivity by defining a context- and content-specific ratio of output to input (ratio of company turnover to employee cost [80], individual/team performance [35,36,81], etc.) and gathering data in a real environment [14]; (iii) evaluating performance based on specifically designed tests (neurobehavioral tasks such as logical/comprehension/numerical/visual/memory etc. [82] or perceptual motor tasks [83]). The main drawbacks of assessing indoor environmental conditions by measuring productivity come from the environmental effects not being directly reflected on the subjects' task performance [84], due to partial adaptation to the stressing agents [51] and individual motivation [85], which can offset the effects on task performance. In fact, productivity is known to be influenced by many factors, among which are personal, social, organizational, and environmental factors [14].

1.3.2. Attention and Distraction

Another performance measure used is *selective attention* and its counterpart, distraction. To overcome its limited processing resources and reduce the amount of sensory information it needs to comprehend [86], the brain uses various filtering mechanisms. It therefore constantly oscillates between states of deep focus—as we momentarily disengage from the real world when we are highly concentrated on a task [87]—and awareness of our surroundings—when we are distracted by environmental stimuli [86]. The oscillations between bursts of attention can therefore be closely linked to our behavior [88], providing indications on the effects the surrounding environments have on us. Specifically, increased cognitive engagement (attention) is known to produce decreased sensitivity to visual events [87] with slower reaction times, decreased situation awareness [89], and specific eye movements such as spontaneously closing the eyes or looking away [90]. The neural oscillations leading to selective attention, alternating periods of either heightened or diminished perceptual sensitivity, have been studied through human behavioral studies [88,91,92], but also using physiological measures as electroencephalography (EEG) [88] and oculomotor capture [87]. Researching the impact of selective attention can be a complex task, as can be seen by the multiple attempts to collect and interpret data from different sources:

- Detection of foreseeable behavior and movements in the test subjects: detection of head and neck movement using a series of sensors [93,94], detection of eye-movement data [95], gait analysis [96,97];
- Detection of specific biosignals, such as measuring cortical activity related to externally cued auditory events using event-related potentials (using EEG, electrooculogram (EOG) and electromyogram (EMG) signal extraction) [98–100] or using other measures as oculomotor control [101,102].

1.3.3. Mental Workload

Similarly to selective attention, *mental workload* or mental fatigue, defined as "the mental resources devoted to the tasks of an individual" [103], is used to measure a lack in performance typically characterized by higher error rates, decreasing efficiency and alertness, and effort disinclination [104], resulting in some cases also in detrimental health effects on long timeframes [105,106]. While measurements of productivity are limited and offset by individual motivation and ability to maintain the performance, mental workload can be assessed using physiological measures, such as electroencephalography (EEG) on subjects while being asked to perform tasks specifically designed to activate typical cognitive functions [2]. Other commonly used measures of physiological metrics include event-related potentials, magnetoencephalography, positron emission tomography, electrooculograms, cardiovascular measures, pupillometry, respiratory measures, and electrodermal measures [107].

1.3.4. Considerations on Performance-Based Metrics

Summarizing the reviewed types of performance-based metrics, only selective attention and mental workload appear to enable an objective measurement of the effect of the surrounding environment on users through the use of physiological measurements. These systems measuring cortical activity are on one hand accurate, providing a good option as a quantitative approach but present some intrinsic limitations: (i) datasets are very complex to read, and in some cases (as selective attention) do not provide a specific EEG signature; (ii) datasets can potentially be affected by many other factors [2]; (iii) the potentially high number of false positives due to interference from other body functions and movements. Reliability of the results can be theoretically achieved by adopting a multimodal approach [89], combining two or more physiological detection methods.

As far as this review could find, these methods are commonly used in the field of neuroscience [108], psychiatry [109], computer science and biomedical informatics [110], as well as architecture, computing and engineering [89], but very few attempts have been undertaken in the field of building physics [2,111].

As discussed in the beginning of the paper, another aspect that part of building physics studies focuses on, apart from comfort (or *sensation* and *preference*), is mental workload and productivity. While both terms have a similar individual complexity, they also have a similar timeframe: both comfort and productivity are momentary states. However, these are also some of the most important factors in building operations and have a direct impact on the economy and use of resources. This argument is significant on a strategic level. While comfort studies prioritize optimizing building energy use and user satisfaction, and studies on productivity bring a more direct economic impact; both constellations forgo the health implications as they are not visible at the same timescales.

1.4. Physiological Metrics

Physiological aspects refer to the autonomic responses of the bodily parts of an organism to a stimulus, to withstand changes in environmental conditions that are outside their optimal range [27], or to trigger a behavioral reaction in response to a threat [28].

1.4.1. Discomfort

As previously discussed, comfort perception is the main key measure used in building physics to assess the quality of indoor environments, although the concept is known to be strongly impacted by psychological factors and the data are potentially biased. The sensation of *discomfort* on the other hand is the complementary entity of comfort [112] and is more broadly researched in fields such as ergonomics [112,113], medicine [114–116] and psychology [33,117,118]. Discomfort is described as a mental or physical uneasiness or annoyance [119], resulting in a natural response of avoidance or reduction in the source of the discomfort [114]. While in many studies physical discomfort is used as a synonym of pain [33,116], not every discomfort can be attributed to pain [114,115]. In most research, discomfort is identified and measured using self-report [112], assessing the severity, frequency, and duration of work-related body-part discomfort [113,116], or the observation of physiological and behavioral outcomes [114,116]. Generally, literature agrees on the subjective nature of the sensation of discomfort [114,116–118]. In physiological terms, there are no receptors to measure comfort (which is defined as lack of discomfort), while discomfort does in fact leave a mark.

1.4.2. Stress

Stress is the prototypical response to an internal or external event, force, or condition (the stressor), triggering a cascade of processes to help the body to adapt [25]. The effects are behavioral, physiological, and psychological [120], testifying how the stress experience relates both to the objective perception and the subjective evaluation of an event [56]. Physiological stress response is largely involuntary and is mediated by the autonomic nervous system (ANS) [56] as neurotransmitters and hormones signal action to the body [121]. Examples of the physiological stress response are palpitations, sweating, dry mouth, shortness of breath, fidgeting, accelerated speech, augmentation of negative emotions (if already being experienced), and longer duration of stress fatigue [25]. Although the concept is often negatively connotated, under certain circumstances, stress can also have positive adaptive effects such as arousal [56,122] and enhancing immune function (e.g., preparing the immune system for challenges such as wounds or infection) [121]. The impact on health and performance has also been shown to vary depending on the person's individual mindset regarding the stress conditions [123]. Damaging, long-term consequences of stress are essentially encountered as the reaction becomes a chronic condition [29,124–127]. Stress has been suggested to have four main components impacting health [128]: exposure, reactivity, recovery, and restoration. (i) The number of stressors experienced (exposure); (ii) the strength of the individual physiological reaction (reactivity); (iii) time of recovery from the stress reaction (recovery); (iv) healing processes of the organism (restoration), which might be hindered by the stressful condition.

In other words, stress is the physiological reaction of the body to a necessity of adaptation, triggering the body's homeostatic functions to regulate (temperature, heart activity, blood pressure, respiration, and glucose levels) [56,129,130]. As long as the range of adaptation required is within the capacity of the organism [131], the dominant perception is of comfort. If the stimuli exceed the capacity of adaptation, the resulting sensation is of discomfort. Stress can therefore become an interesting measure in the field of building physics as a source indicator for physiological change in conditions. Especially as, contrary to the case of mental workload, stress has many known biophysical signatures, allowing for better data precision and reliability [89].

1.4.3. Considerations on Physiological Metrics

Physiological metrics appear to have a good potential to sustain the current mainstream data collection methods in the field of building engineering, potentially offering reliable and objective methods to measure environmental stressors. Specifically, recent advancements in sensing technology and time-sensitive biomedical data acquisition are opening up new possibilities to use biomedical signals as indicators of quality. The promise of parsing physiologically relevant data is that the impacts of additional variables, such as adaptive behavior, tolerance, psychology, exposure time, can be studied against a constant, in a situation where almost all other parameters vary.

1.5. Research Gap

More objective means are needed in the field of building physics to measure the effects of indoor environments on users as the mainstream parameters of indoor environmental quality, comfort, and discomfort are widely accepted as subjective sensations and data collection methods are prone to be biased. Following a review of methods to measure human wellbeing and health, the use of physiological signals to detect stress is seen as a promising alternative.

The use of these methods is however relatively new to the field of building and climate engineering, where solid theoretical foundations need to be established before being systematically used. While biosignal patterns are regularly researched for psychological, social, and mental stress conditions, comparatively few attempts were found to be made in the field of building physics to investigate the patterns of individual biosignal features triggered by stress factors originating from the built environment.

Within this context, this review reports on a broader spectrum of available and advanced biosensing techniques used in the fields of building engineering, human physiology, neurology, and psychology. The aim of this review is to offer a systematic and comprehensive insight on the current capacity to detect stress from the built environment using biosignal measures. Emphasis is put on the efficiency, robustness, and consistency of biosignals as a data source. Specifically, the paper will:

1. Review the current state of knowledge in neighboring fields of study that can be of use in the field of building physics;
2. Review the current parameters and measures used to describe indoor environments;
3. List and assess relevant biosignals highlighting their effectiveness and reliability with regard to stress detection;
4. Establish the relationship between the multitude of biosignal features and their corresponding behavior under different environmental conditions;
5. Establish reliable biosignal (and multimodal biosignal) indices that reveal the underlying physiological mechanisms of the stress response;
6. Discuss existing limits and solutions of the methods reviewed.

2. Materials and Methods

The paper explores the interaction and interdependence between (i) indoor environmental parameters affecting humans and (ii) human biosignals that respond to the environmental stressors by activating a physiological response. As such, the review was developed in subsequent phases of research and analysis. The topic is addressed from a holistic point of view, defining the relevant features through a qualitative and multidisciplinary research approach. State-of-art knowledge in the relevant fields was reviewed by searching online available databases: ScienceDirect, Scopus, MDPI and ResearchGate and through the university libraries of the authors' home university.

A total of 246 sources were reviewed throughout this paper. A large majority of the publications, 82.5% of the total amount, were published after 2000, and 93.1% were published after 1990. Figure 1 shows a histogram based on publication years.

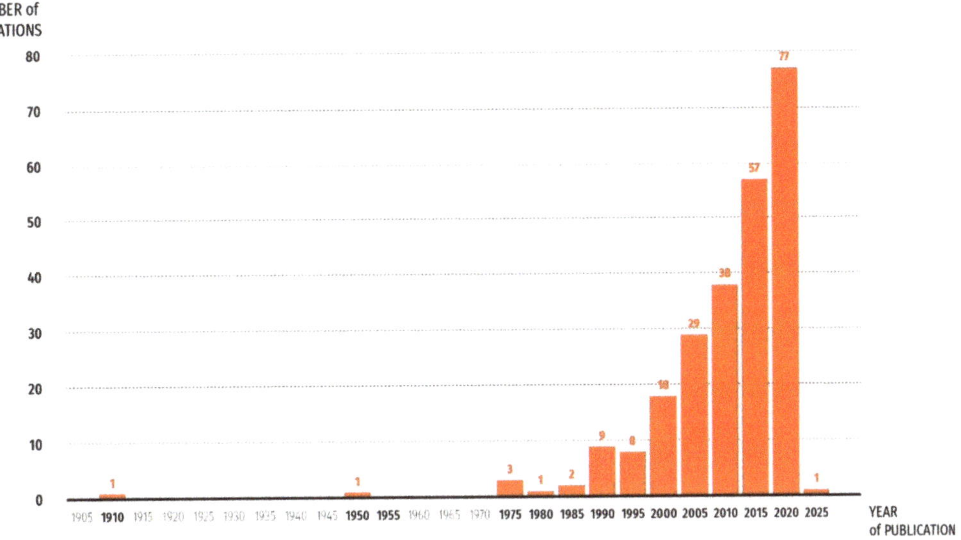

Figure 1. Histogram of publication years of sources.

Out of these 246 sources, journal papers seem to be the absolute majority when looking at the types of the publications. Figure 2 shows the breakdown of the entire bibliography based on the publication types.

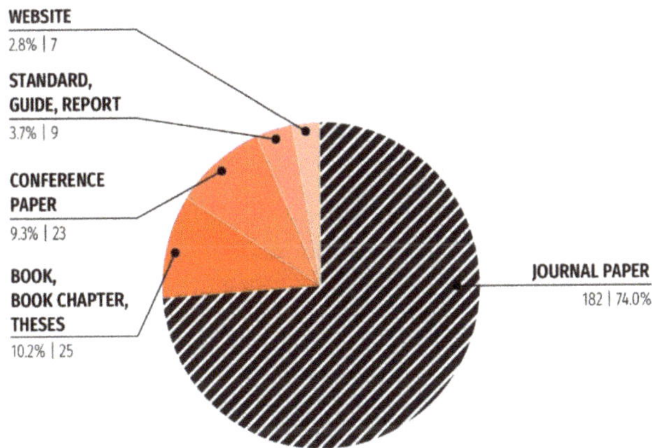

Figure 2. Breakdown of publication types.

Due to the interdisciplinary nature of the research topic and aiming to achieve a broad overview of the methods used to measure human health and wellbeing, the research reviewed appears to be extremely broad and diverse. Table A1 categorizes the publications based on their authors' original research fields and Figure 3 illustrates their distribution. As can be seen in both Table A1 (please see Appendix A) and Figure 3, nearly half of the reviewed material comes from medical fields (general medicine, neuroscience, biomedical engineering, biology, psychology) and 31% originates from fields related to built environments (architecture, building engineering, building physics). The third prominent discipline seems to be computer sciences with 11%, including relevant sensing and data acquisition technologies or data analytics.

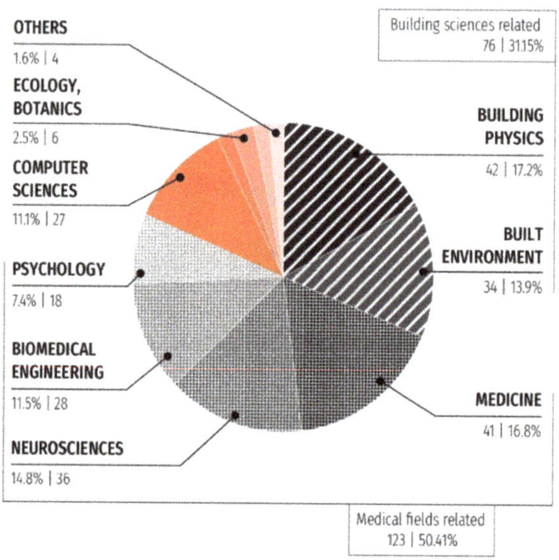

Figure 3. Distribution of main research fields of authors.

2.1. Methods for Measuring Human Wellbeing and Health in Indoor Environments

For the first step, the available methods commonly used, according to literature, to measure human wellbeing and health in the indoor environment were briefly summarized. The databases were searched with the keywords "health, well-being, indoor environment, self-selection, assessment, questionnaire, survey, satisfaction, performance, comfort, productivity, attention, mental workload, fatigue, task performance, cognitive performance, physiological, discomfort, stress, biosignal".

For this part, 157 sources were reviewed in total: 10 books, book chapters and theses, 18 conference papers, 121 journal papers, 4 standards, guides or reports and 4 websites. The publication dates range from 1950 to 2021, with 130 of the publications dated after 2000.

The majority of the research came from the fields of built environment (28/157) and indoor environment (building physics and engineering, 24/157) as well as medicine (27/157). Table A2 shows the sources used in this chapter based on the authors' main research fields in further detail (please see Appendix A).

Table A3 provides a more specific grouping for the publications contributing to the review and discussion of definitions of human aspects. "Stress" comes forward as the most repeated term amongst these publications, mostly with medical origins (25 publications out of 38) and computer sciences and engineering background (12 out of 38), with no occurrences in built environment or building engineering originated publications. However, the term "comfort" is the most common in these two fields (with 23 out of 36), with only 6 from a medical background. Finally, "health" comes forward as a shared term, as it appears in the publications from a medical background 21 times out of 36, and 14 times in the sources with a built environment background.

2.2. State of Art: Indoor Environmental Quality (IEQ) Parameters

For the second step, indoor environmental parameters typically used in the field of building physics were summarized in order to give a structured overview of the aspects that are usually taken into consideration in the design and environmental regulation of indoor spaces. The databases were searched using different combinations of keywords, such as "indoor environmental quality, IEQ, air quality, IAQ, thermal IEQ, thermal comfort, visual IEQ, visual comfort, visual health, acoustic IEQ, acoustic comfort, acoustic health, indoor health, comfort, occupant comfort, user comfort". The literature available on IEQ parameters appears to be very broad as the concepts are well-established in the field of building physics. This is the reason why the literature search was focused on identifying relevant reviews, proceeding in a second step to find specific articles that focused on single aspects of interest.

As a result, 90 sources were reviewed in total, amongst which 50 of these sources were also used in the previous summary of measures for human health and wellbeing. The search includes recent as well as well-established sources on different indoor environmental condition theories between the years 1973 and 2021, and 75 of these were published after 2000. As can be seen in further detail in Table A4, the majority of publications (72 out of 90) are journal papers (please see Appendix A).

The majority of the reviewed material came from the field of building related research and building physics (in total 58 out of 90). Table A5 shows a catalogue of IEQ-related publications reviewed in this paper, based on their authors' main research fields.

Looking at the cross relation of human factors and IEQ parameters, it seems that "health" and "comfort" come forward as frequent terms used in reviewed IEQ research, followed by "productivity", "performance, mental workload". However, comparing Table A6 to Table A3, where the sources on human aspects were categorized based on their focus, it was revealed that the literature on "stress" was the most prominent. However, in Table A6, it becomes clear that research on stress in relation to IEQ is not equally prominent; with "health" (33/90) and "comfort" (25/90) having higher priorities amongst reviewed publications on IEQ, while "stress" being found as the focus point of a single publication in this category.

Looking at the distribution of subcategories of IEQ parameters, thermal IEQ research presents itself as most prevalent, with 26 publications out of 90, while acoustic IEQ research comes last with 9 publications. Table A7 provides an overview of sources based on their relative IEQ parameter focuses.

Aspects that are relevant in terms of human health and comfort or otherwise useful for further bridging with biosensing techniques were identified and summarized. More specifically, all IEQ parameters were discussed in terms of:

- Threats to human health and wellbeing;
- Variables and sub-parameters;
- Solutions and strategies to improve the specific IEQ parameters and achieve healthy ranges for the indoor conditions;
- Parameter measurements, units in use, methods of measurement, limits to the methods.

2.3. Background Research on Physiological Signatures

For the third step, an extensive investigation of published studies on available biosensing techniques, used to measure human stress response, was performed. Broad research criteria were initially adopted in order to take into consideration a larger pool of potential biosignals. Databases were therefore scanned using different combinations of a number of keywords, among which were "biosignals, biologic signals, biomedical signals, biosensors, neurophysiology, psychophysiology, electroencephalogram (EEG), electrocardiogram (ECG), heart rate (HR), electrodermal activity (EDA), galvanic skin response (GSR), physiology of comfort, physiology of stress, stress recognition, mental workload, stress hormones", and the research was expanded to the sources cited by the selected literature. The search was subsequently narrowed down to more recent publications, focusing on research and technologies achieved within the last 20 years (with the exception of publications that were of specific relevance).

As a result, a total of 110 publications were reviewed; the majority (78/110) being journal papers. Table A8 shows the sources based on their publication types.

When checked against the authors' main research fields (Table A9), it was clear that medical fields take the lead, followed by computer sciences and engineering in sensing and data acquisition technologies. Amongst the reviewed sources, research from built environment or building physics seems to have weak direct correlation to physiologic or biologic methods in terms of human aspects (health, comfort, wellbeing) in the context of indoor environmental quality.

2.4. Selection of Biosensing Techniques

Relevant biosensing techniques were subsequently selected from the larger pool of physiological signatures based on their relevance in the context of measuring potential indoor environmental stress sources.

Originally, the list of biosignals retrieved from the literature review was as shown below (see Figure 4):

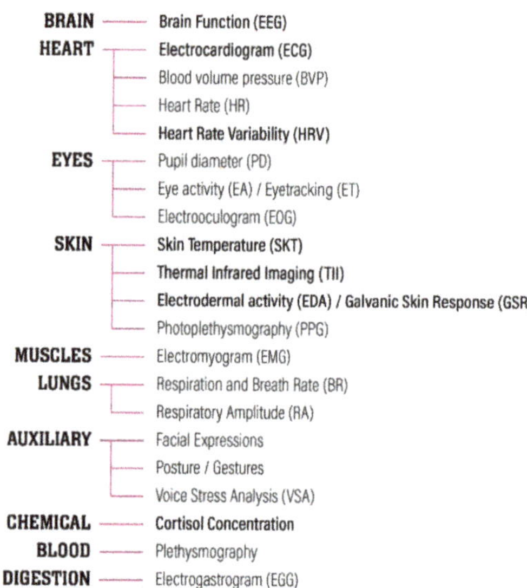

Figure 4. List of biosignals gathered from the literature review. Biosignals marked as bold are examined in this paper.

Selected biosensing techniques were then organized according to hierarchy and relevance following their prevalence in stress literature, but more importantly in building engineering stress literature, efficacy and accuracy of the results, and robustness of the system. The selected biosensing techniques were then examined in means of their physiological processes and relevance to the stress cues as investigated in this paper.

Finally, limits and potentials of these techniques described in the literature were further mentioned and discussed for all selected biosignals.

3. Results

3.1. Indoor Environmental Parameters

The factors influencing an individual's perception of the environment of a built space include a variety of aspects characterizing on one hand the physical setting itself and on the other the person's reactions to the context. The parameters in the physical space that can trigger a (conscious or subconscious) reaction in the users essentially include (i) environmental (air quality, climate, noise, light), (ii) functional/spatial (disturbances, interruptions, distance from work, resources, plan, layout, ergonomics) and (iii) psychological factors (privacy, territoriality, aesthetics) [35,36]. This review focuses on the factors pertaining to the physical environment, and as such only the aspects under point (i) are further explored.

The physical parameters and conditions of an indoor space are mostly straightforward and objective. These can be measured according to state-of-the-art procedures and with broadly accepted units. Ranges of acceptability for human health conditions are set through national laws and standards, although in most domains research still does not agree with precision on the ranges of acceptability and healthiness [16,47]. A common way to describe the physical indoor conditions is through the assessment of the indoor environmental quality (IEQ). This system is adopted in many green building systems—the United States Green Building Council (USGBC), the Leadership in Energy and Environmental Design (LEED) and the Comprehensive Assessment System for Building Environmental Efficiency (CASBEE) [14].

Indoor environmental quality (IEQ) describes the environmental conditions inside a building in relation to the health and wellbeing of its occupants and is determined through air quality, thermal, visual, and acoustic parameters [2,132,133]. In comparison to indoor air quality (IAQ) and thermal IEQ, both visual and acoustic IEQ have received way less attention [43], as can also be seen by the number of sources reviewed in each section (see Table A7).

3.1.1. Indoor Air Quality (IAQ)

Variables influencing indoor air quality (IAQ) include airborne contaminants from indoor and outdoor sources (gases and particles from equipment and furnishings, cleaning products, building materials, pollutants, etc.), ventilation rate, humidity and the type of indoor activities and the occupants' adaptive opportunities (manually intervening to adjust the conditions) [133]. Perceived IAQ is an umbrella of reported descriptors such as odor/smell, stuffy air, and dry and wet (humid) air [134].

The quality of indoor air has a direct impact on users' health and comfort [21,42,135]. Poor IAQ can directly cause severe respiratory and cardiovascular diseases, allergies, asthma, and sick building syndrome (SBS) [11,12,14,21]. As indoor air has been found to be in some cases up to five times as polluted as outdoor air [136] and new buildings tend to be increasingly tightly insulated from the external environment to achieve better energy balance [12,15,21], greater attention needs to be given to achieving good IAQ.

- Airborne Contaminants

Airborne contaminants from indoor and outdoor sources, such as gases and particles from equipment and furnishings, cleaning products, building materials, pollutants, etc., substantially affect IAQ. Common airborne contaminants include volatile organic compounds (VOCs, such as formaldehyde and benzene), microbiological volatile organic compounds (MVOCs), nitrogen oxides (NO and NO2), polycyclic aromatic hydrocarbons (PAHs) and carbon monoxide (CO) [135,137]. Carbon dioxide (CO2) is mostly not considered as a pollutant as its major source (in non-industrial indoor environments) is the human metabolism itself, but rather it is considered as a contribution to lesser air quality [53]. Finally, a great number of chemicals and materials used in everyday environments are also known to impact IAQ. A (non-exhaustive) list includes:

1. Polychlorinated biphenyls (PCBs) used in electrical equipment, caulking, paints and surface coatings;
2. Chlorinated and brominated flame retardants, used in electronics, furniture, and textiles;
3. Pesticides used to control insects, weeds, and other pests in agriculture, lawn maintenance, and the built environment;
4. Phthalates used in vinyl, plastics, fragrances, and other products;
5. Alkylphenols used in detergents, pesticide formulations, and polystyrene plastics;
6. Parabens used to preserve products such as lotions and sunscreens.

As a result of weak regulatory requirements for chemical safety testing, only limited toxicity data are available for many new chemicals that are being developed on a daily basis. Mechanisms of action, adverse effects, and dose–response relationships for many of these chemicals are poorly understood and no systematic screening of common chemicals for endocrine disrupting effects is currently underway, so questions remain as to the health impacts of these exposures [9]. Over the past 15 years, some chemical classes commonly used in building materials, furnishings, and consumer products have been shown to be endocrine disrupting chemicals, meaning they interfere with the action of endogenous hormones [9].

Solutions to contrast the accumulation of these pollutants include raising the ventilation rate, controlling or reducing indoor human activities such as smoking and cooking, and controlling the safety of building materials (paints, preservatives, etc.) and indoor elements (furniture, equipment, cleansers, etc.) [135,138]. Indoor greenery has been found to have positive biofiltering effects taking up toxic agents as benzene, formaldehyde, trichloroethylene [139], particulate matter (PM, fine solid particles suspended in a gas) [33] and CO2 from the air during the photosynthetic process [14,78]. Additionally, studies have suggested that buildings exhibit decreasing tendencies of VOCs in indoor environments over time, which is why old buildings tend to have lower VOC levels than new buildings [14].

Measuring of IAQ is a complex task as it has many potential components [14]. The most common methods for indoor air sampling and analyses are toxic organic (TO) methods, air pollutants in indoor air IP-methods (compendium of methods for the determination of air pollutants in indoor air) and air-phase petroleum hydrocarbon (APH) methods [140]. The emission rate of air pollution is measured in olf (from "olfactory"), where 1 olf is the emission rate of pollutants from one standard person (an average adult working in a non- industrial workplace, sedentary and in thermal comfort with a hygienic standard equivalent to 0.7 baths per day) [141]. Decipol (from "pollution") is used to represent the level of perceived air quality, where 1 decipol is the pollution caused by one standard person (one olf) ventilated by 10 l/s of unpolluted air [141]. The decipol scale ranges from 0.1 decipol (outdoors in a city) to 10 decipol (sick building). [141]. The main difficulties and limits to the measurement and mapping of VOCs are linked to their very broad diversity in physio-chemical properties [14]. This entails varying degrees of sensitivity to the diverse compounds and subjective reactions in individuals, which in turn brings difficulties in developing standard measures for sampling and analysis [14].

- Ventilation Rate

Indoor spaces are ventilated to reduce contaminants in the air and exchange the exhausted indoor air. While higher ventilation rates give better IAQ values [14], they also result in thermal exchanges between indoor and outdoor areas and therefore in higher energy consumption as the air entering needs to be treated to meet the indoor thermal comfort requirements [45]. Researchers however argue that achieving better health and productivity of occupants with higher IAQ levels results in substantial financial returns in comparison to the annual energy and maintenance costs of the building [14].

The ventilation rate is expressed as the outdoor air flow into a building per unit of time divided by the number of people in the building, and thus in liters per second (l/s) per person or per square meter.

- Humidity

Recommended levels of indoor humidity are between 40 and 60% [142]. Higher rates of humidity in indoor spaces have shown to improve sleep quality and reduce effects on the vocal cord on one hand [143], but also to create unhealthy conditions that can bring microbial growth—fungal, bacterial and mold [133,142]—and affect the rate of outgassing of formaldehyde from indoor building materials, the rate of formation of acids and salts from sulfur and nitrogen dioxide, and the rate of formation of ozone [142]. Low humidity on the other hand can favor the survival and transmission of many influenza viruses as well as aggravate the eye tear film stability and physiology, and the osmolarity of the upper airways [143]. "Dry air", which is a very common complaint in indoor spaces [134], is a sensation most likely caused by an exposure to sensory irritants such as indoor air pollutants [48,144]. Literature recommends distinguishing between elevated moisture in construction materials, elevated relative humidity (RH) resulting in condensation on surfaces, and RH in the breathing and ocular zone [143].

Relative humidity (RH) is measured as a percentage of water vapor in the room air, relative to the amount of vapor that the air could contain at a given temperature and is usually measured with a hygrometer. Absolute humidity (AH) measures the water in grams per kg of air at a defined pressure and is used in some cases for comparison and identification of associations, also considering sometimes better correlation between outdoor and indoor AH [143].

3.1.2. Thermal IEQ

Thermal comfort is one of the main factors affecting occupants' perceived environmental satisfaction [2] and IEQ-productivity belief [6], influencing their health and well-being [60,132]. Uncomfortable thermal indoor conditions have been reported to affect cognitive performance and productivity [2,14,51] and to be in extreme cases responsible for sick building flu-like symptoms (eye, nose, and throat irritation) [2]. The range of indoor temperatures deemed acceptable also has a strong bearing on building energy requirements [51]. It is therefore understandable that the opportunity of achieving important cuts in operational CO2 emissions with the energy efficiency upgrade of the existing building stock [145] has driven much research to focus on the thermal aspect.

- Definition

Thermal comfort as we refer to in the field of building physics is traditionally defined through people's psychophysical responses, rather than purely biophysical processes. Thermal comfort is a subjective state of mind where each individual, influenced by physical, physiological and psychological factors, expresses a judgement of satisfaction with the thermal environment [14,47,49]. While thermal comfort is a highly subjective condition, thermal sensation is more objective and is therefore used to describe the human response to thermal comfort [146]. Other indicators used include thermal acceptability (the degree of an occupant's approval of the environment, which is subjective and directly related to the individual's expectation) and thermal preference (the expressed ideal thermal state of the environment) [30]. Variables that influence the comfort sensation are multiple. On one hand are the physical factors defining the thermal state of the environment: mean radiant temperature, relative humidity, air velocity and air temperature [14]. On the other are the factors influencing the human perception and preference towards the thermal conditions: individual factors such as age, gender sex, metabolism rate, mood, etc. [34,147]; dynamic factors such as clothing, activity patterns, posture (sedentary or steady conditions) [58]; and contextual climatic conditions as geographical factors, weather, and time of the year [14].

- Thermal Comfort Models

Achieving optimal indoor thermal comfort conditions is a complex task that has been at the center of scientific debate for about fifty years [50,62]. The outcome has been two approaches for defining thermal comfort.

The "single temperature optimum" model is based on climate chamber data, on the heat balance theory and on thermoregulation physiology [59]. This analytical model defines the predicted mean vote (PMV) index and the predicted percentage of dissatisfied occupants (PPD) through physical parameters (air temperature, mean radiant temperature, air velocity and relative humidity) and human variables (clothing insulation and activity level) [62]. Thermal comfort is therefore expressed around a single optimum temperature for any given combination of comfort parameters [148]. In this approach, any deviation from the thermal optimum anticipates a loss in comfort, performance, and productivity [51].

Biosignals (short for biologic signals or biomedical signals) are signals that are used to extract information on a biologic system to be examined [171]. In biomedical applications, they are a critical part of diagnosis [172]. The signal source can be at the molecular, cell, systemic or organ level. Extraction of a biosignal can go from something as simple as a physician feeling the pulse of a patient to understand the heart rate, to the use of an electrocardiogram or to measure the electrical activity of the heart more precisely and continuously, with the help of electrodes placed on the chest, arms, and legs. Overall, biosignals can be:

1. Spatial (mono-dimensional biosignals, e.g., electrocardiogram (ECG) associated with heart muscle contractions measures heart activity by detecting changes);
2. Temporal (two-dimensional biosignals, e.g., functional magnetic resonance changes associated with blood flow imaging (fMRI, functional magnetic resonance imaging measures brain activity by detecting); or
3. Spatio-temporal (three-dimensional biosignals, e.g., a medical ultrasound movement by detecting changes in the reflection of measured surfaces or internal organs structural sound waves on the tissues) records of a biological event [173].

Biosignals can be categorized in several ways: by their existence (permanent, i.e., EEG, ECG vs. induced, i.e., plethysmography, where an artificial current is induced in the tissue), by the observed time-frame changes (dynamic, i.e., heart rate or static, i.e., core temperature) or by their physical nature [174]. The latest is the most frequent classification method, categorized and elaborated as follows:

1. Bioelectric signals: Bioelectric signals are the most common and well-known biosignals. Bioelectric phenomena have had scientific value for the past 200 years in terms of modern medicine [175]. They convey the electrical activity created by nerve and muscle cells. Well known examples can be listed as electroencephalogram (EEG), electrocardiogram (ECG), electroretinogram (ERG), electrooculogram (EOG), electrogastrogram (EGG), electroneurogram (ENG), electromyogram (EMG), galvanic skin response (GSR) [176].
2. Bioimpedance signals: Bioimpedance signals are useful for estimating body composition through the amount of electric impedance passing through the body. Using bioimpedance signals, parameters can be figured such as: body cell mass, extracellular mass, fat-free mass, fat mass or total body water [177,178].
3. Biomagnetic signals: Several organs produce weak magnetic fields, as a result of their electric activity. For instance, the source for the magnetocardiogram (MCG) or magnetoencephalogram (MEG) is the electric activity of the cardiac muscle or nerve cells, respectively, as it is the source of the electrocardiogram (ECG) and electroencephalogram (EEG) [175].
4. Biomechanical signals: Results from the mechanical functions of the body, such as pressure, tension, motion. Examples can be listed as blood pressure data, human movement data via accelerometer sensors in Parkinson's disease patients, gait, balance and pose (Parkinson's disease, mobile applications, fitness). Biomechanical signals are particularly of interest in sports science, or physical rehabilitation processes [179].
5. Bioacoustic signals: Several physiological activities make noise and can be captured as acoustic data when amplified. Examples are cardiac sounds (phonocardiography) to examine heart valves' closure strength and stiffness, recording snoring in order to investigate sleep apnea, listening to respiratory sounds to detect pulmonary disorders. Apart from the medical field, use of bioacoustic data had been an important tool for animal researchers, identifying animal behavioral patterns [180,181].
6. Biochemical signals: Provide information about concentration of various chemical agents in the body. Common examples are glucose level data for diabetes control, blood oxygen level data for asthma, obstructive pulmonary disease, or heart and kidney failure detection. Biochemical signals, in general, are deemed as amongst the highest accuracy signals to detect stress levels in the human body, particularly via urine, saliva, or blood samples [182].

7. Bio-optical signals: Bio-optical signals are naturally occurring or induced optical functions of the examined biologic system. Examples of use include estimating blood oxygenation by measuring transmitted vs. backscattered light from a tissue, using dye dilution and monitoring the bloodstream to observe cardiac output, or controlling fluorescence characteristics of the amniotic fluid to acquire information about the health of the fetus [171].

All of these types of biosignals have different levels of sensitivity, accuracy, timescale, and types of dedicated sensors [183]. The use of these sensors suggests another important classification relevant to the focus of this research: the level of invasiveness. This factor is mainly critical in order to eliminate any psychological bias on one hand, by ensuring that the test subjects are not aware of the experiment, but also relevant due to the applicability of the biosensing experiments outside laboratory environments on the other.

The level of invasiveness is a direct correlation between the biosignal sensor and its relation to its biologic host; naturally, the least invasive biomedical data collection methods would be through the use of no contact sensors at all (i.e., thermal infrared imaging). Minimally invasive methods can be directly in contact with the skin (i.e., measurements obtained through biopotential electrodes such as electrocardiogram, electroencephalography, electrodermal activity), whereas the most invasive methods would implement sensors within the body (i.e., rectal, oral thermometers), puncturing the skin (i.e., use of needle electrodes for electromyographic signal acquisition) or even surgically placing the sensor (i.e., a blood pressure sensor placed in an artery, vein, or in the heart) [29].

3.2.2. Use of Biosignals in the Field of Building Engineering

The human body constantly tries to self-regulate against environmental changes, trying to keep the state of homeostasis and, as defined by Selye [184], stress is "a state of biological activation triggered by the person interacting with external agents that force her or his capacity to adapt". According to Schneiderman et al. [130], following a stressful event, a cascade of changes in the nervous, cardiovascular, endocrine, and immune systems take place. As previously discussed, these changes are traceable and quantifiable through sensing of biosignals.

Stress literature is well researched and has a long history. From a medical point of view, it goes back to the beginning of the 20th century [185] and gained even more traction in fundamental and clinical neuroscience research in the 50s, after Hans Selye's stress theory [186]. Since then, stress has been a research topic in many specific fields such as psychology, economics, ergonomics, sociology, endocrinology, complementary medicine, animal breeding, etc. It has been a research topic in the domain of building physics as well, regarding occupant stress (mental or physical, in terms of *thermal stress*, *heat stress* or *cold stress*), as reviewed in the first chapter.

However, the use of biosignals as part of stress-research is relatively new, even more so in the field of building engineering. As indicated in Section 2, Materials and Methods, the list of publications researched for this review mostly tried to correlate IEQ parameters to comfort sensation and mental workload, while none, to our knowledge, investigated a long-term effect.

3.2.3. Limitations

Albeit promising, the use of biosignals in the field of building engineering has still several limitations, most of which pose setbacks for robust real-world applications so far. The majority of the studies using biosignals have been conducted in test chambers, laboratories or well-controlled environments, and the gap in research to transfer the know-how from the lab setting to real-world conditions still exists [120]. While this approach is beneficial in establishing working protocols with minimal noise and more comparable data, it is also well-known that people's tolerance limits may drastically vary from test environments to real-world settings in which there will be many unforeseen stressors, influencing the overall stress [122,187]. However, both the sensing and data analysis technologies are rapidly developing, and the possibility of the use of biosignals to further develop our know-how in realistic settings does not seem unrealistic.

The following are the most critical obstacles, according to our literature review.

1. Technical problems

    a. Noise:

    The physiological processes that produce biosignals, most of the time, cannot be reduced to an unequivocal course of events. Motion of the entire body, or parts of it, creates noise in the data. While there are several methods to improve signal-to-noise ratio, and higher performing sensors already eliminate noise to a certain degree, this still poses a problem. While de Santos et al. [188] remarks that "the subjects were strongly indicated not to move during the experiment procedure, in order to avoid noise in physiological signal acquisition", it significantly deviates from a realistic setting, ergo realistic conclusions, since it not only is irrational to expect test subjects to be that restricted while doing their mundane tasks, it is also effective on the strain this command causes on the subjects, meaning the results will not be representative for real-world settings.

    Similarly, Schmidt et al. [189] states that certain biosignals, such as electrooculography or electromyogram, are prone to producing a lot of noise in real world settings, and therefore are not very suitable for any use outside the laboratory conditions.

    b. Level of invasiveness, mobility, or wearability of sensors:

    The last decade has seen a great development in wearable technologies. The wearables used in the field can be watch-like, chest-belt, stationary devices, or recently flexible sensor patches and sensors that can be integrated into fabric. These technologies offer an increased wearing comfort, potential new measurement positions, and are less intrusive [189]. Even biochemical signals such as saliva and sweat, which are sought after as direct stress measurements, can be analyzed non-invasively through recently developed wearable electrochemical sensors [190].

    However, there are still problems associated with mobile technologies: once stationary devices, EEGs now have mobile alternatives with wearable headsets, with wireless data transfer technology. Nevertheless, they are still not fully unobtrusive; in some experiments, it was reported that continuous wearing of said headset caused headaches [2], making it less suitable for prospective long-term experiments in real settings.

    c. Accuracy of sensors:

    Another topic with mobile and wearable sensors as an emerging concern has been the level of accuracy that can be obtained during the testing period [191]. Continuing with the example of mobile EEGs, mechanical setbacks still exist; several studies [192,193] have explored the drop in signal quality and accuracy after a certain time of usage, due to sweat on the skin changing the bioimpedance of electrodes.

2. Problems with data acquisition:

    Depending on country-specific regulations, data privacy is strictly controlled and regulated. This means that data collection methods, particularly when video and/or speech recording is involved, might have to be modified or restricted accordingly [29].

3. Need for self-reporting:

    Sharma and Gedeon [194] and Hernandez et al. [195] state that most methods that use biosensing techniques (alone or in combination) still use questionnaires to validate induced stressors. This eliminates the potential avoidance of psychological bias in the resulting data. On the other hand, Hernandez et al. [196] suggests that self-reports, when used as ground-truths, result in more accurate stress detection in person-specific models.

4. Need for multi-modal biosignals for better insight:

    Generally speaking, multimodal systems perform better when compared to the accuracy reached by unimodal systems [197]. However, Kyriakou et al. [120] make a critical point by stressing the importance of combining the appropriate signals and signal processing algorithms. A multimodal data collection is also practical in order to study the cross-modal effects that one can expect to observe in a real-world setting [198].

Thereby also increasing the applicability problems parabolically, making it even more complicated under the real-world conditions.

3.2.4. State of the Art

In this chapter, relevant biosignals used to assess environmental stressors in the built environment are reviewed and classified. While there is a plethora of biosignals, the ones presented in this chapter are selected on the basis of their occurrence in stress studies directly related to IEQ parameters.

The presented biosignals are simply classified regarding their position in the human body.

- Brain: Electroencephalogram (EEG)

Since the early 20th century [199,200], brain activities can be recorded by electrodes. The system is called electroencephalogram, or EEG for short; by "electro = electrical", "encephalo = brain" and "gram = record" and used for brain function study [201]. EEG is the electrical recording of brain activity, represented as voltage fluctuations resulting from ionic current flows within the neurons of the brain [202]. EEG records brain activity at the millisecond level, and therefore is considered to be a strong tool to provide a direct measure of the dynamic interaction between the brain and other stimuli in real-time [203]. It is considered to be a non-invasive method as it can be recorded by electrodes of varying numbers placed on the scalp. However, the level of obtrusiveness can be discussed.

The amplitude of the EEG signals ranges between 10–200 V, with a frequency falling in the range 0.5–40 Hz. There are five frequency bands:

1. Delta (γ): 0.5–4 Hz in frequency. Delta waves are the slowest EEG waves, normally detected during deep and unconscious sleep.
2. Theta (θ): 4–8 Hz in frequency. Theta waves are observed during some states of sleep and quiet focus.
3. Alpha (α): 8–12 Hz in frequency. Alpha waves originate during periods of relaxation with eyes closed but still awake.
4. Beta (β): 12–25 Hz in frequency. Beta waves originate during normal consciousness and active concentration and are associated with increase in alertness and arousal.
5. Gamma (γ): above 25 Hz in frequency. Gamma waves are known to have stronger electrical signals in response to visual stimulation [202,204,205].

It is also important to note that the precise range assigned to these bands can vary across studies [206] but apart from the exact thresholds, the classification stands.

The importance of these bands is that they are associated with particular cognitive processes. For example, Nyhus and Curran [207] states that theta and gamma bands are correlated to memory processes such as retrieval and encoding, Jensen et al. [208] mentions the connection between alpha and gamma waves and visual processing and Doesburg et al. [209] addresses the whole scalp gamma frequency synchronization in association with consciousness. According to Hamid et al. [210], the presence of stress has been considered to be responsible for an increase in the EEG beta band power.

EEG is a spatio-temporal biosignal, meaning the data not only vary with time but also in location. Where the signal originates at or the comparison of activities in different hemispheres can carry valuable information. An example relevant to stress studies is frontal alpha asymmetry (FAA), which is defined as "the difference between right and left alpha activity over frontal regions of the brain" [211]. It is speculated that the greater left frontal activity is associated with reacting to positive stimuli while the greater right frontal activity is associated with the tendency to withdraw from responding to negative stimuli. The extent of asymmetry has been suggested to vary under conditions of chronic stress and alpha asymmetry is suggested as a potential biomarker for stress classification [212].

Another use of spatial EEG data is when mental workload or performance are presented as evaluation parameters. In task-based experiments, the researchers look at the frontal lobe since as mental demand increases, so does the theta band activity [105]. By recording and comparing the rise in activity, it is possible to quantify the physiological impact of the stressor on the human body. The use of EEG signal to identify attention and distraction is relatively common, as seen in distraction studies in the automotive industry [89,98].

There are methods reported quantifying human acute stress in response to induced stressors (such as impromptu speech, examination, mental task, public speaking, and the cold pressor test) using EEG signal recordings, but the literature lacks a classification of long-term stress using EEG [212]. Even more so, very few of the acute stressors in the literature are related to IEQ parameters (i.e., [2]).

However, studies exist outside the realm of building physics, as [213] study the EEG signals of cold and warm sensation on skin, using 15 healthy subjects. During the experiments, subjects were partially exposed to a temperature range of 14 to 48 °C, with different intervals and intensities. While the temperatures, particularly in the warmer range, are not usual from an IEQ point of view, the study provides valuable insight as to how brain activities react to thermal stimuli.

Finally, it is worth mentioning that EEGs, similar to many other biosensing methods, are prone to noise and artefacts that may obscure the data and their interpretation. These artefacts may be caused by the test subjects' movement (eye movements, shivering, coughing, hiccupping, breathing) or due to the faults in the equipment (electrode popping, cable movements, electrical or electromagnetic interferences) [214].

- Heart: Electrocardiogram (ECG) and Heart Rate Variability (HRV)

An electrocardiogram (ECG) measures the electrical manifestation of the ionic potential of the heart, via numerous electrodes placed on the body surface near to the particular organ (e.g., chest, hands, and legs) [215].

Each cardiac cycle in the ECG is characterized by successive waveforms, known as P wave, QRS complex (including Q, R and S waves occurring in rapid succession) and T wave (Figure 5). These waveforms represent the depolarization and repolarization activities in the heart's cells of atrium and ventricle [216].

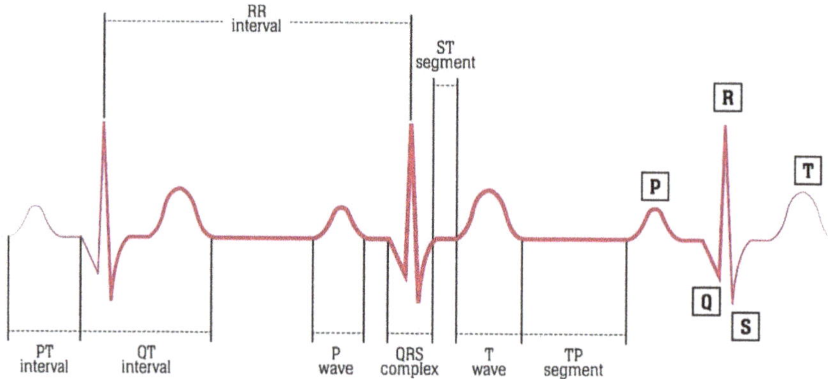

Figure 5. A typical electrocardiogram (ECG) signal that includes three heartbeats and the information lying in the P, Q, R, S, and T waves [after 215].

Studies using the heart as a biosignal data source either use heart rate, or heart rate variability. Heart rate is the measurement of heart beats per minute (bpm), while heart rate variability is the fluctuation of the length of consecutive heartbeat intervals [217], also known as R-R intervals [218] or N-N intervals.

These intervals are not periodic; however, the variation is not random either. The oscillations of a healthy heart are deemed complex and dynamically changing, depending on the "extrinsic protocols imposed on the heart" [219], since the cardiovascular system plays a vital role in reacting to stressors and maintaining the state of homeostasis [220]. Due to this reason, HRV is not only a factor of the heart, but also a rich data source with information on actions of the nervous system in response to stress factors [221].

Accordingly, the use of HRV signals plays a vital role in stress assessment research [183], with several studies being published not only generally in stress literature, but also in the field of building engineering, since the actions of the nervous system in the control of cardiac activities are sensitive to changes in temperature.

As Yao et al. [201] explains, environmental temperature can have an impact on the vagal and sympathetic nerve activity; as under thermally uncomfortable situations, sympathetic activity prevails, while in comfortable situations, the vagal activity overcomes. In humans, the vagus nerve is for the protection of the body, while the sympathetic nerve is triggered for stress response, including thermal stress. Thermoregulation of the human body as a system is controlled by the sympathetic nerve.

The way activities of vagal or sympathetic nerves are presented in HRV are via high or low frequency components. The vagus nerve is thought to excite the high frequency (HF) component of HRV while the low frequency (LF) is thought to be of both. As a result, any physiological reaction concerning thermoregulation triggers the use of a sympathetic nerve, resulting in a higher LF power and lower HF power, and the ratio of LF/HF increases. For this reason, studying the LF/HF ratio data becomes a reliable parameter to understand thermal sensation and physiological comfort in humans [222].

Studies conducted by [201,222–224] have all come up with similar results, contributing to the findings that "higher LF/HF yields unpleasant thermal sensation and discomfort". In Nkurikiyeyezu et al. [225], the researchers had 17 subjects doing light office work under cold, neutral, and hot environments while collecting data on statistical, spectral, and nonlinear HRV indices. With help from machine learning classification algorithms, the study concludes that it is possible to predict people's thermal states in a reliable manner, which was found to be up to 93.7% accuracy. There are no real-world studies on IEQ parameters so far; however, experiments focusing on mental stress using HRV data exist with varying success rates [226,227].

- Skin: Skin Temperature (SKT), Thermal Infrared Imaging (TII), Electrodermal activity (EDA)/Galvanic Skin Response (GSR)

Human skin is a significant source of data when it comes to understanding physiological reactions of the body in relation to its environment. It is the primary sensory organ and anatomical interface the humans have with their immediate environments. The sensations on the skin reflect both the biological processes happening intrinsically, or how the body is impacted by the outer stimuli [228]. To illustrate the skin's adaptability, the operational temperature ranges can be compared. As previously mentioned, the core of the human body requires a rather narrow temperature range, between ~36 to ~40 °C, with normal core temperature being ~37 °C, and it has been established that the skin temperature can vary between ~15 to ~42 °C without sensation of pain [228].

Similar to HRV, skin is part of the thermoregulation system, which is part of the autonomic nervous system (ANS). Therefore, skin-originating biosignals (skin temperature, electrodermal activity or galvanic skin response, photoplethysmograms) can provide variable insights on ANS activity [229].

Amongst these biosignals, skin temperature has been the one investigated most extensively, as it provides a direct understanding in human thermal sensation and comfort estimation, while the sensors are quite inexpensive [230]. Use of contact thermometers to acquire skin temperature has proven successful as seen in [231–235], constant thermocouples in different forms and both in real-world and laboratory settings seem to function and the results seem to be good indicators in predicting thermal sensation. However, thermocouple sensors require electrodes on the skin, making the data collection process slightly obtrusive [188,230].

As an alternative, sensing skin temperature remotely is a method being utilized by several researchers. Infrared imaging technologies are widely available today, and medical infrared has utilized the heat signature of skin to map skin temperature since the 1960s [171]. Current technology presents low-cost no-contact and non-intrusive sensors; however, there are reported application problems: Li et al. [230] states infrared thermometers have a handicap of having such narrow field of views, meaning the thermometers need to be placed very close (few centimeters) to the test subjects. Alternatively, the use of thermographic (thermal) cameras can be placed away from test subjects, while in that case, data accuracy significantly drops in comparison to thermocouple sensors or even infrared thermometers. Nevertheless, with correct data analysis methods, improved data still seem to have been proven to be a robust alternative for real-time thermal comfort detection and even prediction.

One final biosignal of interest uses the skin's thermoregulatory response as the data source. Electrodermal activity (EDA), also known as galvanic skin response (GSR), measures the changes in electrical properties of the skin caused by eccrine sweat gland activity, which is a key factor in the thermoregulatory process [236]. The idea dates back to the very beginning of the 20th century to Carl Jung, who for the first time mentioned electrodermal activity in connection to emotions in a psychoanalysis book [237].

Several studies successfully link EDA data to stress detection and quantification [238–240]. Recently, the use of wearables is becoming more reliable in the field, leading the way to long-term EDA monitoring as well [240].

- Chemical: Cortisol

There are two main pathways in the human body to relay the stress from the brain to the body. The first way is via the sympathetic nervous system, through which cardiovascular and skin-related biosignals were used as stress signifiers. The second way is through the hypothalamic–pituitary–adrenal (HPA) axis, the hormonal route. Stress triggers the HPA axis, a neuroendocrine system that regulates central and peripheral homeostatic adaptive responses to stress [126,129,241,242].

Biochemical samples, primarily urine, saliva, and blood samples, and particularly analysis of hormones such as cortisol and alpha-amylase, are amongst the primary measures used to identify the impacts of stress on the body in conventional psychology methods and long-term stress studies [183,212,243].

While these analyses are deemed to provide high accuracy in stress detection, data collection can pose difficulties in regard to the intrusiveness of obtaining the samples, or the time interval of the data acquisition. Additionally, Lee et al. [244] stress the importance of the timing of the sampling, as cortisol levels in blood also vary diurnally, increasing during early morning and decreasing towards the night. Another important aspect is that while the biochemical signals from fluid analysis may correlate with acute stress, to understand the long-term chronic stress impacts some researchers have suggested looking at extraction of cortisol from hair fibers [245]. In order to overcome the limitations of the intrusive nature of data collection, the use of wearables in the forms of sweat patches, wristbands, or epidermal sensors is becoming more available. As Seshadri et al. [190] remark, while this enables researchers to collect biologic data continuously and is truly non-intrusive, the majority of these devices are only available as market products and have not been clinically tested and validated yet.

4. Discussion

4.1. Summary and Research Gap

Our understanding of indoor environmental quality is as good as the methods we use to assess it. A brief review of the state-of-the-art has highlighted how little consensus there is concerning the actual measures used, both between radically different fields and within fields of research. Recurrent criticisms are mostly focused on two main aspects:

(i) the effect against which quality (or lack thereof) is measured; (ii) the reliability of the collected data. In the first case, human health and wellbeing are commonly implied to be the end-objective, whether the effects measured vary broadly and include self-selection, performance-oriented or physiological metrics. The use of different measures raises on one hand questions concerning their relevance and appropriateness in reflecting health parameters; on the other, it makes comparisons between studies very difficult. In the second case, the importance of taking into account psychological and individual factors in the equation is on one hand underlined by many researchers as not being adequately considered; on the other hand, these components are widely accepted as having a strong potential for data bias.

This gap in terms of measuring efficiency substantially affects the mental framework of all professionals dealing with the built environment, from designers to policymakers, with end effects not only on human health and wellbeing, but also on the overall energetic and environmental building performance.

Within this context, recent advancements in sensing technology and time-sensitive biomedical data acquisition have opened up new possibilities to use biomedical signals as an increasingly reliable objective methodology to measure environmental stressors. The promise of parsing *physiologically* relevant data is so that the impacts of additional variables, such as adaptive behavior, tolerance, psychology, exposure time, can be studied against a constant, in a situation where, as previously explained, almost all other parameters vary. The use of these methods is however relatively new to the field of building and climate engineering, where the established scientific practice largely uses self-selection or performance-based assessments (questionnaires and interviews) to measure the quality range of measured indoor conditions.

The main hypothesis supported by this paper is that introducing data collection based on biomedical signals to measure human stress in indoor environments has the potential to give new insights, more data reliability and enable comparisons between studies in the field of building physics. Hence, the review aims to contribute with a systematic and comprehensive insight on the current capacity to detect stress from the built environment using biosignal measures, in order to set a foundation for the development of this method of data collection in the field of building physics.

4.2. Approach

This paper reviews the state-of-the-art knowledge concerning (i) indoor environmental parameters affecting humans in terms of IEQ and (ii) human biosignals that respond to the environmental IEQ stressors and the physiological response they activate, and then maps the interaction and interdependence between the two. More than 240 scientific publications were analyzed, published between 1906 and 2021, dealing with topics ranging from built indoor environment, medicine and neuroscience, biomedical engineering, sensing technology, psychology, ecology, and economy.

4.3. Main Findings

Indoor environmental parameters commonly used in building engineering and indoor climate research were reviewed. Indoor environmental quality (IEQ) is determined by assessing air quality, thermal, visual, and acoustic parameters. Although the use of these parameters is well-established throughout the reviewed literature, this brief review was necessary in order to establish a common ground between current practice and the introduction of biosignal measures. This will allow bridging the new methods introduced and the state-of-art knowledge on the features of the indoor environment affecting humans, but also to set a basis for comparison with the existing research in the field.

Moreover, discussing IEQ from a health-oriented rather than a comfort-oriented perspective, relevant sub-parameters were identified, which are, in the current state of research, not taken into consideration. This is the case of visual IEQ, where the range of the natural light spectrum should be taken into consideration also for its health-related properties (as to trigger the production of vitamin D) rather than only for its optical properties.

Parallelly, human biosignals were reviewed. Biosignal definitions and classifications were introduced in order to provide a systematic framework for professionals dealing with the built environment. The physical nature as well as the level of invasiveness of the biosignals are discussed. The physical nature of biosignals directly relates to the physical IEQ aspects of the indoor environments, while the invasiveness of the technology enables an easy and unbiased data collection that does not interfere with the study subjects' behavior patterns or awareness of being observed.

Successively, the review reported on the main limitations connected to the use of biosignal measures, being mainly of technical nature (noise, wearability, and accuracy), country-specific legal issues involved with data collection and storage and the use in some contexts of self-reporting as validation. The necessity of relying on multimodal data acquisition is also highlighted as a possible limit; however, if systematically integrated in the research method, it remains a valid method for cross validating the acquired data. The possibility of performing this kind of analysis can therefore also be seen as a strength, yielding potentially high quality and complete sets of data.

Finally, the review reports on the specific biosignals that relate to the IEQ parameters. These are classified into four categories depending on the biophysical aspect measured: brain, heart, skin and chemical.

4.4. Limitations

The introduction of biosignals as measures of indoor environmental quality still appears to have a number of limitations, in spite of the potential benefits highlighted by this review.

Defining the metrics of environmental quality remains a complex task due to a number of intrinsic complexities:

a. Multifaceted conditions of the indoor environment (temperature, light, air quality, humidity, etc.);

b. Elaborate human psychophysiological reactions affecting the regulation of environmental conditions for conditioned buildings (individual preferences, personal methods of adaptive behavior, i.e., different levels of clothing, different body mass index (BMIs), different sense of comfort, etc.);

c. Complex human psychophysiological reactions to the environmental conditions (stress, health, wellbeing, etc.);

d. The many-to-many relationships between the factors mentioned above, and finally;

e. The above-mentioned relationships being dependent on time factors (expectations changing together with the outdoor conditions and changing of the seasons, history of individual acclimatization, time of exposure to certain climatic conditions, etc.).

Additionally, a lack of knowledge and training from the side of building professionals in terms of the medical and technical background necessary to correctly use some biosensor measures, to collect and to interpret the data is a main limit to the effective use of these technologies. This problem is especially evident with some biosignals, such as the EEG, where possibly specialists from other fields need to be involved. Although the interdisciplinarity of these types of studies has the potential to contribute with innovative practices and relevant findings, it poses important limits in terms of time and resources that can discourage the use of these methods in regular ordinary research.

Finally, the study has addressed "indoor environments" as a whole, not distinguishing further between the type of environments, in the aim of presenting the method as an overarching tool that can be successively adapted to more specific case-studies. From the analysis of the reviewed literature, a great majority of the research has up to today focused on office spaces, a few on housing indoors and another few on healthcare contexts.

Intrinsic limitations of the review itself, in spite of the high interdisciplinarity of the sources reviewed, might come from the authors having similar cultural and professional backgrounds. Additionally, the studies reviewed were exclusively written in the English language, with the potential exclusion of relevant sources published in other languages.

4.5. Future Directions

This review aims to set a foundation for the introduction of biosignals for the assessment of environmental parameters in the built environment, but is in that sense only a first step within a long process. Follow-up studies to effectively establish the biosignal data collection methods into current practice include:

1. The application of the identified biosignal measures in indoor environmental research, specifically starting with their use in test chamber lab experiments before staging exploratory studies in real-life indoor contexts and outdoor environments;
2. Parallels with the existing research in the field of comfort studies need to be further deepened, both in order to use stress research with biosignals to support comfort studies and to build up knowledge concerning stress related biosignals supported by the existing knowledge from comfort studies;
3. Implications in terms of building energy efficiency and indoor health are other parallel lines of research that open up, aiming to update the existing design practice as well as building regulations.

4.6. Conclusions

This review has highlighted how the relationship between daily human indoor habitats and human health needs to be researched further. While in current practice, comfort studies prioritize optimizing building energy use and user satisfaction, and studies on productivity bring a more direct economic impact, both constellations forgo the health implications as they are not visible at the same timescales. In other words, a balance needs to be found between the right dose of a higher overall human resilience and building resilience [246].

The use of biosignal measures to detect environment-related stress conditions is put forward as a promising and reliable method to effectively focus on short and long-term human health aspects. The review is a first stage within a longer process of translation of a method that is already established in other fields of study using stress research in the context of the built environment.

Author Contributions: Conceptualization, S.G.L.P., B.K., S.C.K. and T.A.; methodology, S.G.L.P. and B.K.; formal analysis, S.G.L.P., B.K. and S.C.K.; investigation, S.G.L.P., B.K. and S.C.K.; resources, S.G.L.P., B.K. and T.A.; data curation, B.K. and S.C.K.; writing—original draft preparation, S.G.L.P., B.K. and S.C.K.; writing—review and editing, S.G.L.P., B.K. and S.C.K.; visualization, B.K.; supervision, S.G.L.P., B.K. and T.A.; project administration, S.G.L.P., B.K. and T.A.; funding acquisition, S.G.L.P., B.K. and T.A. All authors have read and agreed to the published version of the manuscript.

Funding: This research was funded through contributions of the ALEXANDER VON HUMBOLDT STIFTUNG foundation, Bonn, Germany, through a research grant to Sandra G. L. Persiani, grant number ITA 1211263 HFST-P and of the TUM SEED FUND RESEARCH through a research grant to Bilge Kobas.

Institutional Review Board Statement: Not applicable.

Informed Consent Statement: Not applicable.

Data Availability Statement: Data sharing not applicable.

Conflicts of Interest: The authors declare no conflict of interest. The funders had no role in the design of the study; in the collection, analyses, or interpretation of data; in the writing of the manuscript, or in the decision to publish the results.

Abbreviations

AH	absolute humidity
ANS	autonomic nervous system
APH	air-phase petroleum hydrocarbon
BMI	body mass index
BRI	building-related illnesses
BVP	blood volume pressure
ECG	electrocardiogram
EDA	electrodermal activity
EEG	electroencephalogram
EGG	electrogastrogram
EMG	electromyogram
ENG	electroneurogram
EOG	electrooculogram
ERG	electroretinogram
GSR	galvanic skin response
HF	high frequency
HPA	hypothalamic–pituitary–adrenal
HR	heart rate
HRV	heart rate variability
IAQ	indoor air quality
IEQ	indoor environmental quality
IP	air pollutants in indoor air
LW	low frequency
MCG	magnetocardiogram
MEG	magnetoencephalogram
MVOC	microbiological volatile organic compounds
NO	nitrogen oxides
PAH	polycyclic aromatic hydrocarbons
PCB	polychlorinated biphenyls
PMV	predicted mean vote
PPD	predicted percentage of dissatisfied occupants
RH	relative humidity
SAD	seasonal affective disorder
SBS	sick building syndrome
SKT	skin temperature
SPL	sound pressure levels
SWB	subjective wellbeing
SWD	shift work disorder
SWL	sound power
TNZ	thermoneutral zone
TO	toxic organic
TTI	thermal infrared imaging
VOC	volatile organic compounds

Appendix A

Table A1. Sources based on their main research fields.

Origin	Number (Total: 246)	References
Building physics	42	[9,11,13,21,32,39,48,53,61,71,72,112,132–137,140,141,143,144,146–148,150,153,154,156,161–199,201,226,230,246].
Medicine	41	[3–5,7,10,19,20,31,38,40,62,65–70,73,79,90,98,107,114–116,121,122,124,127,142,149,151,152,155,162,184,219,222,241–245].

Table A1. *Cont.*

Origin	Number (Total: 246)	References
Neurosciences	36	[21,28,86,88,91,92,101,102,104–106,108,125,126,128,129,157,185–188,192,193,196,198,200,203,205–209,214,216,220,221].
Built environment	34	[1,2,6,14–17,30,42–47,49–52,57–60,64,74–77,81,82,85,103,111,131,145].
Biomedical engineering, biology	28	[24,29,95,138,170–183,195,204,215,217,218,223,227,228,231,232].
Computer sciences, engineering	27	[56,89,93,94,96,97,110,120,189–191,194,197,202,210,212–225,229,233–236,238–240].
Psychology	18	[8,22,25,37,38,63,87,99,100,109,117,118,123,125,126,130,158,160,211,237].
Economy, business	4	[35,36,55,80].
Ecology, botanics	6	[12,27,33,78,159,169].
Ergonomics	4	[34,83,84,113].
Others	4	[18,26,54,119].

Table A2. Sources about comfort, wellbeing, stress, and other definitions, based on their main research fields.

Origin	Number (Total: 157)	References
Built environment	28	[2,6,14–17,30,42,43,46,47,49–51,57,59,60,64,74–77,81,82,85,103,111,131].
Medicine	27	[3,7,19,20,31,38,40,62,65,66,73,79,90,98,107,114–116,121,122,124,127,149,151,184,222,241–243].
Building physics	24	[9,11,13,32,39,48,53,61,71,72,112,134,141,147,148,150,156,161,166–168,199,201,246].
Neurosciences	24	[21,28,86,88,91,92,101,102,104–106,108,125,126,128,129,157,185,186,188,189,196,205,206].
Computer sciences, engineering	17	[56,89,93,94,96,97,110,120,190,191,194,210,224,234,235,238,240].
Psychology	13	[8,22,25,37,38,63,87,100,117,118,123,125,126,130,158].
Biomedical engineering, biology	9	[24,29,95,183,195,223,227,228,231].
Economy, business	4	[35,36,55,80].
Ergonomics	4	[34,83,84,113].
Others	3	[26,54,119].
Ecology, botanic	2	[33,159].

Table A3. Sources about human aspects, categorized based on their focus.

Terms	Number (Total: 157)	References
Stress	38	[29,56,65,110,120–130,183–186,188–191,194–196,210,224,227,234,235,238,240–243].
Comfort	36	[15,16,30,34,35,42,43,46–51,57,59,61,62,66,71,72,111–119,131,147–149,159,201,222,223,231,246].
Health	36	[3,8,9,13,15–17,19,38,39,48,53,66,79,82,114–116,121,124,126–130,134,151,156,157,168,190,199,241–243,246].
Performance	21	[2,31,53,74–77,82–84,103–107,111,120,122,161,190,206].
Behaviour	17	[16,23,24,28,49–51,55,71,112,113,117,125,131,147,150,246].
Attention	16	[86–98,100–102,206].
Productivity	14	[13,14,33,35,36,76,77,79–81,85,89,158,168].
Wellness, wellbeing	13	[19,20,22,37,38,40,82,84,118,158,187,205,243].
Preference	8	[6,7,55,60,61,63,111,166].
Satisfaction	3	[30,32,64].

Table A4. Sources about indoor environmental quality (IEQ), based on their publication types.

Type	Number (Total: 90)	References
Books, book chapters, theses	6	[36,52,74,76,146,156].
Conference paper	5	[50,51,77,110,226].
Journal paper	72	[2,4,5,8,9,11–16,21,31–33,35,42,44–49,53,60–62,66–71,75,78,79,81–83,108,111,131–135,137–139,141–144,150,152–155,158,159,161,162,165–167,169,199,225,228,230,232,233].
Standards, guides, reports	6	[10,58,136,148,149,164].
Website	1	[145].

Table A5. Sources about IEQ, based on their main research fields.

Origin	Number (Total: 90)	References
Building physics	34	[9,11,13,21,32,48,53,61,71,132–137,139–141,143,144,146,148,150,153,154,156,161,164–167,199,226,230].
Built environment	24	[2,14–16,42,44–47,49–52,58,60,74–77,81,82,111,131,145].
Medicine	15	[4,5,10,31,62,66–70,79,142,152,155,162].
Ecology, botanics	5	[12,33,78,159,169].
Biomedical engineering, biology	3	[138,228,232].
Computer sciences, engineering	3	[110,225,233].
Psychology	2	[8,158].
Economy, business	2	[35,36].
Neurosciences	1	[108].
Ergonomics	1	[83].

Table A6. Focus points of the publications on IEQ.

Terms	Number (Total: 90)	References
Health	33	[4,8–10,12,13,15,16,21,48,53,66–70,78,79,82,134,135,138,139,142–144,153–156,162,164,199].
Comfort	25	[15,16,35,42,44–51,58,61,62,66,71,111,131,148,159,165,225,226,230].
Productivity	12	[13,14,33,35,36,52,76–79,81,158].
Performance	11	[2,31,53,74–77,82,83,111,161].
Behaviour	8	[16,49–51,71,78,131,150].
Preference	5	[60,61,111,132,166].
Wellness, wellbeing	2	[82,158].
Stress	1	[110].
Satisfaction	1	[32].

Table A7. Types of IEQ parameters in publications on IEQ.

Parameter	Number (Total: 90)	References
Thermal	26	[2,42,45,46,48,50,51,58,62,66–71,83,111,131,146,148,199,225,226,228,230,232,233].
Indoor air quality	19	[5,9,11,12,42,45,50,53,77,135,137–144,165].
IEQ	18	[13–16,21,32,36,47,49,52,60,75,76,78,79,82,132,133].
Visual	12	[8,33,44,150,152–156,158,159,199].
Acoustic	9	[31,61,161,162,164–167,169].

Table A8. Sources about physiologic signatures, based on their publication types.

Type	Number (Total: 110)	References
Books, book chapters, theses	13	[116,170,171,173–176,180,182,196,198,203,237].
Conference paper	17	[29,95,98,110,121,172,179,181,194,195,210,226,234–236,238,239].
Journal paper	78	[7,8,17,28,56,65–70,86–88,91,99,101,102,104,105,108,109,120,122,142,149,151,163,177,178,183,185,187–193,197,199–202,204–209,211–225,227–233,240–245].
Standards, guides, reports	2	[10,107].

Table A9. Sources about physiologic signatures, based on their main research fields.

Origin	Number (Total: 110)	References
Neurosciences	27	[28,86,88,91,101,102,104,105,108,185,187,188,192,193,196,198,200,203,205–209,214,216,220,221].
Biomedical engineering, biology	26	[29,95,170–183,195,204,215,217,218,223,227,228,231,232].
Medicine	23	[7,10,65–70,98,107,116,121,122,142,149,151,219,222,241–245].
Computer sciences, engineering	22	[56,110,120,189–191,194,197,202,210,212,213,224,225,229,233–236,238–240].
Psychology	6	[8,87,99,109,211,237].
Building physics	5	[163,199,201,226,230].
Built environment	1	[17].

References

1. United Nations. Department of Economic and Social Affairs. 2018. Available online: https://www.un.org/development/desa/en/news/population/2018-revision-of-world-urbanization-prospects.html (accessed on 29 November 2020).
2. Wang, X.; Li, D.; Menassa, C.C.; Kamat, V.R. Investigating the effect of indoor thermal environment on occupants' mental workload and task performance using electroencephalogram. *Build. Environ.* **2019**, *158*, 120–132. [CrossRef]
3. Diffey, B. An overview analysis of the time people spend outdoors. *Br. J. Dermatol.* **2011**, *164*, 848–854. [CrossRef] [PubMed]
4. Brasche, S.; Bischof, W. Daily time spent indoors in German homes—Baseline data for the assessment of indoor exposure of German occupants. *Int. J. Hyg. Environ. Health* **2005**, *208*, 247–253. [CrossRef] [PubMed]

5. Klepeis, N.; Nelson, W.; Ott, W.; Robinson, J.; Tsang, A.M.; Switzer, P.; Behar, J.V.; Hern, S.C.; Engelmann, W.H. The National Human Activity Pattern Survey (NHAPS): A resource for assessing exposure to environmental pollutants. *J. Expo. Sci. Environ. Epidemiol.* **2001**, *11*, 231–252. [CrossRef]

6. Chen, C.F.; Yilmaz, S.; Pisello, A.L.; De Simone, M.; Kim, A.; Hong, T.; Bandurski, K.; Bavaresco, M.V.; Liu, P.-L.; Zhu, Y. The impacts of building characteristics, social psychological and cultural factors on indoor environment quality productivity belief. *Build. Environ.* **2020**, *185*, 107–189. [CrossRef]

7. Lee, E.; Kim, M. Light and Life at Night as Circadian Rhythm Disruptors. *Chronobiol. Med.* **2019**, *1*, 95–102. [CrossRef]

8. Bedrosian, T.; Nelson, R. Timing of light exposure affects mood and brain circuits. *Transl. Psychiatry* **2017**, *7*. [CrossRef] [PubMed]

9. Rudel, R.A.; Perovich, L.J. Endocrine disrupting chemicals in indoor and outdoor air. *Atmos. Environ.* **2009**, *43*, 170–181. [CrossRef] [PubMed]

10. Scientific Committee on Emerging and Newly Identified Health Risks (SCENIHR). *Health Effects of Artificial Light*; European Union: Brussels, Belgium, 2012; ISBN 978-92-79-26314-9.

11. Mentese, S.; Mirici, N.A.; Elbir, T.; Palaz, E.; Mumcuoğlu, D.T.; Cotuker, O.; Bakar, C.; Oymak, S.; Otkun, M.T. A long-term multi-parametric monitoring study: Indoor air quality (IAQ) and the sources of the pollutants, prevalence of sick building syndrome (SBS) symptoms, and respiratory health indicators. *Atmos. Pollut. Res.* **2020**, in press. [CrossRef]

12. Brilli, F.; Fares, S.; Ghirardo, A.; de Visser, P.; Catatyud, V.; Munoz, A.; Annesi-Maesano, I.; Sebastiani, F.; Alivemini, A.; Varriale, V.; et al. Plants for Sustainable Improvement of Indoor Air Quality. *Trends Plant Sci.* **2018**, *23*, 507–512. [CrossRef]

13. Fisk, W.J. Estimates of potential nationwide productivity and health benefits from better indoor environments: An update. In *Indoor Air Quality Handbook*; Berkeley Lab: Berkeley, CA, USA, 2000; Volume 4, ISBN 9780071414845.

14. Al Horr, Y.; Arif, M.; Haushik, A.; Mazroei, A.; Katafygiotou, M.; Elsarrag, E. Occupant productivity and office indoor environment quality: A review of the literature. *Build. Environ.* **2016**, *105*, 369–389. [CrossRef]

15. Ortiz, M.A.; Kurvers, S.R.; Bluyssen, P.M. A review of comfort, health, and energy use: Understanding daily energy use and wellbeing for the development of a new approach to study comfort. *Energy Build.* **2017**, *152*, 323–335. [CrossRef]

16. Šujanová, P.; Rychtáriková, M.; Sotto Mayor, T.; Hyder, A. A Healthy, Energy-Efficient and Comfortable Indoor Environment, a Review. *Energies* **2019**, *12*, 1414. [CrossRef]

17. Van Marken Lichtenbelt, W.; Hanssen, M.; Pallubinsky, H.; Kingma, B.; Schellen, L. Healthy excursions outside the thermal comfort zone. *Build. Res. Inf.* **2017**, *45*, 819–827. [CrossRef]

18. WHO. Constitution. Online Article. Available online: https://www.who.int/about/who-we-are/constitution#:~{}:text=Health%20is%20a%20state%20of,belief%2C%20economic%20or%20social%20condition (accessed on 29 November 2020).

19. Cronin de Chavez, A.; Backett-Milburn, K.; Parry, O.; Platt, S. Understanding and researching wellbeing: Its usage in different disciplines and potential for health research and health promotion. *Health Educ. J.* **2005**, *64*, 70–87. [CrossRef]

20. Lindert, J.; Bain, P.A.; Kubzansky, L.D.; Stein, C. Well-being measurement and the WHO health policy Health 2010: Systematic review of measurement scales. *Eur. J. Public Health* **2015**, *25*, 731–740. [CrossRef]

21. Serrano-Jiménez, A.; Lizana, J.; Molina-Huelva, M.; Barrios-Padura, Á. Indoor Environmental Quality in Social Housing with Elderly Occupants in Spain: Measurement Results and Retrofit Opportunities. *Build. Eng.* **2020**, *30*. [CrossRef]

22. Yetton, B.D.; Revord, J.; Margolis, S.; Lyubomirsky, S.; Seitz, A.R. Cognitive and physiological measures in well-being science: Limitations and lessons. *Front. Psychol.* **2019**, *10*, 1630. [CrossRef]

23. Steimer, T. The biology of fear- and anxiety-related behaviors. *Dialogues Clin. Neurosci.* **2002**, *4*, 231–249. [CrossRef]

24. Levitis, D.A.; Lidicker, W.A., Jr.; Freund, G. Behavioural biologists do not agree on what constitutes behaviour. *Anim. Behav.* **2009**, *78*, 103–110. [CrossRef]

25. American Psychological Association (APA). Dictionary of Psychology. Available online: https://dictionary.apa.org/ (accessed on 26 November 2020).

26. Oxford Lexico. Mannerism, Oxford English and Spanish Dictionary. Available online: https://www.lexico.com/definition/mannerism (accessed on 7 November 2020).

27. Elliott, M.; Quintino, V. *The Estuarine Quality Paradox Concept, Encyclopedia of Ecology*, 2nd ed.; Elsevier: Amsterdam, The Netherlands, 2019; ISBN 978-0-444-64130-4.

28. Jansen, A.; Nguyen, X.; Karpitsky, V.; Mettenleiter, M. Central Command Neurons of the Sympathetic Nervous System: Basis of the Fight-or-Flight Response. *Sci. Mag.* **1995**, *5236*, 644–646. [CrossRef]

29. Nkurikiyeyezu, K.; Yokokubo, A.; Lopez, G. Importance of individual differences in physiological-based stress recognition models. In Proceedings of the 15th International Conference on Intelligent Environments (IE), Rabat, Morocco, 24–27 June 2019; pp. 37–43. [CrossRef]

30. Langevin, J.; Wen, J.; Gurian, P.L. Modeling thermal comfort holistically: Bayesian estimation of thermal sensation, acceptability, and preference distributions for office building occupants. *Build. Environ.* **2013**, *69*, 206–226. [CrossRef]

31. Jahncke, H.; Halin, N. Performance, fatigue and stress in open-plan offices: The effects of noise and restoration on hearing impaired and normal hearing individuals. *Noise Health* **2012**, *14*, 260. [CrossRef]

32. Frontczak, M.; Schiavon, S.; Goins, J.; Arens, E.; Zhang, H.; Wargocki, P. Quantitative relationships between occupant satisfaction and satisfaction aspects of indoor environmental quality and building design. *Indoor Air* **2012**, *22*, 119–131. [CrossRef]

33. Lohr, V.I.; Pearson-Mims, C.H.; Goodwin, G.K. Interior plants may improve worker productivity and reduce stress in a windowless environment. *J. Environ. Hortic.* **1996**, *14*, 97–100. [CrossRef]

34. Macpherson, R. Thermal stress and thermal comfort. *Ergonomics* **1973**, *16*, 611–622. [CrossRef]
35. Feige, A.; Wallbaum, H.; Janser, M.; Windlinger, L. Impact of sustainable office buildings on occupant's comfort and productivity. *J. Corp. Real Estate* **2013**, *15*, 7–34. [CrossRef]
36. Oseland, N. *Environmental Factors Affecting Office Worker Performance: A Review of Evidence*; CIBSE: London, UK, 1999; ISBN 0900953950.
37. Evans, G.W.; Wells, N.M.; Moch, A. Housing and Mental Health: A Review of the Evidence and a Methodological and Conceptual Critique. *J. Soc. Issues* **2003**, *59*, 475–500. [CrossRef]
38. Evans, G.W.; Wells, N.M.; Chan, H.Y.E.; Saltzmand, H. Housing quality and mental health. *J. Consult. Clin. Psychol.* **2000**, *68*, 526–530. [CrossRef]
39. Ulrich, R.S. Effects of interior design on wellness: Theory and recent scientific research. *J. Health Care Inter. Des.* **1991**, *3*, 97–109.
40. Hancock, T. Lalonde and beyond: Looking back at "a new perspective on the health of Canadians". *Health Promot. Int.* **1986**, *1*, 93–100. [CrossRef] [PubMed]
41. Lalonde, M. *A New Perspective on the Health of Canadians: A Working Document*; Government of Canada: Ottawa, ON, Canada, 1974.
42. Ma, N.; Aviv, D.; Guo, H.; Braham, W.W. Measuring the right factors: A review of variables and models for thermal comfort and indoor air quality. *Renew. Sustain. Energy Rev.* **2021**, *135*. [CrossRef]
43. Andargie, M.S.; Touchie, M.; O'Brien, W. A review of factors affecting occupant comfort in multi-unit residential buildings. *Build. Environ.* **2019**, *160*, 106–182. [CrossRef]
44. Preto, S.; Gomes, C.C. Lighting in the Workplace: Recommended Illuminance (lux) at Workplace Environs. In *Advances in Design for Inclusion. AHFE 2018. Advances in Intelligent Systems and Computing*; Di Bucchianico, G., Ed.; Springer: Berlin/Heidelberg, Germany, 2019. [CrossRef]
45. Rim, D.; Schiavon, S.; Nazaroff, W.W. Energy and Cost Associated with Ventilating Office Buildings in a Tropical Climate. *PLoS ONE* **2015**, *10*. [CrossRef]
46. Kingma, B.R.M.; Frijns, A.J.H.; Schellen, L.; Lichtenbelt, W.D.V.M. Beyond the classic thermoneutral zone Including thermal comfort. *Temperature* **2014**, *1*, 142–149. [CrossRef]
47. Chappells, H.; Shove, E. Debating the future of comfort: Environmental sustainability, energy consumption and the indoor environment. *Build. Res. Inf.* **2005**, *33*, 32–40. [CrossRef]
48. Fang, L.; Wyon, D.P.; Clausen, G.; Fanger, P.O. Impact of indoor air temperature and humidity in an office on perceived air quality, SBS symptoms and performance. *Indoor Air* **2004**, *14*, 74–81. [CrossRef]
49. Brager, G.S.; de Dear, R.J. Thermal Adaptation in the Built Environment: A Literature Review. *Energy Build.* **1998**, *27*, 83–96. [CrossRef]
50. Brager, G.S.; de Dear, R.J. Climate, Comfort & Natural Ventilation: A new adaptive comfort standard for ASHRAE Standard 55. In Proceedings of the Moving Thermal Comfort Standards into the 21st Century 2012, Windsor, UK, 5 January 2016.
51. De Dear, R.; Zhang, F. Dynamic Environment, Adaptive Comfort, and Cognitive Performance. In Proceedings of the 7th International Buildings Physics Conference 2018, Syracuse, NY, USA, 23–26 September 2018; pp. 1–6. [CrossRef]
52. Heerwagen, J.H.; Orians, G.H. Humans, habitats, and aesthetics. In *The Biophilia Hypothesis*; Kellert, S.R., Wilson, E.O., Eds.; Island Press: Washington, DC, USA, 1993; pp. 138–172.
53. Zhang, X.; Wargocki, P.; Lian, Z.; Thyregod, C. Effects of exposure to carbon dioxide and bioeffluents on perceived air quality, self-assessed acute health symptoms, and cognitive performance. *Indoor Air* **2017**, *27*, 47–64. [CrossRef]
54. Merriam-Webster. Self-Select, Merriam-Webster Dictionary. Available online: https://www.merriam-webster.com/dictionary/self-select (accessed on 26 November 2020).
55. Heckman, J.J. *Selection Bias and Self-Selection*; Durlauf, S.N., Blume, L.E., Eds.; Microeconometrics; The New Palgrave Economics Collection, Palgrave Macmillan: London, UK, 2010. [CrossRef]
56. Giannakakis, G.; Grigoriadis, D.; Giannakaki, K.; Simantiraki, O.; Roniotis, A.; Tsiknakis, M. Review on psychological stress detection using biosignals. *IEEE Trans. Affect. Comput.* **2019**. [CrossRef]
57. Wang, J.; Wang, Z.; de Dear, R.; Luo, M.; Ghahramani, A.; Lin, B. The uncertainty of subjective thermal comfort measurement. *Energy Build.* **2018**, *181*, 38–49. [CrossRef]
58. *ASHRAE Standard 55. Thermal Environmental Conditions for Human Occupancy*; American Society of Heating, Refrigerating and Air Conditioning Engineers: Atlanta, GA, USA, 2017.
59. Djongyang, N.; Tchinda, R.; Njomo, D. Thermal comfort: A review paper. *Renew. Sustain. Energy Rev.* **2010**, *14*, 2626–2640. [CrossRef]
60. Wong, L.T.; Mui, K.W.; Hui, P.S. A multivariate-logistic model for acceptance of indoor environmental quality (IEQ) in offices. *Build. Environ.* **2008**, *43*, 1–6. [CrossRef]
61. Pellerin, N.; Candas, V. Effects of steady-state noise and temperature conditions on environmental perception and acceptability. *Indoor Air* **2004**, *14*, 129–136. [CrossRef]
62. Fanger, P.O. Assessment of man's thermal comfort in practice. *Br. J. Ind. Med.* **1973**, *30*, 313–324. [CrossRef]
63. Lawless, H.T.; Heymann, H. Acceptance and Preference Testing. In *Sensory Evaluation of Food*; Springer: Boston, MA, USA, 1999. [CrossRef]
64. Brager, G.; Baker, L. Occupant satisfaction in mixed-mode buildings. *Build. Res. Inf.* **2009**, *37*, 369–380. [CrossRef]
65. Liu, T.K.; Chen, Y.P.; Hou, Z.Y.; Wang, C.C.; Chou, J.H. Noninvasive evaluation of mental stress using a refined rough set technique based on biomedical signals. *Artif. Intell. Med.* **2014**, *61*, 97–103. [CrossRef]

66. Johnson, F.; Mavrogianni, A.; Ucci, M.; Wardle, J. Could increased time spent in a thermal comfort zone contribute to population increases in obesity? *Obes. Rev.* **2011**, *12*, 543–551. [CrossRef]

67. Keith, S.; Redden, D.T.; Katzmarzyk, P.; Boggiano, M.; Hanlon, E.; Benca, R.; Ruden, D.; Pietrobelli, A.; Barger, J.; Fontaine, K.; et al. Putative contributors to the secular increase in obesity: Exploring the roads less traveled. *Int. J. Obes.* **2006**, *30*, 1585–1594. [CrossRef]

68. Hansen, J.C.; Gilman, A.P.; Øyvind Odland, J. Is thermogenesis a significant causal factor in preventing the "globesity" epidemic? *Med. Hypotheses* **2010**, *75*, 250–256. [CrossRef]

69. Moellering, D.R.; Smith, D.L., Jr. Ambient Temperature and Obesity. *Curr. Obes. Rep.* **2012**, *1*, 26–34. [CrossRef] [PubMed]

70. Hanssen, M.J.; Hoeks, J.; Brans, B.; van der Lans, A.; Schaart, G.; van den Driessche, J.J.; Jörgensen, J.; Boekschoten, M.; Hesselink, M.; Havekes, B.; et al. Short-term cold acclimation improves insulin sensitivity in patients with type 2 diabetes mellitus. *Nat. Med.* **2015**, *21*, 863–865. [CrossRef] [PubMed]

71. Nicol, J.F.; Humphreys, M.A. Adaptive thermal comfort and sustainable thermal standards for buildings. *Energy Build.* **2002**, *34*, 563–572. [CrossRef]

72. Herrera-Limones, R.; Millán-Jiménez, A.; López-Escamilla, Á.; Torres-García, M. Health and Habitability in the Solar Decathlon University Competitions: Statistical Quantification and Real Influence on Comfort Conditions. *J. Environ. Res. Public Health* **2020**, *17*, 5926. [CrossRef] [PubMed]

73. Mueller, E.T.; Landes, R.D.; Kowal, B.P.; Yi, R.; Stitzer, M.L.; Burnett, C.A.; Bickel, W.K. Delay of smoking gratification as a laboratory model of relapse: Effects of incentives for not smoking, and relationship with measures of executive function. *Behav. Pharm.* **2009**, *20*, 461–473. [CrossRef]

74. Tullis, T.; Albert, B. Performance Metrics. In *Measuring the User Experience. Collecting, Analyzing, and Presenting Usability Metrics, Interactive Technologies*, 2nd ed.; Tullis, T., Albert, B., Eds.; Morgan Kaufmann: Burlington, VT, USA, 2013; Chapter 4; pp. 63–97. [CrossRef]

75. O'Brien, W.; Gaetani, I.; Carlucci, S.; Hoes, P.-J.; Hensen, J.L.M. On occupant-centric building performance metrics. *Build. Environ.* **2017**, *122*, 373–385. [CrossRef]

76. Seppanen, O.; Fisk, W.J. Title A model to estimate the cost effectiveness of the indoor environment improvements in office work. In *Creating the Productive Workplace*, 2nd ed.; Clements-Croome, D., Ed.; Taylor & Francis: Abingdon, UK, 2006. [CrossRef]

77. Djukanovic, R.; Wargocki, P.; Fanger, P.O. Cost-benefit analysis of improved air quality in an office building. *Proc. Indoor Air* **2002**, *1*, 808–813.

78. Armijos Moya, T.; van den Dobbelsteen, A.; Ottelé, M.; Bluysse, P.M. A review of green systems within the indoor environment. *Indoor Built Environ.* **2019**, *28*, 298–309. [CrossRef]

79. Gray, T.; Birrell, C. Are Biophilic-Designed Site Office Buildings Linked to Health Benefits and High Performing Occupants? *Int. J. Environ. Res. Public Health* **2014**, *11*, 12205–12221. [CrossRef]

80. Oseland, N.; Bartlett, P. *Improving Office Productivity: A Guide for Business and Facilities Managers*; Longman: London, UK, 1999; ISBN 978-0582357488.

81. Heerwagen, J. Green Buildings, organizational success and occupant productivity. *Build. Res. Inf.* **2000**, *28*, 353–367. [CrossRef]

82. Vimalanathan, K.; Babu, T.R. The effect of indoor office environment on the work performance, health and well-being of office workers. *J. Environ. Health Sci. Eng.* **2014**, *12*. [CrossRef] [PubMed]

83. Ramsey, J.D. Task performance in heat: A review. *Ergonomics* **1995**, *38*, 154–156. [CrossRef] [PubMed]

84. Paas, F.G.W.C.; Van Merriënboer, J.J.G. The efficiency of instructional conditions: An approach to combine mental effort and performance measures. *Hum. Factors J. Hum. Factors Erg. Soc.* **1993**, *35*, 737–743. [CrossRef]

85. Lan, L.; Lian, Z.; Pan, L.; Ye, Q. Neurobehavioral approach for evaluation of office workers' productivity: The effects of room temperature. *Build. Environ.* **2009**, *44*, 1578–1588. [CrossRef]

86. Wais, P.E.; Rubens, M.T.; Boccanfuso, J.; Gazzaley, A. Neural Mechanisms Underlying the Impact of Visual Distraction on Retrieval of Long-Term Memory. *J. Neurosci.* **2010**, *30*, 8541–8550. [CrossRef] [PubMed]

87. Buetti, S.; Lleras, A. Distractibility is a Function of Engagement, Not Task Difficulty: Evidence from a New Oculomotor Capture Paradigm. *J. Exp. Psychol. General.* **2016**, *145*, 1382–1405. [CrossRef]

88. Fiebelkorn, I.C.; Pinsk, M.A.; Kastner, S. Namic Interplay within the Frontoparietal Network Underlies Rhythmic Spatial Attention. *Neuron* **2018**, *99*, 842–853. [CrossRef]

89. Pizzamiglio, S.; Naeem, U.; ur Réhman, S.; Sharif, M.S.; Abdalla, H.; Turner, D.L. A Multimodal Approach to Measure the Distraction Levels of Pedestrians using Mobile Sensing. *Procedia Comput. Sci.* **2017**, *113*, 89–96. [CrossRef]

90. Glenberg, A.M.; Schroeder, J.L.; Robertson, D.A. Averting the gaze disengages the environment and facilitates remembering. *Mem. Cogn.* **1998**, *26*, 651–658. [CrossRef]

91. Song, K.; Meng, M.; Chen, L.; Zhou, K.; Luo, H. Behavioral oscillations in attention: Rhythmic α pulses mediated through θ band. *J. Neurosci. Off. J. Soc. Neurosci.* **2014**, *34*, 4837–4844. [CrossRef]

92. Van Rullen, R.; Carlson, T.; Cavanagh, P. The blinking spotlight of attention. *Proc. Natl. Acad. Sci. USA* **2017**, *104*, 19204–19209. [CrossRef]

93. Gallahan, S.L.; Golzar, G.F.; Jain, A.P.; Samay, A.E.; Trerotola, T.J.; Weisskopf, J.G.; Lau, N. Detecting and Mitigating Driver Distraction with Motion Capture Technology: Distracted Driving Warning System. In Proceedings of the IEEE Systems and Information Engineering Design Symposium, Charlottesville, VA, USA, 26 April 2013; pp. 76–81. [CrossRef]

94. Hamaoka, H.; Hagiwara, T.; Masahiro, T.A.D.A.; Munehiro, K. A study on the behavior of pedestrians when confirming approach of right/left-turning vehicle while crossing a crosswalk. In Proceedings of the IEEE Intelligent Vehicles Symposium (IV), Gold Coast, QLD, Australia, 23–26 June 2013; pp. 106–110. [CrossRef]

95. Mizoguchi, F.; Nishiyama, H.; Iwasaki, H. A new approach to detecting distracted car drivers using eye-movement data. In Proceedings of the IEEE 13th International Conference on Cognitive Informatics and Cognitive Computing, London, UK, 18–20 August 2014; pp. 266–272. [CrossRef]

96. Zaki, M.H.; Sayed, T. Exploring walking gait features for the automated recognition of distracted pedestrians. *IET Intell. Transp. Syst.* **2016**, *10*, 106–113. [CrossRef]

97. Uemura, Y.; Kajiwara, Y.; Shimakawa, H. Estimating Distracted Pedestrian from Deviated Walking Considering Consumption of Working Memory. In Proceedings of the International Conference on Computational Science and Computational Intelligence (CSCI), Las Vegas, NV, USA, 15–17 December 2016; pp. 1164–1167. [CrossRef]

98. Killane, I.; Browett, G.; Reilly, R.B. Measurement of attention during movement: Acquisition of ambulatory EEG and cognitive performance from healthy young adults. In Proceedings of the Annual International Conference of the IEEE Engineering in Medicine and Biology Society, Osaka, Japan, 3–7 July 2013; pp. 6397–6400. [CrossRef]

99. Almahasneh, H.; Chooi, W.T.; Kamel, N.; Malik, A.S. Deep in thought while driving: An EEG study on drivers' cognitive distraction. *Transp. Res. Part F Traffic Psychol. Behav.* **2014**, *26*, 218–226. [CrossRef]

100. Bigliassi, M.; Karageorghis, C.I.; Nowicky, A.V.; Wright, M.J.; Orgs, G. Effects of auditory distraction on voluntary movements: Exploring the underlying mechanisms associated with parallel processing. *Psychol. Res.* **2018**, *82*, 720–733. [CrossRef] [PubMed]

101. Casteau, S.; Smith, D.T. Covert attention beyond the range of eye-movements: Evidence for a dissociation between exogenous and endogenous orienting. *Cortex* **2020**, *122*, 170–186. [CrossRef] [PubMed]

102. Husain, M.; Kennard, C. Visual neglect associated with frontal lobe infarction. *J. Neurol.* **1996**, *243*, 652–657. [CrossRef] [PubMed]

103. Aryal, A.; Ghahramani, A.; Becerik-Gerber, B. Monitoring fatigue in construction workers using physiological measurements. *Autom. Constr.* **2017**, *82*, 154–165. [CrossRef]

104. Zhao, C.; Zheng, C.; Zhao, M.; Tu, Y.; Liu, J. Multivariate autoregressive models and kernel learning algorithms for classifying driving mental fatigue based on electroencephalographic. *Expert Syst. Appl.* **2011**, *38*, 1859–1865. [CrossRef]

105. Holm, A.; Lukander, K.; Korpela, J.; Sallinen, M.; Müller, K.M.I. Estimating brain load from the EEG. *Sci. World J.* **2009**, *9*, 639–651. [CrossRef]

106. Boksem, M.A.S.; Tops, M. Mental fatigue: Costs and benefits. *Brain Res. Rev.* **2008**, *59*, 125–139. [CrossRef]

107. Kramer, A.F. *Physiological Metrics of Mental Workload: A Review of Recent Progress*; Navy Personnel Research and Development Center: California, CA, USA, 1990; ISBN 92152-6800. Available online: https://pdfs.semanticscholar.org/475f/074528e18cda794 77ce02eb50fc1463fe56a.pdf (accessed on 28 November 2020).

108. Bokiniec, P.; Zampieri, N.; Lewin, G.R.; Poulet, J.F. The neural circuits of thermal perception. *Curr. Opin. Neurobiol.* **2018**, *52*, 98–106. [CrossRef] [PubMed]

109. Sur, S.; Sinha, V.K. Event-related potential: An overview. *Ind. Psychiatry J.* **2009**, *18*, 70–73. [CrossRef] [PubMed]

110. Giannakakis, G.; Marias, K.; Tsiknakis, M. A stress recognition system using HRV parameters and machine learning techniques. In Proceedings of the 8th International Conference on Affective Computing and Intelligent Interaction Workshops and Demos (ACIIW), Cambridge, UK, 3–6 September 2019; pp. 269–272. [CrossRef]

111. Chang, T.Y.; Kajackaite, A. Battle for the thermostat: Gender and the effect of temperature on cognitive performance. *PLoS ONE* **2019**, *14*. [CrossRef] [PubMed]

112. Zhang, L.; Helander, M.G.; Drury, C.G. Identifying Factors of Comfort and Discomfort in Sitting. *Hum. Factors* **1996**, *38*, 377–389. [CrossRef]

113. Cameron, J.A. Assessing work-related body-part discomfort: Current strategies and a behaviorally oriented assessment tool. *Int. J. Ind. Ergon.* **1996**, *18*, 389–398. [CrossRef]

114. Ashkenazy, S.; DeKeyser Ganz, F. The Differentiation between Pain and Discomfort: A Concept Analysis of Discomfort. *Pain Manag. Nurs.* **2019**, *20*, 556–562. [CrossRef]

115. Stanghellini, V. Review Article: Pain versus discomfort—Is differentiation clinically useful? *Aliment. Pharmacol. Ther.* **2001**, *15*, 145–149. [CrossRef]

116. Eliav, E.; Gracely, R.H. Measuring and assessing pain. In *Orofacial Pain and Headache*; Benoliel, R., Sharav, Y., Eds.; Elsevier Ltd.: Amsterdam, The Netherlands, 2008; Chapter 3; pp. 45–56. [CrossRef]

117. Grabisch, M.; Duchêne, J.; Lino, F.; Perny, P. Subjective Evaluation of Discomfort in Sitting Positions. *Fuzzy Optim. Decis. Mak.* **2002**, *1*, 287–312. [CrossRef]

118. Zillmann, D.; Rockwell, S.; Schweitzer, K.; Sundar, S.S. Does humor facilitate coping with physical discomfort? *Motiv. Emot.* **1993**, *17*, 1–21. [CrossRef]

119. Merriam-Webster. Discomfort, Merriam-Webster Dictionary. Available online: https://www.merriam-webster.com/dictionary/ discomfort (accessed on 29 November 2020).

120. Kyriakou, K.; Resch, B.; Sagl, G.; Petutschnig, A.; Werner, C.; Niederseer, D.; Liedlgruber, M.; Wilhelm, F.; Osborne, T.; Pykett, J. Detecting Moments of Stress from Measurements of Wearable Physiological Sensors. *Sensors* **2019**, *19*, 3805. [CrossRef]

121. Dhabhar, F.S.; Satoskar, A.R.; Bluethmann, H.; David, J.R.; McEwen, B.S. Stress-induced enhancement of skin immune function: A role for γ interferon. *Proc. Natl. Acad. Sci. USA* **2000**, *97*, 2846–2851. [CrossRef] [PubMed]

122. Healey, J.A.; Picard, R.W. Detecting stress during real-world driving tasks using physiological sensors. *IEEE Trans. Intell. Transp. Syst.* **2005**, *6*, 156–166. [CrossRef]

123. Crum, A.J.; Salovey, P.; Achor, S. Rethinking stress: The role of mindsets in determining the stress response. *J. Personal. Soc. Psychol.* **2013**, *104*. [CrossRef]

124. McEwen, B.S. Protective and damaging effects of stress mediators. *N. Engl. J. Med.* **1998**, *338*, 171–179. [CrossRef] [PubMed]

125. Peters, A.; McEwen, B.S.; Friston, K. Uncertainty and stress: Why it causes diseases and how it is mastered by the brain. *Prog. Neurobiol.* **2017**, *156*, 164–188. [CrossRef] [PubMed]

126. Marques, A.H.; Silverman, M.N.; Sternberg, E.M. Evaluation of stress systems by applying noninvasive methodologies: Measurements of neuroimmune biomarkers in the sweat, heart rate variability and salivary cortisol. *Neuroimmunomodulation* **2010**, *17*, 205–208. [CrossRef]

127. McEwen, B.S. Central effects of stress hormones in health and disease: Understanding the protective and damaging effects of stress and stress mediators. *Eur. J. Pharmacol.* **2008**, *583*, 174–185. [CrossRef]

128. Uchino, B.; Smith, T.; Holt-Lunstad, J.; Campo, R.; Reblin, M. Stress and Illness. In *Handbook of Psychophysiology*; Cacioppo, J.L., Tassinary, L.G., Berntson, G.G., Eds.; Cambridge University Press: Cambridge, UK, 2007; pp. 608–632.

129. Cacha, L.A.; Poznanski, R.R.; Latif, A.Z.A.; Ariff, T.M. Psychophysiology of chronic stress: An example of mind-body interaction. *NeuroQuantology* **2019**, *17*, 53–63. [CrossRef]

130. Schneiderman, N.; Ironson, G.; Siegel, S.D. Stress and health: Psychological, behavioral, and biological determinants. *Annu. Rev. Clin. Psychol.* **2005**, *1*, 607–628. [CrossRef]

131. Baker, N.; Standeven, M. Thermal comfort for free-running buildings. *Energy Build.* **1996**, *23*, 175–182. [CrossRef]

132. Lai, A.C.K.; Mui, K.W.; Wong, L.T.; Law, L.Y. An evaluation model for indoor environmental quality (IEQ) acceptance in residential buildings. *Energy Build.* **2009**, *41*, 930–936. [CrossRef]

133. Larsen, T.S.; Rohde, L.; Trangbæk Jønsson, K.; Rasmussen, B.; Lund Jensen, R.; Knudsen, H.N.; Witterseh, T.; Bekö, G. IEQ-Compass—A tool for holistic evaluation of potential indoor environmental quality. *Building Environ.* **2020**, *172*. [CrossRef]

134. Bluyssen, P.M.; Roda, C.; Mandin, C.; Fossati, S.; Carrer, P.; de Kluizenaar, Y.; Mihucz, V.G.; de Oliveira Fernandes, E.; Bartzis, J. Self-reported health and comfort in 'modern' office buildings: First results from the European OFFICAIR study. *Indoor Air* **2016**, *26*, 298–317. [CrossRef] [PubMed]

135. Lizana, J.; Almeida, S.M.; Serrano-Jiménez, A.; Becerra, J.A.; Gil-Báez, M.; Barrios-Padura, A.; Chacartegui, R. Contribution of Indoor Microenvironments to the Daily Inhaled Dose of Air Pollutants in Children. *Importance Bedrooms. Environ.* **2020**, *183*, 107188. [CrossRef]

136. U.S. Environmental Protection Agency. *The Total Exposure Assessment Methodology (TEAM) Study: Summary and Analysis*; EPA/600/6-87/002-a; Office of Research and Development U.S. Environmental Protection Agency: Washington, DC, USA, 1987.

137. Geiss, O.; Giannopoulos, G.; Tirendi, S.; Barrero-Moreno, J.; Larsen, B.R.; Kotzias, D. TheAIRMEXstudy—VOC measurements in public buildings and schools/kindergartens in eleven European cities: Statistical analysis of the data. *Atmos. Environ.* **2011**, *45*, 3676–3684. [CrossRef]

138. Panagiotaras, D.; Nikolopoulos, D.; Petraki, E.; Kottou, S.; Koulougliotis, D.; Yannakopoulos, P.; Kaplanis, S. Comprehensive experience for indoor air quality assessment: A review on the determination of volatile organic compounds (VOCs). *J. Phys. Chem. Biophys.* **2014**, *4*. [CrossRef]

139. Wolverton, B.C.; Mcdonald, R.C.; Watkins, E.A. Foliage plants for removing indoor air-pollutants from energy efficient homes. *Econ. Bot.* **1984**, *38*, 224–228. [CrossRef]

140. WSC POLICY #02-430. *Indoor Air Sampling and Evaluation Guide*; Commonwealth of Massachusetts Executive Office of Environmental Affairs, Department of Environmental Protection: Boston, MA, USA, 2002. Available online: https://www.mass.gov/doc/wsc-02-430-indoor-air-sampling-and-evaluation-guide-0/download (accessed on 29 October 2020).

141. Fanger, P.O.; Fcibase Fashrae, D. Sc. Olf and decipol: New units for perceived air quality. *Build. Serv. Eng. Res. Technol.* **1988**, *9*, 155–157. [CrossRef]

142. Arundel, A.V.; Sterling, E.M.; Biggin, J.H.; Sterling, T.D. Indirect health effects of relative humidity in indoor environments. *Environ. Health Perspect.* **1986**, *65*, 351–361. [CrossRef]

143. Wolkoff, P. Indoor air humidity, air quality, and health—An overview. *Int. J. Hyg. Environ. Health* **2018**, *221*, 376–390. [CrossRef]

144. Sundell, J.; Lindvall, T. Indoor air humidity and the sensation of dryness as risk indicators of SBS. *Indoor Air* **1993**, *3*, 382–390. [CrossRef]

145. European Commission. Building Stock Characteristics, EU Buildings Factsheets Topics Tree, Energy. Available online: https://ec.europa.eu/energy/eu-buildings-factsheets-topics-tree/building-stock-characteristics_en (accessed on 29 November 2020).

146. Hensen, J.L.M. *On the Thermal Interaction of Building Structure and Heating and Ventilating System*; Technische Universiteit Eindhoven: Eindhoven, The Netherlands, 1991; ISBN 90-386-0081-X.

147. Berglund, L. Mathematical models for predicting thermal comfort response of building occupants. *Ashrae J. Am. Soc. Heat. Refrig. Air Cond. Eng.* **1977**, *19*, 17–91.

148. 140 International Standard 7730. *Moderate Thermal Environments—Determination of the PMV and PPD Indices and Specification of the Conditions of Thermal Comfort*; International Standards Organization: Geneva, Switzerland, 2005.

149. Pallubinsky, H.; Schellen, L.; Lichtenbelt, W.D.V.M. Exploring the human thermoneutral zone—A dynamic approach. *J. Therm. Biol.* **2019**, *79*, 199–208. [CrossRef] [PubMed]

150. Butler, D.L.; Biner, P.M. Preferred Lighting Levels: Variability among Settings, Behaviors, and Individuals. *Environ. Behav.* **1987**, *19*, 695–721. [CrossRef]
151. Stevens, R.G.; Hansen, J.; Costa, G.; Haus, E.; Kauppinen, T.; Aronson, K.J.; Castaño-Vinyals, G.; Davis, S.; Frings-Dresen, M.H.; Fritschi, L.; et al. Considerations of circadian impact for defining 'shift work' in cancer studies: IARC Working Group Report. *Occup. Environ. Med.* **2011**, *68*, 154–162. [CrossRef]
152. Almutawa, F.; Vandal, R.; Wang, S.Q.; Lim, H.W. Current status of photoprotection by window glass, automobile glass, window films, and sunglasses. *Photodermatol. Photoimmunol. Photomed.* **2013**, *29*, 65–72. [CrossRef]
153. Mottram, V.; Middleton, B.; Williams, P.; Arendt, J. The impact of bright artificial white and 'blue enriched' light on sleep and circadian phase during the polar winter. *J. Sleep Res.* **2011**, *20*, 154–161. [CrossRef]
154. Rea, M.; Figueiro, M.; Bullough, J. Circadian photobiology: An emerging framework for lighting practice and research. *Lighting Res. Technol.* **2002**, *34*, 177–187. [CrossRef]
155. Van Bommel, W.; van den Beld, G. Lighting for work: A review of visual and biological effects. *Lighting Res. Technol.* **2004**, *36*, 255–266. [CrossRef]
156. Aries, M.B.C. *Human Lighting Demands: Healthy Lighting in an Office Environment*; Technische Universiteit Eindhoven: Eindhoven, The Netherlands, 2005. [CrossRef]
157. Walker, W.H.; Walton, J.C.; DeVries, A.C. Circadian rhythm disruption and mental health. *Transl. Psychiatry* **2020**, *10*. [CrossRef]
158. Kwallek, N.; Lewis, C.M.; Robbins, A.S. Effects of office interior color on workers' mood and productivity. *Percept. Motor Skills* **1988**, *66*, 123–128. [CrossRef]
159. Chang, C.-Y.; Chen, P.-K. Human response to window views and indoor plants in the workplace. *HortScience* **2005**, *40*, 1354–1359. [CrossRef]
160. Kaplan, R.; Kaplan, S. *The Experience of Nature: A Psychological Perspective*; CUP Archive: Cambridge, UK, 1989; ISBN 978-0521341394.
161. Ayr, U.; Cirillo, E.; Fato, I.; Martellotta, F. A new approach to assessing the performance of noise indices in buildings. *Appl. Acoust.* **2003**, *64*, 129–145. [CrossRef]
162. Stansfeld, S.A.; Matheson, M.P. Noise pollution: Non-auditory effects on health. *Br. Med Bull.* **2003**, *68*, 243–257. [CrossRef] [PubMed]
163. Evans, G.W.; Bullinger, M.; Hygge, S. Chronic noise exposure and physiological response: A prospective study of children living under environmental stress. *Psychol. Sci.* **1998**, *9*, 75–77. [CrossRef]
164. World Health Organization. *Environmental Noise Guidelines for the European Region*; WHO Regional Office for Europe: Copenhagen, Denmark, 2018; ISBN 978-92-890-5356-3.
165. Field, C.D.; Digerness, J. Acoustic design criteria for naturally ventilated buildings. *J. Acoust. Soc. Am.* **2008**, *123*, 9269–9273. [CrossRef]
166. Mui, K.; Wong, L. A method of assessing the acceptability of noise levels in air-conditioned offices. *Building Serv. Eng. Res. Technol.* **2006**, *27*, 249–254. [CrossRef]
167. Payne, S.R. The production of a perceived restorativeness soundscape scale. *Appl. Acoust.* **2013**, *74*, 255–263. [CrossRef]
168. World Green Building Council (WGBC). Health, Wellbeing & Productivity in Offices, World Green Building Council. 2014. Available online: https://www.worldgbc.org/sites/default/files/compressed_WorldGBC_Health_Wellbeing__Productivity_Full_Report_Dbl_Med_Res_Feb_2015.pdf (accessed on 29 November 2020).
169. Azkorra, Z.; Perez, G.; Coma, J.; Cabeza, L.F.; Bures, S.; Alvaro, J.E.; Erkoreka, A.; Urrestarazu, M. Evaluation of green walls as a passive acoustic insulation system for buildings. *Appl. Acoust.* **2015**, *89*, 46–56. [CrossRef]
170. Ermes, M. *Methods for the Classification of Biosignals Applied to the Detection of Epileptiform Waveforms and to the Recognition of Physical Activity*; Tampere University of Technology: Tampere, Finland, 2009; Available online: http://www.vtt.fi/publications/index.jsp (accessed on 29 November 2020).
171. Bronzino, J.D. Biomedical Engineering: A Historical Perspective. In *Biomedical Engineering, Introduction to Biomedical Engineering*, 3rd ed; Enderle, J.D., Bronzino, J.D., Eds.; Academic Press: Cambridge, MA, USA, 2012; Chapter 1; pp. 1–33. ISBN 9780123749796. [CrossRef]
172. Fong, S.; Lan, K.; Sun, P.; Mohammed, S.; Fiaidhi, J. A time-series pre-processing methodology for biosignal classification using statistical feature extraction. In Proceedings of the IASTED International Conference on Biomedical Engineering (BioMed 2013), Innsbruck, Austria, 13–15 February 2013; pp. 207–214. [CrossRef]
173. Escabí, M. Biosignal Processing. In *Biomedical Engineering, Introduction to Biomedical Engineering*, 3rd ed.; Enderle, J.D., Bronzino, J.D., Eds.; Academic Press: Cambridge, MA, USA, 2012; Chapter 11; pp. 667–746. ISBN 9780123749796. [CrossRef]
174. Kaniusas, E. *Biomedical Signals and Sensors I, Biological and Medical Physics, Biomedical Engineering*; Springer: Berlin/Heidelberg, Germany, 2012; ISBN 978-3-642-24842-9.
175. Malmivuo, J.; Plonsey, R. *Bioelectromagnetism: Principles and Applications of Bioelectric and Biomagnetic Fields*; Oxford University Press: Oxford, UK, 1995. [CrossRef]
176. Towe, B.C. Bioelectricity and its Measurement. In *Standard Handbook of Biomedical Engineering & Design*; Kutz, M., Ed.; The McGraw-Hill Companies, Inc.: New York, NY, USA, 2009; Volume 1, Chapter 20; pp. 481–528. ISBN 9780071498388.
177. Walter-Kroker, A.; Kroker, A.; Mattiucci-Guehlke, M.; Glaab, T. A practical guide to bioelectrical impedance analysis using the example of chronic obstructive pulmonary disease. *Nutr. J.* **2011**, *10*. [CrossRef]

178. Kyle, U.G.; Bosaeus, I.; De Lorenzo, A.D.; Deurenberg, P.; Elia, M.; Gómez, J.M.; Heitmann, B.L.; Kent-Smith, L.; Melchior, J.C.; Pirlich, M.; et al. Bioelectrical impedance analysis-part I: Review of principles and methods. *Clin. Nutr.* **2004**, *23*, 1226–1243. [CrossRef] [PubMed]

179. Callejas-Cuervo, M.; Alvarez, J.C.; Alvarez, D. Capture and analysis of biomechanical signals with inertial and magnetic sensors as support in physical rehabilitation processes. In Proceedings of the IEEE 13th International Conference on Wearable and Implantable Body Sensor Networks (BSN), San Francisco, CA, USA, 14–17 June 2016; pp. 119–123. [CrossRef]

180. Hadjileontiadis, L.J.; Rekanos, I.T.; Panas, S.M. *Bioacoustic Signals*; Akay, M., Ed.; Wiley Encyclopedia of Biomedical Engineering: Hoboken, NJ, USA, 2006. [CrossRef]

181. Pourhomayoun, M.; Dugan, P.; Popescu, M.; Risch, D.; Lewis, H.; Clark, C. Classification for Big Dataset of Bioacoustic Signals Based on Human Scoring System and Artificial Neural Network. In Proceedings of the ICML 2013 Workshop on Machine Learning for Bioacoustics, Atlanta, GA, USA, 16–21 June 2010.

182. Karthikeyan, P.; Murugappan, M.; Yaacob, S. Detection of human stress using short-term ECG and HRV signals. *J. Mech. Med. Biol.* **2013**, *13*. [CrossRef]

183. Neuman, M.R. Biomedical Sensors. In *Sensors, Nanoscience, Biomedical Engineering, and Instruments (The Electrical Engineering Handbook)*, 3rd ed.; Dorf, R.C., Ed.; CRC Press: Boca Raton, FL, USA, 2006; Chapter 8; pp. 8.1–8.11. ISBN 9780849373466.

184. Selye, H. Stress and the general adaptation syndrome. *Br. Med J.* **1950**, *1*, 1383–1392. [CrossRef] [PubMed]

185. Godoy, L.D.; Rossignoli, M.T.; Delfino-Pereira, P.; Garcia-Cairasco, N.; de Lima Umeoka, E.H. A Comprehensive Overview on Stress Neurobiology: Basic Concepts and Clinical Implications. *Front. Behav. Neurosci.* **2018**, *12*. [CrossRef]

186. Tan, S.Y.; Yip, A. Hans Selye (1907–1982): Founder of the stress theory. *Singap. Med J.* **2018**, *59*, 170–171. [CrossRef]

187. Picard, R.W. Automating the Recognition of Stress and Emotion: From Lab to Real-World Impact. *IEEE Multim.* **2016**, *23*, 3–7. [CrossRef]

188. De Santos, A.; Sánchez-Avila, C.; Guerra-Casanova, J.; Bailador-Del Pozo, G. Real-Time Stress Detection by Means of Physiological Signals. In *Recent Application in Biometrics*; Yang, J., Poh, N., Eds.; IntechOpen: London, UK, 2011; pp. 23–44. Available online: https://www.intechopen.com/books/recent-application-in-biometrics/hand-biometrics-in-mobile-devices (accessed on 29 November 2020). [CrossRef]

189. Schmidt, P.; Reiss, A.; Dürichen, R.; Laerhoven, K.V. Wearable-Based Affect Recognition-A Review. *Sensors* **2019**, *19*, 4079. [CrossRef]

190. Seshadri, D.R.; Li, R.T.; Voos, J.E.; Rowbottom, J.R.; Alfes, C.M.; Zorman, C.A.; Drummond, C.K. Wearable sensors for monitoring the physiological and biochemical profile of the athlete. *NPJ Digit. Med.* **2019**, *2*. [CrossRef]

191. Can, Y.S.; Chalabianloo, N.; Ekiz, D.; Ersoy, C. Continuous stress detection using wearable sensors in real life: Algorithmic programming contest case study. *Sensors* **2019**, *19*, 1849. [CrossRef]

192. Kappenman, E.S.; Luck, S.J. The effects of electrode impedance on data quality and statistical significance in ERP recordings. *Psychophysiology* **2010**, *47*, 888–904. [CrossRef]

193. Cruz-Garza, J.G.; Brantley, J.A.; Nakagome, S.; Kontson, K.; Megjhani, M.; Robleto, D.; Contreras-Vidal, J.L. Deployment of Mobile EEG Technology in an Art Museum Setting: Evaluation of Signal Quality and Usability. *Front. Hum. Neurosci.* **2017**, *11*. [CrossRef] [PubMed]

194. Wijsman, J.; Grundlehner, B.; Liu, H.; Penders, J.; Hermens, H. Wearable physiological sensors reflect mental stress state in office-like situations. In Proceedings of the 2013 Humaine Association Conference on Affective Computing and Intelligent Interaction (ACII), Geneva, Switzerland, 2–5 September 2013; pp. 600–605. [CrossRef]

195. Sharma, N.; Gedeon, T. Hybrid Genetic Algorithms for Stress Recognition in Reading, Evolutionary Computation. In Proceedings of the 11th European conference on Evolutionary Computation, Machine Learning and Data Mining in Bioinformatics, Vienna, Austria, 3–5 April 2013. [CrossRef]

196. Hernandez, J.; Morris, R.; Picard, R.W. Call Center Stress Recognition with Person-Specific Models. In *ACII 2011, Part I, LNCS 6974*; D´Mello, S., Ed.; Springer: Berlin/Heidelberg, Germany, 2011; pp. 125–134. [CrossRef]

197. D'mello, S.K.; Kory, J. A Review and Meta-Analysis of Multimodal Affect Detection Systems. *ACM Comput. Surv.* **2015**, *47*, 43. [CrossRef]

198. Chinazzo, G.; Wienold, J.; Andersen, M. Daylight affects human thermal perception. *Sci. Rep.* **2019**, *9*, 1–15. [CrossRef] [PubMed]

199. Luck, S.J. *An Introduction to the Event-Related Potential Technique*; The MIT Press: Cambridge, MA, USA, 2005; ISBN 978-0-262-12277-1.

200. Stone, J.L.; Hughes, J.R. Early History of Electroencephalography and Establishment of the American Clinical Neurophysiology Society. *J. Clin. Neurophysiol.* **2013**, *30*, 28–44. [CrossRef] [PubMed]

201. Yao, Y.; Lian, Z.; Liu, W.; Jiang, C.; Liu, Y.; Lu, H. Heart rate variation and electroencephalograph—The potential physiological factors for thermal comfort study. *Indoor Air* **2009**, *19*, 93–101. [CrossRef] [PubMed]

202. Abo-Zahhad, M.; Ahmed, S.; Abbas, S.N. A New EEG Acquisition Protocol for Biometric Identification Using Eye Blinking Signals. *Int. J. Intell. Syst. Appl.* **2015**, *7*, 48–54. [CrossRef]

203. Cohen, M.X. *Analyzing Neural Time Series Data: Theory and Practice*; The MIT Press: Cambridge, MA, USA, 2014; ISBN 9780262019873.

204. Thakor, N.V.; Tong, S. Advances in quantitative electroencephalogram analysis methods. *Annu. Rev. Biomed. Eng.* **2004**, *6*, 453–495. [CrossRef]

205. Wang, X.; Nie, D.; Lu, B. Emotional state classification from EEG data using machine learning approach. *Neurocomputing* **2014**, *129*, 94–106. [CrossRef]

206. Klimesch, W. EEG alpha and theta oscillations reflect cognitive and memory performance: A review and analysis. *Brain Res. Rev.* **1999**, *29*, 169–195. [CrossRef]
207. Nyhus, E.; Curran, T. Functional role of gamma and theta oscillations in episodic memory. *Neurosci. Biobehav. Rev.* **2010**, *34*, 1023–1035. [CrossRef]
208. Jensen, O.; Gips, B.; Bergmann, T.O.; Bonnefond, M. Temporal coding organized by coupled alpha and gamma oscillations prioritize visual processing. *Trends Neurosci.* **2014**, *37*, 357–369. [CrossRef] [PubMed]
209. Doesburg, S.M.; Roggeveen, A.B.; Kitajo, K.; Ward, L.M. Large-scale gamma-band phase synchronization and selective attention. *Cereb. Cortex* **2008**, *18*, 386–396. [CrossRef] [PubMed]
210. Hamid, N.H.A.; Sulaiman, N.; Murat, Z.H.; Taib, M.N. Brainwaves stress pattern based on perceived stress scale test. In Proceedings of the IEEE 6th Control and System Graduate Research Colloquium (ICSGRC), Shah Alam, Malaysia, 10–11 August 2015; pp. 135–140. [CrossRef]
211. Davidson, R.J.; Ekman, P.; Saron, C.D.; Senulis, J.A.; Friesen, W.V. Approach-withdrawal and cerebral asymmetry: Emotional expression and brain physiology. *J. Personal. Soc. Psychol.* **1990**, *58*, 330–341. [CrossRef]
212. Saeed, S.M.U.; Anwar, S.M.; Khalid, H.; Majid, M.; Bagci, U. EEG Based Classification of Long-Term Stress Using Psychological Labeling. *Sensors* **2020**, *20*, 1886. [CrossRef] [PubMed]
213. Mulders, D.; De Bodt, C.; Lejeune, N.; Courtin, A.; Liberati, G.; Verleysen, M.; Mouraux, A. Dynamics of the perception and EEG signals triggered by tonic warm and cool stimulation. *PLoS ONE* **2020**, *15*. [CrossRef] [PubMed]
214. Islam, M.K.; Rastegarnia, A.; Yang, Z. Methods for artifact detection and removal from scalp EEG: A review. *Neurophysiol. Clin. Clin. Neurophysiol.* **2016**, *46*, 287–305. [CrossRef]
215. Singh, Y.N.; Singh, S.K.; Ray, A.K. Bioelectrical Signals as Emerging Biometrics: Issues and Challenges. *ISRN Signal Process.* **2012**, 1–13. [CrossRef]
216. Manriquez, A.; Zhang, Q.; Médigue, C.; Papelier, Y.; Sorine, M. Multi-lead T wave end detection based on statistical hypothesis testing. *IFAC Proc. Vol.* **2006**, *39*, 93–98. [CrossRef]
217. Malik, M.; Camm, A.J. Heart Rate Variability. *Clin. Cardiol.* **1990**, *13*, 570–576. [CrossRef]
218. McCraty, R.; Shaffer, F. Heart Rate Variability: New Perspectives on Physiological Mechanisms, Assessment of Self-regulatory Capacity, and Health risk. *Glob. Adv. Health Med.* **2015**, *4*, 46–61. [CrossRef]
219. Acharya, U.R.; Joseph, K.P.; Kannathal, N.; Lim, C.; Suri, J. Heart rate variability: A review. *Med Biol. Eng. Comput.* **2006**, *44*, 1031–1051. [CrossRef] [PubMed]
220. Shaffer, F.; Ginsberg, J.P. An Overview of Heart Rate Variability Metrics and Norms. *Front. Public Health* **2017**, *5*, 1–17. [CrossRef]
221. Faurholt-Jepsen, M.; Kessing, L.V.; Munkholm, K. Heart rate variability in bipolar disorder: A systematic review and meta-analysis. *Neurosci. Biobehav. Rev.* **2017**, *73*, 68–80. [CrossRef] [PubMed]
222. Zhu, H.; Wang, H.; Liu, Z.; Li, D.; Kou, G.; Li, C. Experimental study on the human thermal comfort based on the heart rate variability (HRV) analysis under different environments. *Sci. Total Environ.* **2018**, *616–617*, 1124–1133. [CrossRef] [PubMed]
223. Liu, W.; Lian, Z.; Liu, Y. Heart rate variability at different thermal comfort levels. *Eur. J. Appl. Physiol.* **2008**, *103*, 361–366. [CrossRef] [PubMed]
224. Hjortskov, N.; Rissén, D.; Blangsted, A.K.; Fallentin, N.; Lundberg, U.; Søgaard, K. The effect of mental stress on heart rate variability and blood pressure during computer work. *Eur. J. Appl. Physiol.* **2004**, *92*, 84–89. [CrossRef] [PubMed]
225. Nkurikiyeyezu, K.N.; Suzuki, Y.; Lopez, G.F. Heart rate variability as a predictive biomarker of thermal comfort. *J. Ambient Intell. Humaniz. Comput.* **2018**, *9*, 1465–1477. [CrossRef]
226. Fernandeza, S.; Lázaroa, I.; Arnaiza, A.; Calis, G. Application of heart rate variability for thermal comfort in office buildings in real-life conditions Santiago. In Proceedings of the Creative Construction Conference, Ljubljana, Slovenia, 30 June–3 July 2018; pp. 798–805. [CrossRef]
227. Melillo, P.; Bracale, M.; Pecchia, L. Nonlinear Heart Rate Variability features for real-life stress detection. Case study: Students under stress due to university examination. *Biomed. Eng. Online* **2011**, *10*, 96. [CrossRef]
228. Filingeri, D. Neurophysiology of Skin Thermal Sensations. *Compr. Physiol.* **2016**, *6*, 1429. [CrossRef]
229. Cho, D.; Ham, J.; Oh, J.; Park, J.; Kim, S.; Lee, N.K.; Lee, B. Detection of stress levels from biosignals measured in virtual reality environments using a kernel-based extreme learning machine. *Sensor* **2017**, *17*, 2435. [CrossRef]
230. Li, D.; Menassa, C.C.; Kamat, V.R. Non-intrusive interpretation of human thermal comfort through analysis of facial infrared thermography. *Energy Build.* **2018**, *176*, 246–261. [CrossRef]
231. Yao, Y.; Lian, Z.; Liu, W.; Shen, Q. Experimental Study on Skin Temperature and Thermal Comfort of the Human Body in a Recumbent Posture under Uniform Thermal Environments. *Indoor Built Environ.* **2007**, *16*, 505–518. [CrossRef]
232. Choi, J.; Loftness, V. Investigation of human body skin temperatures as a bio-signal to indicate overall thermal sensations. *Build. Environ.* **2012**, *58*, 258–269. [CrossRef]
233. Sim, S.Y.; Koh, M.J.; Joo, K.M.; Noh, S.; Park, S.; Kim, Y.H.; Park, K.S. Estimation of Thermal Sensation Based on Wrist Skin Temperatures. *Sensors* **2016**, *16*, 420. [CrossRef] [PubMed]
234. Zhai, J.; Barreto, A.; Chin, C.; Li, C. Realization of stress detection using psychophysiological signals for improvement of human-computer interactions. In Proceedings of the IEEE SoutheastCon, Ft. Lauderdale, FL, USA, 8–10 April 2005; pp. 415–420. [CrossRef]

235. Angus, F.; Zhai, J.; Barreto, A. Front-end analog pre-processing for real-time psychophysiological stress measurements. In Proceedings of the 9th World Multi-Conference on Systemics, Cybernetics and Informatics (WMSCI 05), Orlando, FL, USA, 10–13 July 2005; pp. 218–221.

236. Topoglu, Y.; Watson, J.; Suri, R.; Ayaz, H. Electrodermal activity in ambulatory settings: A narrative review of literature. *Adv. Intell. Syst. Comput.* **2020**, *953*, 91–102. [CrossRef]

237. Jung, C.G. *Studies in Word-association: Experiments in the Diagnosis of Psychopathological Conditions carried out at the Psychiatric Clinic of the University of Zurich*; Moffat, Yard & Company: New York, NY, USA, 1906. [CrossRef]

238. Bakker, J.; Pechenizkiy, M.; Sidorova, N. What's your current stress level? Detection of stress patterns from GSR sensor data. In Proceedings of the 2011 IEEE 11th International Conference on Data Mining Workshops (ICDMW), Washington, DC, USA, 11–14 December 2011; pp. 573–580. [CrossRef]

239. Johannes, S.; Rüdiger, P.; Marc, S.; Manfred, R. Towards Flexible Mobile Data Collection in Healthcare. In Proceedings of the 29th IEEE International Symposium on Computer-Based Medical Systems (CBMS), Dublin, Ireland, 20–24 June 2016; pp. 181–182. [CrossRef]

240. Zangróniz, R.; Martínez-Rodrigo, A.; Pastor, J.M.; López, M.T.; Fernández-Caballero, A. Electrodermal activity sensor for classification of calm/distress condition. *Sensors* **2017**, *17*, 2324. [CrossRef] [PubMed]

241. Stephens, M.A.; Wand, G. Stress and the HPA axis: Role of glucocorticoids in alcohol dependence. *Alcohol Res. Curr. Rev.* **2012**, *34*, 468–483, Corpus ID: 142581511.

242. Smith, S.M.; Vale, W.W. The role of the hypothalamic-pituitary-adrenal axis in neuroendocrine responses to stress. *Dialogues Clin. Neurosci.* **2006**, *8*, 383–395. [CrossRef]

243. Oswald, L.M.; Zandi, P.; Nestadt, G.; Potash, J.B.; Kalaydjian, A.E.; Wand, G.S. Relationship between cortisol responses to stress and personality. *Neuropsychopharmacology* **2006**, *31*, 1583–1591. [CrossRef]

244. Lee, D.Y.; Kim, E.; Choi, M.H. Technical and clinical aspects of cortisol as a biochemical marker of chronic stress. *BMB Rep.* **2015**, *48*, 209–216. [CrossRef]

245. Stalder, T.; Kirschbaum, C. Analysis of cortisol in hair—State of the art and future directions. *Brainbehav. Immun.* **2012**, *26*, 1019–1029. [CrossRef] [PubMed]

246. Schweiker, M. Rethinking resilient comfort—Definitions of resilience and comfort and their consequences for design, operation, and energy use. In *11th Windsor Conference on Thermal Comfort 2020: Resilient Comfort*; Roaf, S., Nicol, F., Finlayson, W., Eds.; Ecohouse Initiative Ltd.: Oxford, UK, 2020; pp. 34–46.

Article

Effect of Thermal, Acoustic and Air Quality Perception Interactions on the Comfort and Satisfaction of People in Office Buildings

Leonidas Bourikas [1,*] , Stephanie Gauthier [2], Nicholas Khor Song En [2] and Peiyao Xiong [2]

[1] School of Architecture, Imagination Lancaster, LICA, Lancaster University, Lancaster LA1 4YW, UK
[2] Sustainable Energy Research Group, Faculty of Engineering and Physical Sciences, University of Southampton, Southampton SO17 1BJ, UK; s.gauthier@soton.ac.uk (S.G.); nsek1g17@soton.ac.uk (N.K.S.E.); px1n19@soton.ac.uk (P.X.)
* Correspondence: l.bourikas@lancaster.ac.uk

Abstract: Current research on human comfort has identified a gap in the investigation of multi-domain perception interactions. There is a lack of understanding the interrelationships of different physio-socio-psychological factors and the manifestation of their contextual interactions into cross-modal comfort perception. In that direction, this study used data from a post occupancy evaluation survey (n = 26), two longitudinal comfort studies (n = 1079 and n = 52) and concurrent measurements of indoor environmental quality factors (one building) to assess the effect of thermal, acoustic and air quality perception interactions on comfort and satisfaction of occupants in three mixed-mode university office buildings. The study concluded that thermal sensation (TSV) is associated with both air quality (ASV) and noise perception (NSV). The crossed effect of the interaction of air quality and noise perception on thermal sensation was not evident. The key finding was the significant correlation of operative temperature (T_{op}) with TSV as expected, but also with noise perception and overall acoustic comfort. Regarding the crossed main effects on thermal sensation, a significant effect was found for the interactions of (1) T_{op} and (2) sound pressure levels ($SPL30$) with air quality perception respectively. Most importantly, this study has highlighted the importance of air quality perception in achieving occupants' comfort and satisfaction with office space.

Keywords: human thermal perception; comfort; multi-domain interactions; indoor air quality; noise sensation; cross-modal perception

Citation: Bourikas, L.; Gauthier, S.; Khor Song En, N.; Xiong, P. Effect of Thermal, Acoustic and Air Quality Perception Interactions on the Comfort and Satisfaction of People in Office Buildings. *Energies* **2021**, *14*, 333. https://doi.org/10.3390/en14020333

Received: 1 December 2020
Accepted: 18 December 2020
Published: 9 January 2021

Publisher's Note: MDPI stays neutral with regard to jurisdictional claims in published maps and institutional affiliations.

1. Introduction

Energy use of buildings as well as the health and wellbeing of a building's occupants have been repeatedly associated with the interaction between occupants and their indoor environment as they respond to environmental cues to achieve comfort [1–4]. Indoor environmental quality (IEQ) is often evaluated by assessing thermal, acoustic, visual and air quality factors [1,5]. A recent literature review, however, has identified a gap in studies on the combined effects of these factors on overall comfort [6], as most studies take place in controlled environments (e.g., climate chambers) and/or focus only on single effects.

Comfort research has historically been approached with siloed, discipline-specific approaches and methods that fail to provide a holistic interpretation of the physio-socio-psychological relationships and their evolution over time and external forcing. In addition, differences between individuals [3,7], and the yet unknown mechanisms of how stimuli trigger sensations, add complexity when evaluating comfort and applying comfort models to building design and the operation of building systems [8]. From a trans-modal perspective [9], the interactions between different comfort domains (i.e., cross-modal effects [10–12]) are interpreted as comfort and satisfaction within a physio-psychological context.

Current research has investigated the relationships and mechanisms that link the level of comfort with the perception of indoor environment quality and its contextual manifestation. In general, comfort dimensions are distinguished as expectation (non-sensory stimuli), sensation (sensory stimuli) and relative perception (satisfaction and preference) [13]. This interpretation of comfort is linked to the "affect" defined in Heydarian et al.'s (2020) [14] review of behavioural theories and to the holistic review into the drivers of thermal perception [15]. In that direction, Schweiker et al. (2020) examined the concept of "seasonal alliesthesia" that approaches comfort through the view of thermal pleasure and suggests the existence of long-term experiences and expectations, which affect contextual thermal perception [13,16]. Studies on multi-modal perception interactions and their effect on comfort remain limited. Tang et al. (2020) in a controlled laboratory experiment identified relationships between indoor air quality and both thermal satisfaction and sound pressure levels [17]. In another controlled experiment, Yang and Moon (2019) investigated multisensory interactions between thermal, acoustic and illuminance conditions, and their study has revealed that indoor sound levels have a larger impact on thermal comfort than illuminance, whereas interestingly sound levels had significant effects on both the overall indoor environment and visual comfort [18]. Similarly, in an experimental laboratory study, satisfaction with thermal conditions has been shown to influence satisfaction with other IEQ factors, but most importantly thermal satisfaction has created "comfort expectations" that subsequently affect an occupant's evaluation of other surveyed parameters [19]. These important findings all come from studies in controlled environments and quite often with student participants. Most studies usually assess the combined effects of IEQ factors, and they evaluate the relative effect and influence of factors, instead of explicitly looking at interaction effects in terms of the examined factors. This paper addresses those limitations by presenting results from "real life" office environments, and we analyze the multi-modal interactions (i.e. cross-modal) between thermal, acoustic and air quality perception and commonly monitored environmental IEQ factors.

Over the last few decades, before the SARS COVID pandemic, open-space offices prevailed as an office type, mainly due to the reduction in facility costs that this working environment offers. Several studies have shown that open-space design has adversely affected productivity, health and wellbeing in the workspace [20,21]. Amount of space, noise, visual intrusiveness and lack of privacy are commonly identified by employees as the key factors for their (lack of) satisfaction with their workspace environment [22,23]. More recently, studies have shown that despite the important roles of factors related to the indoor environment, there are additional socio-psychological parameters that could influence thermal perception and comfort, such as personal mood, aesthetics, likable architecture, office layout and employer's policies related to wellbeing [24–27]. Post Occupancy Evaluation (POE) methodologies have been increasingly used to identify issues that might affect a building's performance and occupant satisfaction. POE has gained attention from the industry and has been included as a requirement in popular "green" building certifications and sustainability assessment methods. Reportedly, though, high scores in such certification schemes often do not translate into occupants' satisfaction with the indoor environment [28,29]. Nevertheless, POE results offer valuable insights into the aspects that require attention and cause dissatisfaction among the buildings' users.

This study combines results from a POE and comfort surveys to evaluate the mixed effects of indoor environment parameters and multi-domain comfort on thermal comfort perception. The present work aims to provide new evidence on the cross-modal comfort perception in "real-living" conditions. The research questions addressed by this study are as follows:

(1) Is there a cross-modal effect of thermal, acoustic and air quality perception on occupants' comfort?

(2) How does indoor environmental quality affect thermal perception?

2. Materials and Methods

2.1. Study Design

This research was carried out in three stages, described as follows.

Stage 1: In the first stage, a Post Occupancy Evaluation (POE) was carried out following the Building Use Studies (BUS) methodology [30,31]. The aim of POE was to gain an understanding of the buildings users' satisfaction with their living environment. BUS is an established POE methodology that has been widely used and reported in relevant literature [29,32–34]. On top of the expected benefits, BUS offers the opportunity to compare occupants' satisfaction and building performance with other similar buildings worldwide.

The case study building was a mixed-mode office building at the University of Southampton (referred to as B1). This pilot study had 26 participants who completed the BUS online questionnaire survey during August 2020. The survey includes 16 sections, collecting demographic data such as age, gender, working hours and days at the building and at their work area. It is also designed to collect feedback regarding the building's environmental performance, building services and overall design. The analysis of the POE focused on the reported satisfaction with typical indoor environmental quality (IEQ) (air quality, thermal conditions, noise and lighting) factors and the occupants' thermal perception and comfort in winter and summer. The survey applied a 7-point scale to assess the satisfaction level from "1" representing "very dissatisfied" to "7" representing "very satisfied". Similarly, a 7-point scale was used to assess thermal comfort from "1" representing "uncomfortable" to "7" representing "comfortable".

Stage 2: In the second stage, building upon the findings from the POE, a weekly "right-here right-now" comfort survey took place over a period of a year (from 25 July 2017 to 25 June 2018) to review the relationship among thermal sensation (TSV), thermal preference (TPV), acoustic perception (NSV) and perceived air quality (ASV). The thermal sensation vote (TSV), thermal preference vote (TPV), acoustic (i.e., noise) perception vote (NSV) and air quality perception vote (ASV) were collected from the questionnaire responses, as shown in Table 1. Four different scales were used following the smart controls and thermal comfort project (SCATs) survey method [35]. Thermal sensation (TSV) was assessed with the ASHRAE 7-point scale [36], and thermal preference, acoustic and air quality perception were evaluated with a relevant 7-point scale, respectively (Table 1). The survey responses were collected from participants in two mixed-mode office buildings at the University of Southampton (referred to as B2 and B3). The number of valid, completed surveys and consequently the sample size is n = 1079 responses from 116 individual participants. In addition to the surveys, the researchers interviewed the respective building managers to understand any daily operation issues and common grievances from the occupants.

Table 1. Sampled questions related to the investigated variables from the weekly survey questionnaires.

Thermal Perception "How Do You Feel Right Now?" (TSV)						
Cold (coded: −3)	Cool	Slightly cool	Neutral (coded: 0)	Slightly warm	Warm	Hot (coded: +3)
Thermal Preference "At this Moment, Would You Prefer to be … ?" (TPV)						
Much cooler (coded: −3)	Cooler	Slightly cooler	Without change (coded: 0)	Slightly warmer	Warmer	Much warmer (coded: +3)
Noise Perception "How Do You Find the Background Noise Level?" (NSV)						
Very Noisy (coded: −3)	Noisy	Slightly Noisy	Neither noisy nor quiet (coded: 0)	Slightly quiet	Quiet	Very quiet (coded: +3)
Air Quality Perception "How Do You Find the Air Quality?" (ASV)						
Very bad (coded: −3)	Bad	Slightly Bad	Neither bad nor good (coded: 0)	Slightly good	Good	Excellent (coded: +3)

Stage 3: In Stage 3 of the research design, a small and focused study in building B1 was undertaken to assess thermal, acoustic and air quality perception with a comfort survey and concurrent environmental conditions monitoring (i.e., sound pressure levels, air temperature and relative humidity), between January and March 2020. Typical indi-

vidual control opportunities were monitored such as (1) private office door open/close, (2) window open/close, (3) air conditioning on/off, (4) ventilation on/off and (5) heating on/off. The sample comprised 12 participants, and most worked in an open plan office (N = 9) in B1 or individual offices (N = 3) at the same floor level. Following a similar method as Stage 2, the 12 participants in the Stage 3 survey were asked to complete three surveys; an initial background survey that collected contextual metadata and demographic information, a weekly survey that assessed their comfort perception and productivity, and a final feedback survey. The questionnaire followed the thermal comfort informative for subjective evaluation in the ASHRAE 55 [36] and EN 15251 (Annex H) [37]. The three investigated variables (*TSV, NSV, ASV*) were assessed with the same as in Stage 2 perceptual scales. Non-intrusive environmental sensors were installed in selected locations that represent the average ambient environment. Although the participants from B1 were different from the participants at B2 and B3, they all worked in similar buildings at the same location, and they did similar jobs. In addition, the buildings were maintained and operated by the same organization, they all had mixed-mode ventilation and followed the same operation schedule.

2.2. Case Study Buildings

The three case study buildings have been commissioned recently and they are located at the Highfield and Bolderwood Campuses of the University of Southampton, UK. All three buildings have mixed-mode ventilation (both mechanical and natural ventilation) providing heating in winter and peak cooling only during extreme heat events in summer. Figure 1 shows the three buildings' typical floor plans which all encompass cellular offices and open plan office spaces next to a central atrium. The atriums are key features of the natural ventilation strategy.

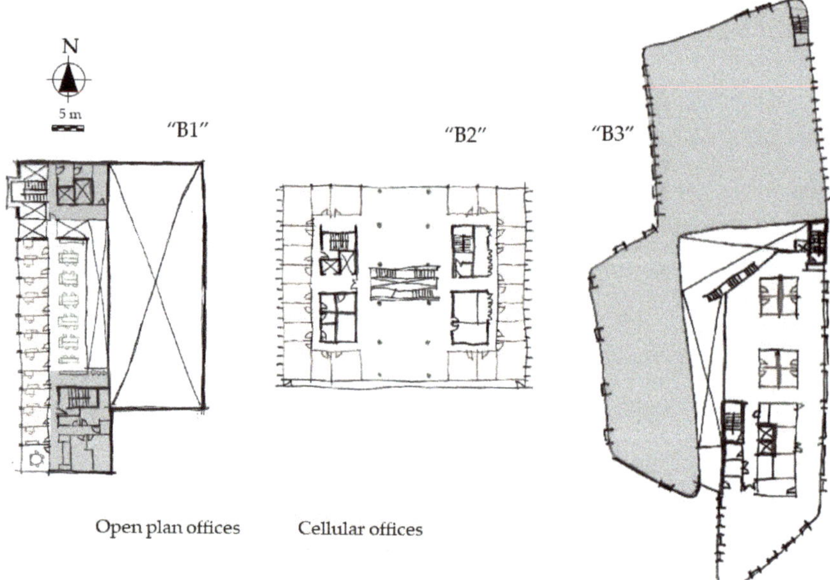

Figure 1. Floor plans of the open (green) and cellular (orange) office spaces in B1 (left), B2 (middle) and B3 (right) buildings at the University of Southampton Bolderwood (B1 and B2) and Highfield (B3) campuses.

Traffic noise levels as represented with $LA_{eq,16h}$ were estimated around the 60 dB level for all the buildings. $LA_{eq,16h}$ is defined as the annual average noise level (in dB) for the 16-h period between 07:00–23:00 and it is typically assessed at a receptor's height of

4 m above ground. B1 floors are connected with an atrium that starts from the common, multi-functional space at the ground floor and extends upwards to the roof of the building. Each floor has an open plan office and a mezzanine style area open to the atrium and the corridor next to the private, individual offices (Figure 1).

2.3. Indoor Environmental Quality (IEQ) Data Collection

The dataloggers used for the IEQ data collection in Stage 3 of this study were installed whenever possible at heights and locations that would be representative of the occupants' working conditions while seated at a desk. Air temperature (T) and relative humidity (RH) were monitored with 5 min frequency. Sound pressure levels were monitored every second. The specifications of the data loggers and the sampling rates are shown in mboxtabreftabref:energies-1040727-t002. In the beginning of Stage 3, the reverberation time (RT) was measured with a Brüel & Kjær 2250 handheld analyser in unoccupied conditions at different places of B1, all with a potential to impact the acoustic conditions of the surveyed offices. The wide band reverberation times were averaged across the 400 Hz to 1.25 kHz frequencies and the acoustic conditions were assessed based on the RT60 (*T30*) measurements. The sound test at the atrium was the only one not surprisingly to return high RT longer than 1.2 s while the other tested locations had RT values lower than 1 s.

Table 2. Environmental monitoring data logger specifications.

Device	Model	Sampling Rate	Measure	Accuracy
Data-logging sound level meter	Reed Instruments SD-4023. Fast weighting. (200 ms) 'A' frequency weighting.	1 s	Sound pressure level, L_P (dB)	1 kHz $\pm$ 1.4 dB; 250 Hz $\pm$ 1.9 dB; 500 Hz $\pm$ 1.9 dB; 125 Hz $\pm$ 2.0 dB; 63 Hz $\pm$ 2.5 dB; 2 kHz $\pm$ 2.6 dB; 31.5 Hz $\pm$ 3.5 dB; 4 kHz $\pm$ 3.6 dB; 8 kHz $\pm$ 5.6 dB
Temperature and relative humidity sensor	MadgeTech RHTemp101A	5 min	Temperature (°C) Relative humidity (%)	$\pm$0.5 °C, $\pm$3.0%
Reverberation time analyser	Brüel & Kjær 2250 handheld analyser	NA	Wide band reverberation time, T_{60} (s)	0.1–0.7 s (Min)

Three sound pressure level (SPL) meters were used to assess background noise conditions during the working and after work hours. The sound level meters were positioned at heights and locations to represent the sound heard by the participants while at their typical seating positions (about 1.0 m above floor level). The sampling time was 1 s to allow for the detection of short impulse sound sources. The collected sound data were used to compute time-averaged acoustic parameters such as the equivalent continuous sound level (LA_{eq}) and LA_{10}, a statistical noise level measure commonly used to evaluate if the sound pressure levels exceed the acceptable threshold (i.e., 50 dB [38]) for 10% of the study's duration.

2.4. Building Services and Controls

Individual, cellular offices in B1 have operable windows, thermostats and passive infrared sensors that control the mechanical ventilation and cooling through a roof mounted air-handling unit. The heating is provided with a wet system through radiators with thermostatic valve (TRV) control. Open plan offices have no direct access to windows. Ventilation and cooling are provided through a roof-mounted, centralised air-handling unit.

B2 has operable windows with external shades whereas mechanical ventilation and cooling are provided by centralised air handling units through floor grilles across the span of the floor. Trench heating is provided with radiators around the perimeter of the floor, next to the windows, through grilles at the floor. There are thermostats in each individual office and at different zones of the large open plan office space.

B3 has operable windows too. Mechanical ventilation, heating and cooling are provided by centralised air handling units and distributed by wall and roof mounted grilles. There are thermostats in the individual offices and in different zones at the open plan office space.

In most cases the thermostats at all studied buildings are connected with the building management systems (BMS) and the temperature of each zone is pre-set and controlled by the building managers and the estates and facilities personnel. Individual access and control of the temperature is restricted.

2.5. Participants

The number of participants in the sample at each stage and the response rates are shown in Table 3. The survey duration refers to the number of weeks that comfort survey questionnaires were sent to each participant. In general, the response rate was acceptable for web-based surveys [39] and comparable with previous relevant studies when sample size is considered [40].

Table 3. Sample size and response rate at the different stages of the study.

Stage	Sample Size	Duration (Weeks)	Response Rate	Type
Stage 1	26	1	65%	Post Occupancy Evaluation (POE)
Stage 2	116	48	20%	Comfort-longitudinal
Stage 3	12	5	90%	Comfort-longitudinal

The sample at all stages comprised administrative staff, academic staff and postgraduate researchers that work regularly and have a permanent workspace in the three case study buildings. All participants had consented to participate in the questionnaire surveys. All responses and data were pseudonymized, and any data were collected and managed in accordance with the approved ethics.

3. Results

3.1. Stage 1—Post Occupancy Evaluation and Satisfaction with Building (B1, n = 26 Valid Questionnaires)

The BUS POE results in general revealed high level of satisfaction with the building design and environmental conditions. The overall comfort was overwhelmingly voted as satisfactory (light shades of grey in Figure 2) and similar results were obtained for the satisfaction with the thermal conditions in the heating (winter) and cooling (summer) seasons. In the heating season especially more than 75% of the responses indicated satisfaction with thermal comfort. This level of satisfaction was less prominent during summer, pointing to issues that likely have to do with the ventilation and cooling of the office space during hot spells (Figure 3).

In relation to this study, it is interesting that the results from the POE, despite the high levels of satisfaction with most aspects of the building's environment and use, showed a rather widespread dissatisfaction with the overall noise levels (Figure 2). That result was expected as background noise has been repeatedly identified as a cause of dissatisfaction in open plan offices [41,42], and the individual offices in the sample have visitors regularly. The POE analysis did not assess the relationship between noise dissatisfaction and interruption frequency.

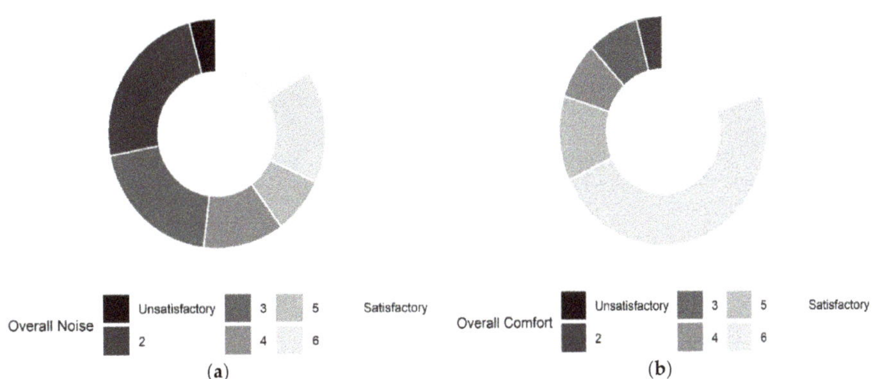

Figure 2. Level of satisfaction with: (**a**) Overall Noise; (**b**) Overall Comfort from the POE results in Stage 1 of the study.

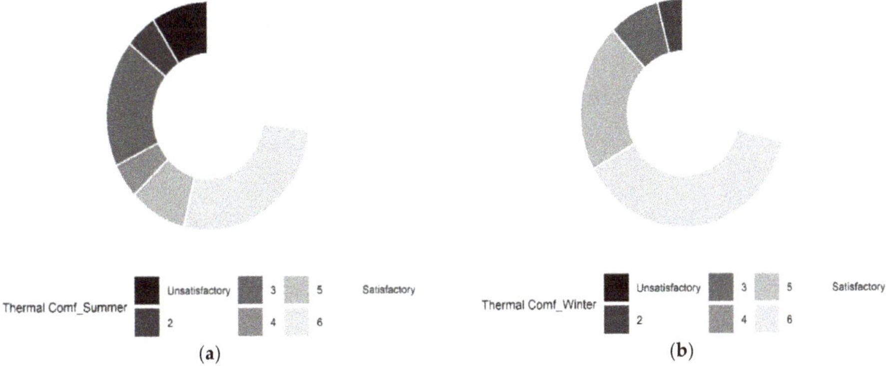

Figure 3. Satisfaction vote for the thermal conditions in the office: (**a**) in summer; (**b**) in winter.

The review of satisfaction levels for different room occupancy (i.e., single, twin or multi) revealed a clear pattern. Individual office occupants were likely to be dissatisfied with the frequency of interruptions but not noise. Whereas, in open plan and shared office spaces, users were mostly dissatisfied with the overall noise rather than the interruptions (Figure 4).

Following that rationale, the next hypothesis was that background noise is related to productivity changes and at the same time noise perception and productivity are likely to be affected by the hours at work. This means busy, "long" days in the office will increase the dissatisfaction with noise and decrease the productivity. The Goodman Kruskal's gamma test was used to investigate if there was a correlation of the Overall Noise with Interruption Frequency and with Productivity Change at the office spaces occupied by the participants. Goodman Kruskal's gamma (GKG) is a test of association between two ordinal variables such as in this case where responses are in the Likert scale. GK gamma (G) takes values in the space $(-1, 1)$ with $G = -1$ and $G = 1$ denoting a strong negative and positive association respectively. The productivity measurement in this study is based on a self-assessment of the survey respondents, which is completely subjective and might be biased by the occupants' comfort at the time of response. In Figure 5, the responses were grouped by the number of hours the participants reported that they spent working in the office every day.

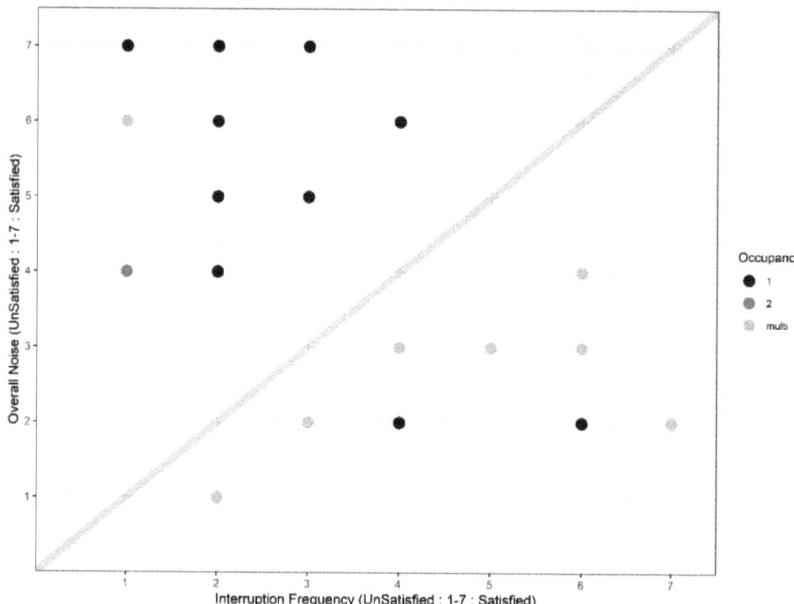

Figure 4. Scatterplot of the Overall Noise satisfaction and satisfaction with Interruption Frequency grouped by different room occupancy (1: single office; 2: twin office; multi: open plan multi occupancy).

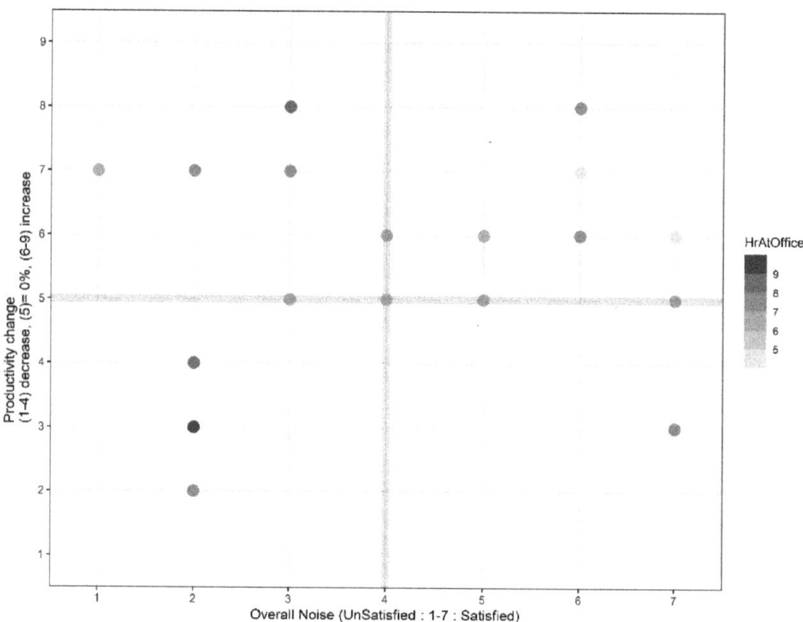

Figure 5. Scatterplot of the Overall Noise satisfaction and Productivity Change grouped by the duration of occupants' workdays (HrAtOffice: hours spent at work).

The productivity change was reported with a scale from "1" to "9" in 10% increments and "5" denoting the central, "no change", point. The satisfaction with overall noise is shown as before with a 7-point scale where "4" marks the "neither noisy nor quiet" point.

The GKG test of the association showed a significant but moderate negative correlation between Overall Noise and Interruption Frequency satisfaction (G = −0.55, confidence intervals (CI (−0.84, −0.27), p = 0.003 < 0.01). As overall noise satisfaction increases, the interruption frequency satisfaction decreases. In other words, as the noise level increases the participant may be less sensitive to interruption disturbances. The relationship between Overall Noise satisfaction and Productivity Change was not significant (G = 0.165, CI (−0.26, 0.59), p = 0.352 > 0.05). However, Figure 5 shows that participants working longer hours at the office are likely to be more unsatisfied with overall noise level (7.7 h on average for participants reported being unsatisfied (levels 1–3) and 6.9 h on average for participants reported being satisfied (levels 5–7)). A previous study has found that the working hours threshold between office occupants who are satisfied and those who are dissatisfied with the IEQ is as low as 20 h/week [43].

3.2. Stage 2—Comfort Survey (B2 and B3, n = 1079 Valid Questionnaires)

This stage builds on the results of the POE to examine the relationship among thermal sensation (TSV), thermal preference (TPV), acoustic perception (NSV) and perceived air quality (ASV). Stage 1 has identified that noise is the only category where satisfaction levels were rated from average to low. The hypothesis at this stage of the study was that there is a relationship among the three examined comfort domains and their interaction terms. In this context, the study assessed the GKG to examine the existence of correlations among TSV, NSV and ASV and investigate the effect of noise perception, indoor air quality and their interactions on thermal perception. Initial descriptive analysis did not indicate the existence of a linear relationship between noise perception (NSV) and thermal sensation (TSV) (Figure 6). The distribution of responses in the graph shown in Figure 6, suggests that most responses are in the neutral to "warm" (TSV 0, 1) and the "neither noisy nor quiet" to "slightly noisy" (NSV 0, −1) area of the graph.

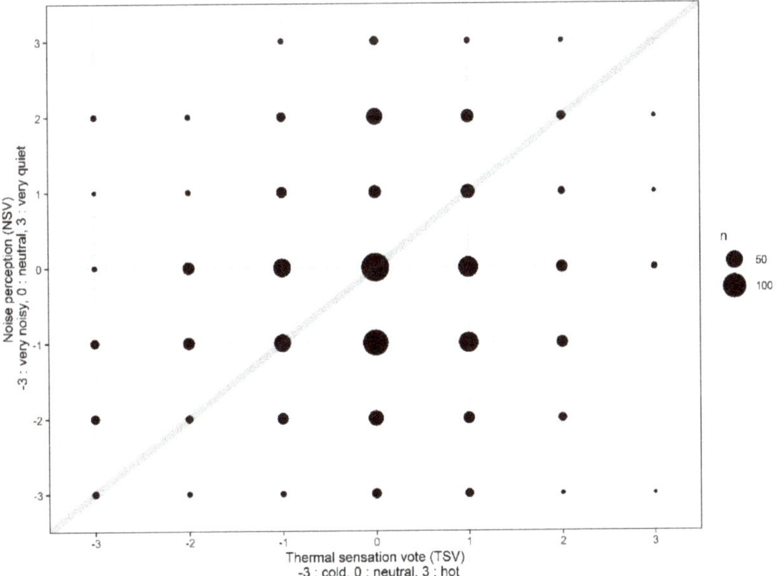

Figure 6. Comparison of agreement between thermal and noise perception vote results for Stage 2 (the number of responses n is proportional to the area of circles).

With regard to the association between air quality perception and thermal perception, Figure 7 indicates that most responses are on and around the central point denoting neutral vote. Interestingly, "slightly bad" to "bad" air quality perception (ASV −1 to −2) was associated with a warm perception of the thermal environment. This result raises questions about the role of air movement and relative humidity levels which are usually responsible for the feeling of "stuffiness" that is typically assessed through carbon dioxide concentration monitoring [44]. Following this finding, Figure 8 describes the relationship of air movement sensation (*AMS*) and air quality sensation (*ASV*).

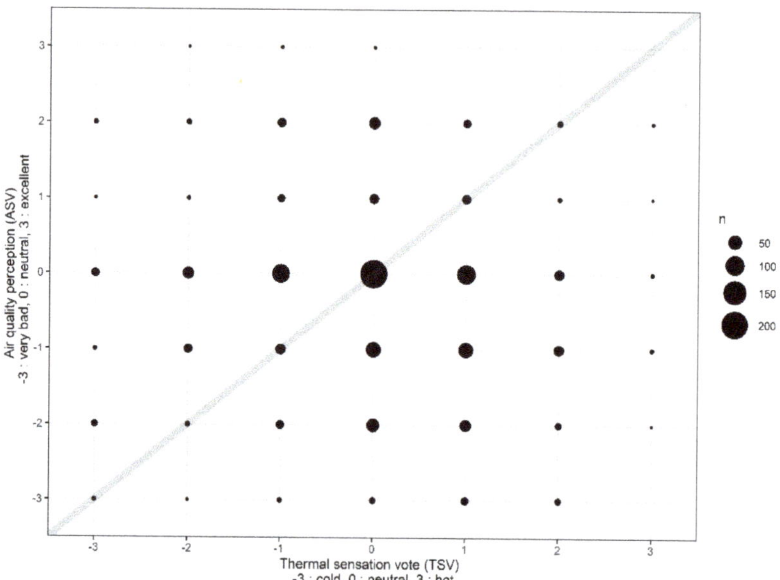

Figure 7. Comparison of agreement between thermal and air quality perception vote results for Stage 2 (the number of responses n is proportional to the area of circles).

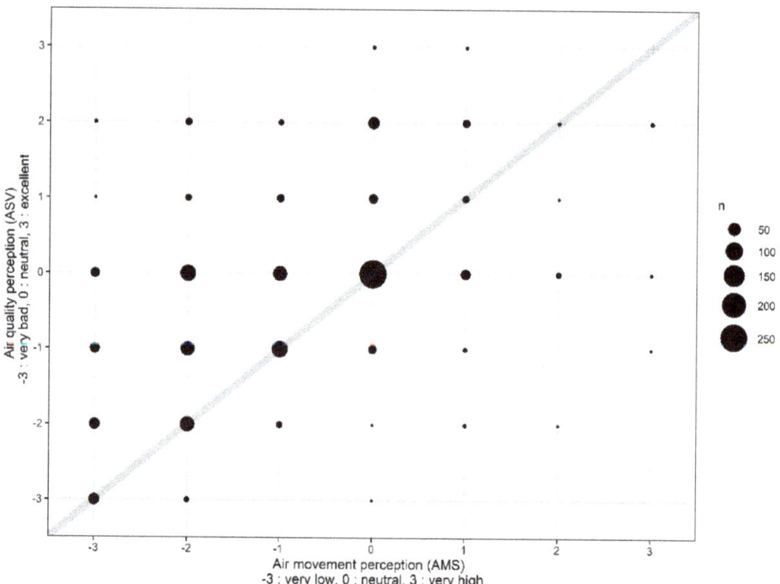

Figure 8. Comparison of agreement between air movement and air quality perception vote results for Stage 2 (the number of responses n is proportional to the area of circles).

The results, as expected, show low air movement perception associated with "bad" air quality perception. Most importantly, the questionnaire results suggest that air movement is perceived as being low (*AMS* 0 to -3) and the air quality is perceived as rather poor (*ASV* 0 to -3).

The GKG test revealed a significant but weak negative correlation between *TSV* and *ASV* ($G = -0.114$, CI (-0.18, -0.04), $p = 0.026 < 0.05$). Similarly, *TSV* was found to be positively but weakly correlated with *NSV* ($G = -0.109$, CI (0.04, 0.18), $p = 7.966 \times 10^{-05} < 0.01$). As expected, there was also a significant correlation between *ASV* and *NSV* ($G = 0.258$, CI (0.18, 0.33), $p = 2.2 \times 10^{-16} < 0.01$).

The last test examined the combined effect of air quality perception (*ASV*) and noise perception (*NSV*) on thermal perception (*TSV*) and the effect of their interactions. For this analysis, mixed effect logistic regression models were assessed, following the methodology described by Schweiker & Wagner, 2018 [45]. The analysis used the R [46] packages lme4 [47] and lmerTest [48]. The mixed effect models were compared with the maximum likelihood estimation (ML). The stepwise backward elimination process (function step.model in R) was used to evaluate the significance of the effects of interactions and optimise the final model. The participants' unique ID was used as the random effect variable. The final model had the formula (*TSV* ~ *ASV* + *NSV* + (1 | *ID*), where (1 | *ID*) is the random effect variable. The effect of the interaction term *ASV*:*NSV* on the *TSV* was not significant ($p = 0.83 > 0.05$).

To follow the modelling analysis, the study explored the relationship between *TSV*, *TPV* and *ASV*. It is important to note that the highest levels of thermal sensation ('hot' coded "3") and thermal preference ('much cooler' coded "-3" and 'much warmer' coded "3") should not be taken into consideration as the sample sizes were small and the confidence intervals were large. As shown in Figure 9, when participants felt cooler (*TSV*) or preferred to be warmer (*TPV*) there was a slight adverse effect on perceived air quality (*ASV*). However, when participants felt warmer (*TSV*) or preferred to be cooler (*TPV*) there was a strong adverse effect on perceived air quality. In summary, the respondents gave the most favourable assessment of air quality when they felt thermally comfortable and preferred no change in their thermal environment.

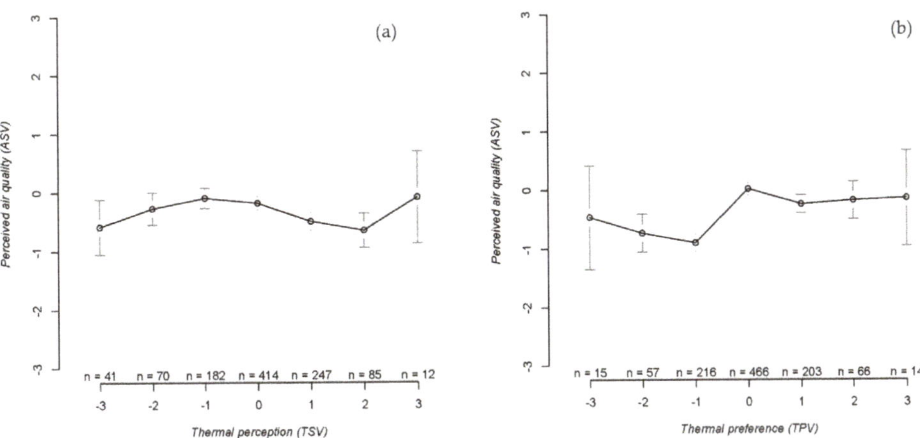

Figure 9. Perceived air quality for: (**a**) thermal perception votes; (**b**) for thermal preference votes, for Stage 2 (n shows the number of responses).

3.3. Stage 3—Comfort Study in Focus Group (B1, n = 52 Valid Questionnaires)

3.3.1. Environmental Conditions

A summary of common descriptive statistics of the environmental variables monitored in B1 during the third stage of this study is shown in Table 4. The mean and median values are almost equal indicating symmetrical distribution of the variables. Relatively small standard deviation in combination with the small variability (Max-Min) suggest that the building's environment was quite stable.

Table 4. Descriptive statistics of the main environmental parameters monitored at Stage 3.

Environmental Parameters	Min	Max	Mean	Median	Std Dev
$SL5$	39 dB	58 dB	46 dB	45 dB	4 dB
$SL30$	40 dB	53 dB	47 dB	47 dB	3 dB
LA_{eq}	41 dB	51 dB	47 dB	48 dB	2 dB
T_a	22.2 °C	24.7 °C	24.2 °C	24.4 °C	0.5 °C
RH	26%	41%	33%	33%	4%
T_{out}	6 °C	14 °C	10 °C	10 °C	2 °C
RH_{out}	56%	98%	82%	83%	11%

An example of the noise level measurements from one of the three sensors in the open plan office is shown in Figure 10. The red-dashed line shows the 50 dB acoustic comfort limit suggested for office spaces by building design guidelines [38].

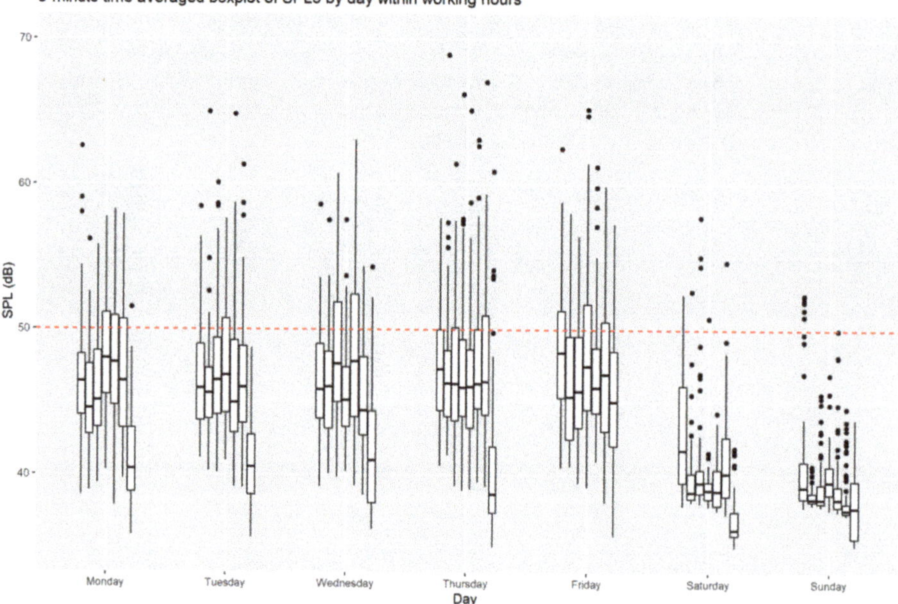

Figure 10. Boxplots of the 5 min averaged observations of sound pressure levels (*SPL*) collected from the open plan office in B1 during working hours (weekend included as baseline) for Stage 3. Red-dashed line denotes the recommended acoustic comfort threshold [38] for office space in the UK.

The 5-min averaged sound pressure level observations pointed out that during the working days the sound exceeds the recommended levels in about 25% of the observations (Q4 of boxplots) within the working hours. That result could confirm the existence of issues with acoustic comfort and noise satisfaction as it was shown by the POE at Stage 1 of the analysis in this paper (note: same office but different sample).

3.3.2. Correlation Analysis and Effects of Interactions

The first part of the correlation analysis investigates the relationship between the current noise perception when the questionnaire was answered (NSV), the acoustic comfort (NSC) and the noise levels perception in the hours before the survey (NSB). The main hypothesis was that the acoustic comfort will be related both to the previous and current sound levels. The Goodman Kruskal's gamma was calculated for the NSV, NSB and NSC pairs to reveal the strength and direction of any associations. The test revealed a strong negative ($G = -0.735$, CI (-1, -0.4)) and significant association ($p = 0.02 < 0.05$) between the current noise perception (NSV) and acoustic comfort (NSC). As expected, when noise perception changed from "very noisy" (-3) to "very quiet" (3) the acoustic comfort vote changed from "uncomfortable" (5) to "comfortable" (1). Interestingly, the analysis also found significant strong associations between the noise preconditions (NSB) with current noise perception (NSV) ($G = 0.796$, CI (0.6, 0.99), $p = 2.03 \times 10^{-7} < 0.01$) and with the acoustic comfort vote (NSC) ($G = -0.912$, CI (-1, -0.776), $p = 3.53 \times 10^{-6} < 0.01$) respectively.

The second part of the analysis at this stage, looked at the relationship between current noise perception (NSV), with thermal sensation (TSV), thermal comfort (TCV) and the perception of indoor air quality (ASV). Results from Stage 2 had found that TSV is associated with both the ASV and the NSV but there was not any significant effect of their interaction. It is difficult to assess the impact of the sample size effect on the results and the sample of Stage 3 is small. Therefore, Kendall's correlation was chosen for the analysis in order to build confidence on the previous results. The Kendall's correlation matrix of results (Table 5) confirmed this study's findings that thermal perception is negatively correlated with air quality perception (ASV) ($TSV \sim ASV$, $\tau_b = -0.363$, $p = 0.003 < 0.01$) but there was not conclusive evidence on an association between TSV and noise perception (NSV).

Table 5. Kendall's correlation results (τ_b(p)) matrix for Stage 3 of the study design.

Variables	NSV	NSC	TCV	ASV
TSV	−0.225 (0.06)	−0.091 (0.48)	0.239 (0.06)	−0.363 (0.003)
ASV	0.066 (0.57)	0.176 (0.17)	−0.134 (0.29)	1
NSC	−0.47 (0.0001)	1	−0.106 (0.43)	0.176 (0.17)
NSV	1	−0.47 (0.0001)	−0.176 (0.15)	0.066 (0.57)

This finding lead to the investigation of direct associations of TSV with sound and temperature environmental factors (Table 6) instead of the perception votes used until this point. The 30 min averaged sound pressure level observations ($SPL30$) were negatively associated with noise perception but the association with acoustic comfort was not significant. Most importantly, the operative temperature (T_{op}) was found to be significantly associated with all thermal (TSV), noise (NSV) perception and acoustic comfort (NSC).

Table 6. Kendall's correlation results (τ_b(p)) for two monitored IEQ factors.

Variables	TSV	TCV	NSV	NSC	ASV
SPL30	−0.042 (0.70)	−0.024 (0.83)	−0.210 (0.046)	0.136 (0.23)	−0.038 (0.73)
T_{op}	0.237 (0.03)	0.094 (0.41)	−0.267 (0.01)	0.232 (0.046)	−0.07 (0.53)

Lastly, mixed effect logistic regression analysis was used, following the same methods with stage 2, to evaluate the effect of any interactions between thermal, noise and air quality parameters on their impact on thermal sensation. Different combinations of predictors and random effect variables were examined with thermal sensation (TSV) as the dependent variable. The independent fixed effects variables were the operative temperature (T_{op}), the air quality perception (ASV), the 30 min average sound pressure level ($SPL30$), the daily average outdoor temperature (T_{out_d}), and the outdoor and indoor temperature at the time of the survey (T_{in/out_s}). The continuous environmental variables were standardized, scaled based on their means and standard deviations. The mixed effect models were compared with the maximum likelihood estimation (ML) and a stepwise backward elimination process (function step.model in R) was used to evaluate the significance of the effects of interactions and optimise the final model.

The results of the mixed effect logistic regression with TSV as dependent variable and NSV as random effect variable, revealed that the effect of the interactions of (1) T_{op} with ASV and (2) $SPL30$ with ASV are both significant ($p < 0.05$) whereas the independent environmental variables related with the outdoor and indoor temperature and their interactions are not. Findings about the interaction terms were consistent across cases where thermal (TCV) and acoustic (NSC) comfort were allowed to vary randomly (i.e., set as random effect variable). The final model had the formula ($TSV \sim T_{op} + SPL30 + ASV + (1 \,|\, NSV) + T_{op}{:}ASV + SPL30{:}ASV$) where ($1 \,|\, NSV$) is the random effect variable and $T_{op}{:}ASV + SPL30{:}ASV$ are the interaction terms for T_{op} and $SPL30$ with ASV. The confidence intervals are shown in Table 7 (standardized variables).

Table 7. Confidence intervals for the independent variables (standardized) with significant effect on TSV.

Confidence Intervals	2.5%	97.5%
Intercept	0.346	0.740
Top	0.264	0.701
SPL30	−0.769	−0.209
ASV	−0.521	−0.184
Top:ASV	−0.408	−0.051
SPL30:ASV	0.117	0.491

4. Discussion

The POE results identified potential issues with building acoustics and noise in the case study building B1. Individual, cellular office occupants were mostly dissatisfied with the frequency of interruptions, whereas the occupants of the open plan office were mostly dissatisfied with background noise. Open plan offices are often preferred as a solution because they maximize density, but their design should include provisions for noise control [41,42]. The marked contrast between the cause of dissatisfaction in single/twin and multi occupancy offices in an academic context could also be a result of the role and responsibilities of individuals and the hierarchy in ranking and space. Senior academics are more likely to have individual offices and frequent interruptions from visitors. Open plan office space is typically occupied by research students, researchers and administration staff. In general, the POE showed that noise was the only category that occupants were not satisfied with the building and the indoor environmental quality (IEQ). In relation to this proposition of indoor noise nuisance, it was found that the overall noise perception was negatively associated with the frequency of interruptions. Most importantly it was shown that noise did not have a significant effect on productivity. This result contradicts previous studies that concluded background noise has adverse effects on productivity [49–52]. Relevant research has shown that the factors associated with productivity and the magnitude of their effect are related to the office types, and they include, as expected, IEQ parameters and demographics but also building design features and organizational arrangements (e.g., working hours, position and workplace arrangements) [27,53].

Adding to the current literature on IEQ and its association with health and comfort, it was shown that bad air quality is generally associated with a "warm" thermal sensation response. This could be describing the feeling of "stuffiness", a condition often attributed to low air movement, high RH and a high concentration of carbon dioxide [44,54]. Looking into the association of related IEQ components and their interactions, this study pointed out that air quality (ASV) and noise perception (NSV) are both correlated with thermal perception (TSV), but there is no evidence for the effect of their interaction on TSV. Air quality perception was correlated with both TSV and NSV, showing once again the complexity of occupants' perception and comfort. With regard to the relationship between air quality and thermal comfort, participants who were feeling uncomfortable (in particular if feeling warm and preferring to be cooler) perceived the air-quality worse than participants who felt comfortable. This finding is consistent with results from previous studies [55].

Building on the findings from the first two stages, the third stage of the analysis in this study looked at noise perception and its relationship to background noise exposure duration (based on sound pressure levels in the hours before the survey was taken). Interestingly, the noise perception in the hours before the survey was associated with both the noise perception at the time of the survey (NSV) and overall acoustic comfort (NSC). Then a combined effect analysis of air quality and noise perception on the thermal perception was undertaken through a different method in order to validate the findings from stage 2. The results confirmed the weak negative correlation between TSV and ASV but could not conclude on the relationship of TSV with noise perception. The contradictory results lead to the investigation of the effect of the monitored environmental parameters directly on the thermal and noise perception. The results indicated that operative temperature, T_{op}, was correlated with thermal sensation as expected, but it was also correlated with noise perception and acoustic comfort. In conclusion, thermal perception in this last analysis was found to be a function of the T_{op}, the sound pressure levels ($SPL30$) and air quality perception (ASV), but there was also a significant effect found of the interaction terms of (1) T_{op} with ASV and (2) $SPL30$ with ASV. A finding that confirmed the importance of good air quality to achieve comfort in office buildings.

5. Conclusions

This study concluded that thermal sensation is associated with both air quality and noise perception. An effect of the interaction of air quality and noise perception on thermal sensation was not evident. A key finding was the significant correlation of operative temperature (T_{op}) with TSV as expected, but also a correlation of operative temperature with noise perception and overall acoustic comfort. Regarding the crossed effect of IEQ parameters on the thermal sensation, a significant effect was found for the interaction of (1) T_{op} with air quality perception and (2) sound pressure levels ($SPL30$) with air quality perception. Most importantly, the findings highlighted the importance of air quality perception in achieving comfort and high satisfaction within the space where people work. It is expected that air quality in office buildings will play an important role in the future design of buildings and building services systems. The conclusions of this study aim to contribute to the research on cross-modal comfort perception and its application to the development of models and systems for adaptive buildings.

5.1. Internal and External Validity

The research design could be replicated in studies with non-university office buildings to examine the consistency of the findings. The differences between the POE and "right-here-right-now" surveys might not allow a direct comparison of results from the surveys in this study. In addition, the variation in the sample size and survey duration might introduce a "seasonal effect" on the questionnaires' results due to the number of valid questionnaires used in the analysis and their distribution across the seasons. The impact of different data collection methods on the final sample size and on the quality of responses should be also examined.

The results may apply only to buildings that are similar to the buildings in this study in terms of use and characteristics (i.e., mixed mode with concurrent or change-over mode of operation). Any findings might not be transferable to other building types that would have different adaptive opportunities. The generalization of the results might be further restricted by the sample size and the participants' characteristics, especially their environmental awareness and energy-saving attitudes, as they all are highly educated and they work in relevant disciplines in an academic environment. The effects of education, gender, age and possibly of any relevant beliefs and social roles, as result of their profession, should be carefully investigated in future work.

5.2. Future Research

This study has identified the importance of indoor environmental quality, and in particular air quality perception and operative temperature, in the cross-modal perception of comfort in office buildings. It is suggested that longitudinal, POE-comfort mixed surveys are a useful tool to investigate the effect of building design and environmental factors on occupants' comfort and behaviour. Further research is required to explore the trans-modal perspective of comfort through studying the integration of cross-modal comfort perception with contextual preferences and behaviours for different occupant characteristics. This trans-modal comfort approach could be used to develop adaptive, reactive and dynamic systems for buildings with occupant-specific zonal conditions that proactively respond to external forcing (including different degrees of environmental discomfort), the preferences of present occupants and the use of personalized controls.

It has been suggested that the ways we work and live may have changed permanently due to the SARS COVID pandemic. These changes will inevitably affect the places where we live and work, especially as work and home spaces become similar. Under these propositions, research about comfort and IEQ in office spaces will also become relevant in the context of homeworking. It will likely become more important to monitor and control the IEQ and comfort parameters in buildings that were designed and built with different adaptive opportunities, specifications, systems, occupancy assumptions and use in mind.

Future work will focus on the study of air quality and IEQ-related parameters in both office and home-office spaces. Future research needs to further investigate the variability in the impact of different office design characteristics (between work and home) on cross-modal comfort. The findings could be also used in the research about an office design's effect on the occupants' physical and mental wellbeing and the necessary retrofits and adjustments to office space both at work and home, as we adapt in new ways to future living.

Author Contributions: L.B. and S.G. contributed to the conception of the study design; L.B., S.G., N.K.S.E. and P.X. contributed to field work of the case study buildings; L.B., S.G., N.K.S.E. and P.X. contributed to the quality assurance and processing of the case study raw data; L.B., S.G., N.K.S.E. and P.X. contributed to the analysis and interpretation of data; L.B and S.G. contributed to drafting the article and revising it critically; All authors contributed to the approval of the final version. All authors have read and agreed to the published version of the manuscript.

Funding: This research received no external funding.

Institutional Review Board Statement: The study was approved by the Ethics Committee of the University of Southampton (ERGO 28671, approved on 3 July 2017, ERGO 53502, approved on 29 November 2019 and ERGO 58356, approved on 4 August 2020).

Informed Consent Statement: Informed consent was obtained from all subjects involved in the study.

Data Availability Statement: Data not available. The data are not publicly available due to ethical restrictions.

Acknowledgments: This study was conducted within the framework of the International Energy Agency—Energy in Buildings and Communities Program (IEA-EBC) Annex 69-Strategy and Practice of Adaptive Thermal Comfort in Low Energy Buildings and Annex 79-Occupant-Centric Building Design and Operation. L.B. and S.G. would like to thank the Sustainable Energy Research Group at the University of Southampton for supporting this work (www.energy.soton.ac.uk). Part of the work of L.B. was supported by the Lancaster School of Architecture and Imagination Lancaster (imagination.lancaster.ac.uk) at LICA, Lancaster University.

Conflicts of Interest: The authors declare no conflict of interest.

References

1. Ortiz, M.A.; Kurvers, S.R.; Bluyssen, P.M. A review of comfort, health, and energy use: Understanding daily energy use and wellbeing for the development of a new approach to study comfort. *Energy Build.* **2017**, *152*, 323–335. [CrossRef]
2. Al Horr, Y.; Arif, M.; Kaushik, A.; Mazroei, A.; Katafygiotou, M.; Elsarrag, E. Occupant productivity and office indoor environment quality: A review of the literature. *Build. Environ.* **2016**, *105*, 369–389. [CrossRef]
3. Mahdavi, A.; Berger, C. Predicting Buildings' Energy Use: Is the Occupant-Centric "Performance Gap" Research Program Ill-Advised? *Front Energy Res.* **2019**, *7*, 124. [CrossRef]
4. Manfren, M.; Nastasi, B.; Piana, E.; Tronchin, L. On the link between energy performance of building and thermal comfort: An example. *AIP Conf. Proc.* **2019**, *21*, 23.
5. Antoniadou, P.; Papadopoulos, A.M. Occupants' thermal comfort: State of the art and the prospects of personalized assessment in office buildings. *Energy Build.* **2017**, *153*, 136–149. [CrossRef]
6. Schweiker, M.; Ampatzi, E.; Andargie, M.S.; Andersen, R.K.; Azar, E.; Barthelmes, V.M.; Berger, C.; Bourikas, L.; Carlucci, S.; Chinazzo, G.; et al. Review of multi-domain approaches to indoor environmental perception and behaviour. *Build. Environ.* **2000**, 106804. Available online: https://www.sciencedirect.com/science/article/abs/pii/S0360132320301621 (accessed on 6 January 2021). [CrossRef]
7. Wang, Z.; de Dear, R.; Luo, M.; Lin, B.; He, Y.; Ghahramani, A.; Zhu, Y. Individual difference in thermal comfort: A literature review. *Build. Environ.* **2018**, *138*, 181–193. [CrossRef]
8. Carlucci, S.; Bai, L.; de Dear, R.; Yang, L. Review of adaptive thermal comfort models in built environmental regulatory documents. *Build. Environ.* **2018**, *137*, 73–89. [CrossRef]
9. Nordvall, M.; Arvola, M. Perception, Meaning and Transmodal Design. *DRS2016 Futur Think.* **2016**, *3*, 1–12.
10. Yang, W.; Moon, H.J. Cross-modal effects of noise and thermal conditions on indoor environmental perception and speech recognition. *Appl. Acoust.* **2018**, *141*, 1–8. [CrossRef]
11. Chinazzo, G.; Wienold, J.; Andersen, M. Influence of indoor temperature and daylight illuminance on visual perception. *Light. Res. Technol.* **2019**, *52*, 350–370. [CrossRef]
12. Lam, C.K.C.; Yang, H.; Yang, X.; Liu, J.; Ou, C.; Cui, S.; Kong, X.; Hang, J. Cross-modal effects of thermal and visual conditions on outdoor thermal and visual comfort perception. *Build. Environ.* **2020**, *186*, 107297. [CrossRef]
13. Schweiker, M.; Schakib-Ekbatan, K.; Fuchs, X.; Becker, S. A seasonal approach to alliesthesia. Is there a conflict with thermal adaptation? *Energy Build.* **2020**, *212*, 109745. [CrossRef]
14. Heydarian, A.; McIlvennie, C.; Arpan, L.; Yousefi, S.; Syndicus, M.; Schweiker, M.; Jazizadeh, F.; Rissetto, R.; Pisello, A.L.; Piselli, C.; et al. What drives our behaviors in buildings? A review on occupant interactions with building systems from the lens of behavioral theories. *Build. Environ.* **2020**, *179*, 106928. [CrossRef]
15. Schweiker, M.; Huebner, G.M.; Kingma, B.R.M.; Kramer, R.; Pallubinsky, H. Drivers of diversity in human thermal perception—A review for holistic comfort models. *Temperature* **2018**, *5*, 308–342. [CrossRef] [PubMed]
16. Schweiker, M.; Rissetto, R.; Wagner, A. Thermal expectation: Influencing factors and its effect on thermal perception. *Energy Build.* **2020**, *210*, 109729. [CrossRef]
17. Tang, H.; Ding, Y.; Singer, B. Interactions and comprehensive effect of indoor environmental quality factors on occupant satisfaction. *Build. Environ.* **2020**, *167*, 106462. [CrossRef]
18. Yang, W.; Moon, H.J. Combined effects of acoustic, thermal, and illumination conditions on the comfort of discrete senses and overall indoor environment. *Build. Environ.* **2019**, *148*, 623–633. [CrossRef]
19. Geng, Y.; Ji, W.; Lin, B.; Zhu, Y. The impact of thermal environment on occupant IEQ perception and productivity. *Build. Environ.* **2017**, *121*, 158–167. [CrossRef]
20. Barnes, J.; Wineman, J.; Adler, N. Open Office Space: The Wave of the Future for Academic Health Centers? *Acad. Med.* **2020**, *95*, 52–58. [CrossRef]
21. Richardson, A.; Potter, J.; Paterson, M.; Harding, T.; Tyler-Merrick, G.; Kirk, R.; Reid, K.; McChesney, J. Office design and health: A systematic review. *N. Z. Med. J.* **2017**, *130*, 39–49. [PubMed]
22. Frontczak, M.; Schiavon, S.; Goins, J.; Arens, E.; Zhang, H.; Wargocki, P. Quantitative relationships between occupant satisfaction and satisfaction aspects of indoor environmental quality and building design. *Indoor Air* **2012**, *22*, 119–131. [CrossRef] [PubMed]
23. Kim, J.; de Dear, R. Nonlinear relationships between individual IEQ factors and overall workspace satisfaction. *Build. Environ.* **2012**, *49*, 33–40. [CrossRef]
24. Castaldo, V.L.; Pigliautile, I.; Rosso, F.; Cotana, F.; De Giorgio, F.; Pisello, A.L. How subjective and non-physical parameters affect occupants' environmental comfort perception. *Energy Build.* **2018**, *178*, 107–129. [CrossRef]
25. Kwon, M.; Remøy, H.; van den Dobbelsteen, A.; Knaack, U. Personal control and environmental user satisfaction in office buildings: Results of case studies in the Netherlands. *Build. Environ.* **2019**, *149*, 428–435. [CrossRef]
26. Ma, N.; Aviv, D.; Guo, H.; Braham, W.W. Measuring the right factors: A review of variables and models for thermal comfort and indoor air quality. *Renew. Sustain. Energy Rev.* **2021**, *135*, 110436. [CrossRef]
27. Chen, C.F.; Yilmaz, S.; Pisello, A.L.; De Simone, M.; Kim, A.; Hong, T.; Bandurski, K.; Bavaresco, M.V.; Liu, P.L.; Zhu, Y. The impacts of building characteristics, social psychological and cultural factors on indoor environment quality productivity belief. *Build. Environ.* **2020**, *185*, 107189. [CrossRef]

28. Pastore, L.; Andersen, M. Building energy certification versus user satisfaction with the indoor environment: Findings from a multi-site post-occupancy evaluation (POE) in Switzerland. *Build. Environ.* **2019**, *150*, 60–74. [CrossRef]
29. Byrd, H.; Rasheed, E.O. The productivity paradox in green buildings. *Sustainability* **2016**, *8*, 347. [CrossRef]
30. Leaman, A. Usable Buildngs Trust, The Partner Network Building Use Studies (BUS) Methodology 2006. Available online: https://busmethodology.org.uk/ (accessed on 6 January 2021).
31. Leaman, A.; Stevenson, F.; Bordass, B. Building evaluation: Practice and principles. *Build. Res. Inf.* **2010**, *38*, 564–577. [CrossRef]
32. Khoshbakht, M.; Baird, G.; Rasheed, E.O. The influence of work group size and space sharing on the perceived productivity, overall comfort and health of occupants in commercial and academic buildings. *Indoor Built Environ.* **2020**. [CrossRef]
33. Baird, G.; Chun, S.; Gandhi, A.; Gjerde, M. Assessment of changes in building performance from the users' point of view: A before-and-after case study of an academic department. In Proceedings of the 53rd International Conference of Architectural Science Association, Roorkee, Uttrakhand, India, 28–30 November 2019.
34. Rasheed, E.O.; Khoshbakht, M.; Baird, G. Does the number of occupants in an office influence individual perceptions of comfort and productivity?-new evidence from 5000 office workers. *Buildings* **2019**, *9*, 73. [CrossRef]
35. McCartney, K.J.; Fergus Nicol, J. Developing an adaptive control algorithm for Europe. *Energy Build.* **2002**, *34*, 623–635. [CrossRef]
36. ASHRAE Standing Standard Project Committee 55. *ASHRAE Standard 55-2010. Thermal Environmental Conditions for Human Occupancy*; ASHRAE: Atlanta, GA, USA, 2010.
37. CEN. *Standard EN15251. Indoor Environmental Input Parameters for Design and Assessment of Energy Performance of Buildings Addressing Indoor Air Quality, Thermal Environment, Lighting and Acoustics*; CEN: Brussels, Belgium, 2007.
38. CIBSE. Guide A: Environmental Design. *Chart. Inst. Build. Serv. Eng.* **2015**. [CrossRef]
39. Poynton, T.A.; DeFouw, E.R.; Morizio, L.J. A Systematic Review of Online Response Rates in Four Counseling Journals. *J. Couns. Dev.* **2019**, *97*, 33–42. [CrossRef]
40. Langevin, J.; Gurian, P.L.; Wen, J. Tracking the human-building interaction: A longitudinal field study of occupant behavior in air-conditioned offices. *J. Environ. Psychol.* **2015**, *42*, 94–115. [CrossRef]
41. Lee, Y.; Aletta, F. Acoustical planning for workplace health and well-being: A case study in four open-plan offices. *Build. Acoust.* **2019**, *26*, 207–220. [CrossRef]
42. Banbury, S.P.; Berry, D.C. Office noise and employee concentration: Identifying causes of disruption and potential improvements. *Ergonomics* **2005**, *48*, 25–37. [CrossRef]
43. Bae, S.; Asojo, A.O.; Martin, C.S. Impact of occupants' demographics on indoor environmental quality satisfaction in the workplace. *Build. Res. Inf.* **2020**, *48*, 301–315. [CrossRef]
44. Ramalho, O.; Mandin, C.; Ribéron, J.; Wyart, G. Air Stuffiness and Air Exchange Rate in French Schools and Day-Care Centres. *Int. J. Vent.* **2013**, *12*, 175–180. [CrossRef]
45. Schweiker, M.; Wagner, A. Interactions between thermal and visual (dis-)comfort and related adaptive actions through cluster analyses. In Proceedings of the BauSIM2018—7. Deutsch-Österreichische IBPSA-Konferenz Tagungsband, Karlsruhe, Germany, 26–28 September 2018; pp. 204–215.
46. R Core Team R: A Language and Environment for Statistical Computing 2020. Available online: http://www.r-project.org/ (accessed on 6 January 2021).
47. Bates, D.; Mächler, M.; Bolker, B.M.; Walker, S.C. Fitting linear mixed-effects models using lme4. *J. Stat. Softw.* **2015**, *67*, 1–48. [CrossRef]
48. Kuznetsova, A.; Brockhoff, P.B.; Christensen, R.H.B. lmerTest Package: Tests in Linear Mixed Effects Models. *J. Stat. Softw.* **2017**, *82*, 1–26. [CrossRef]
49. Göçer, Ö.; Candido, C.; Thomas, L.; Göçer, K. Differences in occupants' satisfaction and perceived productivity in high- and low-performance offices. *Buildings* **2019**, *9*, 199. [CrossRef]
50. Varjo, J.; Hongisto, V.; Haapakangas, A.; Maula, H.; Koskela, H.; Hyönä, J. Simultaneous effects of irrelevant speech, temperature and ventilation rate on performance and satisfaction in open-plan offices. *J. Environ. Psychol.* **2015**, *44*, 16–33. [CrossRef]
51. Roskams, M.J.; Haynes, B.P. Testing the relationship between objective indoor environment quality and subjective experiences of comfort. *Build. Res. Inf.* **2020**. [CrossRef]
52. Cahyani, A.A. Influence of Work Environment Noise to Productivity of Employee Performance of Sidoarjo District. *J. Public Heal. Sci. Res.* **2020**, *1*, 12–17. [CrossRef]
53. Göçer, Ö.; Hua, Y.; Göçer, K. Completing the missing link in building design process: Enhancing post-occupancy evaluation method for effective feedback for building performance. *Build. Environ.* **2015**, *89*, 14–27. [CrossRef]
54. Reinikainen, L.M.; Aunela-Tapola, L.; Jaakkola, J.J.K. Humidification and perceived indoor air quality in the office environment. *Occup. Environ. Med.* **1997**, *54*, 322–327. [CrossRef]
55. Humphreys, M.A.; Nicol, J.F.; Mccartney, K.J. An Analysis of Some Subjective Assessments of Indoor Air-quality in Five European Countries. *Indoor Air* **2002**, *5*, 86–91.

MDPI

St. Alban-Anlage 66

4052 Basel

Switzerland

Tel. +41 61 683 77 34

Fax +41 61 302 89 18

www.mdpi.com

Energies Editorial Office

E-mail: energies@mdpi.com

www.mdpi.com/journal/energies